Graduate Texts in Mathematics 238

Graduate Texts in Mathematics

1 TAKEUTI/ZARING. Introduction to Axiomatic Set Theory. 2nd ed.
2 OXTOBY. Measure and Category. 2nd ed.
3 SCHAEFER. Topological Vector Spaces. 2nd ed.
4 HILTON/STAMMBACH. A Course in Homological Algebra. 2nd ed.
5 MAC LANE. Categories for the Working Mathematician. 2nd ed.
6 HUGHES/PIPER. Projective Planes.
7 J.-P. SERRE. A Course in Arithmetic.
8 TAKEUTI/ZARING. Axiomatic Set Theory.
9 HUMPHREYS. Introduction to Lie Algebras and Representation Theory.
10 COHEN. A Course in Simple Homotopy Theory.
11 CONWAY. Functions of One Complex Variable I. 2nd ed.
12 BEALS. Advanced Mathematical Analysis.
13 ANDERSON/FULLER. Rings and Categories of Modules. 2nd ed.
14 GOLUBITSKY/GUILLEMIN. Stable Mappings and Their Singularities.
15 BERBERIAN. Lectures in Functional Analysis and Operator Theory.
16 WINTER. The Structure of Fields.
17 ROSENBLATT. Random Processes. 2nd ed.
18 HALMOS. Measure Theory.
19 HALMOS. A Hilbert Space Problem Book. 2nd ed.
20 HUSEMOLLER. Fibre Bundles. 3rd ed.
21 HUMPHREYS. Linear Algebraic Groups.
22 BARNES/MACK. An Algebraic Introduction to Mathematical Logic.
23 GREUB. Linear Algebra. 4th ed.
24 HOLMES. Geometric Functional Analysis and Its Applications.
25 HEWITT/STROMBERG. Real and Abstract Analysis.
26 MANES. Algebraic Theories.
27 KELLEY. General Topology.
28 ZARISKI/SAMUEL. Commutative Algebra. Vol.I.
29 ZARISKI/SAMUEL. Commutative Algebra. Vol.II.
30 JACOBSON. Lectures in Abstract Algebra I. Basic Concepts.
31 JACOBSON. Lectures in Abstract Algebra II. Linear Algebra.
32 JACOBSON. Lectures in Abstract Algebra III. Theory of Fields and Galois Theory.
33 HIRSCH. Differential Topology.
34 SPITZER. Principles of Random Walk. 2nd ed.
35 ALEXANDER/WERMER. Several Complex Variables and Banach Algebras. 3rd ed.
36 KELLEY/NAMIOKA et al. Linear Topological Spaces.
37 MONK. Mathematical Logic.
38 GRAUERT/FRITZSCHE. Several Complex Variables.
39 ARVESON. An Invitation to C^*-Algebras.
40 KEMENY/SNELL/KNAPP. Denumerable Markov Chains. 2nd ed.
41 APOSTOL. Modular Functions and Dirichlet Series in Number Theory. 2nd ed.
42 J.-P. SERRE. Linear Representations of Finite Groups.
43 GILLMAN/JERISON. Rings of Continuous Functions.
44 KENDIG. Elementary Algebraic Geometry.
45 LOÈVE. Probability Theory I. 4th ed.
46 LOÈVE. Probability Theory II. 4th ed.
47 MOISE. Geometric Topology in Dimensions 2 and 3.
48 SACHS/WU. General Relativity for Mathematicians.
49 GRUENBERG/WEIR. Linear Geometry. 2nd ed.
50 EDWARDS. Fermat's Last Theorem.
51 KLINGENBERG. A Course in Differential Geometry.
52 HARTSHORNE. Algebraic Geometry.
53 MANIN. A Course in Mathematical Logic.
54 GRAVER/WATKINS. Combinatorics with Emphasis on the Theory of Graphs.
55 BROWN/PEARCY. Introduction to Operator Theory I: Elements of Functional Analysis.
56 MASSEY. Algebraic Topology: An Introduction.
57 CROWELL/FOX. Introduction to Knot Theory.
58 KOBLITZ. p-adic Numbers, p-adic Analysis, and Zeta-Functions. 2nd ed.
59 LANG. Cyclotomic Fields.
60 ARNOLD. Mathematical Methods in Classical Mechanics. 2nd ed.
61 WHITEHEAD. Elements of Homotopy Theory.
62 KARGAPOLOV/MERLZJAKOV. Fundamentals of the Theory of Groups.
63 BOLLOBAS. Graph Theory.

(continued after index)

Martin Aigner

A Course in Enumeration

With 55 Figures and 11 Tables

Martin Aigner
Freie Universität Berlin
Fachbereich Mathematik und Informatik
Institut für Mathematik II
Arnimallee 3
14195 Berlin, Germany
aigner@math.fu-berlin.de

Mathematics Subject Classification (2000): 05-01

ISSN 0072-5285
ISBN 978-3-642-07253-6
e-ISBN 978-3-540-39035-5

Springer is a part of Springer Science+Business Media

springer.com

Softcover reprint of the hardcover 1st edition 2007

Cover design: WMX Design GmbH, Heidelberg

Printed on acid-free paper 46/3180/YL - 5 4 3 2 1 0

Preface

Counting things is probably mankind's earliest mathematical experience, and so, not surprisingly, combinatorial enumeration occupies an important place in virtually every mathematical field. Yet apart from such time-honored notions as binomial coefficients, inclusion-exclusion, and generating functions, combinatorial enumeration is a young discipline. Its main principles, methods, and fields of application evolved into maturity only in the last century, and there has been an enormous growth in recent years. The aim of this book is to give a broad introduction to combinatorial enumeration at a leisurely pace, covering the most important subjects and leading the reader in some instances to the forefront of current research.

The text is divided into three parts: Basics, Methods, and Topics. This should enable the reader to understand what combinatorial enumeration is all about, to apply the basic tools to almost any problem he or she may encounter, and to proceed to more advanced methods and some attractive and lively fields of research.

As prerequisites, only the usual courses in linear algebra and calculus and the basic notions of algebra and probability theory are needed. Since graphs are often used to illustrate a particular result, it may also be a good thing to have a text on graph theory at hand. For terminology and notation not listed in the index the books by R. Diestel, Graph Theory, Springer 2006, and D.B. West, Introduction to Graph Theory, Prentice Hall 1996, are good sources. Given these prerequisites, the book is best suited for a senior undergraduate or first-year graduate course.

It is commonplace to stress the importance of exercises. To learn enumerative combinatorics one simply must do as many exercises as possible. Exercises appear throughout the text to illustrate some points and entice the reader to complete proofs or find generalizations. There are 666 exercises altogether. Many of them contain hints, and for those marked with ▷ you will find a solution in the appendix. In each section, the exercises appear in two groups, divided by a horizontal line. Those in the first part should be doable with modest effort, while those in the second half require a little more work. Each chapter closes with a special highlight, usually a famous and attractive problem illustrating the foregoing material, and a short list of references for further reading.

I am grateful to many colleagues, friends, and students for all kinds of contributions. My special thanks go to Mark de Longueville, Jürgen Schütz, and Richard Weiss, who read all or part of the book in its initial stages; to Margrit Barrett and Christoph Eyrich for the superb technical work and layout; and to David Kramer for his meticulous copyediting.

It is my hope that by the choice of topics, examples, and exercises the book will convey some of the intrinsic beauty and intuitive mathematical pleasure of the subject.

Berlin, Spring 2007 Martin Aigner

Contents

Part II: Methods

Introduction

Enumerative combinatorics addresses the problem how to count the number of elements of a finite set given by some combinatorial conditions. We could ask, for example, how many pairs the set $\{1,2,3,4\}$ contains. The answer is 6, as everybody knows, but the result is not really exciting. It gives no hint how many pairs the sets $\{1,2,\ldots,6\}$ or $\{1,2,\ldots,100\}$ will contain. What we really want is a formula for the number of pairs in $\{1,2,\ldots,n\}$, for *any* n.
A typical problem in enumerative combinatorics looks therefore as follows: We are given an *infinite* family of sets S_n, where n runs through some index set I (usually the natural numbers), and the problem consists in determining the *counting function* $f : I \to \mathbb{N}_0$, $f(n) = |S_n|$. The sets may, of course, have two or more indices, say $S_{i,j}$, with $f(i,j) = |S_{i,j}|$.
There is no straightforward answer as to what "determining" a counting function means. In the example of the number of pairs, $f(n) = \frac{n(n-1)}{2}$, and everybody will accept this as a satisfactory answer. In most cases, however, such a "closed" form is not attainable. How should we proceed then?

Summation.
Suppose we want to enumerate the *fixed-point-free* permutations of $\{1,2,\ldots,n\}$, that is, all permutations σ with $\sigma(i) \neq i$ for all i. Let D_n be their number; they are called the *derangement numbers*, since after permuting them no item appears in its original place. For $n = 3$, $\binom{1\,2\,3}{2\,3\,1}$ and $\binom{1\,2\,3}{3\,1\,2}$ are the only fixed-point-free permutations, hence $D_3 = 2$. We will later prove $D_n = n! \sum_{k=0}^{n} \frac{(-1)^k}{k!}$; the counting function is expressed as a *summation formula.*

Recurrence.
Combinatorial considerations yield, as we shall see, the recurrence $D_n = (n-1)(D_{n-1} + D_{n-2})$ for $n \geq 3$. From the starting values $D_1 = 0$, $D_2 = 1$, one obtains $D_3 = 2$, $D_4 = 9$, $D_5 = 44$, and by induction the general formula $D_n = n! \sum_{k=0}^{n} \frac{(-1)^k}{k!}$. Sometimes we might even prefer a recurrence to a closed formula. The *Fibonacci*

numbers F_n are defined by $F_0 = 0$, $F_1 = 1$, $F_n = F_{n-1} + F_{n-2}$ $(n \geq 2)$. We will later derive the formula $F_n = \frac{1}{\sqrt{5}}\left[(\frac{1+\sqrt{5}}{2})^n - (\frac{1-\sqrt{5}}{2})^n\right]$. This expression is quite useful for studying number-theoretic questions about F_n, but the combinatorial properties are much more easily revealed through the defining recurrence.

Generating Function.
An entirely different idea regards the values $f(n)$ as *coefficients* of a power series $F(z) = \sum_{n\geq 0} f(n)z^n$. $F(z)$ is then called the *generating function* of the counting function f. If f has two arguments, then it will be represented by $F(y,z) = \sum_{m,n} f(m,n)y^m z^n$, and similarly for any number of variables. So why should this say anything new about f? The strength of the method rests on the fact that we can perform *algebraic* operations on these series, like sum, product, or derivative, and then read off recurrences or identities as equations between coefficients. As an example, we shall see that for the derangement numbers, $\sum_{n\geq 0} \frac{D_n}{n!} z^n = \frac{e^{-z}}{1-z}$, and this generating function encodes all the information about the numbers D_n.
The first two chapters lay the ground for these ideas. They contain, so to speak, the vocabulary of combinatorial enumeration, the fundamental coefficients, and types of generating functions with which all counting begins.
In Part II we are going to apply the elementary combinatorial objects and data sets. In most problems the work to be done is to
- solve recurrences,
- evaluate sums,
- establish identities,
- manipulate generating functions,
- find explicit bijections,
- compute determinants.

In Chapters 3–6 general methods are developed that will allow a systematic, almost mechanical approach to a great variety of concrete enumeration problems.
Finally, in the third part we look at several important and beautiful topics where enumeration methods have proved extremely useful. They range from questions in analysis and algebra to problems in knot theory and models in statistical physics, and will, it is hoped, convince the reader of the power and elegance of combinatorial reasoning.

Part I: Basics

1 Fundamental Coefficients

1.1 Elementary Counting Principles

We begin by collecting a few simple rules that, though obvious, lie at the root of all combinatorial counting. In fact, they are so obvious that they do not need a proof.

Rule of Sum. *If $S = \bigcup_{i=1}^{t} S_i$ is a union of disjoint sets S_i, then $|S| = \sum_{i=1}^{t} |S_i|$.*

In applications, the rule of sum usually appears in the following form: we classify the elements of S according to a set of properties e_i $(i = 1,\dots,t)$ that preclude each other, and set $S_i = \{x \in S : x \text{ has } e_i\}$.

The sum rule is the basis for most recurrences. Consider the following example. A set X with n elements is called an *n-set*. Denote by $S = \binom{X}{k}$ the family of all k-subsets of X. Thus $|S| = \binom{n}{k}$, where $\binom{n}{k}$ is the usual binomial coefficient. For the moment $\binom{n}{k}$ is just a symbol, denoting the size of $\binom{X}{k}$. Let $a \in X$. We classify the members of S as to whether they do or do not contain a: $S_1 = \{A \in S : a \in A\}$, $S_2 = \{A \in S : a \notin A\}$. We obtain all sets in S_1 by combining all $(k-1)$-subsets of $X \smallsetminus a$ with a; thus $|S_1| = \binom{n-1}{k-1}$. Similarly, S_2 is the family of all k-subsets of $X \smallsetminus a$: $|S_2| = \binom{n-1}{k}$. The rule of sum yields therefore the *Pascal recurrence* for binomial coefficients

$$\binom{n}{k} = \binom{n-1}{k-1} + \binom{n-1}{k} \qquad (n \geq k \geq 1)$$

with initial value $\binom{n}{0} = 1$.

Note that we obtain this recurrence without having computed the binomial coefficients.

Rule of Product. *If $S = \prod_{i=1}^{t} S_i$ is a product of sets, then $|S| = \prod_{i=1}^{t} |S_i|$.*

S consists of all t-tuples $(a_1, a_2, \dots, a_t)$, $a_i \in S_i$, and the sets S_i are called the *coordinate sets* .

Example. A sequence of 0's and 1's is called a *word* over $\{0,1\}$, and the number of 0's and 1's the *length* of the word. Since any coordinate set S_i has two elements, the product rule states that there are 2^n n-words over $\{0,1\}$. More generally, we obtain r^n words if the alphabet A contains r elements. We then speak of n-words over the *alphabet* A.

Rule of Bijection. *If there is a bijection between* S *and* T, *then* $|S| = |T|$.

The typical application goes as follows: Suppose we want to count S. If we succeed in mapping S bijectively onto a set T (whose size t is known), then we can conclude that $|S| = t$.

Example. A simple but extremely useful bijection maps the power-set 2^X of an n-set X, i.e., the family of *all* subsets of X, onto the n-words over $\{0,1\}$. Index $X = \{x_1, x_2, \ldots, x_n\}$ in any way, and map $A \subseteq X$ to $(a_1, a_2, \ldots, a_n)$ where $a_i = 1$ if $x_i \in A$ and $a_i = 0$ if $x_i \notin A$. This is obviously a bijection, and we conclude that $|2^X| = 2^n$. The word $(a_1, \ldots, a_n)$ is called the *incidence vector* or *characteristic vector* of A.

The rule of bijection is the source of many intriguing combinatorial problems. We will see several examples in which we deduce by algebraic or other means that two sets S and T have the same size. Once we know that $|S| = |T|$, there exists, of course, a bijection between these sets. But it may be and often is a challenging problem to find in the aftermath a "natural" bijection based on combinatorial ideas.

Rule of Counting in Two Ways. *When two formulas enumerate the same set, then they must be equal.*

This rule sounds almost frivolous, yet it often reveals very interesting identities. Consider the following formula:

$$\sum_{i=1}^{n} i = \frac{(n+1)n}{2}. \qquad (1)$$

We may, of course, prove (1) by induction, but here is a purely combinatorial argument. Take an $(n+1) \times (n+1)$ array of dots, e.g., for $n = 4$:

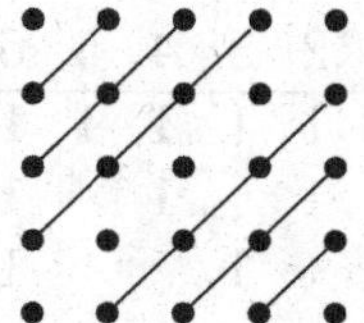

The diagram contains $(n+1)^2$ dots. But there is another way to count the dots, namely by way of diagonals, as indicated in the figure. Clearly, both the upper and lower parts account for $\sum_{i=1}^{n} i$ dots. Together with the middle diagonal this gives $2\sum_{i=1}^{n} i + (n+1) = (n+1)^2$, and thus $\sum_{i=1}^{n} i = \frac{(n+1)n}{2}$.

We even get a bonus out of it: the sum $\sum_{i=1}^{n} i$ enumerates another quantity, the family S of all *pairs* in the $(n+1)$-set $\{0,1,2,\ldots,n\}$. Indeed, we may partition S into disjoint sets S_i according to the *larger* element i, $i = 1,\ldots,n$. Clearly, $|S_i| = i$, and thus by the sum rule $|S| = \sum_{i=1}^{n} i$. Hence we have the following result: the number of pairs in an n-set is $\binom{n}{2} = \frac{n(n-1)}{2}$.

The typical application of the rule of counting in two ways is to consider incidence systems. An *incidence system* consists of two sets S and T together with a relation I. If aIb, $a \in S$, $b \in T$, then we call a and b *incident*. Let $d(a)$ be the number of elements in T that are incident to $a \in S$, and similarly $d(b)$ for $b \in T$. Then

$$\sum_{a \in S} d(a) = \sum_{b \in T} d(b) .$$

The equality becomes obvious when we associate to the system its *incidence matrix* M. Let $S = \{a_1,\ldots,a_m\}$, $T = \{b_1,\ldots,b_n\}$, then $M = (m_{ij})$ is the $(0,1)$-matrix with

$$m_{ij} = \begin{cases} 1 & \text{if } a_i I b_j, \\ 0 & \text{otherwise.} \end{cases}$$

The quantity $d(a_i)$ is then the i-th row sum $\sum_{j=1}^{n} m_{ij}$, $d(b_j)$ is the j-th column sum $\sum_{i=1}^{m} m_{ij}$. Thus we count the total number of 1's once by row sums and the other time columnwise.

Example. Consider the numbers 1 to 8, and set $m_{ij} = 1$ if i divides j, denoted $i \mid j$, and 0 otherwise. The incidence matrix of this divisor relation looks as follows, where we have omitted the 0's:

	1	2	3	4	5	6	7	8
1	1	1	1	1	1	1	1	1
2		1		1		1		1
3			1			1		
4				1				1
5					1			
6						1		
7							1	
8								1

The j-th column sum is the number of divisors of j, which we denote by $t(j)$ thus, e.g., $t(6) = 4, t(7) = 2$. Let us ask how many divisors a number from 1 to 8 has on *average*. Hence we want to compute $\overline{t}(8) = \frac{1}{8}\sum_{j=1}^{8} t(j)$. In our example $\overline{t}(8) = \frac{5}{2}$, and we deduce from the matrix that

n	1	2	3	4	5	6	7	8
$\overline{t}(n)$	1	$\frac{3}{2}$	$\frac{5}{3}$	2	2	$\frac{7}{3}$	$\frac{16}{7}$	$\frac{5}{2}$

How large is $\overline{t}(n)$ for arbitrary n? At first sight this appears hopeless. For prime numbers p we have $t(p) = 2$, whereas for powers of 2, say, an arbitrarily large value $t(2^k) = k+1$ results. So we might expect that the function $\overline{t}(n)$ shows an equally erratic behavior. The following beautiful application of counting in two ways demonstrates that quite the opposite is true!

Counting by columns we get $\sum_{j=1}^{n} t(j)$. How many 1's are in row i? They correspond to the multiples of i, $1 \cdot i, 2 \cdot i, \ldots$, and the last multiple is $\lfloor \frac{n}{i} \rfloor i$. Our rule thus yields

$$\overline{t}(n) = \frac{1}{n}\sum_{j=1}^{n} t(j) = \frac{1}{n}\sum_{i=1}^{n} \lfloor \frac{n}{i} \rfloor \sim \frac{1}{n}\sum_{i=1}^{n} \frac{n}{i} = \sum_{i=1}^{n} \frac{1}{i},$$

where the error going from the second to the third sum is less than 1. The last sum $H_n = \sum_{i=1}^{n} \frac{1}{n}$ is called the n-th *harmonic number*. We know from analysis (by approximating $\log x = \int_1^x \frac{1}{t}\,dt$) that $H_n \sim \log n$, and obtain the unexpected result that the divisor function, though locally erratic, behaves on average extremely regularly: $\overline{t}(n) \sim \log n$.

You will be asked in the exercises and in later chapters to provide combinatorial proofs of identities or recurrences. Usually, this means a combination of the elementary methods we have discussed in this section.

Exercises

1.1 We are given t disjoint sets S_i with $|S_i| = a_i$. Show that the number of subsets of $S_1 \cup \ldots \cup S_t$ that contain at most one element from each S_i is $(a_1+1)(a_2+1)\cdots(a_t+1)$. Apply this to the following number-theoretic problem. Let $n = p_1^{a_1} p_2^{a_2} \cdots p_t^{a_t}$ be the prime decomposition of n then $t(n) = \prod_{i=1}^{t}(a_i + 1)$. Conclude that n is a perfect square precisely when $t(n)$ is odd.

▷ **1.2** In the parliament of some country there are 151 seats filled by 3 parties. How many possible distributions (i, j, k) are there that give no party an absolute majority?

1.3 Use the sum rule to prove $\sum_{k=0}^{n} 2^k = 2^{n+1} - 1$, and to evaluate $\sum_{k=1}^{n}(n-k)2^{k-1}$.

1.4 Suppose the chairman of the math department stipulates that every student must enroll in exactly 4 of 7 offered courses. The teachers give the number in their classes as 51, 30, 30, 20, 25, 12, and 18, respectively. What conclusion can be drawn?

▷ **1.5** Show by counting in two ways that $\sum_{i=1}^{n} i(n-i) = \sum_{i=1}^{n}\binom{i}{2} = \binom{n+1}{3}$.

* * *

1.6 Join any two corners of a convex n-gon by a chord, and let $f(n)$ be the number of pairs of crossing chords, e.g., $f(4) = 1$, $f(5) = 5$. Determine $f(n)$ by Pascal's recurrence. The result is very simple. Can you establish the formula by a direct argument?

1.7 In how many ways can one list the numbers $1, 2, \ldots, n$ such that apart from the leading element the number k can be placed only if either $k-1$ or $k+1$ already appears? Example: 324516, 435216, but not 351246.

▷ **1.8** Let $f(n, k)$ be the number of k-subsets of $\{1, 2, \ldots, n\}$ that do not contain a pair of consecutive integers. Show that $f(n, k) = \binom{n-k+1}{k}$, and further that $\sum_{k=0}^{n} f(n, k) = F_{n+2}$ (Fibonacci number).

1.9 Euler's φ-function is $\varphi(n) = \#\{k : 1 \le k \le n,\ k \text{ relatively prime to } n\}$. Use the sum rule to prove $\sum_{d|n}\varphi(d) = n$.

1.10 Evaluate $\sum_{i=1}^{n} i^2$ and $\sum_{i=1}^{n} i^3$ by counting configurations of dots as in the proof of $\sum_{i=1}^{n} i = \frac{n(n+1)}{2}$.

▷ **1.11** Let $N = \{1, 2, \ldots, 100\}$, and $A \subseteq N$ with $|A| = 55$. Show that A contains two numbers with difference 9. Is this also true for $|A| = 54$?

1.2 Subsets and Binomial Coefficients

Let N be an n-set. We have already introduced the *binomial coefficient* $\binom{n}{k}$ as the number of k-subsets of N. To derive a formula for $\binom{n}{k}$ we look first at words of length k with symbols from N.

Definition. A *k-permutation* of N is a k-word over N all of whose entries are distinct.

For example, 1235 and 5614 are 4-permutations of $\{1, 2, \ldots, 6\}$. The number of k-permutations is quickly computed. We have n possibilities for the first letter. Once we have chosen the first entry, there are $n-1$ possible choices for the second entry, and so on. The product rule thus gives the following result:

The number of k-permutations of an n-set equals $n(n-1)\cdots(n-k+1)$ $(n, k \geq 0)$.

For $k = n$ we obtain, in particular, $n! = n(n-1)\cdots 2 \cdot 1$ for the number of n-permutations, i.e., of ordinary permutations of N. As usual, we set $0! = 1$.

The expressions $n(n-1)\cdots(n-k+1)$ appear so frequently in enumeration problems that we give them a special name:

$n^{\underline{k}} := n(n-1)\cdots(n-k+1)$ are the *falling factorials* of length k, with $n^{\underline{0}} = 1$ $(n \in \mathbb{Z},\ k \in \mathbb{N}_0)$.

Similarly,

$n^{\overline{k}} := n(n+1)\cdots(n+k-1)$ are the *rising factorials* of length k, with $n^{\overline{0}} = 1$ $(n \in \mathbb{Z},\ k \in \mathbb{N}_0)$.

Now, every k-permutation consists of a unique k-subset of N. Since every k-subset can be permuted in $k!$ ways to produce a k-permutation, counting in two ways gives $k!\binom{n}{k} = n^{\underline{k}}$, hence

$$\binom{n}{k} = \frac{n^{\underline{k}}}{k!} = \frac{n(n-1)\cdots(n-k+1)}{k!} \qquad (n, k \geq 0), \tag{1}$$

where, of course, $\binom{n}{k} = 0$ for $n < k$.

Another way to write (1) is

$$\binom{n}{k} = \frac{n!}{k!(n-k)!} \quad (n \geq k \geq 0), \tag{2}$$

from which $\binom{n}{k} = \binom{n}{n-k}$ results.

Identities and formulas involving binomial coefficients fill whole books; Chapter 5 of Graham–Knuth–Patashnik gives a comprehensive survey. Let us just collect the most important facts.

Pascal Recurrence.

$$\binom{n}{k} = \binom{n-1}{k-1} + \binom{n-1}{k}, \quad \binom{n}{0} = 1 \quad (n, k \geq 0). \tag{3}$$

We have already proved this recurrence in Section 1.1; it also follows immediately from (1).

Now we make an important observation, the so-called *polynomial method*. The polynomials

$$x^{\underline{k}} = x(x-1)(x-2)\cdots(x-k+1), x^{\overline{k}} = x(x+1)(x+2)\cdots(x+k-1)$$

over $\mathbb{C}$ (or any field of characteristic 0) are again called the falling resp. rising factorials, where $x^{\underline{0}} = x^{\overline{0}} = 1$. Consider the polynomials

$$\frac{x^{\underline{k}}}{k!} \quad \text{and} \quad \frac{(x-1)^{\underline{k-1}}}{(k-1)!} + \frac{(x-1)^{\underline{k}}}{k!}.$$

Both have degree k, and we know that two polynomials of degree k that agree in more than k values are identical. But in our case they even agree for *infinitely* many values, namely for all non-negative integers, and so we obtain the *polynomial* identity

$$\frac{x^{\underline{k}}}{k!} = \frac{(x-1)^{\underline{k-1}}}{(k-1)!} + \frac{(x-1)^{\underline{k}}}{k!} \quad (k \geq 1). \tag{4}$$

Thus, if we set $\binom{c}{k} = \frac{c^{\underline{k}}}{k!} = \frac{c(c-1)\cdots(c-k+1)}{k!}$ for arbitrary $c \in \mathbb{C}$ ($k \geq 0$), then Pascal's recurrence holds for $\binom{c}{k}$. In fact, it is convenient to extend the definition to negative integers k, setting

$$\binom{c}{k} = \begin{cases} \frac{c^{\underline{k}}}{k!} & (k \geq 0) \\ 0 & (k < 0). \end{cases}$$

Pascal's recurrence holds then in general, since for $k < 0$ both sides are 0:

$$\binom{c}{k} = \binom{c-1}{k-1} + \binom{c-1}{k} \quad (c \in \mathbb{C}, k \in \mathbb{Z}). \tag{5}$$

As an example, $\binom{-1}{n} = \frac{(-1)(-2)\cdots(-n)}{n!} = (-1)^n$.

Here is another useful polynomial identity. From

$$(-x)^{\underline{k}} = (-x)(-x-1)\cdots(-x-k+1) = (-1)^k x(x+1)\cdots(x+k-1)$$

we get

$$(-x)^{\underline{k}} = (-1)^k x^{\overline{k}}, \quad (-x)^{\overline{k}} = (-1)^k x^{\underline{k}}. \tag{6}$$

With $x^{\overline{k}} = (x+k-1)^{\underline{k}}$ this gives

$$\binom{-c}{k} = (-1)^k \binom{c+k-1}{k}, \quad (-1)^k \binom{c}{k} = \binom{k-c-1}{k}. \tag{7}$$

Equation (6) is called the *reciprocity law* between the falling and rising factorials.

The recurrence (3) gives the Pascal matrix $P = \left(\binom{n}{k}\right)$ with n as row index and k as column index. P is a lower triangular matrix with 1's on the main diagonal. The table shows the first rows and columns, where the 0's are omitted.

$n \backslash k$	0	1	2	3	4	5	6	7
0	1							
1	1	1						
2	1	2	1					
3	1	3	3	1				
4	1	4	6	4	1			
5	1	5	10	10	5	1		
6	1	6	15	20	15	6	1	
7	1	7	21	35	35	21	7	1

$\binom{n}{k}$

There are many beautiful and sometimes mysterious relations in the Pascal matrix to be discovered. Let us note a few formulas that we will need time and again. First, it is clear that $\sum_{k=0}^{n}\binom{n}{k} = 2^n$, since we are counting all subsets of an n-set. Consider the column-sum of index k down to row n, i.e., $\sum_{i=0}^{n}\binom{i}{k}$. By classifying the $(k+1)$-subsets of $\{1,2,\ldots,n+1\}$ according to the last element $i+1$ $(0 \le i \le n)$ we obtain

$$\sum_{i=0}^{n}\binom{i}{k} = \binom{n+1}{k+1}. \tag{8}$$

Let us next look at the down diagonal from left to right, starting with row m and column 0. That is, we want to sum $\sum_{i=0}^{n}\binom{m+i}{i}$. In the table above, the diagonal with $m = 3$, $n = 3$ is marked, summing to $35 = \binom{7}{3}$. Writing $\sum_{i=0}^{n}\binom{m+i}{i} = \sum_{i=0}^{n}\binom{m+i}{m} = \sum_{k=0}^{m+n}\binom{k}{m}$, this is just a sum like that in (8), and we obtain

$$\sum_{i=0}^{n}\binom{m+i}{i} = \binom{m+n+1}{n}. \tag{9}$$

Note that (9) holds in general for $m \in \mathbb{C}$.

From the reciprocity law (7) we may deduce another remarkable formula. Consider the *alternating* partial sums in row 7: $1, 1-7 = -6, 1-7+21 = 15, -20, 15, -6, 1, 0$. We note that these are precisely the binomial coefficients immediately above, with alternating sign. Let us prove this in general; (7) and (9) imply

$$\sum_{k=0}^{m}(-1)^k\binom{n}{k} = \sum_{k=0}^{m}\binom{k-n-1}{k} = \binom{m-n}{m} = (-1)^m\binom{n-1}{m}. \tag{10}$$

The reader may wonder whether there is also a simple formula for the partial sums $\sum_{k=0}^{m}\binom{n}{k}$ without signs. We will address this question of when a "closed" formula exists in Chapter 4 (and the answer for this particular case will be no).

Next, we note an extremely useful identity that follows immediately from (2); you are asked in the exercises to provide a combinatorial argument:

$$\binom{n}{m}\binom{m}{k} = \binom{n}{k}\binom{n-k}{m-k} \quad (n, m, k \in \mathbb{N}_0). \tag{11}$$

Binomial Theorem.

$$(x+y)^n = \sum_{k=0}^{n} \binom{n}{k} x^k y^{n-k} . \tag{12}$$

Expand the left-hand side, and classify according to the number of x's taken from the factors. The formula is an immediate consequence.

For $y = 1$ respectively $y = -1$ we obtain

$$(x+1)^n = \sum_{k=0}^{n} \binom{n}{k} x^k, \ (x-1)^n = \sum_{k=0}^{n} (-1)^{n-k} \binom{n}{k} x^k , \tag{13}$$

and hence for $x = 1$, $\sum_{k=0}^{n} \binom{n}{k} = 2^n$ and

$$\sum_{k=0}^{n} (-1)^k \binom{n}{k} = \delta_{n,0} , \tag{14}$$

where $\delta_{i,j}$ is the Kronecker symbol

$$\delta_{i,j} = \begin{cases} 1 & i = j, \\ 0 & i \neq j. \end{cases}$$

This last formula will be the basis for the inclusion–exclusion principle in Chapter 5. We may prove (14) also by the bijection principle. Let N be an n-set, and set $S_0 = \{A \subseteq N : |A| \text{ even}\}$, $S_1 = \{A \subseteq N : |A| \text{ odd}\}$. Formula (14) is then equivalent to $|S_0| = |S_1|$ for $n \geq 1$. To see this, pick $a \in N$ and define $\phi : S_0 \to S_1$ by

$$\phi(A) = \begin{cases} A \cup a & \text{if } a \notin A, \\ A \smallsetminus a & \text{if } a \in A. \end{cases}$$

This is a desired bijection.

Vandermonde Identity.

$$\binom{x+y}{n} = \sum_{k=0}^{n} \binom{x}{k} \binom{y}{n-k} \quad (n \in \mathbb{N}_0) . \tag{15}$$

Once again the polynomial method applies. Let R and S be disjoint sets with $|R| = r$ and $|S| = s$. The number of n-subsets of $R \cup S$ is $\binom{r+s}{n}$. On the other hand, any such set arises by combining a k-subset of R with an $(n-k)$-subset of S. Classifying the n-subsets A according to $|A \cap R| = k$ yields

$$\binom{r+s}{n} = \sum_{k=0}^{n} \binom{r}{k}\binom{s}{n-k} \quad \text{for all } r, s \in \mathbb{N}_0 .$$

The polynomial method completes the proof.

Example. We have $\sum_{k=0}^{n} \binom{n}{k}^2 = \sum_{k=0}^{n} \binom{n}{k}\binom{n}{n-k} = \binom{2n}{n}$.

Multiplying both sides of (15) by $n!$ we arrive at a "binomial" theorem for the falling factorials:

$$(x+y)^{\underline{n}} = \sum_{k=0}^{n} \binom{n}{k} x^{\underline{k}} y^{\underline{n-k}} \tag{16}$$

and the reciprocity law (6) gives the analogous statement for the rising factorials:

$$(x+y)^{\overline{n}} = \sum_{k=0}^{n} \binom{n}{k} x^{\overline{k}} y^{\overline{n-k}} . \tag{17}$$

Multisets.

In a set all elements are distinct, in a *multiset* we drop this requirement. For example, $M = \{1,1,2,2,3\}$ is a multiset over $\{1,2,3\}$ of size 5, where 1 and 2 appear with multiplicity 2. Thus the size of a multiset is the number of elements counted with their multiplicities. The following formula shows the importance of rising factorials:

The number of k-multisets of an n-set is

$$\frac{n^{\overline{k}}}{k!} = \frac{n(n+1)\cdots(n+k-1)}{k!} = \binom{n+k-1}{k} . \tag{18}$$

Just as a k-subset A of $\{1,2,\ldots,n\}$ can be interpreted as a *monotone* k-word $A = \{1 \le a_1 < a_2 < \cdots < a_k \le n\}$, a k-multiset is a monotone k-word *with repetitions* $\{1 \le a_1 \le \cdots \le a_k \le n\}$. This interpretation immediately leads to a proof of (18) by the bijection rule. The map $\phi : A = \{a_1 \le a_2 \le \cdots \le a_k\} \longrightarrow A' = \{1 \le a_1 < a_2+1 < a_3+2 < \cdots < a_k+k-1 \le n+k-1\}$ is clearly a bijection, and (18) follows.

Multinomial Theorem.

$$(x_1 + \cdots + x_m)^n = \sum_{(k_1,\cdots,k_m)} \binom{n}{k_1 \ldots k_m} x_1^{k_1} \cdots x_m^{k_m}, \qquad (19)$$

where

$$\binom{n}{k_1 \ldots k_m} = \frac{n!}{k_1! \cdots k_m!}, \quad \sum_{i=1}^{m} k_i = n, \qquad (20)$$

is the *multinomial coefficient.*

The proof is similar to that of the binomial theorem. Expanding the left-hand side we pick x_1 out of k_1 factors; this can be done in $\binom{n}{k_1} = \frac{n!}{k_1!(n-k_1)!}$ ways. Out of the remaining $n - k_1$ factors we choose x_2 from k_2 factors in $\binom{n-k_1}{k_2} = \frac{(n-k_1)!}{k_2!(n-k_1-k_2)!}$ ways, and so on.

A useful interpretation of the multinomial coefficients is the following. The ordinary binomial coefficient $\binom{n}{k}$ counts the number of n-words over $\{x, y\}$ with exactly k x's and $n - k$ y's. Similarly, the multinomial coefficient $\binom{n}{k_1 \ldots k_m}$ is the number of n-words over an alphabet $\{x_1, \ldots, x_m\}$ in which x_i appears exactly k_i times.

Lattice Paths.
Finally, we discuss an important and pleasing way to look at binomial coefficients. Consider the $(m \times n)$-lattice of integral points in $\mathbb{Z}^2$, e.g., $m = 6$, $n = 5$ as in the figure,

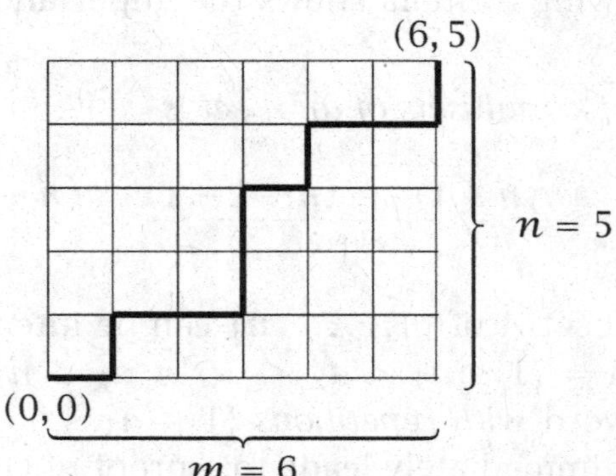

and look at all lattice paths starting at $(0,0)$, terminating at (m,n), with steps one to the right or one upward. We will call the horizontal steps $(1,0)$-steps since the x-coordinate is increased by 1, and similarly, we call the vertical steps $(0,1)$-steps. Let $L(m,n)$ be the number of these lattice paths. The initial conditions are $L(m,0) = L(0,n) = 1$, and classification according to the first step immediately gives

$$L(m,n) = L(m-1,n) + L(m,n-1).$$

This is precisely Pascal's recurrence for $\binom{m+n}{m}$, and we conclude that

$$L(m,n) = \binom{m+n}{m}. \tag{21}$$

Another quick way to see this is by encoding the paths. We assign the symbol E(ast) to a $(1,0)$-step and N(orth) to a $(0,1)$-step. The lattice paths correspond then bijectively to $(m+n)$-words over $\{E,N\}$ with precisely m E's, and this is $\binom{m+n}{m}$. In the example above, the encoding is given by ENEENNENEEN. The lattice path interpretation allows easy and elegant proofs of many identities involving binomial coefficients.

Example. Consider the following variant of the Vandermonde identity: $\sum_{k=0}^{n}\binom{s+k}{k}\binom{n-k}{m} = \binom{s+n+1}{s+m+1}$ $(s,m,n \in \mathbb{N}_0)$. For $n < m$, both sides are 0, so assume $n \geq m$, and look at the $(s+m+1) \times (n-m)$-lattice. The number of paths is $\binom{s+n+1}{s+m+1}$. Now we classify the paths according to the *highest* coordinate $y = k$ where they touch the vertical line $x = s$.

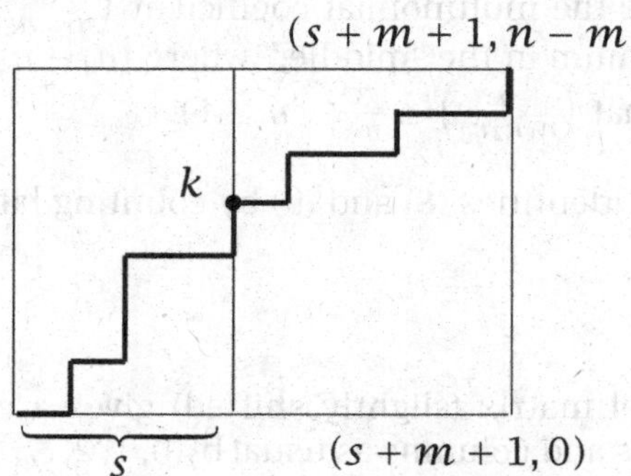

Then the next step is a $(1,0)$-step, and the sum and product rules give

$$\binom{s+n+1}{s+m+1} = \sum_{k=0}^{n-m} \binom{s+k}{k}\binom{m+(n-m-k)}{m} = \sum_{k=0}^{n} \binom{s+k}{k}\binom{n-k}{m}.$$

Exercises

1.12 Prove $\binom{n}{m}\binom{m}{k} = \binom{n}{k}\binom{n-k}{m-k}$ $(n \geq m \geq k \geq 0)$ by counting pairs of sets (A, B) in two ways, and deduce $\sum_{k=0}^{m}\binom{n}{k}\binom{n-k}{m-k} = 2^m\binom{n}{m}$.

▷ **1.13** Use the previous exercise to show that $\binom{2n}{2k}\binom{2n-2k}{n-k}\binom{2k}{k} = \binom{2n}{n}\binom{n}{k}^2$ for $n \geq k \geq 0$.

1.14 Show that $\binom{n}{k} = \frac{n}{k}\binom{n-1}{k-1}$, $\binom{n}{k} = \frac{k+1}{n-k}\binom{n}{k+1}$, and use this to verify the *unimodal* property for the sequence $\binom{n}{k}$, $0 \leq k \leq n$: $\binom{n}{0} < \binom{n}{1} < \cdots < \binom{n}{\lfloor n/2 \rfloor} = \binom{n}{\lceil n/2 \rceil} > \cdots > \binom{n}{n}$.

1.15 Show that that the sum of right-left diagonals in the Pascal matrix ending at $(n, 0)$ is the Fibonacci number F_{n+1}, i.e., $F_{n+1} = \sum_{k \geq 0}\binom{n-k}{k}$.

1.16 Show that $r^{\underline{k}}(r - \frac{1}{2})^{\underline{k}} = \frac{(2r)^{\underline{2k}}}{2^{2k}}$ $(r \in \mathbb{C}, k \in \mathbb{N}_0)$, and deduce $\binom{-1/2}{n} = (-\frac{1}{4})^n\binom{2n}{n}$, $\binom{-3/2}{n} = (-\frac{1}{4})^n(2n+1)\binom{2n}{n}$.

▷ **1.17** Show that the multinomial coefficient $\binom{n}{n_1 \ldots n_k}$ assumes for fixed n and k its maximum in the "middle," where $|n_i - n_j| \leq 1$ for all i, j. Prove in particular that $\binom{n}{n_1 n_2 n_3} \leq \frac{3^n}{n+1}$ $(n \geq 1)$.

1.18 Prove the identities (8) and (9) by counting lattice paths.

* * *

▷ **1.19** The Pascal matrix (slightly shifted) gives a curious prime number test. Index rows and columns as usual by $0, 1, 2, \ldots$. In row n we insert the $n+1$ binomial coefficients $\binom{n}{0}, \binom{n}{1}, \ldots, \binom{n}{n}$, but shifted to the columns $2n, \ldots, 3n$. In addition, we draw a circle around each of these numbers that is a multiple of n, as in the table.

n \ k	0	1	2	3	4	5	6	7	8	9	10	11	12
0	1												
1			①	①									
2					1	②	1						
3							1	③	③	1			
4									1	④	6	④	1

Show that k is a prime number if and only if all elements in column k are circled. Hint: k even is easy, and for odd k the element in position (n, k) is $\binom{n}{k-2n}$.

1.20 Let $a_n = \frac{1}{\binom{n}{0}} + \frac{1}{\binom{n}{1}} + \cdots + \frac{1}{\binom{n}{n}}$. Show that $a_n = \frac{n+1}{2n} a_{n-1} + 1$ and compute $\lim_{n\to\infty} a_n$ (if the limit exists). Hint: $a_n > 2 + \frac{2}{n}$ and $a_{n+1} < a_n$ for $n \geq 4$.

▷ **1.21** Consider $(m+n)$-words with exactly m 1's and n 0's. Count the number of these words with exactly k runs, where a run is a maximal subsequence of consecutive 1's. Example: 1011100110 has 3 runs.

1.22 Prove the following variants of Vandermonde's identity algebraically (manipulating binomial coefficients) and by counting lattice paths.

a. $\sum_k \binom{r}{m+k}\binom{s}{n-k} = \binom{r+s}{m+n}$, b. $\sum_k \binom{r}{m+k}\binom{s}{n+k} = \binom{r+s}{r-m+n}$,

1.23 Give a combinatorial argument for the identity

$$\sum_k \binom{2r}{2k-1}\binom{k-1}{s-1} = 2^{2r-2s+1}\binom{2r-s}{s-1}, r, s \in \mathbb{N}_0 .$$

▷ **1.24** Consider the $(m \times n)$-lattice in $\mathbb{Z}^2$. A *Delannoy path* from $(0,0)$ to (m,n) uses steps $(1,0)$, $(0,1)$ and diagonal steps $(1,1)$ from (x,y) to $(x+1, y+1)$. The number of these paths is the *Delannoy number* $D_{m,n}$. Example for $D_{2,1} = 5$:

.

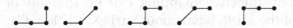

Prove that $D_{m,n} = \sum_k \binom{m}{k}\binom{n+k}{m}$. Hint: Classify the paths according to the number of diagonal steps.

1.25 Prove the identity $\sum_k \binom{m-r+s}{k}\binom{n+r-s}{n-k}\binom{r+k}{m+n} = \binom{r}{m}\binom{s}{n}$, $m, n \in \mathbb{N}_0$. Hint: Write $\binom{r+k}{m+n} = \sum_i \binom{r}{m+n-i}\binom{k}{i}$, and apply (11).

1.26 Prove that $\frac{2^{2n}}{2\sqrt{n}} \leq \binom{2n}{n} \leq \frac{2^{2n}}{\sqrt{n}}$ for $n \geq 1$. Hint: For the upper bound prove the stronger result $\binom{2n}{n} \leq 2^{2n} / (1 + \frac{1}{n})\sqrt{n}$.

1.3 Set-partitions and Stirling Numbers $S_{n,k}$

Our next combinatorial objects are the partitions of a set N into non-empty disjoint sets A_i, $N = A_1 \dot{\cup} \cdots \dot{\cup} A_k$. Let us denote by $\prod(N)$ the family of all partitions of N. The sets A_i are called the *blocks* of the partition, and a partition into k blocks is a *k-partition.*

Definition. The *Stirling number $S_{n,k}$* (of the second kind) is the number of k-partitions of an n-set, where by definition $S_{0,0} = 1$, $S_{0,k} = 0$ $(k > 0)$. The number of all partitions is the *Bell number* $\text{Bell}(n)$; thus $\text{Bell}(n) = \sum_{k=0}^{n} S_{n,k}$, $\text{Bell}(0) = 1$.

Why the numbers $S_{n,k}$ are called Stirling numbers of the *second* kind has historical reasons that will become clear in the next section. (As expected there are also those of the first kind.) The notation $S_{n,k}$ is the most widely used; in computer science books one also finds $\left\{ {n \atop k} \right\}$.

Example. $N = \{1,2,3,4,5\}$ has fifteen 2-partitions; thus $S_{5,2} = 15$:

1234|5	123|45	145|23
1235|4	124|35	234|15
1245|3	125|34	235|14
1345|2	134|25	245|13
2345|1	135|24	345|12

Stirling numbers occur quite naturally when we count mappings between sets N and R. Let us denote by $\text{Map}(N,R)$ the set of all mappings from N to R, by $\text{Inj}(N,R)$ the *injective* mappings, and by $\text{Surj}\,(N,R)$ the *surjective* mappings.

Suppose $|N| = n$, $|R| = r$. Clearly, $|\text{Map}(N,R)| = r^n$, and $|\text{Inj}(N,R)| = r^{\underline{n}}$, since any injective mapping can be thought of as an n-permutation of the set R. What about $\text{Surj}\,(N,R)$? For $f \in \text{Surj}\,(N,R)$ the pre-images $f^{-1}(y)$ $(y \in R)$ form an *ordered* r-partition of N, $f^{-1}(y_1)|f^{-1}(y_2)|\cdots|f^{-1}(y_r)$. Since an ordinary r-partition of N corresponds to $r!$ surjective mappings (by permuting the blocks), we infer

$$|\text{Map}(N,R)| = r^n, \quad |\text{Inj}(N,R)| = r^{\underline{n}}, \quad |\text{Surj}\,(N,R)| = r!S_{n,r}\,. \qquad (1)$$

Now, every $f \in \text{Map}(N,R)$ has a unique image $f(N) = A \subseteq R$ onto which it is mapped surjectively. If we classify $\text{Map}(N,R)$ according to the image we arrive at

$$r^n = |\mathrm{Map}(N,R)| = \sum_{A \subseteq R} |\mathrm{Surj}\,(N,A)| = \sum_{k=0}^{r} \sum_{A:|A|=k} |\mathrm{Surj}\,(N,A)|$$
$$= \sum_{k=0}^{r} \binom{r}{k} k! S_{n,k} = \sum_{k=0}^{r} S_{n,k} r^{\underline{k}}.$$

In conclusion, we have found a formula that combines powers, falling factorials, and Stirling numbers, and our familiar polynomial argument gives the polynomial identity

$$x^n = \sum_{k=0}^{n} S_{n,k} x^{\underline{k}}. \tag{2}$$

Note that we may stop the summation at n, since obviously $S_{n,k} = 0$ for $n < k$.

Polynomial Sequences.
Formula (2) is the first instance in which two polynomial sequences (x^n) and $(x^{\underline{n}})$ are *linearly* connected. Such "connecting" identities are a fertile source for various relations involving combinatorial coefficients. We will elaborate on this theme in Section 2.4; for the moment, let us collect some basic facts.

A *polynomial sequence* is a sequence $p_0(x), p_1(x), p_2(x), \ldots$ of polynomials (over $\mathbb{C}$, or some field K of characteristic 0) with $\deg p_n(x) = n$. Usually, we assume that all $p_n(x)$ have leading coefficient 1. Any such sequence $(p_n(x))$ is a basis of the vector space $K[x]$. Hence, given two such sequences $(p_n(x))$ and $(q_n(x))$ there are unique coefficients $a_{n,k}$ and $b_{n,k}$ with

$$p_n(x) = \sum_{k=0}^{n} a_{n,k} q_k(x), \quad q_n(x) = \sum_{k=0}^{n} b_{n,k} p_k(x). \tag{3}$$

The numbers $a_{n,k}$ respectively $b_{n,k}$ are called the *connecting coefficients* between the sequences $(p_n(x))$ and $(q_n(x))$. They form infinite lower triangular matrices $(a_{n,k})$ and $(b_{n,k})$, since clearly $a_{n,k} = b_{n,k} = 0$ for $n < k$.

The Stirling numbers $S_{n,k}$ are thus the connecting coefficients of the sequence (x^n) expressed in terms of the basis $(x^{\underline{n}})$. As a first application, we use (2) to derive a recurrence for the numbers $S_{n,k}$. We have

$$x^{\underline{k+1}} = x^{\underline{k}}(x-k) = x \cdot x^{\underline{k}} - kx^{\underline{k}},$$

or

$$x \cdot x^{\underline{k}} = x^{\underline{k+1}} + kx^{\underline{k}}.$$

This implies

$$\begin{aligned} x^n = x \cdot x^{n-1} &= \sum_k S_{n-1,k}(x \cdot x^{\underline{k}}) = \sum_k S_{n-1,k}x^{\underline{k+1}} + \sum_k kS_{n-1,k}x^{\underline{k}} \\ &= \sum_k S_{n-1,k-1}x^{\underline{k}} + \sum_k kS_{n-1,k}x^{\underline{k}}. \end{aligned}$$

Comparing this to (2), we have proved algebraically the recurrence

$$S_{n,k} = S_{n-1,k-1} + kS_{n-1,k} \quad (n \geq 1)\,. \tag{4}$$

Of course, we can explain (4) also by a combinatorial argument. Fix $a \in N$, $|N| = n$. The first summand in (4) counts all k-partitions of N in which $\{a\}$ is a singleton block, and the second summand counts those in which a appears in a block of size ≥ 2, since a may be added to any of the k blocks of a k-partition of $N \smallsetminus a$.

Stirling Matrix.
The first rows and columns of the Stirling matrix $(S_{n,k})$ look as follows, where the 0's above the main diagonal are omitted:

n \ k	0	1	2	3	4	5	6	7
0	1							
1	0	1						
2	0	1	1					
3	0	1	3	1				
4	0	1	7	6	1			
5	0	1	15	25	10	1		
6	0	1	31	90	65	15	1	
7	0	1	63	301	350	140	21	1

$S_{n,k}$

A few special values are easily seen: $S_{n,1} = 1$, $S_{n,2} = 2^{n-1} - 1$, $S_{n,n-1} = \binom{n}{2}$, $S_{n,n} = 1$. That $S_{n,n-1} = \binom{n}{2}$ is clear, since any $(n-1)$-partition must contain exactly one pair and otherwise singletons. A partition of N into two blocks is a pair $\{A, N \smallsetminus A\}$ of complementary non-empty subsets; hence $S_{n,2} = \frac{2^n-2}{2} = 2^{n-1} - 1$.

Finally, for the Bell numbers we have the recurrence

$$\mathrm{Bell}(n+1) = \sum_{k=0}^{n} \binom{n}{k} \mathrm{Bell}(k)\,. \tag{5}$$

For the proof fix $a \in N$ and classify the partitions according to the size of the block containing a.

Exercises

▷ **1.27** Show that $\sum_k S_{n+1,k+1} x^{\underline{k}} = (x+1)^n$, and use the polynomial method to prove $S_{n+1,k+1} = \sum_i \binom{n}{i} S_{i,k}$. Verify this last equality also by a combinatorial argument, and deduce again the Bell number recurrence (5).

1.28 Prove the identities $\sum_{i=k}^{n} S_{i,k}(k+1)^{n-i} = S_{n+1,k+1}$, and $\sum_{i=0}^{n} i S_{m+i,i} = S_{m+n+1,n}$, reminiscent of (8), (9) of the previous section.

1.29 Find a formula for $S_{n,3}$.

1.30 Determine the connecting coefficients of (x^n) expressed in terms of the rising factorials $(x^{\overline{n}})$.

* * *

1.31 Verify $S_{n,k} = \sum 1^{a_1-1} 2^{a_2-1} \cdots k^{a_k-1}$, where the sum extends over all solutions of $a_1 + a_2 + \cdots + a_k = n$ in positive integers.

▷ **1.32** Show that $\binom{k+r}{k} S_{n,k+r} = \sum_{i=k}^{n-r} \binom{n}{i} S_{i,k} S_{n-i,r}$.

1.33 Define the polynomials $p_n(x) = \sum_{k=0}^{n} S_{n,k} x^k$. Use the previous exercise to prove the "Stirling binomial theorem"

$$p_n(x+y) = \sum_{k=0}^{n} \binom{n}{k} p_k(x) p_{n-k}(y).$$

1.34 Determine the number $f(n,k)$ of sequences $a_1 a_2 \ldots a_n$ of positive integers such that the largest entry is k, and the first occurrence of i appears before the first occurence of $i+1$ $(1 \le i \le k-1)$. Hint: $f(n,k) = S_{n,k}$.

▷ **1.35** Give a combinatorial argument that the number of partitions of $\{1,\ldots,n\}$ such that no two consecutive numbers appear in the same block is precisely the Bell number $\mathrm{Bell}(n-1)$.

1.36 Show that $(S_{n,0}, S_{n,1}, \ldots, S_{n,n})$ is a unimodal sequence, for every n. More precisely, prove that there is an index $M(n)$ such that

$$S_{n,0} < S_{n,1} < \cdots < S_{n,M(n)} > S_{n,M(n)+1} > \cdots > S_{n,n}$$

or

$$S_{n,0} < S_{n,1} < \cdots < S_{n,M(n)-1} = S_{n,M(n)} > \cdots > S_{n,n},$$

where $M(n) = M(n-1)$ or $M(n) = M(n-1) + 1$.

Hint: Use recurrence (4) and Exercise 1.27.

1.4 Permutations and Stirling Numbers $s_{n,k}$

For a set N we denote by $S(N)$ the set of all permutations of N, and in particular, by $S(n)$ the set of all permutations of $\{1, 2, \ldots, n\}$. Permutations are one of the classical fields of algebra as well as combinatorics.

A permutation $\sigma \in S(n)$ is first of all a bijective mapping whose canonical notation is

$$\sigma = \begin{pmatrix} 1 & 2 & \ldots & n \\ \sigma(1) & \sigma(2) & \ldots & \sigma(n) \end{pmatrix}.$$

With composition, $S(n)$ forms a group, the *symmetric group* of order n. We read a product always from right to left, thus for

$$\sigma = \begin{pmatrix} 123456 \\ 234165 \end{pmatrix}, \tau = \begin{pmatrix} 123456 \\ 134526 \end{pmatrix},$$

we have

$$\tau\sigma = \begin{pmatrix} 123456 \\ 345162 \end{pmatrix} \text{ and } \sigma\tau = \begin{pmatrix} 123456 \\ 241635 \end{pmatrix}.$$

If we fix the domain $\{1, 2, \ldots, n\}$ in increasing order, the second line is a unique n-permutation. We call $\sigma = \sigma(1)\sigma(2)\ldots\sigma(n)$ the *word representation* of σ. Another way to describe σ is by its cycle decomposition. For every i, the sequence $i, \sigma(i), \sigma^2(i), \ldots$ must eventually terminate with, say, $\sigma^k(i) = i$, and we denote by $(i, \sigma(i), \sigma^2(i), \ldots, \sigma^{k-1}(i))$ the *cycle* containing i. Repeating this for all elements, we arrive at the *cycle decomposition* $\sigma = \sigma_1\sigma_2 \cdots \sigma_t$.

Example. $\sigma = \binom{12345678}{35146827}$ has word representation $\sigma = 35146827$ and cycle form $\sigma = (13)(25687)(4)$.

Cycle Decomposition.
Let us discuss first the cycle decomposition. It is clear that we may start a cycle with any element in the cycle; the rest is then determined. For example, (2568) and (5682) describe the same cycle. The cycles of length 1 are the *fixed points* of σ, those of length 2 correspond to *transpositions* $i \leftrightarrow j$. A permutation that consists of one cycle only is called a *cyclic* permutation. By fixing the leading element, we find that there are $(n-1)!$ cyclic permutations of $S(n)$.

The order in which we write down the cycles is clearly irrelevant. We sometimes use the following standard form: Start each cycle with the largest element, and order the cycles with increasing leading elements. The standard form of $\sigma = (13)(25687)(4)$ is therefore $\sigma = (31)(4)(87256)$. Conversely, from any n-permutation $a_1a_2 \dots a_n$ we may uniquely recover the cycle decomposition: $(a_1a_2\dots)$ $(a_i\dots)(a_j\dots)\dots$, where a_i is the first entry larger than a_1, a_j is the first entry after a_i larger than a_i, and so on. Thus, the number of cycles determined in this way equals the number of *left-to-right maxima* of the word $a_1a_2\dots a_n$.

A natural and very useful graphical representation is to interpret $\sigma \in S(n)$ as a directed graph with $i \to j$ if $j = \sigma(i)$. The cycles of σ correspond then to the directed circuits of the graph.

Example. $\sigma = (142)(738)(5)(69)$ corresponds to the graph

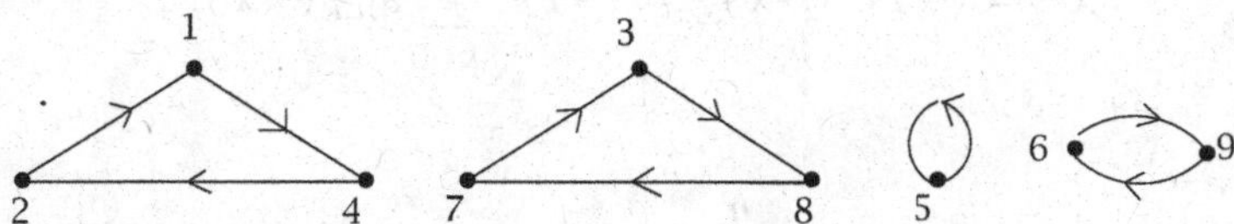

Definition. The *Stirling number* $s_{n,k}$ of the first kind is the number of permutations of an n-set with precisely k cycles, where we set $s_{0,0} = 1$ and $s_{0,k} = 0$ $(k > 0)$ as usual.

Hence $\sum_{k=0}^{n} s_{n,k} = n!$, and the recurrence reads

$$s_{n,k} = s_{n-1,k-1} + (n-1)s_{n-1,k} \quad (n \geq 1)\,. \tag{1}$$

For the proof we classify the permutations in the familiar way according to a fixed element $a \in N$. The first summand counts all permutations of N that have a as fixed point. The second summand counts the remaining permutations, since we may insert a

before each of the $n-1$ elements of any cycle of a permutation of $N \setminus a$ with k cycles.

Recurrence (1) immediately yields the polynomial identity

$$x^{\overline{n}} = \sum_{k=0}^{n} s_{n,k} x^k \,. \tag{2}$$

To show this we note that

$$x^{\overline{n}} = x^{\overline{n-1}}(x+n-1) = x \cdot x^{\overline{n-1}} + (n-1)x^{\overline{n-1}}$$

and hence by induction that

$$\begin{aligned} x^{\overline{n}} &= x \sum_{k\geq 1} s_{n-1,k-1} x^{k-1} + \sum_{k\geq 0} (n-1) s_{n-1,k} x^k \\ &= \sum_{k\geq 1} s_{n-1,k-1} x^k + \sum_{k\geq 0} (n-1) s_{n-1,k} x^k \\ &= \sum_{k\geq 0} s_{n,k} x^k \,. \end{aligned}$$

By the reciprocity law (6) of Section 1.2 we obtain

$$x^{\underline{n}} = (-1)^n (-x)^{\overline{n}} = (-1)^n \sum_{k=0}^{n} s_{n,k} (-x)^k \,,$$

that is,

$$x^{\underline{n}} = \sum_{k=0}^{n} (-1)^{n-k} s_{n,k} x^k \,. \tag{3}$$

The coefficients $(-1)^{n-k} s_{n,k}$ are thus the counterparts of $x^n = \sum_{k=0}^{n} S_{n,k} x^{\underline{k}}$, this time expressing $x^{\underline{n}}$ in terms of the basis (x^n). And this is, of course, the origin of the name Stirling numbers of the *first* and *second* kinds. Incidentally, in some texts $(-1)^{n-k} s_{n,k}$ are called the Stirling numbers of the first kind, and $s_{n,k}$ the *signless* Stirling numbers.

Stirling Matrix.
The table lists the first values of the Stirling matrix $(s_{n,k})$; clearly $(s_{n,k})$ is again lower triangular.

$n \backslash k$	0	1	2	3	4	5	6	7
0	1							
1	0	1						
2	0	1	1					
3	0	2	3	1				
4	0	6	11	6	1			
5	0	24	50	35	10	1		
6	0	120	274	225	85	15	1	
7	0	720	1764	1624	735	175	21	1

$s_{n,k}$

We already know that $s_{n,1} = (n-1)!$ (counting the cyclic permutations). Furthermore, $s_{n,n-1} = \binom{n}{2}$, $s_{n,n} = 1$. To compute $s_{n,2}$, we make use of recurrence (1). Dividing by $(n-1)!$ we obtain

$$\frac{s_{n,2}}{(n-1)!} = \frac{(n-2)!}{(n-1)!} + \frac{(n-1)s_{n-1,2}}{(n-1)!} = \frac{s_{n-1,2}}{(n-2)!} + \frac{1}{n-1},$$

and thus with $s_{2,2} = 1$,

$$\frac{s_{n,2}}{(n-1)!} = \frac{1}{n-1} + \frac{1}{n-2} + \cdots + \frac{1}{2} + \frac{1}{1} = H_{n-1},$$

that is,

$$s_{n,2} = (n-1)!H_{n-1}, \tag{4}$$

where H_{n-1} is the $(n-1)$-st harmonic number.

Word Representation.

Now we look at permutations of $\{1, 2, \ldots, n\}$ in word form $\sigma = a_1 a_2 \ldots a_n$, $a_i = \sigma(i)$. The pair $\{i, j\}$ is called an *inversion* if $i < j$ but $a_i > a_j$. The *inversion number* $\text{inv}(\sigma)$ is the number of inversions of σ, and σ is called *even (odd)* if $\text{inv}(\sigma)$ is even (odd). The *sign* of σ is defined as $\text{sign}(\sigma) = (-1)^{\text{inv}(\sigma)}$. Thus $\text{sign}(\sigma) = 1$ if σ is even and $\text{sign}(\sigma) = -1$ if σ is odd.

The following graphic representation of a permutation σ will prove very useful. We write $1, 2, \ldots, n$ into two rows and join i with $a_i = \sigma(i)$ by an edge. For example, $\sigma = 314265$ has the diagram

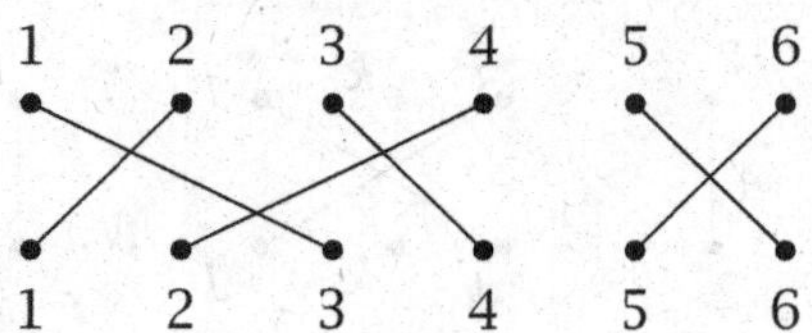

A moment's thought shows that $\{i, j\}$ is an inversion if and only if the edges ia_i and ja_j cross in the diagram. Hence the inversion number $\mathrm{inv}(\sigma)$ equals the number of crossings.

From this setup we may easily deduce a few useful facts. First, it is clear that $\mathrm{inv}(\sigma) = 0$ holds exactly for the identity permutation $\sigma = \mathrm{id}$. Next we see that the diagram read from the bottom up gives the permutation σ^{-1}; thus $\mathrm{inv}(\sigma) = \mathrm{inv}(\sigma^{-1})$, $\mathrm{sign}\,(\sigma) = \mathrm{sign}\,(\sigma^{-1})$. To represent a product $\tau\sigma$ we attach another row to the diagram corresponding to τ. Suppose $\sigma = 314265$, $\tau = 513642$; then we obtain

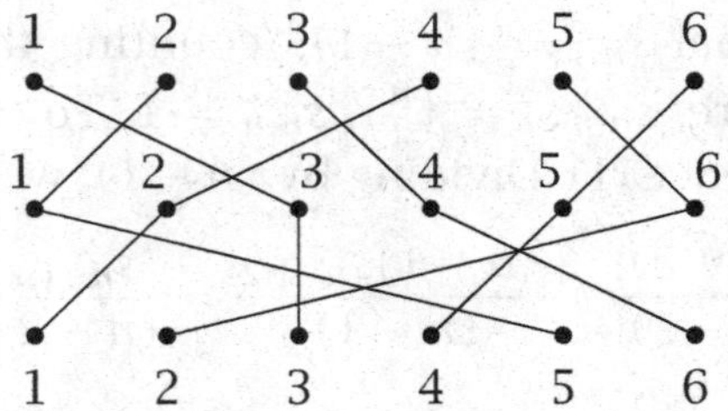

Following the paths of length 2 from top to bottom we get $\tau\sigma = 356124$. Furthermore, we see that $\mathrm{inv}(\sigma) + \mathrm{inv}(\tau) \equiv \mathrm{inv}(\tau\sigma) \pmod 2$. Indeed, $\{i, j\}$ is an inversion in $\tau\sigma$ if and only if the paths starting at i and j cross in the first half but not in the second, or they cross in the second half but not in the first. If they cross in both, as for 5 and 6, then the crossings cancel out, and $\{5, 6\}$ is not an inversion. In summary, we have $\mathrm{sign}\,(\tau\sigma) = \mathrm{sign}\,(\tau)\mathrm{sign}\,(\sigma)$. Thus sign: $S(n) \to \{1, -1\}$ is a homomorphism, from which it follows that for $n \geq 2$ half of the permutations are even and half are odd.

Suppose $\sigma = \sigma_1\sigma_2 \cdots \sigma_t$ is the cycle decomposition of σ. Regarding a cycle $(i_1, i_2, \ldots, i_k)$ as a permutation of $\{1, 2, \ldots, n\}$ by keeping all elements not in the cycle fixed, we see that $\sigma = \sigma_1\sigma_2 \cdots \sigma_t$ is, in fact, a product of permutations. Hence $\mathrm{sign}\,(\sigma) = \prod_{i=1}^{t} \mathrm{sign}\,(\sigma_i)$, so to compute $\mathrm{sign}\,(\sigma)$ it suffices to determine the sign for cycles. But this is easy. First we note that any transposition (i, j) is odd. Look at the diagram

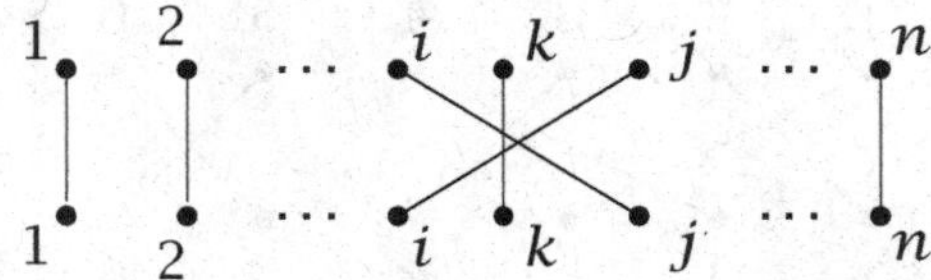

Any edge kk with $k < i$ or $k > j$ contributes no crossing, while the edges kk with $i < k < j$ add 2. Hence modulo 2 there remains the one crossing induced by the inversion $\{i,j\}$, and we get $\text{sign}\,(i,j) = -1$. Now let $(i_1,\ldots,i_\ell)$ be a cycle of length $\ell \geq 3$. Then it is easily checked that the product

$$(i_{\ell-1},i_\ell)(i_{\ell-2},i_{\ell-1})\cdots(i_1,i_2)(i_1,i_2,\ldots,i_\ell)$$

is the identity permutation. Hence $(-1)^{\ell-1}\text{sign}\,(i_1,\ldots,i_\ell) = 1$, and we obtain the following result:

A cyclic permutation of length ℓ *has* $\text{sign} = (-1)^{\ell-1}$.

Exercises

1.37 Use the polynomial method to show that $s_{n+1,k+1} = \sum_{i=0}^{n}\binom{i}{k}s_{n,i}$. Can you find a combinatorial proof?

1.38 What is the expected number of fixed points when all $n!$ permutations of $S(n)$ are equally likely?

1.39 A permutation $\sigma \in S(n)$ is an *involution* if $\sigma^2 = \text{id}$, that is, if all cycles have length 1 or 2. Prove for the number i_n of involutions the recurrence $i_{n+1} = i_n + n i_{n-1}$ $(i_0 = 1)$. What is the number of fixed-point-free involutions?

▷ **1.40** Let $i_n^{(r)}$ be the number of permutations of $\{1,\ldots,n\}$ with no cycles of length greater than r. Prove the recurrence $i_{n+1}^{(r)} = \sum_{k=n-r+1}^{n} n^{\underline{n-k}} i_k^{(r)}$, generalizing the previous exercise.

1.41 Let $\ell > \frac{n}{2}$. Show that the number of permutations $\sigma \in S(n)$ that have a cycle of length ℓ equals $\frac{n!}{\ell}$. What is the proportion $t(n)$ of $\sigma \in S(n)$ that contain a cycle of length $> \frac{n}{2}$ when all permutations are equally likely? Compute $\lim_{n\to\infty} t(n)$.

▷ **1.42** Let $I_{n,k}$ be the number of permutations in $S(n)$ with exactly k inversions, $k = 0,1,\ldots,\binom{n}{2}$. Prove: a. $I_{n,0} = 1$, b. $I_{n,k} = I_{n,\binom{n}{2}-k}$, c. $I_{n,k} = I_{n-1,k} + I_{n,k-1}$ for $k < n$. Is this also true for $k = n$? d. $\sum_{k=0}^{\binom{n}{2}}(-1)^k I_{n,k} = 0$ for $n \geq 2$. The $I_{n,k}$ are called *inversion numbers*.

1.43 Let $\sigma = a_1a_2\ldots a_n \in S(n)$ be given in word form, and denote by b_j the number of elements to the left of j that are larger than j (thus they form an inversion with j). The sequence $b_1b_2\ldots b_n$ is called the *inversion table* of σ. Show that $0 \leq b_j \leq n-j$ $(j = 1,\ldots,n)$ and prove, conversely, that every sequence $b_1 b_2 \dots b_n$ with $0 \le b_j \le n - j$ is the inversion table of a unique permutation.

1.44 A permutation $\sigma \in S(n)$ is called *connected* if for any k, $1 \le k < n$, $\{\sigma(1), \sigma(2), \dots, \sigma(k)\} \ne \{1, \dots, k\}$. Prove $\sum_{i=1}^{n} c(i)(n-i)! = n!$, where $c(i)$ is the number of connected partitions in $S(i)$.

* * *

1.45 Define the *type* of $\sigma \in S(n)$ to be the formal expression $1^{c_1} 2^{c_2} \dots n^{c_n}$, where c_i is the number of cycles in σ of length i; thus $\sum_{i=1}^{n} ic_i = n$. Show that the number of $\sigma \in S(n)$ with type $1^{c_1} 2^{c_2} \dots n^{c_n}$ equals $\frac{n!}{1^{c_1} c_1! 2^{c_2} c_2! \cdots n^{c_n} c_n!}$.

▷ **1.46** Let $\sigma \in S(n)$ have type $t(\sigma) = 1^{c_1} \dots n^{c_n}$. Show that the number of $\pi \in S(n)$ that commute with σ, i.e., $\pi\sigma = \sigma\pi$, is $1^{c_1} c_1! 2^{c_2} c_2! \cdots n^{c_n} c_n!$ (where this is now a real product). Hint: The graph representation helps.

1.47 Prove the following identities for the Stirling numbers $s_{n,k}$:

a. $\sum_{i=k}^{n} s_{i,k} n^{\underline{n-i}} = n! \sum\limits_{i=k}^{n} \frac{s_{i,k}}{i!} = s_{n+1,k+1}$, b. $\sum_{i=0}^{n} (m+i) s_{m+i,i} = s_{m+n+1,n}$,

c. $\sum_k s_{n+1,k+1} \binom{k}{m} (-1)^{k-m} = s_{n,m}$, d. $\sum_k s_{k,\ell} s_{n-k,m} \binom{n}{k} = s_{n,\ell+m} \binom{\ell+m}{m}$ $(\ell, m, n \in \mathbb{N}_0)$.

▷ **1.48** Let $\sigma = a_1 a_2 \dots a_n \in S(n)$ be given in word form. A *run* in σ is a largest increasing subsequence of consecutive entries. The *Eulerian number* $A_{n,k}$ is the number of $\sigma \in S(n)$ with precisely k runs or equivalently with $k-1$ descents $a_i > a_{i+1}$. Thus, e.g., $A_{n,1} = A_{n,n} = 1$ with $12 \dots n$ respectively $n\ n-1 \dots 1$ as the only permutations. Prove the recurrence $A_{n,k} = (n-k+1) A_{n-1,k-1} + k A_{n-1,k}$ for $n, k \ge 1$ with $A_{0,0} = 1$, $A_{0,k} = 0$ $(k > 0)$. Use induction to prove $x^n = \sum_{k=0}^{n} A_{n,k} \binom{x+n-k}{n}$ and deduce the formula $A_{n,k} = \sum_{i=0}^{k} (-1)^i \binom{n+1}{i} (k-i)^n$.

1.49 Use the previous exercise to show further: a. $\sum_{k=1}^{n} A_{n,k} = n!$, b. $A_{n,k} = A_{n,n+1-k}$, c. $\sum_{k=1}^{n} k A_{n,k} = \frac{1}{2}(n+1)!$. What is the expected number of runs when all permutations are equally likely?

1.50 When all permutations are equally likely, what is the probability that k specified elements are in the same cycle? Hint: Consider the standard cycle representation and use $\sum_{i=0}^{n-1} i^{\underline{k-1}} = \frac{n^{\underline{k}}}{k}$, which will be proved in Chapter 5.

▷ **1.51** With all permutations equally likely, what is the expected number of cycles?

1.52 Derive by a combinatorial argument the following recurrence for the derangement numbers: $D_n = (n-1)(D_{n-1} + D_{n-2})$ $(n \ge 2)$, $D_0 = 1$,

$D_1 = 0$. Deduce from this $D_n = nD_{n-1} + (-1)^n$. Can you find a direct argument for this latter recurrence?

1.5 Number-Partitions

After partitions of sets we consider now partitions of positive integers into integer summands. A *number-partition* of n is $n = \lambda_1 + \lambda_2 + \cdots + \lambda_k$, where we assume $\lambda_1 \geq \lambda_2 \geq \cdots \geq \lambda_k \geq 1$. The summands λ_i are called the *parts* of n. We write $\lambda = \lambda_1\lambda_2\ldots\lambda_k$ for short and set $|\lambda| = n$ if $n = \sum_{i=1}^{k} \lambda_i$ is the number partitioned by λ. If λ has k parts, then λ is called a *k-partition*.

$\mathrm{Par}(n)$ denotes the set of partitions of n, $\mathrm{Par}(n;k)$ the set of k-partitions of n, with $p(n) = |\mathrm{Par}(n)|$, $p(n;k) = |\mathrm{Par}(n;k)|$. By definition, $\mathrm{Par}(0)$ consists of the empty partition, $p(0) = 1$.

Example. For $n = 5$ we get the following seven partitions:

$$5, \quad 41, \quad 32, \quad 311, \quad 221, \quad 2111, \quad 11111;$$

thus $p(5) = 7$.

Note that we deal with *unordered* partitions. When the order matters, then, e.g., 311,131, and 113 are distinct *ordered* partitions. A nice application of the bijection principle yields

The number of ordered k-partitions of n equals $\binom{n-1}{k-1}$.

Map the ordered partition $n_1n_2\ldots n_k$ of n to the $(k-1)$-set $\{n_1, n_1+n_2, \ldots, n_1+\cdots+n_{k-1}\}$. Since $n_i \geq 1$, this is a subset of $\{1, 2, \ldots, n-1\}$, and the map is easily seen to be bijective. Ordered partitions are also called *compositions*.

The ordered k-partitions of n are thus the solutions of the equation $x_1 + x_2 + \cdots + x_k = n$ in positive integers. It is interesting to note that the solutions in *non-negative* integers correspond to n-multisets of $\{1, 2, \ldots, k\}$ (the x_i's being the multiplicities). Hence their number is $\binom{n+k-1}{n}$.

Partition Numbers.
Back to ordinary unordered partitions. We have introduced the notation $\mathrm{Par}(n)$, $\mathrm{Par}(n;k)$. Similarly, we use $\mathrm{Par}(n;\leq k)$ for the set

of all partitions of n with at most k parts, and set $p(n;\leq k) = |\mathrm{Par}(n;\leq k)|$. For the partition numbers $p(n;k)$ there is no recurrence of order 2. Instead, we have

$$p(n;k) = p(n-k;\leq k) = p(n-k;1) + \cdots + p(n-k;k)\,. \qquad (1)$$

Indeed, the mapping $\phi : \mathrm{Par}(n;k) \to \mathrm{Par}(n-k;\leq k)$ with $\phi(\lambda_1 \ldots \lambda_k) = \lambda_1 - 1, \lambda_2 - 1, \ldots, \lambda_k - 1$ is clearly a bijection, where we omit possible 0's at the end.

Formula (1) immediately implies the recurrence

$$p(n;k) = p(n-1;k-1) + p(n-k;k)\,. \qquad (2)$$

For small values we have

$p(n;k)$

n \ k	0	1	2	3	4	5	6	7
0	1							
1	0	1						
2	0	1	1					
3	0	1	1	1				
4	0	1	2	1	1			
5	0	1	2	2	1	1		
6	0	1	3	3	2	1	1	
7	0	1	3	4	3	2	1	1

n	0	1	2	3	4	5	6	7
$p(n)$	1	1	2	3	5	7	11	15

An extremely useful graphic description of number-partitions is provided by the so-called *Ferrers diagram*. Let $\lambda = \lambda_1\lambda_2\ldots\lambda_k$ be a partition of n. We put λ_1 dots in row 1, λ_2 dots in row 2, and so on, starting in the same column.

Example. $\lambda = 64221$ has the diagram

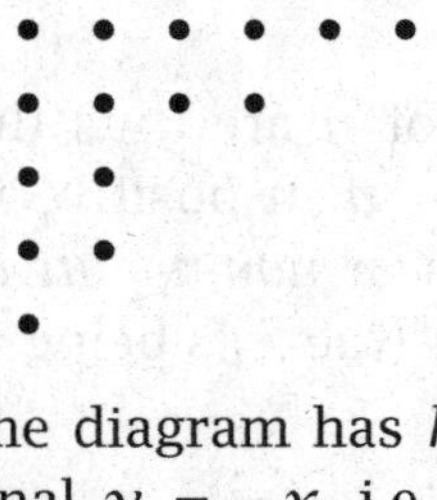

• • • • • •
• • • •
• •
• •
•

Thus for $\lambda \in \mathrm{Par}(n;k)$ the diagram has k rows. Reflecting the diagram at the main diagonal $y = -x$, i.e., interchanging rows and columns, we obtain the *conjugate partition* λ^*. In our example this is $\lambda^* = 542211$:

```
• • • • •
• • • •
• •
• •
•
•
```

Clearly, $\lambda \mapsto \lambda^*$ is an involution of the set $\mathrm{Par}(n)$. Note further that $\lambda_i^* = \#\{j : \lambda_j \geq i\}$, and that the number of parts of λ^* is equal to the highest summand λ_1 of λ. Let us therefore introduce this parameter into our notation. $\mathrm{Par}(n;k;m)$ is the set of all partitions of n with k parts and highest summand m. $\mathrm{Par}(n;k;\leq m)$ and $\mathrm{Par}(n;\leq k;\leq m)$ are similarly defined, with sizes $p(n;k;m)$, $p(n;k;\leq m)$, $p(n;\leq k;\leq m)$. Finally, $p(;\leq k;\leq m) = |\mathrm{Par}(;\leq k;\leq m)|$, with no restriction on n.

Example. For $k = 3$, $m = 2$ the set $\mathrm{Par}(;\leq 3;\leq 2)$ consists of the following partitions:

$$\begin{array}{llll} 222 & 22 & 2 & \emptyset \\ 221 & 21 & 1 & \\ 211 & 11 & & \\ 111 & & & \end{array} \qquad p(;\leq 3;\leq 2) = 10\,.$$

The involution $\lambda \mapsto \lambda^*$ implies that we may interchange the second and third parameters:

$$\begin{aligned} p(n;k;m) &= p(n;m;k), \\ p(n;k;\leq m) &= p(n;\leq m;k), \\ p(;\leq k;\leq m) &= p(;\leq m;\leq k)\,. \end{aligned} \tag{3}$$

The partitions $\lambda = \lambda_1\lambda_2\ldots\lambda_k$, where all parts are *distinct*, i.e., $\lambda_1 > \lambda_2 > \cdots > \lambda_k$, will play a special role. These sets shall be denoted by $\mathrm{Par}_d(n)$, $\mathrm{Par}_d(n;k;m)$, etc. with cardinalities $p_d(n)$, $p_d(n;k;m)$, etc. As a first result we have the following relations:

$$\begin{aligned} p_d(n;k;\leq m) &= p(n - \binom{k+1}{2};\leq k;\leq m-k), \\ p_d(;k;\leq m) &= p(;\leq k;\leq m-k)\,. \end{aligned} \tag{4}$$

This time we use the map $\phi : \mathrm{Par}_d(n;k;\leq m) \to \mathrm{Par}(n - \binom{k+1}{2};\leq k;\leq m-k)$ defined by $\phi(\lambda_1\ldots\lambda_k) = \lambda_1 - k, \lambda_2 - (k-1), \ldots, \lambda_k - 1,$

omitting possible 0's at the end. This is the desired bijection, also for the second equality.

Lattice Paths.
To end these introductory remarks about number-partitions we discuss an important and perhaps unexpected connection to binomial coefficients via lattice paths. Remember that $\binom{m+n}{m}$ counts the lattice paths from $(0,0)$ to (m,n). The following figure demonstrates a natural bijection between these paths and the set $\text{Par}(;\le n;\le m)$, by interpreting the part above the lattice path as a Ferrers diagram.

Example.

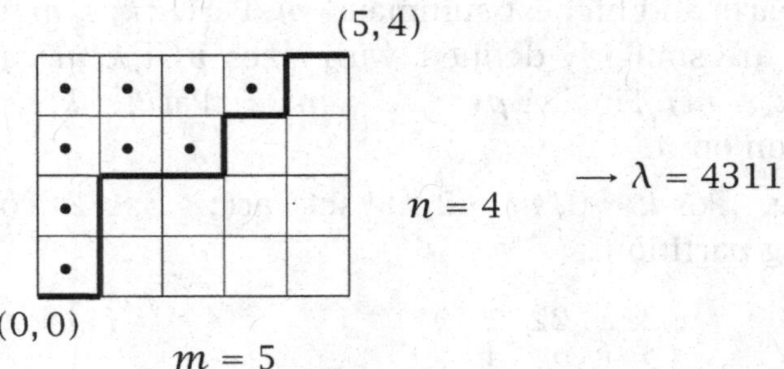

Hence

$$p(;\le n;\le m) = \binom{m+n}{m}. \tag{5}$$

Note that the path that goes up to $y = n$ and then horizontally to (m,n) corresponds to the empty partition. In our example above, $n = 3$, $m = 2$, we obtain $p(;\le 3;\le 2) = \binom{2+3}{2} = 10$.

Summary.
An easy way to remember some of the fundamental coefficients we have encountered so far is to interpret them as distributions of n balls into r boxes. Let N be the set of balls, and R the boxes. Any mapping $f : N \to R$ is then a distribution of the balls, and injective or surjective mappings have the usual meaning. Now suppose the balls cannot be distinguished, but the boxes can. In this case, the different mappings correspond to sequences $x_1x_2\dots x_r$ with $x_1 + \cdots + x_r = n$, where x_j is the number of balls in box j. Hence these mappings correspond to n-multisets of R, the injective mappings to n-subsets of R, and the surjective mappings to ordered r-partitions of n. The reader may easily complete the other rows of the following table.

$f: N \longrightarrow R$	arbitrary	injective	surjective
N dist. R dist.	r^n	$r^{\underline{n}}$	$r!S_{n,r}$
N indist. R dist.	$\frac{r^{\overline{n}}}{n!} = \binom{r+n-1}{n}$	$\frac{r^{\underline{n}}}{n!} = \binom{r}{n}$	$\binom{n-1}{r-1}$
N dist. R indist.	$\sum_{k=0}^{r} S_{n,k}$	0 or 1	$S_{n,r}$
N indist. R indist.	$\sum_{k=1}^{r} p(n;k)$	0 or 1	$p(n;r)$

Exercises

▷ **1.53** Show that the number of self-conjugate partitions, i.e., $\lambda^* = \lambda$, equals the number of partitions of n with all summands odd and distinct.

1.54 Show that $p(n;n-t) = p(t)$ if and only if $n \geq 2t$.

1.55 Prove $p_d(n;k) = p(n - \binom{k}{2}; k)$.

1.56 Verify the recurrence

$$p_d(n;k) = p_d(n-k;k) + p_d(n-k;k-1), 2 \leq k \leq \lfloor \tfrac{n}{2} \rfloor, n \geq 5,$$

with starting values

$$p_d(n;1) = 1, \quad p_d(n;k) = 0 \text{ for } n < \binom{k+1}{2}, \quad p_d\left(\binom{k+1}{2};k\right) = 1.$$

1.57 Express the following quantities in terms of Fibonacci numbers: a. The number of ordered partitions of n into parts greater than 1; b. the number of ordered partitions of n into parts equal to 1 or 2; c. the number of ordered partitions of n into odd parts.

1.58 Show that the number of partitions of n with all parts ≥ 2 equals $p(n) - p(n-1)$.

▷ **1.59** Determine the number of solutions of $x_1 + \cdots + x_k \leq n$ in positive integers; in non-negative integers.

* * *

1.60 Let $e(n), o(n)$, and $sc(n)$ denote, respectively, the number of partitions of n with an even number of even parts, with an odd number of even parts, and that are self-conjugate. Show that $e(n) - o(n) = sc(n)$.

1.61 Prove that for $n \geq 2$, exactly half of the partitions of n into powers of 2 have an even number of parts. For example, when $n = 5$ the partitions are $41, 221, 2111, 11111$.

▷ **1.62** For $\lambda \in \mathrm{Par}(n)$, let $f_m(\lambda)$ be the number of times m appears in λ, and let $g_m(\lambda)$ be the number of *distinct* parts of λ that occur at least m times. Example: $f_2(4333211) = 1$, $g_2(4333211) = 2$. Show that $\sum_\lambda f_m(\lambda) = \sum_\lambda g_m(\lambda)$, where both sums range over $\mathrm{Par}(n)$, m fixed. What is $\sum_{|\lambda|=n} f_1(\lambda)$? Hint: Show that the sums satisfy the same recurrence.

1.63 . Let b_m be the number of pairs $\lambda = \lambda_1\lambda_2\ldots\lambda_k$, $\mu = \mu_1\mu_2\ldots\mu_k$ for some k such that $\lambda_1 > \lambda_2 > \cdots > \lambda_k \geq 1$, $\mu_1 > \mu_2 > \cdots > \mu_k \geq 0$ and $\sum\lambda_j + \sum\mu_j = m$. Prove that $b_m = p(m)$. Hint: Use a clever decomposition of the Ferrers diagram.

▷ **1.64** A partition of n is called *perfect* if it contains precisely one partition for every $m < n$. Thus, if $x_1 + 2x_2 + \cdots + nx_n = n$ $(x_i \geq 0)$, then there is a unique solution of $y_1 + 2y_2 + \cdots + ny_n = m$ with $0 \leq y_i \leq x_i$ for all i and $m < n$. For example, 311, 221, and 11111 are the perfect partitions of 5. Show that the number of perfect partitions of n equals the number of ordered factorizations of $n + 1$ without unit factors.

1.6 Lattice Paths and Gaussian Coefficients

We have discussed several connecting relations between polynomial sequences, the simplest being $x^n = \sum_{k=0}^n \binom{n}{k}(x-1)^k$, with the binomials $\binom{n}{k}$ as connecting coefficients. Now we turn to an important generalization of the coefficient $\binom{n}{k}$.

Definition. The polynomials $g_n(x) = (x-1)(x-q)\cdots(x-q^{n-1})$, $g_0(x) = 1$, are called the *Gaussian polynomials*, where q is an arbitrary complex number or, more generally, an indeterminate.

Writing $x^n = \sum_{k=0}^n \left[{n \atop k}\right]_q g_k(x)$, the connecting coefficients are called the *Gaussian coefficients*. Note that for $q = 1$, $g_k(x) = (x-1)^k$, whence $\left[{n \atop k}\right]_1 = \binom{n}{k}$.

The generalization to arbitrary q turns out to be a fertile source for many enumeration problems. First we note the recurrence

$$\left[{n \atop k}\right]_q = \left[{n-1 \atop k-1}\right]_q + q^k\left[{n-1 \atop k}\right]_q \quad (n \geq k \geq 1), \quad \left[{n \atop 0}\right]_q = 1\,. \tag{1}$$

We certainly have $\left[{n \atop 0}\right]_q = 1$, so assume $k \geq 1$. From $g_k(x) = (x - q^{k-1})g_{k-1}(x)$, we infer

$$\begin{aligned} x^n = x \cdot x^{n-1} &= x \sum_{k\geq 1} \begin{bmatrix} n-1 \\ k-1 \end{bmatrix}_q g_{k-1}(x) = \sum_{k\geq 1} \begin{bmatrix} n-1 \\ k-1 \end{bmatrix}_q x g_{k-1}(x) \\ &= \sum_{k\geq 1} \begin{bmatrix} n-1 \\ k-1 \end{bmatrix}_q g_k(x) + \sum_{k\geq 1} q^{k-1} \begin{bmatrix} n-1 \\ k-1 \end{bmatrix}_q g_{k-1}(x) \\ &= \sum_{k\geq 0} \left(\begin{bmatrix} n-1 \\ k-1 \end{bmatrix}_q + q^k \begin{bmatrix} n-1 \\ k \end{bmatrix}_q \right) g_k(x), \end{aligned}$$

and hence $\left[{n \atop k}\right]_q = \left[{n-1 \atop k-1}\right]_q + q^k \left[{n-1 \atop k}\right]_q$ as claimed.

Now we want to compute the coefficients $\left[{n \atop k}\right]_q$. The guiding idea is the fact that $\left[{n \atop k}\right]_q$ becomes for $q = 1$ the binomial coefficient $\binom{n}{k} = \frac{n!}{k!(n-k)!}$.

Define $[n]_q := \frac{1-q^n}{1-q}$ as an expression in the variable q. Since $\frac{1-q^n}{1-q} = 1 + q + q^2 + \cdots + q^{n-1}$, this becomes the integer n when $q = 1$. Accordingly, we call $[n]_q$ the *q-integer n*. Next, we set

$$[n]_q! = [n]_q[n-1]_q \cdots [1]_q = \frac{(1-q^n)(1-q^{n-1}) \cdots (1-q)}{(1-q)^n} \tag{2}$$

with $[0]_q! = 1$ by definition, which becomes $n!$ for $q = 1$.

Claim.

$$\begin{bmatrix} n \\ k \end{bmatrix}_q = \frac{[n]_q!}{[k]_q![n-k]_q!} \qquad (n \geq k \geq 0). \tag{3}$$

This is certainly true for $n = 0$ or $k = 0$; we proceed by induction on n. From $[n]_q! = \frac{1-q^n}{1-q}[n-1]_q!$ and (1) we obtain

$$\begin{aligned} \begin{bmatrix} n \\ k \end{bmatrix}_q &= \begin{bmatrix} n-1 \\ k-1 \end{bmatrix}_q + q^k \begin{bmatrix} n-1 \\ k \end{bmatrix}_q \\ &= \frac{[n-1]_q!}{[k-1]_q![n-k]_q!} + q^k \frac{[n-1]_q!}{[k]_q![n-1-k]_q!} \\ &= \frac{[n-1]_q!}{[k]_q![n-k]_q!} \left(\frac{1-q^k}{1-q} + \frac{q^k(1-q^{n-k})}{1-q} \right) \\ &= \frac{[n-1]_q!}{[k]_q![n-k]_q!} \cdot \frac{1-q^n}{1-q} = \frac{[n]_q!}{[k]_q![n-k]_q!}. \end{aligned}$$

Note that (3) implies $\left[{n \atop k}\right]_q = \left[{n \atop n-k}\right]_q$. Cancelling terms we may also write

$$\begin{bmatrix} n \\ k \end{bmatrix}_q = \frac{[n]_q[n-1]_q \cdots [n-k+1]_q}{[k]_q!} = \frac{(1-q^n)\cdots(1-q^{n-k+1})}{(1-q^k)\cdots(1-q)}, \tag{4}$$

reminiscent of $\binom{n}{k} = \frac{n^{\underline{k}}}{k!}$ for $q = 1$.

This suggests that we should define the *q-falling* and *q-rising factorial polynomials*

$$x_q^{\underline{n}} = x(x - [1]_q) \cdots (x - [n-1]_q),$$

$$x_q^{\overline{n}} = x(x + [1]_q) \cdots (x + [n-1]_q).$$

Expressing x^n in terms of $x_q^{\underline{k}}$, we call the connecting coefficients the *q-Stirling numbers* $S_{n,k;q}$, and similarly for $x_q^{\overline{n}}$ we obtain the numbers $s_{n,k;q}$. Thus for every q we may develop a calculus of *q-coefficients*, generalizing the previous results (see the exercises).

Lattice Paths.

The most interesting of these "q-generalizations" concerns the interpretation of the binomial coefficients as counting numbers of lattice paths. Recall Section 1.5, where we have demonstrated the remarkable result

$$|\mathrm{Par}(\,;\le n;\le m)| = \binom{m+n}{m},$$

by bijecting lattice paths to the Ferrers diagram above the path. It is natural to classify these paths according to the number i of points in the diagram, i.e., i is the number that is partitioned by λ, $i = 0, 1, \ldots, mn$.

Accordingly, we set

$$A_{m,n}(q) = \sum_{\lambda \in \mathrm{Par}(\,;\le n;\le m)} q^{|\lambda|} = \sum_{i=0}^{mn} a_i q^i, \tag{5}$$

where $a_i = p(i;\le n;\le m)$. Thus $A_{m,n}(1) = p(\,;\le n;\le m) = \binom{m+n}{m}$.

Proposition 1.1. *We have*

$$A_{m,n}(q) = \sum_{i \ge 0} p(i;\le n;\le m) q^i = \begin{bmatrix} m+n \\ m \end{bmatrix}_q. \tag{6}$$

Proof. We show that $A_{m,n}(q)$ and $\left[{m+n \atop m}\right]_q$ satisfy the same initial conditions and the same recurrence. For $m = 0$ or $n = 0$ we have $A_{0,n}(q) = A_{m,0}(q) = 1$ since in this case we get only the empty partition, and also $\left[{0+n \atop 0}\right]_q = \left[{m+0 \atop m}\right]_q = 1$. Assume $m, n \geq 1$; then the recurrence for $\left[{m+n \atop m}\right]_q$ is by (1),

$$\begin{bmatrix} m+n \\ m \end{bmatrix}_q = \begin{bmatrix} m+n-1 \\ m-1 \end{bmatrix}_q + q^m \begin{bmatrix} m+n-1 \\ m \end{bmatrix}_q . \tag{7}$$

For $A_{m,n}(q)$ we split the paths into two classes, depending on whether for the largest summand, $\lambda_1 < m$ or $\lambda_1 = m$. In the first case we obtain $A_{m-1,n}(q)$, and in the second (after deleting the top row containing m dots) $q^m A_{m,n-1}(q)$. Hence

$$A_{m,n}(q) = A_{m-1,n}(q) + q^m A_{m,n-1}(q),$$

and this is precisely recurrence (7). □

As a corollary we can state the q-binomial theorem generalizing $(x+1)^n = \sum_{k=0}^{n} \binom{n}{k} x^k$.

Corollary 1.2. *We have*

$$(1+xq)(1+xq^2)\cdots(1+xq^n) = \sum_{k=0}^{n} \begin{bmatrix} n \\ k \end{bmatrix}_q q^{\binom{k+1}{2}} x^k .$$

Proof. Expanding the left-hand side we obtain

$$(1+xq)\cdots(1+xq^n) = \sum_{k=0}^{n} b_k(q) x^k,$$

where

$$b_k(q) = \sum_{\lambda \in \mathrm{Par}_d(;k;\leq n)} q^{|\lambda|} .$$

Now recall the bijection $\mathrm{Par}_d(;k;\leq n) \to \mathrm{Par}(;\leq k;\leq n-k)$ proved in Section 1.5 (equation (4)) by subtracting k dots from the first row, $k-1$ from the second row, and so on. Taking these $k + (k-1) + \cdots + 1 = \binom{k+1}{2}$ dots into account we obtain

$$b_k(q) = q^{\binom{k+1}{2}} \sum_{\lambda \in \mathrm{Par}(;\leq k;\leq n-k)} q^{|\lambda|},$$

and thus by the proposition $b_k(q) = q^{\binom{k+1}{2}} \left[{n \atop k}\right]_q$. □

As a final application let us generalize the multinomial coefficient $\binom{n}{k_1 \ldots k_m}$. In analogy to $\binom{n}{k_1 \ldots k_m} = \frac{n!}{k_1! \cdots k_m!}$ we define

$$\begin{bmatrix} n \\ k_1 k_2 \ldots k_m \end{bmatrix}_q := \frac{[n]_q!}{[k_1]_q! \cdots [k_m]_q!} = \frac{(1-q^n) \cdots (1-q)}{\prod_{i=1}^{k_1}(1-q^i) \cdots \prod_{i=1}^{k_m}(1-q^i)}, \tag{8}$$

where $\sum_{j=1}^{m} k_j = n$.

We know that the multinomial coefficient counts the set $S(k_1, \ldots, k_m)$ of all n-words $s = s_1 \ldots s_n$ over $\{1, \ldots, m\}$ with k_i i's ($1 \le i \le m$). As usual, we say that $i < j$ ($1 \le i < j \le n$) is an inversion if $s_i > s_j$, $\mathrm{inv}(s) = \#$ inversions is the *inversion number of* s.

Example. $s = 12231133 \in S(3,2,3)$ has inversion number $\mathrm{inv}(s) = 6$.

Note that $S\underbrace{(1,1,\ldots,1)}_{n}$ is exactly the set of permutations $S(n)$, and $\mathrm{inv}(s)$ reduces to the inversion number defined for permutations.

Corollary 1.3. *We have*

$$\begin{bmatrix} n \\ k_1 \ldots k_m \end{bmatrix}_q = \sum_{s \in S(k_1, \ldots, k_m)} q^{\mathrm{inv}(s)}, \quad \sum_{j=1}^{m} k_j = n.$$

Proof. We use induction on m. For $m = 1$, $S(k_1) = S(n)$ consists of the single word $s = 11\ldots1$ with $\mathrm{inv}(s) = 0$, and also $\left[{n \atop n}\right]_q = 1 = q^0$. Consider $m = 2$. By the proposition,

$$\begin{bmatrix} k_1 + k_2 \\ k_1\ k_2 \end{bmatrix}_q = \begin{bmatrix} k_1 + k_2 \\ k_2 \end{bmatrix}_q = \sum_{i \ge 0} a_i q^i$$

with $a_i = p(i; \le k_1; \le k_2)$. Any word s with k_1 1's and k_2 2's corresponds bijectively to a lattice path by going up for 1 and right for 2.

Example.

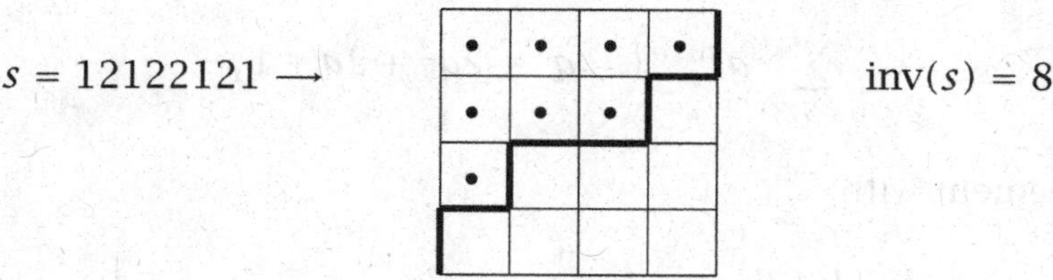

A moment's thought shows that $\mathrm{inv}(s)$ equals precisely the number i of dots above the path, which means that $\sum_{i \ge 0} a_i q^i = \sum_s q^{\mathrm{inv}(s)}$.

Now let $m \ge 3$. To $s \in S(k_1, \dots, k_m)$ we associate a pair (t, u) of words such that $t \in S(k_1, \dots, k_{m-2}, k_{m-1} + k_m)$ arises from s by replacing in s every m by $m - 1$, and where $u \in S(k_{m-1}, k_m)$ is the subword of $m - 1$ and m.

Example. $s = 21331211 \longrightarrow t = 21221211,\ u = 2332.$

Since s can be recovered uniquely from t and u, the map $s \mapsto (t, u)$ is a bijection. Furthermore, we have $\mathrm{inv}(s) = \mathrm{inv}(t) + \mathrm{inv}(u)$, that is, $\sum_s q^{\mathrm{inv}(s)} = \sum_t q^{\mathrm{inv}(t)} \cdot \sum_u q^{\mathrm{inv}(u)}$. With induction this gives

$$\sum_t q^{\mathrm{inv}(t)} = \frac{[n]_q!}{[k_1]_q! \cdots [k_{m-2}]_q! [k_{m-1} + k_m]_q!}$$

$$\sum_u q^{\mathrm{inv}(u)} = \frac{[k_{m-1} + k_m]_q!}{[k_{m-1}]_q! [k_m]_q!},$$

and we conclude that

$$\sum_s q^{\mathrm{inv}(s)} = \sum_t q^{\mathrm{inv}(t)} \cdot \sum_u q^{\mathrm{inv}(u)} = \frac{[n]_q!}{[k_1]_q! \cdots [k_m]_q!}. \qquad \square$$

For $k_1 = k_2 = \cdots = k_n = 1$ we have $\left[{n \atop 1 \dots 1} \right]_q = \frac{[n]_q!}{[1]_q! \cdots [1]_q!} = [n]_q!$, and we infer the following corrollary.

Corollary 1.4.

$$\sum_{\sigma \in S(n)} q^{\mathrm{inv}(\sigma)} = \frac{(1 - q^n)(1 - q^{n-1}) \cdots (1 - q)}{(1 - q)^n}. \tag{9}$$

Example. For $n = 3$, the inversion numbers are $I_{3,0} = 1$, $I_{3,1} = I_{3,2} = 2$, $I_{3,3} = 1$. Thus

$$\sum_{\sigma \in S(3)} q^{\mathrm{inv}(\sigma)} = q^3 + 2q^2 + 2q + 1\,,$$

in agreement with

$$\frac{(1-q^3)(1-q^2)(1-q)}{(1-q)^3} = (1+q+q^2)(1+q) = q^3 + 2q^2 + 2q + 1\,.$$

Exercises

1.65 Verify $\left[{n \atop k}\right]_q = \frac{[n]_q}{[k]_q}\left[{n-1 \atop k-1}\right]_q$, $n \ge k \ge 0$.

▷ **1.66** Prove the recurrence $\left[{n \atop k}\right]_q = q^{n-k}\left[{n-1 \atop k-1}\right]_q + \left[{n-1 \atop k}\right]_q$.

1.67 Verify the reciprocity law $[n]_q^{\underline{k}} = (-1)^k q^{kn+\binom{k}{2}}[-n]_q^{\overline{k}}$.

1.68 Prove $\left[{n \atop m}\right]_q\left[{m \atop k}\right]_q = \left[{n \atop k}\right]_q\left[{n-k \atop m-k}\right]_q$, $n \ge m \ge k \ge 0$.

1.69 The identities $x_q^{\overline{n}} = \sum_{k=0}^{n} s_{n,k;q} x^k$, $x^n = \sum_{k=0}^{n} S_{n,k;q} x_q^{\underline{k}}$ define the q-Stirling numbers. Derive the following recurrences, generalizing the results in Sections 1.3 and 1.4: a. $s_{n,k;q} = s_{n-1,k-1;q} + [n-1]_q s_{n-1,k;q}$, b. $S_{n,k;q} = S_{n-1,k-1;q} + [k]_q S_{n-1,k;q}$.

* * *

▷ **1.70** Prove $\sum_{i=0}^{n} \left[{i \atop k}\right]_q q^{(k+1)(n-i)} = \left[{n+1 \atop k+1}\right]_q$.

1.71 Show that the q-Stirling numbers $S_{n,r;q}$ can be expressed in the following way:

$$S_{n,r;q} = \frac{q^{-\binom{r}{2}}}{[r]_q!}\sum_{k=0}^{r}(-1)^{r-k} q^{\binom{r-k}{2}}\left[{r \atop k}\right]_q [k]_q^n\,.$$

1.72 Prove the q-Vandermonde identity

$$\left[{r+s \atop n}\right]_q = \sum_{k=0}^{n}\left[{r \atop k}\right]_q\left[{s \atop n-k}\right]_q q^{(r-k)(n-k)}\,.$$

▷ **1.73** The sum $G_n = \sum_{k=0}^{n}\left[{n \atop k}\right]_q$ is called a *Galois number*. Prove the recurrence $G_{n+1} = 2G_n + (q^n - 1)G_{n-1}$ $(n \ge 1)$.

1.74 Let V be an n-dimensional vector space over the finite field $GF(q)$, q a prime power. Prove that $\left[{n \atop k}\right]_q$ equals the number of k-dimensional subspaces of V.

▷ **1.75** Let $\phi_{n,k} = q^{\binom{k+1}{2}}\left[{n \atop k}\right]_q$. Prove the identity $\phi_{n,k} = \phi_{n-1,k-1} + \phi_{n-1,k} + (q^{n-k+1} - 1)\phi_{n,k-1}$.

Highlight: Aztec Diamonds

A time-honored topic in combinatorics is to cover a given figure with a set of "bricks" of the same type without overlaps. Any such covering is called a *tiling*, and the task consists in proving that there exists a tiling at all, and if possible to count the number of such tilings. There is a vast and well-developed theory of tilings with regular polygons, as witnessed by ornaments and wallpaper patterns dating back to ancient times.

We consider the simplest type, tilings of a plane lattice figure with the usual 1×2-dominoes. Even for the most natural case, domino tilings of an $m \times n$-rectangle, to determine the number of tilings is no easy task. In Section 10.1 we will derive an intricate and unexpected formula, and for more complicated figures one would expect even greater difficulties.

So it came as a surprise when not long ago, Noam Elkies, Greg Kuperberg, Michael Larsen, and James Propp introduced a plane figure, which they called *Aztec diamond*, for which the answer is as simple as one could hope for.

Take n rows with $2, 4, 6, \ldots, 2n$ cells, stack them on top of each other in the form of a staircase, and reflect this staircase about the x-axis. The resulting configuration is the Aztec diamond $AZ(n)$. The figure shows the diamonds for $n = 1, 2, 3$:

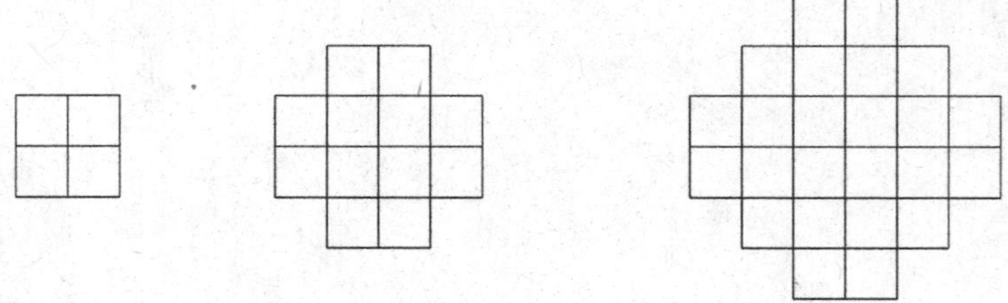

For later reference we note that $AZ(n)$ contains $2n(n+1)$ cells.

Let $A(n)$ be the number of domino tilings of $AZ(n)$. For example, $A(1) = 2$, and $A(2) = 2^3$, $A(3) = 2^6$ are easily seen. The result we want to prove is the following:

Theorem. *We have*

$$A(n) = 2^{\binom{n+1}{2}}. \tag{1}$$

Noam Elkies et al. gave four proofs for this astonishing formula, using various different ideas. One proof stands out in its simplicity and beauty. It uses an ingenious device called domino shuffling, and this is the proof we want to follow.

Domino Shuffling.

Consider the infinite $\mathbb{Z} \times \mathbb{Z}$-chessboard $\mathcal{Z}$, checkered as usual with white and black cells. Every domino thus occupies a black and a white cell. T is called a *partial tiling* of $\mathcal{Z}$ if T covers part (possibly all) of $\mathcal{Z}$. The uncovered cells are called *free cells*, and the uncovered part the *free region*.

Now we introduce the basic operation on a domino tiling, called *domino shuffling*. Horizontal dominoes move one step up or down, vertical dominoes one step to the left or right, as explained in the figure:

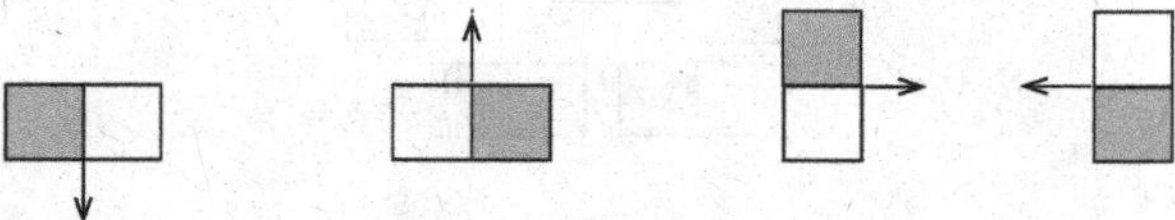

Performing the shuffling operation simultaneously for all dominoes of T, we obtain a new configuration $S(T)$.

A 2×2-square filled with two dominoes and with a black cell in the upper left-hand corner is called a *black block*:

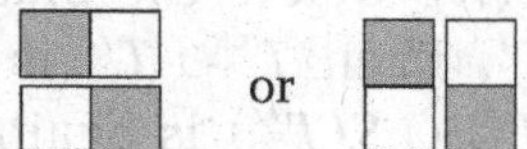

Notice that a black block is left invariant under the shuffling, since the dominoes just change their places. We now come to the crucial definition.

Definition. A partial tiling T is called *reduced* if

a. T contains no black blocks,
b. the free region can be filled with disjoint black blocks.

Main Lemma. *The shuffling operation is an involution on the set of all reduced tilings of* $\mathcal{Z}$.

We postpone the (quite subtle) proof, and show first how the Main Lemma implies (1), or what is the same, the recurrence

$$A(n) = 2^n A(n-1). \tag{2}$$

We place $AZ(n)$ on the infinite board such that the upper left-hand cell is black, and use the partial tiling T'' outside $AZ(n)$ as in the figure:

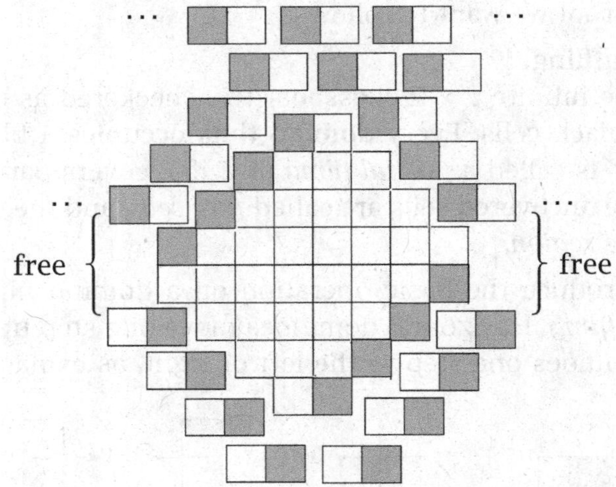

The tiling T'' is complete except for $AZ(n)$ and the two strips right and left of height 2. Note that T'' contains no black blocks, and that the free region outside $AZ(n)$ can be filled with disjoint black blocks. Notice further that the shuffled tiling $S(T'')$ leaves precisely the smaller Aztec diamond $AZ(n-1)$ uncovered. Consider now a complete tiling T of $AZ(n)$, delete the black blocks, and call the resulting partial tiling T'. Then $T' \cup T''$ is a reduced tiling of $\mathcal{Z}$. By the Main Lemma, $S(T') \cup S(T'')$ is again a reduced tiling, with $S(T') \subseteq AZ(n-1)$. Thus we have a bijective correspondence ϕ : $T' \subseteq AZ(n) \mapsto S(T') \subseteq AZ(n-1)$ of reduced tilings of $AZ(n)$ onto reduced tilings of $AZ(n-1)$.

Suppose T' contains t dominoes and k black holes. Then $S(T')$ has again t dominoes and, say, ℓ black holes. Since $|AZ(n)| = 2n(n+1)$, we obtain

$$2n(n+1) - 4k = 2t = 2n(n-1) - 4\ell$$

and thus

$$k = n + \ell\,.$$

Example. The unfilled 2×2-squares correspond to "black holes."

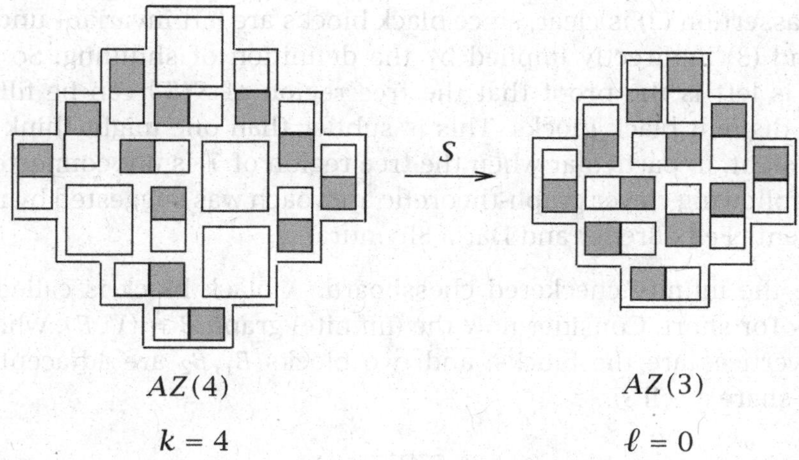

Now any black hole in T' can be filled with dominoes in two ways. Thus to a reduced tiling of $AZ(n)$ with $k = n + \ell$ holes there correspond $2^{n+\ell}$ complete tilings of $AZ(n)$, and similarly there are 2^{ℓ} complete tilings of $AZ(n-1)$ corresponding to $S(T')$. Hence to any given pair $(T', S(T'))$ the number of tilings of $AZ(n)$ is 2^n times the number of tilings of $AZ(n-1)$, and the recurrence $AZ(n) = 2^n AZ(n-1)$, and thus the theorem, follows.

Proof of the Main Lemma.
Let T be a reduced tiling. We have to prove four things:

1. $S(T)$ is a partial tiling,
2. $S(T)$ has no black blocks,
3. $S(S(T)) = T$,
4. $S(T)$ is reduced.

To prove (1) suppose to the contrary that a white cell s is covered twice in $S(T)$ (the case of a black cell is analogous). Then we have in T without loss of generality the following situation:

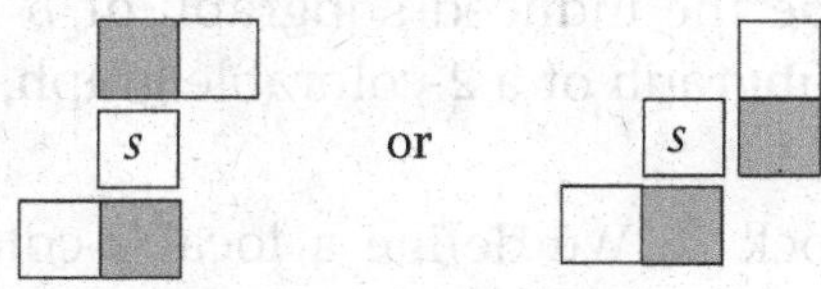

Since T contains no black blocks, s is a free cell of T. But then s could not be in a free black block, which is impossible, since T is reduced.

The assertion (2) is clear, since black blocks are left invariant under S, and (3) is directly implied by the definition of shuffling. So all that is left is the proof that the free region of $S(T)$ can be filled with disjoint black blocks. This is subtler than one might think at first sight, in particular when the free region of T is not connected. The following clever graph-theoretic approach was suggested by my students Felix Breuer and Daria Shymura.

Take the infinite checkered chessboard. A black block is called a *block* for short. Consider now the (infinite) graph $B = (V, E)$, where the vertices are the blocks, and two blocks B_1, B_2 are adjacent if they share a cell s:

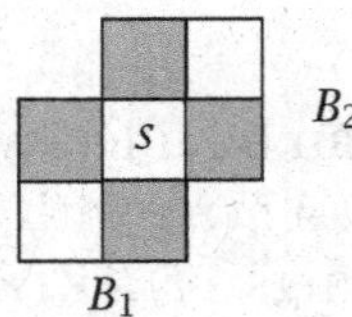

The graph B is bipartite, thus 2-colorable, and 4-regular. Consider a partial tiling T without (black) blocks. A block $A \in V$ is called *rich* (with respect to T) if A contains a complete domino D of T (and therefore exactly one, since T contains no blocks). We say that D is the domino *belonging* to A. Otherwise, A is called *poor.*

Example.

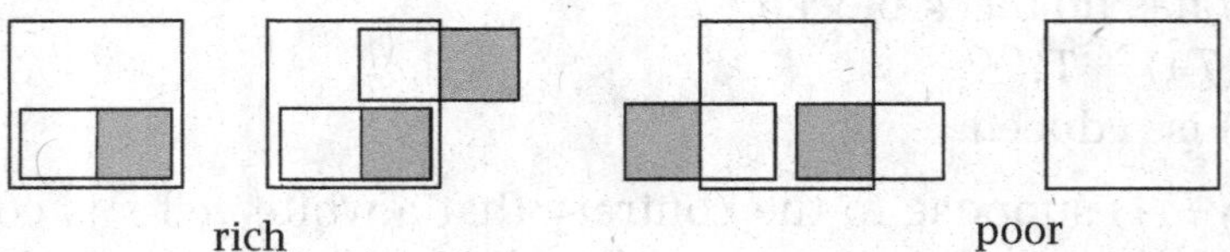

In particular, free blocks that are uncovered by T are always poor. Let $B_T = (V_T, E_T)$ be the induced subgraph of B consisting of the poor blocks. As a subgraph of a 2-colorable graph, B_T is 2-colorable as well.

Consider a rich block A. We define a local 2-coloring of its poor neighbors C as follows: If C contains one-half of the domino belonging to A, we color C red; otherwise, C is colored green.

Example.

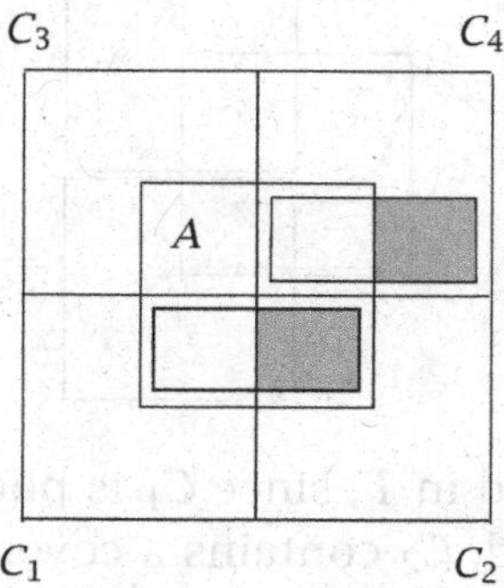

In the figure, C_1, C_2 are colored red, C_3 green, and C_4 is not colored since it is rich.

The following lemma is the main step toward the proof.

Lemma. *Let T be a partial tiling without (black) blocks. Then T is reduced if and only if*

a. *$S(T)$ is a partial tiling,*
b. *there exists a 2-coloring of B_T that coincides with the local 2-colorings of all rich blocks.*

Proof. Suppose T is a reduced tiling. Then we have already seen that (a) holds. Denote by G the set of disjoint blocks that fill up the free region, and by R the poor blocks not in G; thus $V_T = G \dot\cup R$. We color the blocks in G green, and those of R red.

First we show that this is an admissible 2-coloring of B_T. Let C_1 and C_2 be two adjacent blocks in B_T, and s the common cell:

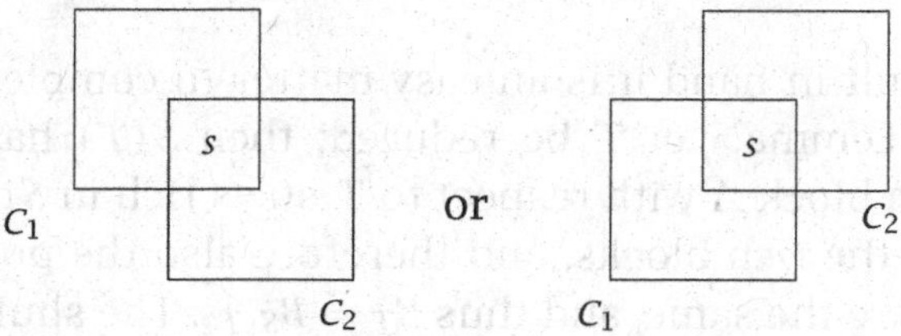

Since C_1 and C_2 are poor blocks, s is not covered in T. But then s is in a green block, so one of C_1 or C_2 is in G, and the other must be in R, since no two green blocks overlap.

It remains to verify that this coloring agrees with all local 2-colorings induced by the rich blocks. Let A be a rich block, and C_1 and C_2 two poor neighbors of A as in the figure:

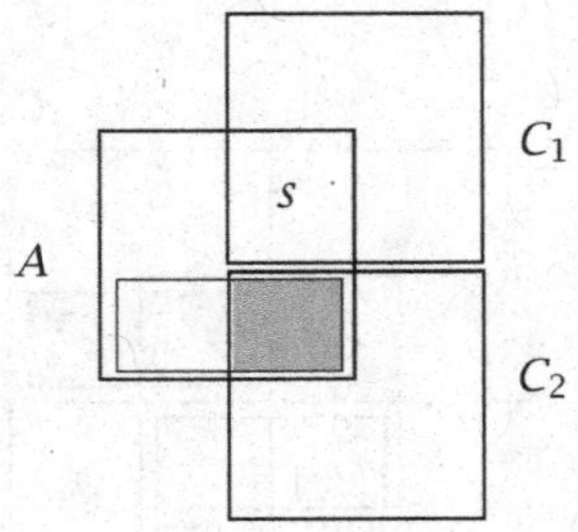

The cell s is not covered in T, since C_1 is poor. It follows that C_1 is in G. On the other hand, C_2 contains a covered cell and hence is in R by definition. So the colorings coincide, and (b) follows.

Now assume (a) and (b), and let $V_T = G \dot{\cup} R$, where $G \dot{\cup} R$ is the 2-coloring of B_T according to (b). The blocks of G are disjoint, and it remains to prove that they fill up the free region of T.

Let s be a free cell, and C_1, C_2 the neighboring blocks containing s. If C_1 and C_2 were both rich, then s would be covered twice in $S(T)$, contradicting (a). If C_1, C_2 are both poor, then one of C_1 or C_2 is in G (and the other in R). Finally, if C_1 is poor and C_2 is rich, then the local 2-coloring of C_2 forces C_1 to be in G. So every free cell is in a block of G. Next we see that no free cell is covered by two green blocks, since no two adjacent blocks of B_T are colored alike. Finally, it is easily seen that no cell that is covered in T belongs to a green block.

In summary, the blocks of G are all contained in the free region of T, and they cover it without overlap, which means that T is reduced. □

With this result in hand it is an easy matter to complete the proof of the Main Lemma. Let T be reduced; then $S(T)$ has no (black) blocks. A rich block A with respect to T stays rich in $S(T)$, and vice versa. Hence the rich blocks, and therefore also the poor blocks of T and $S(T)$, are the same, and thus $B_T = B_{S(T)}$. The shuffling clearly switches the color-classes of the local 2-colorings. Hence (a) and (b) are satisfied for $S(T)$, and the lemma implies that $S(T)$ is reduced.

Notes and References

The material on counting coefficients presented in this chapter forms the established core of combinatorial enumeration, from the classic treatise by MacMahon to the comprehensive monographs by Stanley. For those who are interested in the history of enumeration, the article of Biggs is a valuable source. Good introductions are the books by Riordan and Comtet, and for further reading Chapters 5 and 6 in the book by Graham–Knuth–Patashnik are recommended. The idea of representing combinatorial objects as words or paths has a long history and was in more recent times primarily developed by the French school of combinatorialists. Particularly influential was the article by Foata and Schützenberger. The original proof of the Aztec domino tilings is contained in the paper by Elkies et al., in which they showed unexpected connections to various other branches of mathematics. Their work has sparked a flurry of activity, and indeed there now exist several more proofs drawing from a variety of ideas. Finally, a book that should be on every enumerator's shelf is Sloane's handbook of integer sequences, with a list of several thousand sequences in lexicographic order. If a sequence you have just discovered is known, chances are you will find it there.

1. N.L. Biggs (1979): The roots of combinatorics. *Historia Math.* 6, 109–136.

2. L. Comtet (1974): *Advanced Combinatorics.* Reidel, Dordrecht and Boston.

3. N. Elkies, G. Kuperberg, M. Larsen, and J. Propp (1992): Alternating sign matrices and domino tilings, parts I and II. *Journal Algebr. Comb.* 1, 111–132, 219–234.

4. D. Foata and M. Schützenberger (1970): *Théorie Géometrique des Polynômes Euleriens.* Lecture Notes Math. 138. Springer, Berlin.

5. R.L. Graham, D.E. Knuth, and O. Patashnik (1994): *Concrete Mathematics*, 2nd edition. Addison-Wesley, Reading.

6. P.A. MacMahon (1915): *Combinatory Analysis*, 2 vols. Cambridge Univ. Press, Cambridge; reprinted in one volume by Chelsea, New York, 1960.

7. J. Riordan (1958): *An Introduction to Combinatorial Analysis.* Wiley, New York.
8. N.J.A. Sloane (1994): *The New Book of Integer Sequences.* Springer, Berlin/New York.
9. R.P. Stanley (1997/9): *Enumerative Combinatorics*, 2 vols. Cambridge Univ. Press, Cambridge/New York.

2 Formal Series and Infinite Matrices

We come to the most important idea in enumerative combinatorics, which will allow surprisingly simple proofs of identities and recurrences. Suppose we are interested in the counting function $f : \mathbb{N}_0 \to \mathbb{C}$. We associate to f the *formal series* $F(z) = \sum_{n \geq 0} f(n) z^n$, and say that $F(z)$ is the *generating function* of f.

The adjective "formal" refers to the fact that we regard these series as *algebraic* objects that can be added and multiplied; the symbol z^n is just a mark for where the n-th coefficient $f(n)$ is placed. We do not consider $F(z)$ as a function in the usual sense; that is, we are, with some exceptions, not interested for which values of z the series $F(z)$ converges. Two formal series $F(z)$ and $G(z)$ are simply termed equal if and only if they agree in all coefficients. Whenever we have manipulated two such series algebraically and have obtained $F(z) = G(z)$, then we can read off $f(n) = g(n)$ for all n, and may interpret these identities combinatorially.

2.1 Algebra of Formal Series

If $A(z) = \sum_{n \geq 0} a_n z^n$, then a_n is the n-th *coefficient* of $A(z)$, for which we also write $a_n = [z^n]A(z)$. Sometimes it is useful to consider coefficients with negative index, with the understanding that $a_k = 0$ for $k < 0$. The coefficient a_0 is called the *constant* coefficient, and we also write $a_0 = A(0)$. The set of all formal series over $\mathbb{C}$ shall be denoted by $\mathbb{C}[[z]]$.

Sum and Product.
The *sum* $\sum a_n z^n + \sum b_n z^n$ is defined to be $\sum (a_n + b_n) z^n$, and the *scalar product* $c \sum a_n z^n$ as $\sum (c a_n) z^n$. With these operations $\mathbb{C}[[z]]$ becomes a vector space with 0 as zero-element.

We also have a natural product. For $A(z) = \sum_{n\geq 0} a_n z^n$, $B(z) = \sum_{n\geq 0} b_n z^n$ we set

$$A(z)B(z) = \sum_{n\geq 0}\left(\sum_{k=0}^{n} a_k b_{n-k}\right) z^n . \tag{1}$$

This product is called the *convolution* of $A(z)$ and $B(z)$; it is suggested by the product $z^k z^\ell = z^{k+\ell}$. What is the contribution to z^n in $A(z)B(z)$? We must choose $a_k z^k$ and $b_{n-k}z^{n-k}$, i.e., $a_k b_{n-k}$, and take the sum from $k = 0$ to $k = n$. We see that the convolution is commutative with $A(z) = 1$ as multiplicative identity.

The reader may easily check that these operations satisfy all the usual properties such as associativity and distributivity. Hence $\mathbb{C}[[z]]$ is a commutative ring over $\mathbb{C}$, in fact an integral domain, since $A(z)B(z) = 0$ clearly implies $A(z) = 0$ or $B(z) = 0$. Let us ask next which series $A(z)$ have a multiplicative inverse $B(z)$ with $A(z)B(z) = 1$. This has an easy answer:

$$A(z) = \sum_{n\geq 0} a_n z^n \textit{ has an inverse if and only if } a_0 \neq 0.$$

Since $A(z)B(z) = 1$ implies $a_0 b_0 = 1$, the condition $a_0 \neq 0$ is certainly necessary. Suppose, conversely, $a_0 \neq 0$. We determine the coefficients b_n of $B(z)$ step by step. At the start, $b_0 = a_0^{-1}$. Suppose $b_0, b_1, \ldots, b_{n-1}$ are already (uniquely) determined. Then it follows from $0 = \sum_{k=0}^{n} a_k b_{n-k} = a_0 b_n + \sum_{k=1}^{n} a_k b_{n-k}$, that $b_n = -a_0^{-1}\sum_{k=1}^{n} a_k b_{n-k}$ is uniquely determined.

Example. The geometric series $\sum_{n\geq 0} z^n$ has as inverse $1 - z$ by (1); hence we write $\sum_{n\geq 0} z^n = \frac{1}{1-z}$. Let $A(z) = \sum_{n\geq 0} a_n z^n$. Then by convolution,

$$\frac{A(z)}{1-z} = \sum_{n\geq 0} a_n z^n \cdot \sum_{n\geq 0} z^n = \sum_{n\geq 0}\left(\sum_{k=0}^{n} a_k\right) z^n , \tag{2}$$

thus $\frac{A(z)}{1-z}$ adds the coefficients up to n. In particular, we have

$$\frac{1}{(1-z)^2} = \sum_{n\geq 0}(n+1)z^n .$$

The multiplication $z^m A(z)$, $m \geq 0$, corresponds to a shift of the index by $-m$:

$$z^m \sum_{n\geq 0} a_n z^n = \sum_{n\geq 0} a_n z^{n+m} = \sum_{n\geq 0} a_{n-m} z^n. \tag{3}$$

For example, $\frac{z}{(1-z)^2} = \sum_{n\geq 0} nz^n$.

Composition.

Next, we consider the *composition* of two series. For $A(z) = \sum a_n z^n$, $B(z) = \sum b_n z^n$, we set

$$A(B(z)) = \sum_{n\geq 0} a_n (B(z))^n.$$

Expanding the right-hand side,

$$a_0 + a_1(b_0 + b_1 z + b_2 z^2 + \cdots) + a_2(b_0 + b_1 z + b_2 z^2 + \cdots)^2 + \cdots,$$

we find that the composition is at any rate well defined if $A(z)$ is a polynomial. But when $A(z)$ has infinitely many nonzero coefficients, we must assume that $b_0 = 0$, since otherwise we would obtain an infinite sum $a_0 + a_1 b_0 + a_2 b_0^2 + \cdots$ as constant coefficient. If on the other hand, $b_0 = 0$, then

$$[z^n]A(B(z)) = [z^n] \sum_{k=0}^{n} a_k (b_1 z + b_2 z^2 + \cdots)^k$$

is well defined, since there are no nonzero contributions to $[z^n]$ for $k > n$. In summary, we note:

$A(B(z))$ is well defined when $A(z)$ is a polynomial or $B(0) = 0$.

As for compositional inverses, you are asked in the exercises to prove the following:

Let $A(0) = 0$. Then there exists a unique series $B(z)$ with $B(0) = 0$ such that $A(B(z)) = B(A(z)) = z$ if and only if $a_1 \neq 0$. The series $B(z)$ is called the compositional inverse of $A(z)$, denoted by $A^{\langle -1 \rangle}(z)$. The compositional unit is z.

It is straightforward to verify that sum and product are compatible with composition (whenever defined), that is,

$$\begin{aligned} C(z) &= A(z) + B(z) \text{ implies } C(D(z)) = A(D(z)) + B(D(z)), \\ C(z) &= A(z)B(z) \quad \text{ implies } C(D(z)) = A(D(z)) \cdot B(D(z)); \end{aligned}$$

furthermore, associativity holds. In particular, when $A(z)$ has a compositional inverse and $A(B(z)) = A(C(z))$, then $B(z) = C(z)$, and $A(D(z))^{-1} = A^{-1}(D(z))$, whenever defined.

As examples we have $\sum_{n\geq 0} z^{2n} = \frac{1}{1-z^2}$ and further $\frac{1}{(1-cz)^2} = \sum_{n\geq 0}(n+1)c^n z^n$.

Important Series.

Here is a list of formal series that we will constantly use:

1. $\sum\limits_{n\geq 0} z^n = \frac{1}{1-z}$,
2. $\sum\limits_{n\geq 0} (-1)^n z^n = \frac{1}{1+z}$,
3. $\sum\limits_{n\geq 0} z^{2n} = \frac{1}{1-z^2}$,
4. $\sum\limits_{n\geq 0} \binom{m}{n} z^n = (1+z)^m \quad (m \in \mathbb{Z})$,
5. $\sum\limits_{n\geq 0} \binom{m+n-1}{n} z^n = \frac{1}{(1-z)^m} \quad (m \in \mathbb{Z})$,
6. $\sum\limits_{n\geq 0} \binom{m+n}{n} z^n = \frac{1}{(1-z)^{m+1}} \quad (m \in \mathbb{Z})$,
7. $\sum\limits_{n\geq 0} \binom{n}{m} z^n = \frac{z^m}{(1-z)^{m+1}} \quad (m \in \mathbb{N}_0)$.

We have already seen (1) to (3). When $m \in \mathbb{N}_0$, equation (4) is just the binomial theorem. Suppose $m < 0$. Since $(1+z)^m$ is the inverse of $(1+z)^{-m}$, we have to show that $\sum_{n\geq 0}\binom{m}{n} z^n$ $(m < 0)$ is the inverse of $\sum_{n\geq 0}\binom{-m}{n} z^n$. But this is an immediate consequence of Vandermonde's identity: The n-th coefficient of $\sum\binom{-m}{n} z^n \cdot \sum\binom{m}{n} z^n$ is by convolution

$$\sum_{k=0}^{n}\binom{-m}{k}\binom{m}{n-k} = \binom{0}{n} = \delta_{n,0}\,.$$

Equation (5) (and hence (6)) follows from (4) by the reciprocity law $\binom{m+n-1}{n} = (-1)^n\binom{-m}{n}$. Finally, for (7) we see by index shift that

$$\frac{z^m}{(1-z)^{m+1}} = z^m \sum_{n\geq 0}\binom{m+n}{m} z^n = \sum_{n\geq 0}\binom{n}{m} z^n\,.$$

Equation (4) suggests that we *define*

$$\sum_{n\geq 0}\binom{c}{n} z^n =: (1+z)^c \text{ for arbitrary } c \in \mathbb{C}\,,$$

and similarly we may replace m by $c \in \mathbb{C}$ in (5) and (6). In other words, $(1+z)^c$ is a *symbol* for the formal series on the left-hand side, since we have no definition, in general, of a real or complex power of a formal series. Still, the familiar equality

$$(1+z)^{a+b} = (1+z)^a(1+z)^b \quad (a, b \in \mathbb{C})$$

holds, since the convolution reduces simply to Vandermonde's identity:

$$\binom{a+b}{n} = \sum_{k=0}^{n} \binom{a}{k}\binom{b}{n-k}.$$

Alternatively, we may consider c as an indeterminate and regard $\sum \binom{c}{n} z^n$ as a formal series in $(\mathbb{C}[c])[[z]]$. The coefficient of z^n is then a polynomial in c, and all the expected properties are formally valid.

Let us list three important series known from analysis:

8. $\sum\limits_{n\geq 0} \frac{z^n}{n!} = e^z$,
9. $\sum\limits_{n\geq 1} (-1)^{n-1}\frac{z^n}{n} = \log(1+z)$,
10. $\sum\limits_{n\geq 1} \frac{z^n}{n} = -\log(1-z)$.

Formally, e^z and $\log(1+z)$ are as above just names for the series on the left-hand side. We know from analysis that e^z and $\log z$ are compositional inverses of each other. As we have seen, for formal series such inverses exist only when $a_0 = 0$, $a_1 \neq 0$. Hence we will expect that $\log(1+z)$ is the compositional inverse of $e^z - 1 = \sum_{n\geq 1} \frac{z^n}{n!}$. This is indeed the case, but the proof is a little subtle and will be given in the next chapter. Note that this implies $e^{\log F(z)} = F(z)$ whenever $F(0) = 1$.

Derivative.
Another useful operation that can be defined formally is the derivative. If $F(z) = \sum_{n\geq 0} a_n z^n$, then the *formal derivative* is the series $F'(z) = \sum_{n\geq 1} n a_n z^{n-1} = \sum_{n\geq 0}(n+1)a_{n+1}z^n$. It is easy to check that all familiar properties hold:

$$(F+G)' = F' + G', \quad (FG)' = F'G + FG',$$

$$(F^{-1})' = -\frac{F'}{F^2}, \quad F(G(z))' = F'(G(z)) \cdot G'(z).$$

Furthermore, we see $\sum_{n\geq 0} na_n z^n = zF'(z)$, $\sum_{n\geq 1} na_{n-1}z^{n-1} = (zF(z))'$.

As examples, $(e^z)' = e^z$, $(\log(1+z))' = \sum_{n\geq 1}(-1)^{n-1}z^{n-1} = \frac{1}{1+z}$, and more generally for $F(z)$ with $F(0) = 1$,

$$(\log F(z))' = \frac{F'(z)}{F(z)}. \tag{4}$$

The inverse operation is, of course, formal indefinite integration:

$$\int \sum_{n\geq 0} a_n z^n = \sum_{n\geq 0} a_n \frac{z^{n+1}}{n+1} + \text{constant}. \tag{5}$$

Example. Consider $F(z) = \sum_{n\geq 0} \binom{2n}{n} z^n$, $a_n = \binom{2n}{n}$. We have $a_n = \binom{2n}{n} = \frac{2n(2n-1)}{n^2} a_{n-1}$, and thus $na_n = 4na_{n-1} - 2a_{n-1}$. Comparing coefficients for z^{n-1} we get

$$F'(z) = 4(zF(z))' - 2F(z),$$

hence

$$\begin{aligned} F'(z) &= 4F(z) + 4zF'(z) - 2F(z) \\ &= 4zF'(z) + 2F(z). \end{aligned}$$

This, in turn, yields

$$(\log F(z))' = \frac{F'(z)}{F(z)} = \frac{2}{1-4z} = -\frac{1}{2}(\log(1-4z))',$$

and thus by indefinite integration

$$\log F(z) = -\frac{1}{2}\log(1-4z),$$

since both sides have vanishing constant term. Accepting the familiar fact $m \log A(z) = \log(A(z)^m)$ for $A(0) = 1$, which will be proved later (see Exercise 2.12), we obtain $F(z) = (1-4z)^{-1/2}$, that is,

$$F(z) = \sum_{n\geq 0} \binom{2n}{n} z^n = \frac{1}{\sqrt{1-4z}}. \tag{6}$$

Now that we know the result let us check it using Vandermonde. According to Exercise 1.16,

$$\binom{-1/2}{n} = \left(-\frac{1}{4}\right)^n \binom{2n}{n},$$

and hence by Vandermonde's identity,

$$(-1)^n = \binom{-1}{n} = \sum_{k=0}^{n} \binom{-1/2}{k}\binom{-1/2}{n-k}$$

$$= \left(-\frac{1}{4}\right)^n \sum_{k=0}^{n} \binom{2k}{k}\binom{2(n-k)}{n-k},$$

or

$$\sum_{k=0}^{n} \binom{2k}{k}\binom{2(n-k)}{n-k} = 4^n \quad (n \ge 0)\,. \tag{7}$$

But this is equivalent to $F(z)^2 = \frac{1}{1-4z}$, and we obtain again $F(z) = \frac{1}{\sqrt{1-4z}}$.

Remark. We know from analysis that if two series $F(z)$ and $G(z)$ agree as *functions* in a neighborhood of the origin, then they have identical coefficients, i.e., they are equal as formal series. In other words, if the computations we perform are formally justified, and if the resulting series have as functions nonzero radius of convergence, then we have an identity as formal series. But even if a series has radius of convergence 0 such as $\sum n!z^n$, we may still manipulate the series formally in a meaningful way.

Exercises

2.1 Let $F(z) = \sum_{n\ge 0} a_n z^n$. Show that

$$\tfrac{F(z)+F(-z)}{2} = \sum_{n\ge 0} a_{2n} z^{2n}, \; \tfrac{F(z)-F(-z)}{2} = \sum_{n\ge 0} a_{2n+1} z^{2n+1}\,.$$

The series $F(z)$ is called *even (odd)* if all a_n with odd (even) index are 0. Prove: $F(z)$ is even if and only if $F(z) = F(-z)$, and odd if and only if $F(z) = -F(-z)$.

▷ **2.2** Find the unique sequence (a_n) of real numbers with $\sum_{k=0}^{n} a_k a_{n-k} = 1 \; (n \ge 0)$.

2.3 Prove $e^{(a+b)z} = e^{az}e^{bz} \; (a, b \in \mathbb{C})$, and deduce $e^{mz} = (e^z)^m \; (m \in \mathbb{Q})$.

2.4 Let $F(z) = \sum_{n\ge 0} a_n z^n$ with $a_0 = 0$. Show that $F(z)$ has a compositional inverse $F^{\langle -1\rangle}(z) = \sum_{n\ge 0} b_n z^n$ with $b_0 = 0$ if and only if $a_1 \ne 0$.

2.5 Consider the double series $F(y,z) = \sum_{n\ge 0}\sum_{k\ge 0}\binom{n}{k} y^k z^n$. Determine $F(y,z)$, and also $G(y,z) = \sum_{n\ge 0}\sum_{k\ge 0}\binom{n}{k} y^k \frac{z^n}{n!}$.

▷ **2.6** Find the generating function $F(z)$ with $[z^n]F(z) = \sum_k \binom{r}{k}\binom{r}{n-2k}$.

* * *

2.7 Give a combinatorial argument for $\sum_{k=0}^{n}\binom{2k}{k}\binom{2(n-k)}{n-k} = 4^n$.

▷ **2.8** Determine $\sum_{n\ge 0}\binom{2n+1}{n} z^n$ and $\sum\limits_{n\ge 0}\binom{n}{\lfloor n/2\rfloor} z^n$.
Hint: Use $\sum_{n\ge 0}\binom{2n}{n} z^n = (1-4z)^{-1/2}$.

2.9 Compute the coefficients of the series $F(z) = \sqrt{\frac{1+z}{1-z}}$.

2.10 Verify the associativity $F\big(G(H(z))\big) = F(G(z))(H(z))$ whenever defined.

▷ **2.11** When $F(0) = 1$, show that $(\log F(z))' = \frac{F'(z)}{F(z)}$.

2.12 Prove $m\log(1+z) = \log((1+z)^m)$ for $m \in \mathbb{Q}$.

2.2 Types of Formal Series

We can represent a sequence (a_n) not only by an *ordinary* generating function $\sum_{n\ge 0} a_n z^n$, but also by a so-called *exponential* generating function $\sum_{n\ge 0} a_n \frac{z^n}{n!}$, which is sometimes better adapted to the problem at hand. One example is $\sum_{n\ge 0}\frac{z^n}{n!} = e^z$, which corresponds to the sequence $(1,1,1,\ldots)$. Hence e^z plays a similar role for exponential series as $\sum_{n\ge 0} z^n = \frac{1}{1-z}$ does for ordinary series.

What natural types appear is the subject of this section. We start with the following setup. Let $q_1, q_2, q_3, \ldots$ be a sequence of nonzero complex numbers with $q_1 = 1$. We set $Q_n = q_1 q_2 \cdots q_n$ with $Q_0 = 1$, and associate to the counting function $f : \mathbb{N}_0 \to \mathbb{C}$ the Q-series

$$F(z) = \sum_{n\ge 0} f(n)\frac{z^n}{Q_n}. \tag{1}$$

Thus when $q_n = 1$ for all n, we obtain ordinary series, while for $q_n = n$, i.e., $Q_n = n!$, exponential series result.

The q-Derivative.
Next we define the *q-derivative operator* Δ as follows: Δ is a linear operator on $\mathbb{C}[[z]]$ defined by

$$\begin{aligned} \Delta z^0 &= 0 \\ \Delta z^n &= q_n z^{n-1} \quad (n \geq 1). \end{aligned} \tag{2}$$

By linearity we find for $F(z) = \sum_{n \geq 0} a_n \frac{z^n}{Q_n}$,

$$\Delta F(z) = \sum_{n \geq 0} a_n \frac{\Delta z^n}{Q_n} = \sum_{n \geq 1} a_n \frac{z^{n-1}}{Q_{n-1}} = \sum_{n \geq 0} a_{n+1} \frac{z^n}{Q_n}.$$

Thus Δ shifts the sequence $(a_0, a_1, a_2, \ldots)$ to $(a_1, a_2, a_3, \ldots)$.

Consider the product $C(z) = A(z)B(z)$ as Q-series. Comparing coefficients for z^n we obtain

$$\frac{c_n}{Q_n} = \sum_{k=0}^{n} \frac{a_k}{Q_k} \frac{b_{n-k}}{Q_{n-k}}.$$

The convolution formula for Q-series reads therefore

$$c_n = \sum_{k=0}^{n} \frac{Q_n}{Q_k Q_{n-k}} a_k b_{n-k} = \sum_{k=0}^{n} \begin{bmatrix} n \\ k \end{bmatrix} a_k b_{n-k}, \tag{3}$$

where

$$\begin{bmatrix} n \\ k \end{bmatrix} = \frac{Q_n}{Q_k Q_{n-k}} = \frac{q_n q_{n-1} \cdots q_{n-k+1}}{q_k q_{k-1} \cdots q_1}, \qquad \begin{bmatrix} n \\ 0 \end{bmatrix} = 1. \tag{4}$$

The symbol $\left[\begin{smallmatrix} n \\ k \end{smallmatrix}\right]$ satisfies $\left[\begin{smallmatrix} n \\ k \end{smallmatrix}\right] = \left[\begin{smallmatrix} n \\ n-k \end{smallmatrix}\right]$, and it looks very much like the binomial coefficient. That this is no coincidence will be shortly made clear.

There are few general results on these Q-series. We restrict ourselves to those sequences (q_n) for which Δ satisfies the following product property:

$$\Delta(F(z)G(z)) = (\Delta F(z))G(qz) + F(z)(\Delta G(z))$$

for some fixed q, and (5)

$$\Delta(F(z)G(z)) = \Delta(G(z)F(z)).$$

Looking at (2) we obtain from $z^{m+n} = z^m z^n$,

$$\Delta z^{m+n} = q_{m+n} z^{m+n-1} = (q_m z^{m-1})(q^n z^n) + z^m q_n z^{n-1}.$$

Comparing coefficients and exchanging z^m and z^n gives

$$q_{m+n} = q_m q^n + q_n = q_n q^m + q_m.$$

For $m = 1$, this yields

$$q_{n+1} = q^n + q_n = q \cdot q_n + 1,$$

or

$$q_n = \frac{1-q^n}{1-q}. \tag{6}$$

Conversely, it is easily checked that the sequence ($q_n = \frac{1-q^n}{1-q}$) satisfies (5) for every q. Thus our requirement leads us to the q-calculus discussed in Section 1.6.

From $\Delta z^n = \frac{1-q^n}{1-q} z^{n-1} = \frac{z^n-(qz)^n}{(1-q)z}$, we infer by linearity

$$\Delta F(z) = \frac{F(z) - F(qz)}{(1-q)z}. \tag{7}$$

The Three q-Calculi.
In summary, every $q \in \mathbb{C}$ gives rise to a class of formal Q-series with $Q_n = [n]_q!$. The cases $q = 0$ and 1 play a special role.

A. $q = 0$. Here we have $q_n = 1$ for all n, i.e., $Q_n = 1$, and we obtain the calculus of ordinary series $F(z) = \sum_{n \geq 0} a_n z^n$. The coefficients $\left[{n \atop k}\right]$ are all equal to 1; thus (3) is the ordinary convolution as defined in the previous section. We write $\Delta = D_0$ for the derivative and have by (7),

$$D_0 F(z) = \frac{F(z) - F(0)}{z}. \tag{8}$$

The product rule is

$$D_0(F(z)G(z)) = (D_0 F(z))G(0) + F(z)(D_0 G(z)). \tag{9}$$

B. $q = 1$. Here $\frac{1-q^n}{1-q} = 1 + q + \cdots + q^{n-1} = n$, $Q_n = n!$, as we have noted before. If (a_n) is a sequence, then we denote its exponential generating function by $\hat{F}(z) = \sum_{n\geq 0} a_n \frac{z^n}{n!}$. The convolution coefficients are $\left[{n \atop k}\right] = \frac{n!}{k!(n-k)!} = \binom{n}{k}$, and the product $\hat{C}(z) = \hat{A}(z)\hat{B}(z)$ corresponds to the *binomial convolution*

$$c_n = \sum_{k=0}^{n} \binom{n}{k} a_k b_{n-k}. \tag{10}$$

The derivative is, of course, the formal derivative of the last section, which we denote by $\hat{F}'(z)$ or $D\hat{F}(z)$. The product rule reads as usual, and we obtain for $\hat{F}(z)$,

$$(\hat{F}^k(z))' = k\hat{F}^{k-1}(z)\hat{F}'(z). \tag{11}$$

C. For general q (or q an indeterminate) we get series of the form

$$F(z) = \sum a_n \frac{z^n}{Q_n}, \quad \text{where } Q_n = [n]_q! = \frac{\prod_{i=1}^{n}(1-q^i)}{(1-q)^n}.$$

The coefficients $\left[{n \atop k}\right] = \left[{n \atop k}\right]_q$ are the Gaussian coefficients

$$\left[{n \atop k}\right]_q = \frac{(1-q^n)\cdots(1-q^{n-k+1})}{(1-q^k)\cdots(1-q)},$$

and the convolution for $C(z) = A(z)B(z)$ is

$$c_n = \sum_{k=0}^{n} \left[{n \atop k}\right]_q a_k b_{n-k}.$$

We write D_q for the derivative and have by (7),

$$D_q F(z) = \frac{F(z) - F(qz)}{(1-q)z}.$$

Example. Consider the Stirling series $S_k(z) = \sum_{n\geq 0} S_{n,k} z^n$, $S_0(z) = 1$. The recurrence $S_{n,k} = S_{n-1,k-1} + kS_{n-1,k}$ leads to

$$D_0 S_k(z) = \frac{S_k(z)}{z} = S_{k-1}(z) + kS_k(z) \quad (k \geq 1);$$

thus

$$S_k(z) = \frac{zS_{k-1}(z)}{1-kz}.$$

Iterating, we arrive at

$$S_k(z) = \frac{z^k}{(1-z)(1-2z)\cdots(1-kz)}.$$

Looking at the right-hand side, this gives

$$\begin{aligned} S_{n,k} &= [z^{n-k}]\frac{1}{(1-z)\cdots(1-kz)} = \sum_{\sum b_j = n-k} 1^{b_1}2^{b_2}\cdots k^{b_k} \\ &= \sum 1^{a_1-1}2^{a_2-1}\cdots k^{a_k-1}, \end{aligned}$$

where the sum extends over all solutions of $\sum_{j=1}^{k} a_j = n$ in positive integers (see Exercise 1.31).

Example. A product $\hat{B}(z) = \hat{A}(z)e^z$ for exponential series corresponds to the convolution

$$b_n = \sum_{k=0}^{n} \binom{n}{k} a_k \quad \text{for all } n.$$

Now $\hat{B}(z) = \hat{A}(z)e^z$ holds if and only if $\hat{A}(z) = \hat{B}(z)e^{-z}$, which means for the coefficients with $e^{-z} = \sum_{n\geq 0} \frac{(-1)^n z^n}{n!}$,

$$a_n = \sum_{k=0}^{n} (-1)^{n-k}\binom{n}{k} b_k \quad \text{for all } n,$$

and we have proved the so-called *binomial inversion formula*

$$b_n = \sum_{k=0}^{n}\binom{n}{k}a_k \;\; (\forall n) \iff a_n = \sum_{k=0}^{n}(-1)^{n-k}\binom{n}{k}b_k \;\; (\forall n). \qquad (12)$$

As another example consider the exponential generating function $\hat{D}(z) = \sum_{n\geq 0} D_n \frac{z^n}{n!}$ of the derangement numbers. Classifying the permutations according to their fixed-point set, we clearly have $\sum_{k=0}^{n}\binom{n}{k}D_{n-k} = \sum_{k=0}^{n}\binom{n}{k}D_k = n!$; thus by convolution,

$$\hat{D}(z)e^z = \sum_{n\geq 0}\frac{n!}{n!}z^n = \sum_{n\geq 0} z^n = \frac{1}{1-z},$$

or

$$\hat{D}(z) = \frac{e^{-z}}{1-z}.$$

If we interpret this last identity in terms of ordinary generating functions, we obtain by equation (2) in the previous section our old formula

$$\frac{D_n}{n!} = \sum_{k=0}^{n} \frac{(-1)^k}{k!}.$$

Example. Consider the Gaussian polynomials $g_n(x) = (x-1)\cdots(x-q^{n-1})$. We claim that

$$D_q g_n(x) = \frac{1-q^n}{1-q} g_{n-1}(x). \tag{13}$$

This is certainly true for $n = 0, 1$, so assume $n \geq 2$. The identity $g_n(x) = g_{n-1}(x)(x - q^{n-1})$ and the product rule (5) yield with induction

$$\begin{aligned} D_q g_n(x) &= (D_q g_{n-1}(x))(qx - q^{n-1}) + g_{n-1}(x) \\ &= \frac{1-q^{n-1}}{1-q} g_{n-2}(x)(x - q^{n-2})q + g_{n-1}(x) \\ &= g_{n-1}(x)\Big(\frac{1-q^{n-1}}{1-q}q + 1\Big) \\ &= \frac{1-q^n}{1-q} g_{n-1}(x). \end{aligned}$$

Exercises

2.13 From $B(z) = \frac{A(z)}{1-z} \iff A(z) = B(z)(1-z)$ derive an inversion formula for the coefficients a_n of $A(z)$ and b_n of $B(z)$.

2.14 If $F(0) = 0$, show that $D_0 F^k(z) = F^{k-1}(z)(D_0 F(z))$.

2.15 Show that $(D_q z - z D_q)F(z) = F(qz)$ for q-series.

▷ **2.16** We know the chain rule $\hat{F}(\hat{G}(z))' = \hat{F}'(\hat{G}(z)) \cdot \hat{G}(z)'$ for the exponential calculus ($q = 1$). Extend this to arbitrary q if the inner function is linear, that is, $D_q F(cz) = (D_q F)(cz) \cdot c$, where $(D_q F)(cz)$ means that we compute $D_q F(z)$ and then replace z by cz.

2.17 Prove the chain rule $D_0 F(z^m) = (D_0 F)(z^m) \cdot z^{m-1}$, $m \geq 1$.

▷ **2.18** Define $H_n(x) = \sum_{k=0}^{n} \left[{n \atop k} \right]_q x^k$; $H_n(x)$ is called *q-Hermite polynomial.* Show that $D_q H_n(x) = \frac{1-q^n}{1-q} H_{n-1}(x)$, and further $H_{n+1}(x) = xH_n(x) + H_n(qx)$.

* * *

2.19 Derive an inversion formula from

$$B(z) = \frac{A(z)}{(1-z)^m} \iff A(z) = B(z)(1-z)^m .$$

2.20 Prove the general Leibniz rule in the q-calculus

$$D_q^m(F(z)G(z)) = \sum_{k=0}^{m} \left[{m \atop k} \right]_q (D_q^k F(z))(D_q^{m-k} G(z))(q^k z) .$$

Specialize to D_0 $(q = 0)$.

▷ **2.21** Show that $\sum_{n \geq 0} \left[{m+n-1 \atop n} \right]_q z^n = \frac{1}{(1-z)(1-zq)\cdots(1-zq^{m-1})}$ $(m \in \mathbb{N})$. Hint: Consider $\frac{1}{(1-zq)(1-zq^2)\cdots(1-zq^m)}$ and expand as a series, or apply D_q to both sides.

2.22 Compute the exponential generating function of the polynomials $s_n(x) = \sum_{k=0}^{n} s_{n,k} x^k$ ($s_{n,k}$ Stirling number of the first kind), and derive $s_n(x+y) = \sum_{k=0}^{n} \binom{n}{k} s_k(x) s_{n-k}(y)$.

2.23 Let $\widehat{s}_k(z) = \sum_{n \geq 0} s_{n,k} \frac{z^n}{n!}$. Derive the identity $\widehat{s}_k(z)' = \frac{\widehat{s}_{k-1}(z)}{1-z}$ for $k \geq 1$. What are $\widehat{s}_1(z)$ and $\widehat{s}_2(z)$? Derive an expression for $s_{n,2}$ and $s_{n,3}$ in terms of harmonic numbers.

▷ **2.24** Prove the q-binomial theorem of Section 1.6 by writing $f_n(x) = (1 + xq)(1 + xq^2) \cdots (1 + xq^n)$ in the form $f_n(x) = \sum_{k=0}^{n} q^{\binom{k+1}{2}} a_{n,k} x^k$, and taking the derivative D_q to determine $a_{n,k}$.

2.25 Show that the q-Vandermonde formula in Exercise 1.72 follows from $f_{r+s}(x) = f_r(x) f_s(q^r x)$, where $f_n(x)$ is the polynomial of the previous exercise.

2.3 Infinite Sums and Products

We have defined the sum and product of two formal series, and we may extend this to any finite number of summands or factors. In some instances, especially in discussing series of number-partitions, it is advantageous also to consider countably infinite sums and products.

To treat these problems we introduce the following simple notion of formal convergence. Let $a_0, a_1, a_2, \ldots$ be a sequence of complex numbers (or elements of a field of characteristic 0). We say that the sequence (a_n) converges to b if $a_n = b$ for all but finitely many n. In other words, $a_n = b$ for $n \geq n_0$.

Definition. Suppose $A_0(z), A_1(z), A_2(z), \ldots$ is a sequence of ordinary series, $A_i(z) = \sum_{n\geq 0} a_{i,n} z^n$. We say $(A_i(z))$ *converges* to $F(z) = \sum_{n\geq 0} f_n z^n$ if for every n, the sequence $(a_{i,n} = [z^n]A_i)$ converges to $f_n = [z^n]F$.

The limit series $F(z)$ is obviously unique if it exists. Furthermore, $(A_i(z))$ converges if and only if for every n, the sequence of n-th coefficients becomes eventually constant.

The following facts are immediately established (see Exercise 2.26). Suppose $(A_i(z)) \to F(z)$, $(B_i(z)) \to G(z)$. Then

$$(A_i + B_i) \to F + G, \quad (A_i B_i) \to FG, \quad (A_i') \to F', \tag{1}$$

and if $B_i(0) = 0$ for all i, then

$$\big(A_i(B_i(z))\big) \to F(G(z)). \tag{2}$$

A quick way to observe convergence goes as follows. Define the *degree* of a series $A(z) = \sum_{n\geq 0} a_n z^n \neq 0$ as

$$\deg A = \min k \ \text{ with } a_k \neq 0 .$$

For the 0-series we set $\deg 0 = \infty$. Note that $\deg(AB) = \deg A + \deg B$.

Proposition 2.1. *A sequence $(A_i(z))$ of series converges if and only if for any n,* $\deg\left(A_{i+1}(z) - A_i(z)\right) > n$ *for all but finitely many i. We write this, for short, as* $\lim \deg\left(A_{i+1}(z) - A_i(z)\right) = \infty$.

Proof. Suppose $(A_i(z))$ converges to $F(z)$. Then for fixed k, $a_{i,k} = f_k$ for $i \geq i_k$, and so $a_{i+1,k} - a_{i,k} = 0$ for $i \geq i_k$. Now let n be arbitrary. Then for $k = 0, 1, \ldots, n$,

$$a_{i+1,k} - a_{i,k} = 0 \quad \text{for } i \geq \max(i_0, i_1, \ldots, i_n),$$

which implies

$$\deg(A_{i+1} - A_i) > n \text{ for } i \geq \max(i_0, i_1, \ldots, i_n)\,.$$

Thus $\lim \deg(A_{i+1} - A_i) = \infty$.

Conversely, if $\lim \deg(A_{i+1} - A_i) = \infty$, then given n, we have $\deg(A_{i+1} - A_i) > n$ for $i \geq \ell$. This implies $a_{i+1,n} = a_{i,n}$ for $i \geq \ell$, and so $(A_i(z))$ converges. □

Infinite Sums and Products.

We can now apply this convergence notion to infinite sums and products of formal series. Let $A_0(z), A_1(z), \ldots$ be a sequence of series. Then we define the *sum* $\sum_{k\geq 0} A_k(z)$ as the limit of the partial sums $S_i(z) = \sum_{k=0}^{i} A_k(z)$ if this limit exists. Since $S_{i+1}(z) - S_i(z) = A_{i+1}(z)$, we infer the following result.

Corollary 2.2. *The sum* $\sum_{k\geq 0} A_k(z)$ *exists if and only if* $\lim \deg A_k = \infty$.

Another way to state this is that $\sum_{k\geq 0} A_k(z)$ exists if and only if for any n, there are only finitely many non-zero n-th coefficients among the $A_k(z)$. So our notion of convergence agrees with the natural way to define the sum $F(z) = \sum_{k\geq 0} A_k(z)$ as $f_n = \sum_{k\geq 0} a_{k,n}$ for all n.

Next we turn to infinite *products* $\prod_{k\geq 1} A_k(z)$. To avoid technical details we always assume in such a product that $A_k(0) = 1$ for all k. We define, of course, the infinite product $\prod_{k\geq 1} A_k(z)$ as the limit of the partial products $P_i(z) = \prod_{k=1}^{i} A_k(z)$, if it exists. Again there is a simple criterion when such an infinite product exists as a series, whose proof is left to the exercises.

Proposition 2.3. *Let* $(A_k(z))$ *be a sequence of series with* $A_k(0) = 1$ *for all k. Then the infinite product* $\prod_{k\geq 1} A_k(z)$ *exists if and only if* $\lim \deg(A_k(z) - 1) = \infty$.

As an example, $(1+z)(1+z^2)(1+z^3)\cdots$ is an admissible product; that is, it can be expanded into a formal series, while $(1+z)(1+z^2)(1+z)(1+z^2)\cdots$ is not. Note that the order of the summands in $\sum A_k(z)$ or of the factors in $\prod A_k(z)$ does not matter, since the limit conditions on the degrees and the limit functions are not affected.

Since the constant coefficient of an admissible product $F(z) = \prod_{k\geq 1} A_k(z)$ is 1, $F(z)$ is invertible, and we will expect that $F(z)^{-1} = \prod_{k\geq 1} A_k^{-1}(z)$ holds. This is indeed the case. More generally, we have the following result. First we note the following easy fact (see Exercise 2.29):

$$\deg\left(A(z)B(z)-1\right) \geq \min\Big(\deg\left(A(z)-1\right), \deg\left(B(z)-1\right)\Big) \qquad (3)$$

with equality if $\deg\left(A(z)-1\right) \neq \deg\left(B(z)-1\right)$. Note that $\deg\left(A(z)-1\right) = \deg\left(A^{-1}(z)-1\right)$.

Proposition 2.4. *If $\prod_{k\geq 1} A_k(z)$ and $\prod_{k\geq 1} B_k(z)$ are admissible products, then so is $\prod_{k\geq 1} A_k(z)B_k(z)$, and we have*

$$\prod_{k\geq 1} A_k(z) \cdot \prod_{k\geq 1} B_k(z) = \prod_{k\geq 1} A_k(z)B_k(z)\,. \qquad (4)$$

Proof. From (3) we immediately infer that $\prod_{k\geq 1} A_k(z)B_k(z)$ is again admissible. Furthermore, the factors that make a non-trivial contribution to $[z^n]$ are the same on both sides, so we have in reality reduced the multiplication to finite products. □

Accordingly, if $\prod_{i\geq 1} F_i(z)$ is admissible, then so is $\prod_{i\geq 1} F_i^{-1}(z)$, and further,

$$\prod_{i\geq 1} F_i(z) \cdot \prod_{i\geq 1} F_i^{-1}(z) = 1\,,$$

that is,

$$\left(\prod_{i\geq 1} F_i(z)\right)^{-1} = \prod_{i\geq 1} F_i^{-1}(z)\,. \qquad (5)$$

More generally, if $\prod_{i\geq 1} F_i(z)$ and $\prod_{i\geq 1} G_i(z)$ are both admissible products, then

$$\frac{\prod\limits_{i\geq 1} F_i(z)}{\prod\limits_{i\geq 1} G_i(z)} = \prod_{i\geq 1} \frac{F_i(z)}{G_i(z)}\,. \qquad (6)$$

Example. The following infinite products will be of prime importance when we study generating functions of number-partitions. The product $\prod_{i\geq 1}(1-z^i)$ is admissible, and hence so is $\frac{1}{\prod_{i\geq 1}(1-z^i)}$. Now consider the quotient

$$\frac{\prod_{i\geq 1}(1-z^{2i})}{\prod_{i\geq 1}(1-z^i)}\,.$$

By (6) we have

$$\frac{\prod_{i\geq 1}(1-z^{2i})}{\prod_{i\geq 1}(1-z^i)} = \prod_{i\geq 1}\frac{1-z^{2i}}{1-z^i} = \prod_{i\geq 1}(1+z^i)\,.$$

But on the other hand, by interpreting the numerator as the product $1\cdot(1-z^2)\cdot 1\cdot(1-z^4)\cdot 1\cdot(1-z^6)\cdots$, the factors $1-z^{2i}$ cancel out, and we obtain

$$\frac{\prod_{i\geq 1}(1-z^{2i})}{\prod_{i\geq 1}(1-z^i)} = \frac{1}{\prod_{i\geq 1}(1-z^{2i-1})}\,.$$

In summary, this gives the important (and unexpected) identity

$$\prod_{i\geq 1}(1+z^i) = \frac{1}{\prod_{i\geq 1}(1-z^{2i-1})}\,. \tag{7}$$

We will see in Chapter 3 how infinite products provide stupendously simple proofs for identities between sets of number-partitions.

Exercises

2.26 Prove the assertions in (1) and (2).

2.27 Show that $\sum_{n\geq 0} a_n z^n$ is the limit series of the polynomials $p_i(z) = \sum_{k=0}^{i} a_k z^k$.

▷ **2.28** Show that $\prod_{k\geq 1} A_k(z)$ converges if and only if $\lim \deg(A_k(z)-1) = \infty$.

2.29 Prove $\deg(A(z)B(z)-1) \geq \min\big(\deg(A(z)-1), \deg(B(z)-1)\big)$ with equality if $\deg(A(z)-1) \neq \deg(B(z)-1)$, and $\deg(A(z)-1) = \deg(A^{-1}(z)-1)$ if $A(0)\neq 0$.

2.30 Why is the identity $\frac{1}{1-z} = (1+z)(1+z^2)(1+z^4)\cdots(1+z^{2^k})\cdots$ true?

* * *

▷ **2.31** Let $F(z) = \prod_{i\geq 1} F_i(z)$ be an admissible product. Show that $\frac{F'(z)}{F(z)} = \sum_{i\geq 1} \frac{F_i'(z)}{F_i(z)}$, and compute $\frac{F'(z)}{F(z)}$ for $F(z) = \frac{1}{\prod_{i\geq 1}(1-z^i)}$.

▷ **2.32** We will show in the next section that $\sum_{k=1}^{n} \frac{n!}{k!}\binom{n-1}{k-1}$ $(n \geq 1)$ counts the number of partitions of $\{1,2,\ldots,.n\}$ in which every block is linearly ordered. Example: $n = 3$; 123 gives six orders, 12|3 gives two orders, 1|2|3 gives one order; hence we obtain altogether $6 + 3 \cdot 2 + 1 = 13$. Derive from this the identity $\prod_{i\geq 1} e^{z^i} = e^{\sum_{i\geq 1} z^i} = e^{\frac{z}{1-z}}$. Hint: $e^{\frac{z}{1-z}} = 1 + \sum_{n\geq 1}\left[\sum_{k=1}^{n} \frac{n!}{k!}\binom{n-1}{k-1}\right]\frac{z^n}{n!}$.

2.33 Compute $\sum_{k\geq 1} \frac{z^k}{1-z^k}$.

▷ **2.34** The *Möbius function* in number theory is $\overline{\mu}(1) = 1, \overline{\mu}(n) = 0$ if n is divisible by a square, $\overline{\mu}(n) = (-1)^t$ if n is the product of t distinct primes. Find a simple expression for $F(z) = \prod_{n\geq 1}(1-z^n)^{-\frac{\overline{\mu}(n)}{n}}$. Hint: Use the fact $\log(\prod F_k) = \sum \log F_k$ where $F_k(0) = 1$ and $\sum_{d|n} \overline{\mu}(d) = \delta_{n,1}$. We will consider the function $\overline{\mu}$ in Chapter 5.

2.35 What combinatorial significance has the n-th coefficient in the product $\prod_{k\geq 1}(1+z^k)$?

2.36 Let a_n be the number of partitions of n into powers of 2, where the order of summands does not matter, $a_0 = 1$. Example: $a_3 = 2; 3 = 2+1 = 1+1+1$, $a_4 = 4$ with $4 = 2+2 = 2+1+1 = 1+1+1+1$. Let $b_n = \sum_{k=0}^{n} a_k$. Compute the generating functions $A(z)$ and $B(z)$. What follows for a_n, b_n?

2.4 Infinite Matrices and Inversion of Sequences

There is another algebraic setup that will prove very useful. Let $\mathcal{M}$ be the set of infinite complex matrices, whose rows and columns are indexed by $\mathbb{N}_0$. With entry-wise addition and scalar multiplication, $\mathcal{M}$ becomes a vector space over $\mathbb{C}$. Now let us look at the usual matrix product $A \cdot B$ for two such matrices. In order that this become meaningful, the inner product of the i-th row of A with the j-th column of B must consist of finitely many non-zero summands.

The most important case in which this happens is that either A is a *lower triangular* matrix (i.e., $a_{ij} = 0$ for $i < j$) or B is an *upper triangular* matrix ($b_{ij} = 0$ for $i > j$). Let us denote these sets by $\mathcal{M}^\ell$ respectively $\mathcal{M}^u$.

It is easily checked that $A \in \mathcal{M}^\ell$ has a (unique) inverse A^{-1}, which is again in $\mathcal{M}^\ell$ with $AA^{-1} = A^{-1}A = I$, I the infinite identity matrix, if and only if the main diagonal of A consists of elements $\neq 0$. The analogous result holds for $\mathcal{M}^u$.

Inversion Formulas.
Let us take a closer look at $\mathcal{M}^\ell$. We have introduced in Section 1.3 the idea of connecting two polynomial sequences $(p_n(x))$ and $(q_n(x))$,

$$p_n(x) = \sum_{k=0}^{n} a_{n,k} q_k(x), \quad q_n(x) = \sum_{k=0}^{n} b_{n,k} p_k(x). \tag{1}$$

The simplest example connects (x^n) and $(x-1)^n$,

$$x^n = \sum_{k=0}^{n} \binom{n}{k} (x-1)^k, \quad (x-1)^n = \sum_{k=0}^{n} (-1)^{n-k} \binom{n}{k} x^k, \tag{2}$$

and we also know the Stirling connection

$$x^n = \sum_{k=0}^{n} S_{n,k} x^{\underline{k}}, \quad x^{\underline{n}} = \sum_{k=0}^{n} (-1)^{n-k} s_{n,k} x^k. \tag{3}$$

Since $(p_n(x))$ and $(q_n(x))$ are two bases of the vector space $\mathbb{C}[x]$, the matrices $A = (a_{n,k})$, $B = (b_{n,k})$ are inverses of each other; in fact, they are in $\mathcal{M}^\ell$.

Now let $u = (u_0, u_1, \dots)$, $v = (v_0, v_1, \dots)$ be two sequences of complex numbers. Then

$$u = Av \iff v = Bu.$$

Written out, this means that

$$u_n = \sum_{k=0}^{n} a_{n,k} v_k \ (\forall n) \iff v_n = \sum_{k=0}^{n} b_{n,k} u_k \ (\forall n). \tag{4}$$

We call this an *inversion formula* connecting the sequences u and v. The importance of such a formula is clear: If we know one sequence and the connecting coefficients, then we can compute the other sequence. Any pair $(p_n(x)), (q_n(x))$ of polynomial sequences furnishes therefore an inversion formula. Furthermore, from $B = A^{-1}$ we obtain

$$\sum_{k \geq 0} a_{i,k} b_{k,j} = \delta_{i,j} . \tag{5}$$

The most important inversion formula arises from (2); we have already encountered it in the context of exponential series in Section 2.2.

Binomial Inversion.
The sequences (x^n) and $((x-1)^n)$ give

$$u_n = \sum_{k=0}^{n} \binom{n}{k} v_k \ (\forall n) \iff v_n = \sum_{k=0}^{n} (-1)^{n-k} \binom{n}{k} u_k \ (\forall n) , \tag{6}$$

or in symmetric form,

$$u_n = \sum_{k=0}^{n} (-1)^k \binom{n}{k} v_k \ (\forall n) \iff v_n = \sum_{k=0}^{n} (-1)^k \binom{n}{k} u_k \ (\forall n) .$$

The inverse matrix of the Pascal matrix $(\binom{n}{k})$ is thus $((-1)^{n-k}\binom{n}{k})$.

Example. We have seen in Section 1.3 that for fixed n,

$$r^n = \sum_{k=0}^{n} \binom{r}{k} k! S_{n,k} \quad \text{for all } r .$$

Setting $u_k = k^n$, $v_k = k! S_{n,k}$, binomial inversion gives

$$r! S_{n,r} = \sum_{k=0}^{r} (-1)^{r-k} \binom{r}{k} k^n ,$$

hence

$$S_{n,r} = \frac{1}{r!} \sum_{k=0}^{r} (-1)^{r-k} \binom{r}{k} k^n . \tag{7}$$

Example. An important application of binomial inversion concerns the so-called *Newton representation* of a polynomial $f(x)$. Since the falling factorials $x^{\underline{k}}$ constitute a basis, we may uniquely write

$$f(x) = \sum_{k \geq 0} a_k x^{\underline{k}} = \sum_{k \geq 0} k! a_k \binom{x}{k} .$$

For $x = n$, this gives

$$f(n) = \sum_{k=0}^{n} k!a_k \binom{n}{k}.$$

We set $u_k = f(k)$, $v_k = k!a_k$, and obtain

$$\sum_{k=0}^{n} (-1)^{n-k} \binom{n}{k} f(k) = n!a_n \quad (n \in \mathbb{N}_0). \tag{8}$$

In particular,

$$\sum_{k=0}^{n} (-1)^{n-k} \binom{n}{k} f(k) = 0 \quad \text{when } \deg f < n. \tag{9}$$

As an example consider the sum $\sum_{k=0}^{n}(-1)^k \binom{n}{k}\binom{n+k}{k}$. One way to compute this sum is to use the Vandermonde identity. Since $(-1)^k\binom{n+k}{k} = \binom{-n-1}{k}$, we get

$$\sum_{k=0}^{n} (-1)^k \binom{n}{k}\binom{n+k}{k} = \sum_{k=0}^{n} \binom{n}{n-k}\binom{-n-1}{k} = \binom{-1}{n} = (-1)^n.$$

Alternatively, we may employ (8). The polynomial $f(x) = \binom{x+n}{n}$ written in Newton form $\binom{x+n}{n} = \frac{x^{\underline{n}}}{n!} + \cdots$ has degree n and leading coefficient $a_n = \frac{1}{n!}$. The sum is therefore

$$\sum_{k=0}^{n} (-1)^k \binom{n}{k} f(k) = (-1)^n n!a_n = (-1)^n.$$

Stirling Inversion.
Stirling numbers provide according to (3) another inversion formula:

$$u_n = \sum_{k=0}^{n} S_{n,k} v_k \ (\forall n) \iff v_n = \sum_{k=0}^{n} (-1)^{n-k} s_{n,k} u_k \ (\forall n),$$

and we have $\sum_{k\geq 0} S_{n,k}(-1)^{k-m} s_{k,m} = \delta_{n,m}$.

Going the Other Way.
So far, we have used the recurrence of combinatorial numbers such as the binomial coefficients or the Stirling numbers to infer relations between polyomial sequences. But we may also go the other

way. We begin with polynomial sequences that enjoy certain combinatorial properties, and deduce recurrences for the connecting coefficients.

The bases

$$x^n = (x-0)(x-0)\cdots(x-0),\ x^{\underline{n}} = x(x-1)\cdots(x-n+1),$$
$$x^{\overline{n}} = x(x+1)\cdots(x+n-1)$$

have a property in common: The n-th polynomial arises from the $(n-1)$-st by adjoining a factor $x-a_n$. This suggests the following definition.

Definition. Let $\alpha = (a_1, a_2, a_3, \ldots)$ be a sequence of complex numbers. The *persistent sequence* of polynomials associated with α is

$$p_0^{(\alpha)}(x) = 1,\ p_n^{(\alpha)}(x) = (x-a_1)(x-a_2)\cdots(x-a_n)\quad (n \ge 1)\,.$$

Thus the sequences $x^n, x^{\underline{n}}, x^{\overline{n}}$ are persistent with $\alpha = (0,0,0,\ldots)$, $\alpha = (0,1,2,\ldots)$, and $\alpha = (0,-1,-2,\ldots)$, respectively. Another example is the Gaussian polynomials $g_n(x)$ with $\alpha = (1, q, q^2, \ldots)$.

The following result shows that the connecting coefficients of two persistent sequences always satisfy a 2-term recurrence.

Proposition 2.5. *Let* $p_n^{(\alpha)}(x) = \sum_{k=0}^{n} c_{n,k} p_k^{(\beta)}(x)$ *with* $\alpha = (a_1, a_2, \ldots)$, $\beta = (b_1, b_2, \ldots)$. *Then*

$$\begin{aligned} c_{0,0} &= 1, c_{0,k} = 0\ (k>0), \\ c_{n,0} &= (b_1-a_1)(b_1-a_2)\cdots(b_1-a_n)\ (n \ge 1)\,, \\ c_{n,k} &= c_{n-1,k-1} + (b_{k+1}-a_n)c_{n-1,k}\quad (n,k \ge 1)\,. \end{aligned} \tag{10}$$

Proof. To simplify the notation we set $p_n(x) = p_n^{(\alpha)}(x)$, $q_n(x) = p_n^{(\beta)}(x)$; thus $p_n(x) = \sum_{k=0}^{n} c_{n,k} q_k(x)$.

Clearly, $c_{0,k} = \delta_{0,k}$. Substituting $x = b_1$ we get

$$p_n(b_1) = \sum_{k=0}^{n} c_{n,k} q_k(b_1) = c_{n,0}\,,$$

since b_1 is a root of $q_k(x)$ for $k \ge 1$; thus $c_{n,0} = (b_1-a_1)\cdots(b_1-a_n)$.

Now $p_n(x) = (x - a_n)p_{n-1}(x)$ and $q_{k+1}(x) = (x - b_{k+1})q_k(x)$ imply

$$\begin{aligned} p_n(x) &= \sum_k c_{n-1,k} x q_k(x) - a_n \sum_k c_{n-1,k} q_k(x) \\ &= \sum_k c_{n-1,k} q_{k+1}(x) + \sum_k b_{k+1} c_{n-1,k} q_k(x) - \sum_k a_n c_{n-1,k} q_k(x) \\ &= \sum_k [c_{n-1,k-1} + (b_{k+1} - a_n) c_{n-1,k}] q_k(x), \end{aligned}$$

and the result follows. □

Examples. For $\alpha = (0,0,0,\dots)$, $\beta = (1,1,1,\dots)$, that is, $p_n^{(\alpha)}(x) = x^n$, $p_n^{(\beta)}(x) = (x-1)^n$, we obtain the recurrence

$$c_{n,0} = 1, \ c_{n,k} = c_{n-1,k-1} + c_{n-1,k},$$

and thus $c_{n,k} = \binom{n}{k}$. Similarly, for $a_n = 0$, $b_n = n - 1$ we obtain the recurrence for $S_{n,k}$, while for $a_n = -(n-1)$, $b_n = 0$, the recurrence for the Stirling numbers $s_{n,k}$ results.

The case $x^n = \sum_{k=0}^{n} [{n \atop k}]_q g_k(x)$ corresponds to $a_n = 0$, $b_n = q^{n-1}$, and we get the recurrence $[{n \atop k}]_q = [{n-1 \atop k-1}]_q + q^k [{n-1 \atop k}]_q$ for the Gaussian coefficients.

Exercises

2.37 Show that a lower triangular matrix A has an inverse if and only if the main diagonal consists of nonzero elements.

2.38 Let $P = \left(\binom{n}{k}\right)$ be the Pascal matrix. Verify for the m-th power the identity $P^m(n,k) = \binom{n}{k} m^{n-k}$ $(m \in \mathbb{Z})$.

2.39 Prove an identity of Euler: $\sum_{k=0}^{n} (-1)^k \binom{n}{k} (x-k)^n = n!$.

▷ **2.40** Determine the connecting coefficients $L_{n,k}$ between the sequences $(x^{\overline{n}})$ and $(x^{\underline{n}})$; thus $x^{\overline{n}} = \sum_{k=0}^{n} L_{n,k} x^{\underline{k}}$. You may use either Vandermonde or a counting argument. The numbers $L_{n,k}$ are called *Lah numbers*.

2.41 Derive the 2-term recurrence for the Lah numbers and the Lah inversion formula.

2.42 Show that $(x+1)^{\overline{n}} = \sum_{k=0}^{n} L_{n+1,k+1} (x-1)^{\underline{k}}$, where $L_{n,k}$ are the Lah numbers.

▷ **2.43** Determine the numbers $a_n \in \mathbb{N}_0$ from the identity $n! = a_0 + a_1 n^{\underline{1}} + a_2 n^{\underline{2}} + \cdots + a_n n^{\underline{n}}$ $(n \geq 0)$.

* * *

2.44 Show that $\sum_{k=0}^{n} \binom{n}{k}\binom{m}{k}^{-1} = \frac{m+1}{m+1-n}$ for $m \geq n$, and derive another formula by binomial inversion.

2.45 Show that for the Lah numbers

$$L_{n+1,k+1} = \sum_{i=0}^{n} L_{i,k}(n+k+1)^{\underline{n-i}}.$$

2.46 Use the recurrence for the Lah numbers to prove that $L_{n,k}$ equals the number of k-partitions of $\{1,2,\ldots,n\}$ in which every block is linearly ordered. Example: $L_{3,2} = 6$, with $12|3, 21|3, 13|2, 31|2, 23|1, 32|1$ (see Exercise 2.32).

▷ **2.47** Express the Gaussian polynomial $g_n(x)$ in terms of x^k and derive the Gaussian inversion formula

$$u_n = \sum_{k=0}^{n} \left[{n \atop k}\right]_q v_k \iff v_n = \sum_{k=0}^{n} (-1)^{n-k} q^{\binom{n-k}{2}} \left[{n \atop k}\right]_q u_k .$$

2.48 The following formula is called inversion of Chebyshev type:

$$u_n = \sum_{k=0}^{\lfloor n/2 \rfloor} \binom{n}{k} v_{n-2k} \iff v_n = \sum_{k=0}^{\lfloor n/2 \rfloor} (-1)^k \frac{n}{n-k}\binom{n-k}{k} u_{n-2k} .$$

Hint: Use the polynomial $t_n(x) = \sum_{k=0}^{\lfloor n/2 \rfloor} (-1)^k \frac{n}{n-k}\binom{n-k}{k} x^{n-2k}$.

▷ **2.49** Given $a_n = \sum_{k=0}^{n} \binom{m+k}{k} b_{n-k}$ for $m \in \mathbb{N}_0$, invert the sum to find a formula for b_n in terms of a_k.

2.50 Let $a_1, \ldots, a_n$ be non-negative reals, and set $p(x) = (x+a_1)\cdots(x+a_n)$. Write $p(x) = \sum_{k=0}^{n} c_k x^k$, and show that the sequence $(c_0, \ldots, c_n)$ is unimodal. Hint: Prove the stronger result $c_{k-1}c_{k+1} \leq c_k^2$ for $k = 1, \ldots, n-1$. This latter property is called *logarithmic concavity*. Apply the result to known sequences.

2.5 Probability Generating Functions

In several examples we have considered probability distributions on a set Ω and asked for the expected value of a random variable

$X : \Omega \to \mathbb{N}_0$. For example, in Exercise 1.38 we looked at the set $\Omega = S(n)$ of all permutations of $\{1, \ldots, n\}$ and considered the random variable $X : S(n) \to \mathbb{N}_0$, where $X(\pi)$ is the number of fixed points of π.

Expectation and Variance.
Generating functions often provide a quick way to compute expectation and variance. Let $X : \Omega \to \mathbb{N}_0$ be a random variable, and let $p_n = \text{Prob}(X = n)$ be the probability that X takes the value n. The *probability generating function* is defined as

$$P_X(z) = \sum_{n\geq 0} p_n z^n . \tag{1}$$

Thus the coefficients of $P_X(z)$ are all nonnegative, and $P_X(1) = 1$. Conversely, any such function may be interpreted as a probability generating function.

As usual, the *expected value* of X is declared as

$$\mathbb{E}X = \sum_{n\geq 0} np_n .$$

Hence taking the derivative $P'_X(z) = \sum_{n\geq 1} np_n z^{n-1}$, we obtain

$$\mathbb{E}X = P'_X(1) , \tag{2}$$

if $P'_X(1)$ exists. The *variance*

$$\text{Var}X = \mathbb{E}X^2 - (\mathbb{E}X)^2$$

is computed similarly. From

$$P''_X(z) = \sum_{n\geq 2} n(n-1)p_n z^{n-2} = \sum_{n\geq 2} n^2 p_n z^{n-2} - \sum_{n\geq 2} np_n z^{n-2}$$

we get

$$P''_X(1) = \mathbb{E}X^2 - \mathbb{E}X ,$$

and hence by (2),

$$\text{Var}X = P''_X(1) + P'_X(1) - (P'_X(1))^2 . \tag{3}$$

Remark. In contrast to our formal treatment, we regard $P_X(z)$ and its derivatives as functions and evaluate them at $z = 1$. For polynomials this presents, of course, no difficulties; in the countable case we have to resort to convergence arguments.

Example. Let $\Omega = S(n)$ with the uniform distribution. Consider the random variable $X_n(\pi) = \text{inv}(\pi)$, hence $\text{Prob}(X_n = k) = \frac{I_{n,k}}{n!}$, where $I_{n,k}$ is the number of permutations with exactly k inversions (see Exercise 1.42). The probability generating function is therefore

$$P_{X_n}(z) = \sum_{k \geq 0} \frac{I_{n,k}}{n!} z^k .$$

Classifying the permutations according to the position of n, we obtain $I_{n,k} = I_{n-1,k} + I_{n-1,k-1} + \cdots + I_{n-1,k-n+1}$; hence

$$P_{X_n}(z) = \frac{1 + z + \cdots + z^{n-1}}{n} P_{X_{n-1}}(z) .$$

With $P_{X_1}(z) = 1$ this gives

$$P_{X_n}(z) = \prod_{i=1}^{n} \frac{1 + z + \cdots + z^{i-1}}{i},$$

$$P'_{X_n}(z) = \sum_{i=1}^{n} \left(\prod_{j \neq i} \frac{1 + z + \cdots + z^{j-1}}{j} \right) \frac{1 + 2z + \cdots + (i-1)z^{i-2}}{i} .$$

Setting $z = 1$ we obtain

$$\mathbb{E}X_n = \sum_{i=1}^{n} \frac{i-1}{2} = \frac{n(n-1)}{4} .$$

Using (3), the variance is easily computed as

$$\text{Var}X_n = \frac{(n-1)n(2n+5)}{72} .$$

Random Lattice Walks.
Let us look at an example that is a little more involved. Consider all lattice walks on the integers $\mathbb{Z} = \{\ldots, -2, -1, 0, 1, 2, \ldots\}$ that start at 0 and go at every stage one step to the right or left with equal probability $1/2$. Given an integer $m \geq 1$, what is the probability that a walk reaches m or $-m$? Suppose this probability is 1 (we are going to show this). Then we may consider the random variable X on these walks, where X is the number of steps needed to reach m or $-m$ for the *first* time.

To compute the probability generating function we proceed as follows. Let $G_n^{(m)}$ be the number of walks that reach m or $-m$ for the first time at the n-th step, and set $G^{(m)}(z) = \sum_{n\geq 0} G_n^{(m)} z^n$. Then $\mathrm{Prob}(X = n) = \frac{G_n^{(m)}}{2^n}$, and so

$$P_X(z) = G^{(m)}\left(\frac{z}{2}\right).$$

For the probability of a walk reaching m or $-m$ at all we thus obtain

$$P_X(1) = G^{(m)}\left(\frac{1}{2}\right), \tag{4}$$

and assuming this to be 1, for the expectation

$$\mathbb{E}X = \frac{1}{2} G^{(m)\prime}\left(\frac{1}{2}\right). \tag{5}$$

To find $G^{(m)}(z)$ we solve a few related problems on the way.

A. For $m \geq 1$, let $E_n^{(m)}$ be the number of *positive* walks that return to 0 for the first time at the n-th step, and never reach m. A positive walk is one that moves only along the non-negative integers. Similarly, let $\widetilde{E}_n^{(m)}$ be the number of *all* positive walks that return to 0 at the n-th step, and again do not hit m.

The initial conditions are $E_0^{(m)} = 0$, $\widetilde{E}_0^{(m)} = 1$ for all $m \geq 1$. Classifying the walks according to the first return to 0 we obtain

$$\widetilde{E}_n^{(m)} = \sum_{k=1}^{n} E_k^{(m)} \widetilde{E}_{n-k}^{(m)} \quad (n \geq 1),$$

and thus

$$\widetilde{E}^{(m)}(z) = E^{(m)}(z)\widetilde{E}^{(m)}(z) + 1,$$

that is,

$$\widetilde{E}^{(m)}(z) = \frac{1}{1 - E^{(m)}(z)}. \tag{6}$$

On the other hand, a walk counted by $E_n^{(m)}$ must go first to 1 and at the n-th step from 1 to 0; hence $E_n^{(m)} = \widetilde{E}_{n-2}^{(m-1)}$. This implies by (6),

$$E^{(m)}(z) = z^2 \widetilde{E}^{(m-1)}(z) = \frac{z^2}{1 - E^{(m-1)}(z)}, \quad E^{(1)}(z) = 0. \tag{7}$$

Now set $A_m = E^{(m)}(\frac{1}{2})$. Then $A_m = \frac{1}{4(1-A_{m-1})}$, $A_1 = 0$, and an easy induction on m yields

$$A_m = \frac{m-1}{2m} \quad (m \geq 1). \tag{8}$$

Using again (7) one further computes for $B_m = E^{(m)\prime}(\frac{1}{2})$,

$$B_m = \frac{2(m^2-1)}{3m} \quad (m \geq 1). \tag{9}$$

B. Next we consider $F_n^{(m)}$, the number of positive walks that reach m for the first time at the n-th step. Such a walk may return to 0 before it reaches m, or it may stay strictly positive after the first step. Taking these two possibilities into account, we obtain

$$F_n^{(m)} = \sum_{k=1}^{n} E_k^{(m)} F_{n-k}^{(m)} + F_{n-1}^{(m-1)}, \quad F_0^{(m)} = 0,$$

which translates into

$$F^{(m)}(z) = E^{(m)}(z)F^{(m)}(z) + zF^{(m-1)}(z), \quad F^{(1)}(z) = z.$$

Again we are interested in the quantities $U_n = F^{(m)}(\frac{1}{2})$, $V_m = F^{(m)\prime}(\frac{1}{2})$. The (inductive) proof of the following formulas is left to the exercises:

$$U_m = \frac{1}{m+1}, \quad V_m = \frac{2m(m+2)}{3(m+1)}. \tag{10}$$

Now we are all set to solve the original problem. A walk counted by $G_n^{(m)}$ goes to 1 or -1 on the first step; suppose it goes to 1. After that it may return to 0 before it reaches m (or $-m$), or it may stay strictly positive. With the analogous argument for -1 we arrive at

$$G_n^{(m)} = 2\left(\sum_{k=1}^{n} E_k^{(m)} G_{n-k}^{(m)} + F_{n-1}^{(m-1)}\right), \quad G_0^{(m)} = 0.$$

For the generating functions this means that

$$G^{(m)}(z) = 2(E^{(m)}(z)G^{(m)}(z) + zF^{(m-1)}(z));$$

hence

$$G^{(m)}(z) = \frac{2zF^{(m-1)}(z)}{1 - 2E^{(m)}(z)}. \tag{11}$$

With (8) and (10) the probability of a walk reaching m or $-m$ is

$$P_X(1) = G^{(m)}\left(\frac{1}{2}\right) = \frac{U_{m-1}}{1 - 2A_m} = \frac{1}{m(1 - \frac{m-1}{m})} = 1$$

as announced at the beginning. To compute $\mathbb{E}X$ we have to differentiate (11). An easy calculation yields

$$\mathbb{E}X = \frac{1}{2}G^{(m)\prime}\left(\frac{1}{2}\right) = \frac{(U_{m-1} + \frac{1}{2}V_{m-1})(1 - 2A_m) + U_{m-1}B_m}{(1 - 2A_m)^2},$$

and plugging in the values in (8), (9), and (10) we obtain

$$\mathbb{E}X = m^2 .$$

So we get the nice result that on average, a lattice walk needs m^2 steps until it reaches distance m from the origin for the first time.

Fibonacci Walks.
We close with another random walk problem on $\mathbb{Z}$. Suppose the walker starts at 1 and jumps at each stage with equal probability $\frac{1}{2}$ one step to the left or two steps to the right. What is the probability that the walker ever reaches the origin? We call such walks *Fibonacci walks* for reasons to be explained shortly.

As before, we classify the walks according to the first time they reach 0. Let a_n be the number of Fibonacci walks that hit 0 for the first time at the n-th step, $a_0 = 0$. Setting $A(z) = \sum_{n\geq 0} a_n z^n$, we are thus interested in the probability

$$p = A\left(\frac{1}{2}\right).$$

To determine p we generalize our problem. Let $a_n^{(m)}$ be the number of Fibonacci walks that start at $m \geq 1$ and reach 0 for the first time at the n-th step, $a_n^{(1)} = a_n$. Now, when such a walk starts at $m + 1$, it must eventually go to m (since it moves only one step to the left at each stage), and we obtain the recurrence

$$a_n^{(m+1)} = \sum_{k=0}^{n} a_k a_{n-k}^{(m)} \quad (m \geq 1).$$

In terms of the generating functions this means that $A^{(m+1)}(z) = A(z)A^{(m)}(z)$; hence

$$A^{(m)}(z) = A(z)^m, \quad A^{(0)}(z) = 1. \tag{12}$$

This is nice, but how does it help to compute $A(z)$? Consider the original walk starting at 1. The first step leads to 0 or 3, which means that $a_n = a_{n-1}^{(0)} + a_{n-1}^{(3)}$, that is,

$$A(z) = z + zA(z)^3. \tag{13}$$

Evaluation at $z = \frac{1}{2}$ gives $p = \frac{1}{2} + \frac{1}{2}p^3$; thus p is a root of the equation

$$p^3 - 2p + 1 = (p-1)(p^2 + p - 1) = 0.$$

This gives as possible answers $p = 1$, $p = \frac{-1+\sqrt{5}}{2}$, or $p = \frac{-1-\sqrt{5}}{2}$. The last root is negative and hence impossible, and there remain the possibilities $p = 1$ and $p = \tau - 1$, where $\tau = \frac{1+\sqrt{5}}{2}$ is the *golden section.*

We surmise that $p = \tau - 1 = 0.618$ is the correct answer. To see this we use for once analytical arguments. Clearly, $A(z) = \sum_{n\ge 0} a_n z^n$ is an increasing function in the interval $[0, \frac{1}{2}]$; thus $A'(\frac{1}{2}) > 0$. Now by (13), $A'(z) = 1 + A(z)^3 + 3zA(z)^2 A'(z)$; hence

$$A'(z) = \frac{1 + A(z)^3}{1 - 3zA(z)^2}. \tag{14}$$

If $p = A(\frac{1}{2})$ were equal to 1, then (14) would yield

$$A'(\frac{1}{2}) = \frac{2}{1 - \frac{3}{2}} = -4 < 0,$$

a contradiction. So we have proved $p = \tau - 1$.

Many readers will be familiar with the connection between Fibonacci numbers and the golden section τ (and this is, of course, the reason why we speak of Fibonacci walks). For those who are not, the next section will provide all the necessary details.

Since $A^{(m)}(z) = A(z)^m$, we get as a bonus that for a Fibonacci walk starting at $m \ge 1$ the probability of it ever reaching the origin equals $p_m = (\tau - 1)^m$.

Exercises

▷ **2.51** Suppose the random variable X takes the values $0, 1, \ldots, n-1$ with equal probability $\frac{1}{n}$. Compute $P_X(z)$.

2.52 Let X be the number of heads when a fair coin is tossed n times. Compute $P_X(z)$, and $\mathbb{E}X$, $\mathrm{Var}X$.

2.53 Suppose X and Y are independent random variables, that is, $\mathrm{Prob}(X = m \wedge Y = n) = \mathrm{Prob}(X = m) \cdot \mathrm{Prob}(Y = n)$. Show that $P_{X+Y}(z) = P_X(z)P_Y(z)$.

▷ **2.54** Toss a fair coin until you get heads for the n-th time. Let X be the number of throws necessary. What are $P_X(z)$, $\mathbb{E}X$, and $\mathrm{Var}X$?

2.55 Prove formula (10).

* * *

2.56 Let $H_n^{(m)}$ be the number of lattice walks that reach $m \geq 0$ for the first time at the n-th step, $H^{(m)}(z) = \sum_{n\geq 0} H_n^{(m)} z^n$. Show that $H^{(m)}(\frac{1}{2}) = 1$, and compute the expected number of steps needed. Can you interpret the result?

2.57 The *moments* μ_k of a random variable X are $\mu_k = \mathbb{E}(X^k)$. Show that $1 + \sum_{k\geq 1} \mu_k \frac{y^k}{k!} = P_X(e^y)$.

▷ **2.58** What is the probability generating function for the number of times needed to roll a fair die until all faces have turned up? Find $\mathbb{E}X$.

2.59 How many times do we need to throw a fair coin until we get heads twice in a row? Compute $P_X(z)$, $\mathbb{E}X$, and $\mathrm{Var}X$. Hint: Use that $\sum_{n\geq 0} F_n z^n = \frac{z}{1-z-z^2}$, F_n the n-th Fibonacci number, proved in the next chapter.

2.60 Suppose $2n$ points are arranged around a circle, and that they are pairwise joined by n chords. Clearly, there are $(2n-1)(2n-3)\cdots 3 \cdot 1$ such chord arrangements. Let X be the random variable counting pairs of crossing chords, with all arrangements equally likely. Compute $\mathbb{E}X$ and $\mathrm{Var}X$.

▷ **2.61** Suppose two particles are initially at adjacent vertices of a pentagon. At every step each of the particles moves with equal probability $\frac{1}{2}$ to an adjacent vertex. Let X be the number of stages until both particles are at the same vertex. Find the expected value of X and the variance. What are the results when the two particles start two vertices apart?

Highlight: The Point of (No) Return

Suppose a random walker starts at the origin of the d-dimensional lattice $\mathbb{Z}^d$ and goes at every step with equal probability to one of its $2d$ neighboring lattice points. What is the probability $p^{(d)}$ that he or she eventually returns to the origin? In particular, when is $p^{(d)} = 1$; that is, when does the walker surely return to the origin?

This latter problem was posed and essentially solved by George Pólya with the following beautiful result:

Theorem. *We have*

$$\begin{aligned} p^{(d)} &= 1 \quad \textit{for } d = 1, 2, \\ p^{(d)} &< 1 \quad \textit{for } d \ge 3 . \end{aligned} \tag{1}$$

We proceed in the spirit of the last section. Let $P_n^{(d)}$ be the number of walks of length n that return to 0 for the first time at the n-th step, and $Q_n^{(d)}$ the number of all walks of length n that return to 0 at the n-th step. Thus $P_0^{(d)} = 0$ and $Q_0^{(d)} = 1$ for all d.

Setting $P^{(d)}(z) = \sum_{n\ge 0} P_n^{(d)} z^n$, $Q^{(d)}(z) = \sum_{n\ge 0} Q_n^{(d)} z^n$, we have

$$p^{(d)} = P^{(d)}\left(\frac{1}{2d}\right) = \sum_{n\ge 0} \frac{P_n^{(d)}}{(2d)^n}, \quad q^{(d)} = Q^{(d)}\left(\frac{1}{2d}\right) = \sum_{n\ge 0} \frac{Q_n^{(d)}}{(2d)^n} . \tag{2}$$

Furthermore, classifying the paths counted by $Q_n^{(d)}$ according to the *first* return to 0, we obtain

$$Q_n^{(d)} = \sum_{k=0}^{n} P_k^{(d)} Q_{n-k}^{(d)} ,$$

and so

$$Q^{(d)}(z) = 1 + P^{(d)}(z) Q^{(d)}(z) ,$$

or

$$P^{(d)}(z) = 1 - \frac{1}{Q^{(d)}(z)} .$$

It follows that

$$
\begin{aligned}
p^{(d)} = 1 &\iff q^{(d)} = \sum_{n\geq 0} \frac{Q_n^{(d)}}{(2d)^n} = \infty, \\
p^{(d)} < 1 &\iff q^{(d)} = \sum_{n\geq 0} \frac{Q_n^{(d)}}{(2d)^n} < \infty .
\end{aligned} \tag{3}
$$

In order to prove Pólya's result (1) we thus have to show that the series $\sum_{n\geq 0} \frac{Q_n^{(d)}}{(2d)^n}$ *diverges* for $d = 1$ and 2, and *converges* for all $d \geq 3$. We have chosen the new formulation (3) because the numbers $Q_n^{(d)}$ admit a simple recurrence in d, while the $P_n^{(d)}$'s are hard to handle.

A Recurrence for $Q_n^{(d)}$.

Let us first make a simple observation. Since any walk that returns to the origin must proceed in every dimension equally often in the positive and negative direction, all these walks must be of *even* length.

As a warm-up let us look at $d = 1$. Any walk of length $2n$ must contain n steps to the right and n to the left, hence

$$Q_{2n}^{(1)} = \binom{2n}{n}. \tag{4}$$

Now consider $\mathbb{Z}^{d+1}$. Suppose the walk (of length $2n$) contains $2k$ steps in the first d dimensions and $2n - 2k$ steps in the last dimension. Since we may place these $2k$ steps in $\binom{2n}{2k}$ ways, we obtain the recurrence

$$Q_{2n}^{(d+1)} = \sum_{k=0}^{n} \binom{2n}{2k}\binom{2n-2k}{n-k} Q_{2k}^{(d)}. \tag{5}$$

We may simplify this further.

Claim. *We have*

$$Q_{2n}^{(d)} = \binom{2n}{n} q_{2n}^{(d)}, \tag{6}$$

where the $q_{2n}^{(d)}$'s satisfy the recurrence

$$
\begin{cases}
q_{2n}^{(1)} = 1 & \textit{for all } n, \\
q_{2n}^{(d+1)} = \sum_{k=0}^{n} \binom{n}{k}^2 q_{2k}^{(d)}.
\end{cases} \tag{7}
$$

For $d = 1$, $q_{2n}^{(1)} = 1$ by (4). Induction and (5) yields with the help of Exercise 1.13,

$$Q_{2n}^{(d+1)} = \sum_{k=0}^{n} \binom{2n}{2k}\binom{2n-2k}{n-k}\binom{2k}{k} q_{2k}^{(d)}$$

$$= \binom{2n}{n} \sum_{k=0}^{n} \binom{n}{k}^2 q_{2k}^{(d)},$$

as claimed.

As examples, we obtain $q_{2n}^{(2)} = \sum_{k=0}^{n} \binom{n}{k}^2 = \binom{2n}{n}$, and thus $Q_{2n}^{(2)} = \binom{2n}{n}^2$, and for $d = 3$,

$$q_{2n}^{(3)} = \sum_{k=0}^{n} \binom{n}{k}^2 \binom{2k}{k}. \tag{8}$$

The Cases $d = 1, 2$.
For $d = 1$ we get by (4) and $\binom{2n}{n} \geq \frac{2^{2n}}{2\sqrt{n}}$ (see Exercise 1.26),

$$q^{(1)} = \sum_{n\geq 0} \frac{\binom{2n}{n}}{2^{2n}} \geq 1 + \sum_{n\geq 1} \frac{1}{2\sqrt{n}} \geq 1 + \frac{1}{2} \sum_{n\geq 1} \frac{1}{n},$$

and hence $q^{(1)} = \infty$, since the harmonic series diverges.

Similarly, for $d = 2$ we have $Q_{2n}^{(2)} = \binom{2n}{n}^2$, and thus with the above estimate,

$$q^{(2)} = \sum_{n\geq 0} \frac{\binom{2n}{n}^2}{4^{2n}} \geq 1 + \frac{1}{4} \sum_{n\geq 1} \frac{1}{n},$$

and so again $q^{(2)} = \infty$.

The Case $d \geq 3$.
Suppose we already know that

$$q_{2n}^{(d)} \leq \frac{d^{2n}}{n} \quad \text{for } d \geq 3 \text{ and all } n. \tag{9}$$

Then invoking the upper bound in Exercise 1.26, we obtain

$$q^{(d)} = \sum_{n\geq 0} \frac{\binom{2n}{n}}{(2d)^{2n}} q_{2n}^{(d)} \leq 1 + \sum_{n\geq 1} \frac{2^{2n}}{\sqrt{n}(2d)^{2n}} \frac{d^{2n}}{n}$$

$$= 1 + \sum_{n\geq 1} \frac{1}{n^{3/2}} < \infty ,$$

since we know from analysis that $\sum_{n\geq 1} \frac{1}{n^\alpha}$ converges for any $\alpha > 1$.

To finish the proof we have to verify (9). It is plausible that $d = 3$ is the hardest case, since as the dimension increases we expect that the proportion of random walks that stay away from the origin also increases.

Suppose we have already proved (9) for $d = 3$. Then the recurrence (7) easily extends the inequality (9) to all $d \geq 3$. Indeed, using the simple inequality $\binom{n}{k}^2 \leq \binom{2n}{2k}$ and induction we obtain

$$q_{2n}^{(d+1)} = \sum_{k=0}^{n} \binom{n}{k}^2 q_{2k}^{(d)} \leq 1 + \sum_{k=1}^{n} \binom{2n}{2k} \frac{d^{2k}}{k} ,$$

and it suffices to prove the inequality

$$1 + \sum_{k=1}^{n} \binom{2n}{2k} \frac{d^{2k}}{k} \leq \frac{(d+1)^{2n}}{n} = \frac{1}{n} \sum_{i=0}^{2n} \binom{2n}{i} d^i . \tag{10}$$

But this is easy. First we have $1 \leq \frac{1}{n}\left[\binom{2n}{0} + \binom{2n}{1}d\right] = \frac{1}{n}(1 + 2nd)$. For $1 \leq k \leq n - 1$ we combine the two summands with indices $2k$ and $2k + 1$ on the right of (10) and want to show that

$$\binom{2n}{2k} \frac{d^{2k}}{k} \leq \frac{1}{n}\left[\binom{2n}{2k} d^{2k} + \binom{2n}{2k+1} d^{2k+1}\right] . \tag{11}$$

Now, $\binom{2n}{2k+1} = \binom{2n}{2k} \frac{2n-2k}{2k+1}$; hence (11) reduces to

$$\frac{n}{k} \leq 1 + \frac{2n - 2k}{2k + 1} d , \tag{12}$$

and an easy calculation shows that (12) holds for $k \leq n$ and $d \geq 3$. Finally, for the last term $k = n$ we find that $\binom{2n}{2n} \frac{d^{2n}}{n} \leq \frac{1}{n} \binom{2n}{2n} d^{2n}$ holds with equality.

Thus (10) is true, and it remains to prove (9) for $d = 3$.

A Combinatorial Argument.
Looking at (8), our final task is to prove

$$\sum_{k=0}^{n} \binom{n}{k}^2 \binom{2k}{k} \le \frac{9^n}{n}. \tag{13}$$

One could use estimates such as Stirling's formula, but here is a neater way by interpreting the left-hand side of (13) combinatorially. Consider all words $a_1 \dots a_n b_1 \dots b_n$ of length $2n$ over $\{0, 1, 2\}$, where there are $n - k$ 0's in $a_1 \dots a_n$ and also in $b_1 \dots b_n$, and k 1's (and therefore k 2's) in the remaining $2k$ positions. Clearly, the number of these words is $\binom{n}{k}^2 \binom{2k}{k}$, and therefore $\sum_{k=0}^{n} \binom{n}{k}^2 \binom{2k}{k}$ counts all words $a_1 \dots a_n b_1 \dots b_n$ over $\{0, 1, 2\}$ with

$$\begin{cases} \#0\text{'s in } a_1 \dots a_n = \#0\text{'s in } b_1 \dots b_n, \\ \#1\text{'s} = \#2\text{'s}. \end{cases} \tag{14}$$

Here is the main observation. Given any of the 3^n words $a_1 \dots a_n$ with, say, n_0 0's, n_1 1's, n_2 2's, then the multiplicities of 0, 1, and 2 of any compatible word $b_1 \dots b_n$ are also determined. Indeed, by (14), $b_1 \dots b_n$ must contain n_0 0's, n_2 1's, and n_1 2's. It follows that

$$\sum_{k=0}^{n} \binom{n}{k}^2 \binom{2k}{k} \le 3^n \binom{n}{m_1 m_2 m_3},$$

where $\binom{n}{m_1 m_2 m_3}$ is the largest trinomial coefficient. According to Exercise 1.17 the trinomial coefficients are bounded above by $\frac{3^n}{n+1}$, and so

$$\sum_{k=0}^{n} \binom{n}{k}^2 \binom{2k}{k} \le 3^n \cdot \frac{3^n}{n+1} < \frac{9^n}{n},$$

and the proof is complete.

A final remark: Using analytical methods one can prove that $p^{(3)} = 0.3405$ for the 3-dimensional lattice, and $p^{(d)} \to 0$ for $d \to \infty$, as is to be expected.

Notes and References

The technique of encoding combinatorial sequences as coefficients of a power series has a long history, dating back to Euler and de Moivre. The formal standpoint and formal convergence, however, took a long time to be accepted. The article of Niven was one of the first to attempt a rigorous foundation. For an advanced treatment of rational and algebraic generating functions the books by Stanley are highly recommended. Riordan presents an extensive collection of inverse relations. We have only touched on the subject of probability generating functions. For more on this topic, Chapter 8 in the book by Graham, Knuth, and Patashnik is a good start. Those readers who are interested in an analytical treatment including asymptotics should consult the soon-to-appear book by Flajolet and Sedgewick. For a general background on combinatorics the books by Aigner, Lovász, and van Lint–Wilson are recommended. Lovász's book, in particular, is a veritable treasure trove of combinatorial problems, complete with hints and solutions.

1. M. Aigner (1997): *Combinatorial Theory.* Reprint of the 1979 edition. Springer, Berlin.
2. P. Flajolet and R. Sedgewick (2006): *Analytical Combinatorics.* Preliminary version available on the Web.
3. R.L. Graham, D.E. Knuth, and O. Patashnik (1994): *Concrete Mathematics*, 2nd edition. Addison-Wesley, Reading.
4. L. Lovász (1979): *Combinatorial Problems and Exercises.* North-Holland, Amsterdam.
5. I. Niven (1969): Formal power series. *Amer. Math. Monthly* 76, 871–889.
6. G. Pólya (1921): Über eine Aufgabe der Wahrscheinlichkeitstheorie betreffend die Irrfahrt im Straßennetz. *Math. Annalen* 84, 149–160.
7. J. Riordan (1968): *Combinatorial Identities.* Wiley, New York.
8. R.P. Stanley (1997/9): *Enumerative Combinatorics*, 2 vols. Cambridge Univ. Press, Cambridge/New York.
9. J.H. van Lint and R.M. Wilson (1996): *A Course in Combinatorics.* Cambridge Univ. Press, Cambridge.

Part II: Methods

3 Generating Functions

3.1 Solving Recurrences

The classical application of generating functions is recurrences with constant coefficients. As an introductory example that will serve as a model for the general case, let us consider the simplest recurrence of order 2:

$$F_0 = 0, \quad F_1 = 1, \quad F_n = F_{n-1} + F_{n-2} \quad (n \geq 2).$$

The solution is the sequence F_n of *Fibonacci numbers*. They appear in so many counting problems that a whole journal is dedicated to them. Here is a table of the first numbers:

n	0	1	2	3	4	5	6	7	8	9	10
F_n	0	1	1	2	3	5	8	13	21	34	55

The Four Steps.
How can we compute the n-th Fibonacci number? The following steps are typical for the general case.

Step 1. Write the recurrence as a single equation including the initial conditions.

As always, $F_n = 0$ for $n < 0$, so the recurrence is also correct for $n = 0$. But for $n = 1$ we have $F_1 = 1$, while the right-hand side is 0. The complete recurrence is therefore

$$F_n = F_{n-1} + F_{n-2} + [n = 1]. \tag{1}$$

Here and later we use the "truth"-symbol

$$[P] = \begin{cases} 1 & \text{if } P \text{ is true} \\ 0 & \text{if } P \text{ is false.} \end{cases}$$

Step 2. Express (1) as an equation between generating functions.

We know that lowering the index corresponds to multiplication by a power of z. Thus

$$F(z) = \sum_{n\geq 0} F_n z^n = \sum_{n\geq 0} F_{n-1} z^n + \sum_{n\geq 0} F_{n-2} z^n + \sum_{n\geq 0} [n=1] z^n$$
$$= zF(z) + z^2F(z) + z\,. \tag{2}$$

Step 3. Solve the equation.

This is easy:

$$F(z) = \frac{z}{1-z-z^2}\,. \tag{3}$$

Step 4. Express the right-hand side of (3) as a generating function and compare coefficients.

This is, of course, the only step that requires some work. First we write $1 - z - z^2 = (1 - \alpha z)(1 - \beta z)$, and compute then the partial fractions form

$$\frac{1}{(1-\alpha z)(1-\beta z)} = \frac{a}{1-\alpha z} + \frac{b}{1-\beta z}\,. \tag{4}$$

Now we are through, since

$$F(z) = z\left(\frac{a}{1-\alpha z} + \frac{b}{1-\beta z}\right) = z\left(a \sum_{n\geq 0} \alpha^n z^n + b \sum_{n\geq 0} \beta^n z^n\right)$$
$$= \sum_{n\geq 0} (a\alpha^{n-1} + b\beta^{n-1}) z^n\,,$$

which gives the final result

$$F_n = a\alpha^{n-1} + b\beta^{n-1} \quad (n \geq 0)\,. \tag{5}$$

To get (5) we must therefore compute α and β, and then a and b.

Setting $c(z) = 1 - z - z^2$, we call $c^R(z) = z^2 - z - 1$ the *reflected polynomial*, and we claim that $c^R(z) = (z - \alpha)(z - \beta)$ implies $c(z) = (1 - \alpha z)(1 - \beta z)$. In other words, α and β are the roots of $c^R(z)$.

Let us prove this in full generality. Let $c(z) = 1 + c_1 z + \cdots + c_d z^d$ be a polynomial with complex coefficients of degree $d \geq 1$. The reflected polynomial is then $c^R(z) = z^d + c_1 z^{d-1} + \cdots + c_d$, $c_d \neq 0$, and we clearly have $c(z) = z^d c^R(\frac{1}{z})$. Let $\alpha_1, \ldots, \alpha_d$ be the roots of $c^R(z)$. Then $c^R(z) = (z - \alpha_1) \cdots (z - \alpha_d)$, and therefore

$$c(z) = z^d\left(\frac{1}{z} - \alpha_1\right)\cdots\left(\frac{1}{z} - \alpha_d\right) = (1-\alpha_1 z)\cdots(1-\alpha_d z)$$

as claimed.

For the Fibonacci numbers this gives

$$c^R(z) = z^2 - z - 1 = \left(z - \frac{1+\sqrt{5}}{2}\right)\left(z - \frac{1-\sqrt{5}}{2}\right),$$

$$c(z) = \left(1 - \frac{1+\sqrt{5}}{2}z\right)\left(1 - \frac{1-\sqrt{5}}{2}z\right).$$

The usual notation is $\tau = \frac{1+\sqrt{5}}{2}$, $\hat{\tau} = \frac{1-\sqrt{5}}{2}$, where τ is the famous *golden section* known since antiquity. From $z^2-z-1 = (z-\tau)(z-\hat{\tau})$ we infer $\tau + \hat{\tau} = 1$, $\hat{\tau} = -\tau^{-1}$.

The final step is the partial fractions form:

$$\frac{1}{(1-\tau z)(1-\hat{\tau}z)} = \frac{a}{1-\tau z} + \frac{b}{1-\hat{\tau}z}.$$

Taking the common denominator on the right-hand side we obtain

$$a + b = 1,$$
$$\hat{\tau}a + \tau b = 0,$$

which gives $a = \frac{\tau}{\sqrt{5}}$, $b = -\frac{\hat{\tau}}{\sqrt{5}}$. The final result is therefore by (5),

$$F_n = \frac{1}{\sqrt{5}}(\tau^n - \hat{\tau}^n) = \frac{1}{\sqrt{5}}\left[\left(\frac{1+\sqrt{5}}{2}\right)^n - \left(\frac{1-\sqrt{5}}{2}\right)^n\right]. \tag{6}$$

Since $\left|\frac{1-\sqrt{5}}{2}\right| < 1$, we observe that F_n is the integer nearest to $\frac{1}{\sqrt{5}}\tau^n$.

The following fundamental theorem states that these four steps always work.

Theorem 3.1. *Let $c_1, \ldots, c_d$ be a fixed sequence of complex numbers, $d \geq 1$, $c_d \neq 0$, and set $c(z) = 1 + c_1 z + \cdots + c_d z^d = (1-\alpha_1 z)^{d_1}\cdots(1-\alpha_k z)^{d_k}$, where $\alpha_1, \ldots, \alpha_k$ are the distinct roots of $c^R(z)$. For a counting function $f : \mathbb{N}_0 \to \mathbb{C}$ the following conditions are equivalent:*

(A1) *Recurrence of order d: For all $n \geq 0$,*

$$f(n+d) + c_1 f(n+d-1) + \cdots + c_d f(n) = 0.$$

(A2) *Generating function:*

$$F(z) = \sum_{n\geq 0} f(n)z^n = \frac{p(z)}{c(z)},$$

where $p(z)$ is a polynomial of degree less than d.

(A3) *Partial fractions:*

$$F(z) = \sum_{n\geq 0} f(n)z^n = \sum_{i=1}^{k} \frac{g_i(z)}{(1-\alpha_i z)^{d_i}}$$

for polynomials $g_i(z)$ of degree $< d_i$, $i = 1, \ldots, k$.

(A4) *Explicit form:*

$$f(n) = \sum_{i=1}^{k} p_i(n)\alpha_i^n,$$

where $p_i(n)$ is a polynomial in n of degree less than d_i, $i = 1, \ldots, k$.

Proof. Let us define sets V_i by

$$V_i = \{f : \mathbb{N}_0 \to \mathbb{C} : f \text{ satisfies (Ai)}\},\ i = 1, \ldots, 4.$$

Each V_i is clearly a vector space, since in each of the four cases sum and scalar product respect the condition. Next we observe that each vector space V_i has dimension d. In (A1) the initial values $f(0), \ldots, f(d-1)$ can be chosen arbitrarily, in (A2) the coefficients $p_0, p_1, \ldots, p_{d-1}$ of the polynomial $p(z)$, and in (A3), (A4) the d_i coefficients of g_i respectively p_i, summing to d. Hence if we can show that $V_i \subseteq V_j$, then $V_i = V_j$ holds.

Suppose $f \in V_2$. Then equating coefficients in $c(z)\sum_{n\geq 0} f(n)z^n = p(z)$ for z^{n+d} yields precisely the recurrence (A1). Thus $f \in V_1$, and so $V_1 = V_2$.

Assume $f \in V_3$. Then

$$\sum_{n\geq 0} f(n)z^n = \frac{\sum_{i=1}^{k} g_i(z) \prod_{j\neq i} (1-\alpha_j z)^{d_j}}{\prod_{i=1}^{k} (1-\alpha_i z)^{d_i}} = \frac{p(z)}{c(z)},$$

where $\deg p(z) \leq \max_{1\leq i\leq k}(\deg g_i(z) + \sum_{j\neq i} d_j) < d$. Hence $f \in V_2$, and so $V_1 = V_2 = V_3$.

Finally, we show that $V_3 \subseteq V_4$. Let $f \in V_3$, and look at a summand $\frac{g_i(z)}{(1-\alpha_i z)^{d_i}}$ of $F(z)$. Our list of generating functions in Section 2.1 shows that

$$\frac{1}{(1-\alpha_i z)^{d_i}} = \sum_{n\geq 0} \binom{d_i+n-1}{n} \alpha_i^n z^n = \sum_{n\geq 0} \binom{d_i+n-1}{d_i-1} \alpha_i^n z^n .$$

Multiplication by $g_i(z) = g_0 + g_1 z + \cdots + g_{d_i-1} z^{d_i-1}$ means index shift, that is,

$$\begin{aligned}\frac{g_i(z)}{(1-\alpha_i z)^{d_i}} &= \sum_{n\geq 0} \left[\sum_{j=0}^{d_i-1} g_j \binom{d_i+n-j-1}{d_i-1} \alpha_i^{n-j} \right] z^n \\ &= \sum_{n\geq 0} \left[\sum_{j=0}^{d_i-1} \alpha_i^{-j} g_j \binom{n+d_i-j-1}{d_i-1} \right] \alpha_i^n z^n .\end{aligned}$$

Now we set $p_i(n) = \sum_{j=0}^{d_i-1} \alpha_i^{-j} g_j \binom{n+d_i-j-1}{d_i-1}$; then $p_i(n)$ is a polynomial in n of degree less than or equal to $d_i - 1$. Hence $f \in V_4$, and the proof is complete. □

Readers who have a background in algebra may note that instead of $\mathbb{C}$ we may take any algebraically closed field of characteristic 0.

Example. Suppose we are asked to solve the recurrence

$$a_n = 6a_{n-1} - 9a_{n-2},\ a_0 = 0,\ a_1 = 1 .$$

The associated polynomial is $c(z) = 1 - 6z + 9z^2$, $c^R(z) = z^2 - 6z + 9 = (z-3)^2$. Thus 3 is a root of multiplicity 2. (A4) yields $a_n = (a+bn)3^n$, and with the initial conditions $0 = a_0 = a$, $1 = a_1 = 3b$, $b = \frac{1}{3}$, we obtain the solution $a_n = n3^{n-1}$.

Example. Our task is to compute the eigenvalues of the $n \times n$-matrix

$$M_n = \begin{pmatrix} 0 & 1 & & & 0 \\ 1 & 0 & 1 & & \\ & 1 & 0 & & \\ & & & \ddots & 1 \\ 0 & & & 1 & 0 \end{pmatrix}. \text{ Let } c_n(x) = \det \begin{pmatrix} x & -1 & & & \\ -1 & x & -1 & & 0 \\ & -1 & x & & \\ 0 & & & \ddots & -1 \\ & & & -1 & x \end{pmatrix}$$

be the characteristic polynomial of M_n. Developing the determinant according to the first row, we arrive at the recurrence

$$c_n(x) = xc_{n-1}(x) - c_{n-2}(x), \quad c_0(x) = 1, c_1(x) = x\,.$$

Hence $c(z) = 1 - xz + z^2 = c^R(z)$ with roots $\alpha = \frac{x+\sqrt{x^2-4}}{2}$, $\beta = \frac{x-\sqrt{x^2-4}}{2}$. Step 4 of our familiar procedure gives

$$c_n(x) = \frac{1}{\sqrt{x^2-4}}\left[\left(\frac{x+\sqrt{x^2-4}}{2}\right)^{n+1} - \left(\frac{x-\sqrt{x^2-4}}{2}\right)^{n+1}\right]. \quad (7)$$

If λ is a root of $c_n(x)$, then $\left(\frac{\lambda+\sqrt{\lambda^2-4}}{\lambda-\sqrt{\lambda^2-4}}\right)^{n+1} = 1$; hence

$$\frac{\lambda+\sqrt{\lambda^2-4}}{\lambda-\sqrt{\lambda^2-4}} = e^{\frac{2k\pi i}{n+1}} \quad (k = 1,\ldots,n)\,. \quad (8)$$

Note that $k = 0$, i.e., $\lambda + \sqrt{\lambda^2-4} = \lambda - \sqrt{\lambda^2-4}$, is impossible (why?). Multiplying both sides of (8) by $\lambda + \sqrt{\lambda^2-4}$, we get

$$(\lambda + \sqrt{\lambda^2-4})^2 = 4e^{\frac{2k\pi i}{n+1}}\,;$$

hence $\lambda + \sqrt{\lambda^2-4} = \pm 2e^{\frac{k\pi i}{n+1}}$, $\sqrt{\lambda^2-4} = \pm 2e^{\frac{k\pi i}{n+1}} - \lambda$, and so

$$\lambda^2 - 4 = 4e^{\frac{2k\pi i}{n+1}} \pm 4\lambda e^{\frac{k\pi i}{n+1}} + \lambda^2\,.$$

This, in turn, gives

$$\lambda = \pm\frac{e^{\frac{2k\pi i}{n+1}}+1}{e^{\frac{k\pi i}{n+1}}} = \pm\left(e^{\frac{k\pi i}{n+1}} + e^{-\frac{k\pi i}{n+1}}\right) = \pm 2\cos\frac{k\pi}{n+1}\,.$$

The eigenvalues are therefore $2\cos\frac{k\pi}{n+1}$ $(k = 1,\ldots,n)$, since

$$\cos\frac{k\pi}{n+1} = -\cos\frac{(n+1-k)\pi}{n+1}\,.$$

Simultaneous Recurrences.

The method of generating functions is also successful when we want to solve simultaneous recurrences. In a mathematical competition in 1980 the following problem was posed: Write the number $(\sqrt{2}+\sqrt{3})^{1980}$ in decimal form. What is the last digit before and the first digit after the decimal point?

This appears rather hopeless at first sight, and what has it got to do with recurrences? Consider in general $(\sqrt{2}+\sqrt{3})^{2n}$. We obtain

$(\sqrt{2}+\sqrt{3})^0 = 1$, $(\sqrt{2}+\sqrt{3})^2 = 5+2\sqrt{6}$, $(\sqrt{2}+\sqrt{3})^4 = (5+2\sqrt{6})^2 = 49+20\sqrt{6}$. Are all powers $(\sqrt{2}+\sqrt{3})^{2n}$ of the form $a_n + b_n\sqrt{6}$? With induction we obtain

$$\begin{aligned}(\sqrt{2}+\sqrt{3})^{2n} &= (\sqrt{2}+\sqrt{3})^{2n-2}(\sqrt{2}+\sqrt{3})^2 \\ &= (a_{n-1}+b_{n-1}\sqrt{6})(5+2\sqrt{6}) \\ &= (5a_{n-1}+12b_{n-1}) + (2a_{n-1}+5b_{n-1})\sqrt{6}.\end{aligned}$$

Hence we obtain the simultaneous recurrences

$$\begin{aligned}a_n &= 5a_{n-1} + 12b_{n-1}, \\ b_n &= 2a_{n-1} + 5b_{n-1},\end{aligned} \tag{9}$$

with initial conditions $a_0 = 1$, $b_0 = 0$.

Now we start our machinery.

Step 1. $a_n = 5a_{n-1} + 12b_{n-1} + [n=0]$,
$b_n = 2a_{n-1} + 5b_{n-1}$.

Step 2. $A(z) = 5zA(z) + 12zB(z) + 1$,
$B(z) = 2zA(z) + 5zB(z)$.

Step 3. Solving for $A(z)$, we obtain $A(z) = \frac{1-5z}{1-10z+z^2}$.

Step 4. $c^R(z) = c(z) = z^2 - 10z + 1 = (z-(5+2\sqrt{6}))(z-(5-2\sqrt{6}))$; hence $\alpha_1 = 5+2\sqrt{6}$, $\alpha_2 = 5-2\sqrt{6}$. Partial fractions finally yield

$$a_n = \frac{1}{2}[(5+2\sqrt{6})^n + (5-2\sqrt{6})^n]. \tag{10}$$

Thus we know a_n (and we may also compute b_n in this way), but what does this say about the digits in $(\sqrt{2}+\sqrt{3})^{2n} = a_n + b_n\sqrt{6}$ for $n = 990$? First of all we have $(5+2\sqrt{6})^n = (\sqrt{2}+\sqrt{3})^{2n} = a_n + b_n\sqrt{6}$; hence (10) gives $a_n = \frac{1}{2}(a_n + b_n\sqrt{6} + (5-2\sqrt{6})^n)$, or

$$a_n = b_n\sqrt{6} + (5-2\sqrt{6})^n. \tag{11}$$

Let $\{x\}$ be the fractional part of x, that is, $x = \lfloor x \rfloor + \{x\}$, $0 \le \{x\} < 1$. Since a_n is an integer, we infer from (11) that $\{b_n\sqrt{6}\} + \{(5-2\sqrt{6})^n\} = 1$. The term $(5-2\sqrt{6})^n$ goes to 0. Hence for large n, and certainly for $n = 990$, $(5-2\sqrt{6})^n = 0.000\ldots$, and thus $\{b_n\sqrt{6}\} = 0.99\ldots$ The first digit after the decimal point is therefore 9.

Now let A be the last digit of a_{990} and B that of $b_{990}\sqrt{6}$, i.e., $a_{990} = \ldots A$, $b_{990}\sqrt{6} = \cdot \ldots B.999\ldots$ It follows from (11) that $A \equiv B+1 \pmod{10}$, and so the last digit of $a_{990} + b_{990}\sqrt{6}$ is $A + B \equiv 2A - 1 \pmod{10}$. It remains to determine A, and for this we use the original recurrence (9). The first values (mod 10) are

n	0	1	2	3	4	5
a_n	1	5	9	5	1	5
b_n	0	2	0	8	0	2

and we see that the digits repeat with period 4. In particular, $990 \equiv 2 \pmod 4$; thus $A = 9$, and hence the last digit before the decimal point is $2A - 1 \equiv 7 \pmod{10}$.

Exponential Generating Functions.

In summary, recurrences with constant coefficients can be efficiently dealt with using ordinary generating functions. Whenever the running parameter n is involved in the recurrence, exponential generating functions are usually called for.

As an example, let us consider the number i_n of involutions in $S(n)$. In Exercise 1.39 you were asked to prove combinatorially the recurrence

$$i_{n+1} = i_n + n i_{n-1} \quad (n \geq 0), \quad i_0 = 1\,.$$

Comparing coefficients for z^n means for the exponential generating function $\hat{I}'(z) = \hat{I}(z) + z\hat{I}(z)$; hence

$$(\log \hat{I}(z))' = \frac{\hat{I}'(z)}{\hat{I}(z)} = 1 + z\,,$$

and thus $\log \hat{I}(z) = z + \frac{z^2}{2}$, or $\hat{I}(z) = e^{z + \frac{z^2}{2}}$.

But we can also go the other way, from the generating function to a recurrence. We know that the exponential generating function for the derangement numbers is $\hat{D}(z) = \sum_{n \geq 0} D_n \frac{z^n}{n!} = \frac{e^{-z}}{1-z}$. Taking the derivative we have

$$\hat{D}'(z) = \frac{-e^{-z}(1-z) + e^{-z}}{(1-z)^2} = \frac{z e^{-z}}{(1-z)^2} = \frac{z\hat{D}(z)}{1-z}\,,$$

and thus $(1-z)\hat{D}'(z) = z\hat{D}(z)$. Comparison of coefficients for z^{n-1} now gives $D_n - (n-1)D_{n-1} = (n-1)D_{n-2}$, or

$$D_n = (n-1)(D_{n-1} + D_{n-2}) \quad \text{for } n \geq 1 .$$

Catalan Numbers.
Quite a different kind of recurrence is that of a convolution type. The *Catalan numbers* are defined by

$$C_{n+1} = C_0C_n + C_1C_{n-1} + \cdots + C_nC_0 \quad (n \geq 0), \quad C_0 = 1 . \tag{12}$$

The Catalan numbers are almost as ubiquitous as the binomial coefficients. We will encounter them in many future situations. The first values are

n	0	1	2	3	4	5	6	7	8
C_n	1	1	2	5	14	42	132	429	1430

The classical instance counted by the Catalan numbers is rooted binary plane trees with n vertices (or equivalently, bracketings). From any inner vertex emanates a left or right edge, or both:

$C_1 = 1$

$C_2 = 2$ $C_3 = 5$

In general, the situation at the root of a tree with $n+1$ vertices is

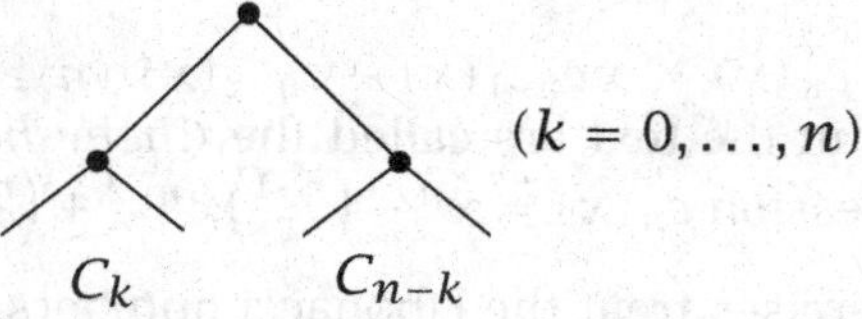

from which $C_{n+1} = \sum_{k=0}^{n} C_kC_{n-k}$ immediately follows.

How can we compute C_n? No problem with generating functions. Let $C(z) = \sum_{n\geq 0} C_nz^n$. Then the convolution (12) translates into $C(z) = zC^2(z) + 1$, or

$$C^2(z) - \frac{1}{z}C(z) + \frac{1}{z} = 0 .$$

Solving this quadratic equation, we get

$$C(z) = \frac{1-\sqrt{1-4z}}{2z} , \tag{13}$$

where the minus sign before the square root must hold since $C(0) = 1$. By Exercise 1.16 we obtain

$$(1-4z)^{1/2} = \sum_{n\ge 0}\binom{1/2}{n}(-4)^n z^n = 1 + \sum_{n\ge 1}\frac{1}{2n}\binom{-1/2}{n-1}(-4)^n z^n$$

$$= 1 - 2\sum_{n\ge 1}\frac{1}{n}\binom{2n-2}{n-1}z^n$$

$$= 1 - 2\sum_{n\ge 0}\frac{1}{n+1}\binom{2n}{n}z^{n+1},$$

or

$$1-\sqrt{1-4z} = 2\sum_{n\ge 0}\frac{1}{n+1}\binom{2n}{n}z^{n+1}.$$

Dividing by $2z$ gives the unexpectedly simple answer

$$C_n = \frac{1}{n+1}\binom{2n}{n}. \tag{14}$$

Note that $C_n = \binom{2n}{n} - \binom{2n}{n-1}$.

Exercises

3.1 The polynomials $c_n(x) = xc_{n-1}(x) - c_{n-2}(x)$ $(n \ge 2)$, $c_0(x) = 1$, $c_1(x) = x$ considered in the text are called the *Chebyshev polynomials.* Prove the explicit expression $c_n(x) = x^n - \binom{n-1}{1}x^{n-2} + \binom{n-2}{2}x^{n-4} \mp \cdots$.

3.2 The following exercises treat the Fibonacci numbers. Prove
a. $\sum_{k=0}^{n} F_k = F_{n+2} - 1$, b. $\sum_{k=1}^{n} F_{2k-1} = F_{2n}$, c. $\sum_{k=0}^{n} F_k^2 = F_n F_{n+1}$.

3.3 Let A be the matrix $\left(\begin{smallmatrix}1&1\\1&0\end{smallmatrix}\right)$. Prove $A^n = \left(\begin{smallmatrix}F_{n+1}&F_n\\F_n&F_{n-1}\end{smallmatrix}\right)$ and derive from this $F_{n+1}F_{n-1} - F_n^2 = (-1)^n$. Show conversely that $|m^2 - mk - k^2| = 1$ for $m, k \in \mathbb{Z}$ implies $m = \pm F_{n+1}$, $k = \pm F_n$ for some n.

▷ **3.4** A subset $A \subseteq \{1, 2, \ldots, n\}$ is called *fat* if $k \ge |A|$ for all $k \in A$. For example, $\{3, 5, 6\}$ is fat, while $\{2, 4, 5\}$ is not. Let $f(n)$ be the number of fat subsets, where $\emptyset$ is fat by definition. Prove: a. $f(n) = F_{n+2}$ (Fibonacci number) and derive from this b. $F_{n+1} = \sum_{k\ge 0}\binom{n-k}{k}$, c. $\sum_{k=0}^{n}\binom{n}{k}F_k = F_{2n}$.

3.5 Let A_n be the number of domino tilings of a $2 \times n$-rectangle. For example, $A_1 = 1, A_2 = 2, A_3 = 3$. Compute A_n.

▷ **3.6** Let $f(n)$ be the number of n-words over the alphabet $\{0,1,2\}$ that contain no neighboring 0's, e.g., $f(1) = 3, f(2) = 8, f(3) = 22$. Determine $f(n)$.

3.7 Compute $\sum_{0<k<n} \frac{1}{k(n-k)}$ using partial fractions or generating functions.

3.8 Decompose a regular n-gon, $n \geq 3$, into triangles by inserting diagonals.
Example: $n = 4$ ⊠, ⊠. 
Show that the number of triangulations is the Catalan number C_{n-2}, $n \geq 3$.

3.9 In an election, exactly n persons vote for candidate A and n people for candidate B. They throw their ballots into the ballot box one after the other. Show that the number of possible ballot lists in which at any stage the number of votes for A is at least as large as that for B equals C_n. Example: $n = 3$, $AAABBB$, $AABABB$, $AABBAB$, $ABAABB$, $ABABAB$.

3.10 Assuming Exercise 3.9, give a bijection proof that C_n equals the number of arrangements of $\{1, 2, \ldots, 2n\}$ in two rows of length n such that the numbers in any row and column appear in increasing order.

▷ **3.11** Let $C(z)$ be the generating function of the Catalan numbers. Prove: $C'(z) = \frac{C^2}{\sqrt{1-4z}}$, and derive from this $(n+1)C_{n+1} = \sum_{k=0}^{n} \binom{2k}{k} C_{n-k+1}$.

3.12 Prove $C(\frac{z}{4z-1}) = 2 - C(z)$ and $C(\frac{z}{1+2z}) = 1 + zC(z^2)$.

* * *

3.13 Solve the recurrence $g_0 = 1, g_n = g_{n-1} + 2g_{n-2} + \cdots + ng_0$ $(n \geq 1)$. Hint: Consider the Fibonacci series $\sum_{n\geq 0} F_{2n} z^n$.

3.14 Compute the ordinary generating function of the harmonic numbers $H_n = 1 + \frac{1}{2} + \cdots + \frac{1}{n}$, and further $\sum_{k=1}^{n-1} H_k H_{n-k}$.

▷ **3.15** Determine the number A_n of ways to fill a $3 \times n$-rectangle with 1×2-dominoes; thus $A_1 = 0$, $A_2 = 3$. Hint: Consider the number B_n of tilings for which the upper left-hand corner is left free, e.g., $B_1 = 1$, $B_2 = 0$, and solve simultaneously .

3.16 In how many ways can one build a $2 \times 2 \times n$-tower with $2 \times 1 \times 1$ bricks?

3.17 Let A_n be the number of ways to tile a $4 \times n$-rectangle using 1×1-squares and copies of L, where L is a 2×2-square with the upper right corner missing (no rotations allowed). Find the generating function of A_n.

▷ **3.18** A row of n light bulbs must be completely turned on; initially they are all off. The first bulb can always be turned on or off. For $i > 1$, the i-th bulb can be switched (turned on or off) only when bulb $i-1$ is on and all earlier bulbs are off. Let a_n be the number of switches needed to turn all bulbs on, and b_n the number needed to turn the n-th bulb on for the first time. Find recurrences for a_n and b_n, and determine these numbers.

3.19 Evaluate the sum $s_n = \sum_{k=0}^{n} \binom{n+k}{2k} 2^{n-k}$ with the following steps: a. generating function $\sum_{n\ge0} s_n z^n$, b. recurrence of second order, c. explicit formula.

▷ **3.20** Determine the generating function $\sum_{n\ge0} p_n(x)z^n$ of the polynomials $p_n(x) = \sum_{k=0}^{n} \binom{n}{k}\binom{2k}{k} x^k$, and consider the special cases $x = -\frac{1}{2}$ and $x = -\frac{1}{4}$. Hint: Use $\sum_{n\ge0} \binom{2n}{n} z^n = (1-4z)^{-1/2}$.

3.21 Prove the formula $\sum_{k=0}^{n} F_k F_{n-k} = \frac{1}{5}(2nF_{n+1} - (n+1)F_n)$, F_n the n-th Fibonacci number.

3.22 Compute the exponential generating function for the numbers g_n defined by $g_0 = 0$, $g_1 = 1$, $g_n = -2ng_{n-1} + \sum_{k=0}^{n} \binom{n}{k} g_k g_{n-k}$ $(n \ge 2)$. Hint: Use Exercise 1.16.

3.23 Let $f(n)$ be the number of cyclic permutations $(a_1, a_2, \ldots, a_n)$ where a_i, a_{i+1} are never consecutive numbers $1,2; 2,3; \ldots; n-1, n; n, 1$. Example: $f(1) = f(2) = 0$, $f(3) = 1$ with $(1,3,2)$ as only possibility. Show that $f(n) + f(n+1) =$ derangement number D_n, and determine the exponential generating function of $f(n)$.

▷ **3.24** Consider the Fibonacci walks of Section 2.5. We have shown that the probability p_m of a walk starting at $m \ge 0$ to reach the origin is $p_m = (\tau - 1)^m$. Compute p_{-m} for $m > 0$. Hint: Establish a recurrence to prove $p_{-m} = 3\tau - 4 + (5 - 3\tau)(F_{m+1} - F_m \tau)$, F_m Fibonacci number. What is $\lim_{m\to\infty} p_{-m}$?

▷ **3.25** Let $C(z)$ be the Catalan generating function. Prove the identity

$$\sum_{n\ge0} \binom{2n+k}{n} z^n = \frac{C(z)^k}{\sqrt{1-4z}}\,.$$

3.26 . Let $a_{m,n} = \sum_{k\ge0} a^k \binom{m}{k}\binom{n}{k}$. Prove $F(y,z) = \sum_{m,n\ge0} a_{m,n} y^m z^n = \frac{1}{1-y-z-(a-1)yz}$.

▷ **3.27** The Delannoy numbers of Exercise 1.24 satisfy $D_{m,n} = D_{m-1,n} + D_{m,n-1} + D_{m-1,n-1}$ $(m, n \ge 1)$ with $D_{m,0} = D_{0,n} = 1$ (clear?). Find from this the generating function $D(y,z) = \sum_{m,n} D_{m,n} y^m z^n$, and deduce from the previous exercise a new formula for $D_{m,n}$.

3.2 Evaluating Sums

Generating functions give us several methods to compute sums, or to establish identities between sums.

The simplest case occurs when $s_n = \sum_{k=0}^{n} a_k$. We then know that $S(z) = \frac{A(z)}{1-z}$ for the corresponding ordinary generating functions.

Example. We want to compute $s_n = \sum_{k=1}^{n} H_k$, where $H_k = 1 + \frac{1}{2} + \cdots + \frac{1}{k}$ are the harmonic numbers. Let $H(z) = \sum_{n\ge1} H_n z^n$ be their generating function. Then

$$H(z) = \frac{\sum_{n\ge1} \frac{1}{n} z^n}{1-z},$$

and thus

$$S(z) = \sum_{n\ge1} s_n z^n = \frac{\sum_{n\ge1} \frac{1}{n} z^n}{(1-z)^2}.$$

With $\frac{1}{(1-z)^2} = \sum_{n\ge0}(n+1)z^n$, convolution yields

$$s_n = \sum_{k=1}^{n} \frac{1}{k}(n-k+1) = (n+1)H_n - n,$$

which can also be written as

$$\sum_{k=1}^{n} H_k = (n+1)(H_{n+1} - 1). \tag{1}$$

Convolution.
A more powerful method is to recognize the sum s_n as a convolution product. In this case $S(z) = A(z)B(z)$, and we may be able to compute the product $A(z)B(z)$ efficiently.

As a start, consider the sum $s_n = \sum_{k=0}^{n} \binom{2k}{k} 4^{-k}$. To make it into a convolution we write it as $s_n = 4^{-n} \sum_{k=0}^{n} \binom{2k}{k} 4^{n-k}$. The inner sum is now the convolution of the two series $\sum_{n\ge0} \binom{2n}{n} z^n = \frac{1}{\sqrt{1-4z}}$ and $\sum_{n\ge0} 4^n z^n = \frac{1}{1-4z}$. The product is $(1-4z)^{-3/2}$, und thus the n-th coefficient is $\binom{-3/2}{n}(-4)^n$. Exercise 1.16 now finishes the work, $s_n = \frac{1}{4^n}(2n+1)\binom{2n}{n}$.

When the sum contains $\binom{n}{k}$, then binomial convolution is called for. Suppose we want to compute $s_n = \sum_{k=1}^{n}(-1)^{k-1}k\binom{n}{k}r^{n-k}$. This is the n-th coefficient of the binomial convolution of $\sum_{n\geq 1}(-1)^{n-1}n\frac{z^n}{n!} = z\sum_{n\geq 0}\frac{(-z)^n}{n!} = ze^{-z}$ and $\sum_{n\geq 0}\frac{(rz)^n}{n!} = e^{rz}$. Now $ze^{(r-1)z} = \sum\frac{(r-1)^{n-1}z^n}{(n-1)!} = \sum\frac{n(r-1)^{n-1}z^n}{n!}$, and we conclude that

$$\sum_{k=1}^{n}(-1)^{k-1}k\binom{n}{k}r^{n-k} = n(r-1)^{n-1} \quad (n \geq 1).$$

"Free Parameter" Method.
Both methods discussed so far have n as limit in the summation. When n does not appear explicitly in the summation, we may consider n as a "free" parameter, treat s_n as a coefficient of $F(z) = \sum s_n z^n$, change the order of the summations on n and k, and try to compute the inner sum. This method (called "snake oil" by H. Wilf) often provides amazingly simple proofs.

Example. We want to compute

$$s_n = \sum_{k\geq 0}\binom{n+k}{m+2k}\binom{2k}{k}\frac{(-1)^k}{k+1} \quad (m, n \in \mathbb{N}_0).$$

We treat n as a "free" parameter, and set

$$F(z) = \sum_{n\geq 0}\left[\sum_{k\geq 0}\binom{n+k}{m+2k}\binom{2k}{k}\frac{(-1)^k}{k+1}\right]z^n.$$

Interchanging summation gives

$$F(z) = \sum_{k\geq 0}\binom{2k}{k}\frac{(-1)^k}{k+1}z^{-k}\sum_{n\geq 0}\binom{n+k}{m+2k}z^{n+k}.$$

Now the inner sum is $\frac{z^{m+2k}}{(1-z)^{m+2k+1}}$ (see the list in Section 2.1). Thus

$$\begin{aligned}
F(z) &= \frac{z^m}{(1-z)^{m+1}} \sum_{k\ge 0} \frac{1}{k+1}\binom{2k}{k}\left(\frac{-z}{(1-z)^2}\right)^k \\
&= \frac{z^m}{(1-z)^{m+1}} \sum_{k\ge 0} C_k \left(\frac{-z}{(1-z)^2}\right)^k \qquad (C_k = \text{Catalan}) \\
&= \frac{z^m}{(1-z)^{m+1}} \frac{1-\sqrt{1+\frac{4z}{(1-z)^2}}}{\frac{-2z}{(1-z)^2}} = \frac{-z^{m-1}}{2(1-z)^{m-1}}\left(1-\frac{1+z}{1-z}\right) \\
&= \frac{z^m}{(1-z)^m} = z\frac{z^{m-1}}{(1-z)^m}.
\end{aligned}$$

But this generating function is again in our list, and we obtain the stupendously simple result

$$s_n = \binom{n-1}{m-1} \text{ for } m \ge 1, \; s_n = [n=0] \text{ for } m = 0.$$

Example. One of the nicest examples in which this approach works concerns the Delannoy numbers $D_{m,n}$. Recall that we noted that $D_{m,n} = \sum_{k\ge 0}\binom{m}{k}\binom{n+k}{m}$ in Exercise 1.24 and found the generating function $\sum_{m,n} D_{m,n} y^m z^n = \frac{1}{1-y-z-yz}$ in Exercise 3.27. Now we want to compute the generating function $D(z) = \sum_{n\ge 0} D_{n,n} z^n$ of the *central Delannoy numbers* $D_{n,n}$.

We have $\binom{n}{k}\binom{n+k}{k} = \binom{2k}{k}\binom{n+k}{2k}$. Thus by our familiar argument,

$$\begin{aligned}
D(z) &= \sum_{n\ge 0}\sum_{k\ge 0}\binom{2k}{k}\binom{n+k}{2k} z^n \\
&= \sum_{k\ge 0}\binom{2k}{k} z^{-k} \sum_{n\ge 0}\binom{n+k}{2k} z^{n+k} \\
&= \sum_{k\ge 0}\binom{2k}{k} z^{-k}\frac{z^{2k}}{(1-z)^{2k+1}} = \frac{1}{1-z}\sum_{k\ge 0}\binom{2k}{k}\left(\frac{z}{(1-z)^2}\right)^k \\
&= \frac{1}{1-z}\frac{1}{\sqrt{1-\frac{4z}{(1-z)^2}}} = \frac{1}{\sqrt{1-6z+z^2}}.
\end{aligned}$$

Once we know the generating function we can try to extract a recurrence out of it.

From $D(z) = (1-6z+z^2)^{-1/2}$ we obtain

$$\frac{D'(z)}{D(z)} = -\frac{1}{2} \cdot \frac{-6+2z}{1-6z+z^2} = \frac{3-z}{1-6z+z^2};$$

thus

$$D'(z)(1-6z+z^2) = D(z)(3-z).$$

Comparing coefficients for z^{n-1}, this translates into

$$nD_{n,n} - 6(n-1)D_{n-1,n-1} + (n-2)D_{n-2,n-2} = 3D_{n-1,n-1} - D_{n-2,n-2},$$

and we obtain the unexpected recurrence

$$nD_{n,n} = (6n-3)D_{n-1,n-1} - (n-1)D_{n-2,n-2} \quad (n \geq 1). \qquad (2)$$

Power Sums.
As a final example of summation techniques let us study the classical problem of evaluating the m-th power sums $P_m(n) = \sum_{k=0}^{n-1} k^m$. The first values are

$$\begin{aligned}
P_0(n) &= n, \\
P_1(n) &= \sum_{k=0}^{n-1} k = \binom{n}{2} = \frac{n^2}{2} - \frac{n}{2}, \\
P_2(n) &= \sum_{k=0}^{n-1} k^2 = \frac{(n-1)(n-\frac{1}{2})n}{3} = \frac{n^3}{3} - \frac{n^2}{2} + \frac{n}{6}, \\
P_3(n) &= \sum_{k=0}^{n-1} k^3 = \binom{n}{2}^2 = \frac{n^4}{4} - \frac{n^3}{2} + \frac{n^2}{4}.
\end{aligned}$$

It appears that $P_m(n)$ is always a polynomial in n of degree $m+1$ and leading coefficient $\frac{1}{m+1}$. To determine $P_m(n)$ we use exponential generating functions. Let

$$\widehat{P}(z,n) = \sum_{m \geq 0} P_m(n) \frac{z^m}{m!}.$$

Substituting for $P_m(n)$ we get

$$\begin{aligned}
\widehat{P}(z,n) &= \sum_{m \geq 0} \left(\sum_{k=0}^{n-1} k^m \right) \frac{z^m}{m!} = \sum_{k=0}^{n-1} \left(\sum_{m \geq 0} \frac{(kz)^m}{m!} \right) \\
&= \sum_{k=0}^{n-1} e^{kz} = \frac{e^{nz}-1}{e^z-1}.
\end{aligned} \qquad (3)$$

Now we set

$$\frac{z}{e^z - 1} = \sum_{n \ge 0} B_n \frac{z^n}{n!} =: \widehat{B}(z). \tag{4}$$

The coefficients B_n are called the *Bernoulli numbers.* Notice that $\frac{z}{e^z-1}$ is well defined, since $e^z - 1$ has constant coefficient 0, whence $\frac{e^z-1}{z}$ is invertible. The equality $\widehat{B}(z)(e^z - 1) = z$ implies by convolution

$$\sum_{k=0}^{n-1} \binom{n}{k} B_k = [n = 1], \tag{5}$$

and the B_n can be successively computed. We obtain for $n = 1$, $B_0 = 1$, and for $n = 2$, $B_0 + 2B_1 = 0$; thus $B_1 = -\frac{1}{2}$. The first Bernoulli numbers are given in the table:

n	0	1	2	3	4	5	6	7	8
B_n	1	$-\frac{1}{2}$	$\frac{1}{6}$	0	$-\frac{1}{30}$	0	$\frac{1}{42}$	0	$-\frac{1}{30}$

It appears that $B_{2n+1} = 0$ for all $n \ge 1$. Indeed, if we add $\frac{z}{2}$ to both sides of (4), the linear term in $\widehat{B}(z)$ vanishes, and we obtain for the left-hand side

$$\frac{z}{e^z - 1} + \frac{z}{2} = \frac{z}{2}\frac{e^z + 1}{e^z - 1}.$$

This is an even function, $F(z) = F(-z)$, and so $B_{2n+1} = 0$ for $n \ge 1$.

The rest is easy. We consider $\widehat{B}(z)(e^{nz} - 1)$ and look at the coefficient $\frac{[z^{m+1}]}{(m+1)!}$. On the one hand, by (4) and (3) we have

$$\widehat{B}(z)(e^{nz} - 1) = z\frac{e^{nz} - 1}{e^z - 1} = z\widehat{P}(z, n);$$

hence this coefficient is $(m+1)P_m(n)$. On the other hand, binomial convolution of $\widehat{B}(z)$ and $(e^{nz} - 1)$ gives $\sum_{k=0}^{m} \binom{m+1}{k} B_k n^{m+1-k}$, and thus the final result

$$\sum_{k=0}^{n-1} k^m = \frac{1}{m+1} \sum_{k=0}^{m} \binom{m+1}{k} B_k n^{m+1-k} \quad (m \ge 0). \tag{6}$$

For $m = 4$ we get from the list of Bernoulli numbers

$$\sum_{k=0}^{n-1} k^4 = \frac{1}{5}\left(n^5 - \frac{5}{2}n^4 + \frac{5}{3}n^3 - \frac{1}{6}n\right).$$

The identity $\sum_{m\ge 0} P_m(n)\frac{z^m}{m!} = \frac{e^{nz}-1}{e^z-1}$ reveals some nice properties of the polynomial $P_m(n)$. First it is clear that $P_m(1) = 0$ for $m \ge 1$, and $P_m(0) = 0$. Now consider $P_m(1-n)$. By multiplying numerator and denominator by e^{-z} we obtain

$$\frac{e^{(1-n)z}-1}{e^z-1} = \frac{e^{-nz}-e^{-z}}{1-e^{-z}} = -\frac{e^{-nz}-1}{e^{-z}-1} + \frac{1-e^{-z}}{1-e^{-z}}$$

and thus

$$P_m(1-n) = (-1)^{m+1}P_m(n) \quad \text{for} \quad m \ge 1\,.$$

In particular, $P_m(\frac{1}{2}) = 0$ for $m \ge 2$ even, and since $P_m(2) = 1$ for $m \ge 1$, we get $P_m(-1) = (-1)^{m+1}$.

Exercises

3.28 Express $\sum_{k=1}^{n-1} H_n H_{n-k}$ in terms of the harmonic numbers.

3.29 Re-prove $\sum_{k=0}^{n} F_k = F_{n+2} - 1$ (F_n Fibonacci number) using generating functions.

3.30 Use generating functions to evaluate:
a. $\sum_{k=0}^{n}(-1)^k\binom{m}{k}\binom{m}{n-k}$, b. $\sum_{k\ge 0} 2k\binom{n}{2k}$, c. $\sum_{k=0}^{n} k\binom{n}{k}^2$.

▷ **3.31** Prove the identity $\sum_{k=0}^{n}\binom{n+k}{k}2^{-k} = 2^n$ and deduce $\sum_{k>n}\binom{n+k}{k}2^{-k} = 2^n$.

3.32 Find $\sum\binom{n}{k}2^{k-n}$ as a sum on k, and also as a sum on n.

3.33 Show that $\sum_{k\ge 0} C_k\binom{n-2k}{\ell-k} = \binom{n+1}{\ell}$, $n, \ell \in \mathbb{N}_0$, and give a combinatorial interpretation (C_k = Catalan number).

3.34 What are the coefficients of $\frac{z}{e^z+1}$? Hint: Use $e^{2z} - 1 = (e^z - 1)(e^z + 1)$.

▷ **3.35** Show that $\sum_{k=0}^{n}\binom{n}{k}(kF_{k-1} - F_k)D_{n-k} = (-1)^n F_n$, where F_n and D_n are the Fibonacci and derangement numbers, respectively.

3.36 Let $s_{n,k}$ be the Stirling numbers of the first kind. Use snake oil to re-prove the identities $s_{n+1,k+1} = \sum_i \binom{i}{k} s_{n,i}$, and $\sum_k s_{n+1,k+1}\binom{k}{m}(-1)^{k-m} = s_{n,m}$.

* * *

3.37 Use the method of free parameters to show that the Chebyshev polynomials $c_n(x) = \sum_k (-1)^k \binom{n-k}{k} x^{n-2k}$ satisfy the recurrence $c_n(x) = xc_{n-1}(x) - c_{n-2}(x)$. Hint: Compute the generating function. What is $c_n(2)$?

3.38 Prove the identity below by the method of free parameters, and give a combinatorial argument that the sum equals the number of k-subsets of $\{1,\dots,n\}$ that contain no three consecutive integers, where $k = n-m+1$:

$$\sum_{i\ge 0} \binom{i}{k-i}\binom{m}{i} = \sum_{j\ge 0} \binom{\lfloor j/2 \rfloor}{k-j}\binom{m-k+\lfloor 3j/2 \rfloor}{j}.$$

3.39 . Find $\sum_k \binom{n}{m+k}\binom{m+k}{2k} 4^k$.

▷ **3.40** We know two expressions for the Delannoy numbers

$$D_{m,n} = \sum_k \binom{m}{k}\binom{n+k}{m} = \sum_k \binom{m}{k}\binom{n}{k} 2^k$$

(see Exercises 1.24, 3.27). Generalize to the following identity, using snake oil:

$$\sum_k \binom{m}{k}\binom{n+k}{m} x^{m-k} = \sum_k \binom{m}{k}\binom{n}{k}(1+x)^k.$$

▷ **3.41** Any generating function $F(z) = \sum_{n\ge 0} a_n z^n$ with $a_0 = 1$, $a_1 \ne 0$, defines a polynomial sequence $(p_n(x))$ by $F(z)^x = \sum_{n\ge 0} p_n(x) z^n$, where $p_n(1) = a_n$ and $p_n(0) = [n = 0]$. Show that $p_n(x)$ has degree n and prove the convolution formulas $p_n(x+y) = \sum_{k=0}^n p_k(x) p_{n-k}(y)$ and $(x+y)\sum_{k=0}^n k p_k(x) p_{n-k}(y) = nx p_n(x+y)$. Hint: $[z^n] e^{x \log F(z)}$ is for $n > 0$ a polynomial $p_n(x)$ in x of degree n with $p_n(0) = 0$. For the second identity use the derivative.

3.42 Apply the preceding exercise to derive the so-called Abel identities:

a. $\sum_k \binom{tk+r}{k}\binom{tn-tk+s}{n-k}\frac{r}{tk+r} = \binom{tn+r+s}{n}$,

b. $\sum_k \binom{n}{k}(tk+r)^k(tn-tk+s)^{n-k}\frac{r}{tk+r} = (tn+r+s)^n$.

Look at some familiar examples for special values of r, s, t.

▷ **3.43** Prove the following identity relating Stirling and Bernoulli numbers:

$$\sum_{j\ge 0} S_{m,j} s_{j+1,k} \frac{(-1)^{j+1-k}}{j+1} = \frac{1}{m+1}\binom{m+1}{k} B_{m+1-k}.$$

What do you get for $k = 1$?

Hint: Express $P_m(n) = \sum_{k=0}^{n-1} k^m$ in terms of $S_{m,j}$, and use $\sum_{k=0}^{n-1} k^{\underline{j}} = \frac{n^{\underline{j+1}}}{j+1}$.

3.3 The Exponential Formula

In the previous section we have studied enumeration problems that can be solved by looking at the product of formal series. Now we turn to composition, in particular composition of exponential generating functions.

Let $f : \mathbb{N}_0 \to \mathbb{C}$ be a counting function, and

$$\widehat{F}(z) = \sum_{n \geq 0} f(n) \frac{z^n}{n!}$$

its exponential generating function. We start with an alternative description of the binomial convolution.

Lemma 3.2. *Let $f, g : \mathbb{N}_0 \to \mathbb{C}$ and $h : \mathbb{N}_0 \to \mathbb{C}$ be defined by*

$$h(|X|) = \sum_{(S,T)} f(|S|)g(|T|),$$

where (S, T) runs through all ordered pairs with $S \cup T = X, S \cap T = \emptyset$. Then

$$\widehat{H}(z) = \widehat{F}(z)\widehat{G}(z).$$

Proof. Let $|X| = n$. There are $\binom{n}{k}$ such pairs (S, T) with $|S| = k$, $|T| = n - k$; hence

$$h(n) = \sum_{k=0}^{n} \binom{n}{k} f(k)g(n-k),$$

and this is precisely binomial convolution. □

We can immediately extend this to k factors. Let $f_1, f_2, \ldots, f_k : \mathbb{N}_0 \to \mathbb{C}$, and

$$h(|X|) = \sum_{(T_1,\ldots,T_k)} f_1(|T_1|) \cdots f_k(|T_k|), \tag{1}$$

where $(T_1, \ldots, T_k)$ runs through all k-tuples with $\bigcup_{i=1}^{k} T_i = X$, $T_i \cap T_j = \emptyset$ $(i \neq j)$. Then we have

$$\widehat{H}(z) = \prod_{i=1}^{k} \widehat{F}_i(z) \tag{2}$$

for the corresponding generating functions.

Theorem 3.3 (Composition Formula). *Let $f, g : \mathbb{N}_0 \to \mathbb{C}$ with $f(0) = 0$, and let $h : \mathbb{N}_0 \to \mathbb{C}$ be defined through*

$$\begin{aligned} h(|X|) &= \sum_{k\geq 1} \sum_{\{B_1,\ldots,B_k\}\in\prod(X)} f(|B_1|)\cdots f(|B_k|)g(k) \quad (|X|>0), \\ h(0) &= g(0), \end{aligned} \tag{3}$$

where the inner sum extends over all k-partitions of X. Then

$$\widehat{H}(z) = \widehat{G}(\widehat{F}(z)).$$

Proof. Let $|X| = n$, and denote by $h_k(n)$ the inner sum of the right-hand side of (3) for fixed k. Since the blocks B_i are nonempty (and therefore distinct), they can be permuted in $k!$ ways. Hence we get by (2) (note that $f(0) = 0$)

$$\widehat{H}_k(z) = \sum_{k\geq 0} h_k(n)\frac{z^n}{n!} = \frac{g(k)}{k!}\widehat{F}(z)^k.$$

Summing over k we obtain

$$\widehat{H}(z) = g(0) + \sum_{k\geq 1} \frac{g(k)}{k!}\widehat{F}(z)^k = \widehat{G}(\widehat{F}(z)). \qquad \square$$

In particular, when $g(k) = 1$ for all k, i.e., $\widehat{G}(z) = e^z$, we obtain the following corollary.

Corollary 3.4 (Exponential Formula). *Let $f : \mathbb{N}_0 \to \mathbb{C}$ with $f(0) = 0$ and $h : \mathbb{N}_0 \to \mathbb{C}$ be defined by*

$$\begin{aligned} h(|X|) &= \sum_{k\geq 1} \sum_{\{B_1,\ldots,B_k\}\in\prod(X)} f(|B_1|)\cdots f(|B_k|) \quad (|X|>0), \\ h(0) &= 1. \end{aligned}$$

Then $\widehat{H}(z) = e^{\widehat{F}(z)}$.

The combinatorial significance rests on the following idea. Many structures, for example graphs, are made up as disjoint union of "connected" substructures. The function g determines the structure on the set of components, and f the inner structure within the individual components. The following examples should make this clear.

Example. In how many ways $h(n)$ can we decompose an n-set into nonempty blocks and choose a linear order on each block? Here $g(n) = 1$ for all n, $f(n) = n!$; hence $\widehat{G}(z) = e^z$, $\widehat{F}(z) = \sum_{n\geq 1} n!\frac{z^n}{n!} = \frac{1}{1-z} - 1 = \frac{z}{1-z}$, and we obtain the result

$$\widehat{H}(z) = e^{\frac{z}{1-z}}.$$

But this should look familiar to you. According to Exercise 2.46 the coefficient of $\frac{z^n}{n!}$ is $\sum_{k=0}^{n} L_{n,k}$, where $L_{n,k}$ are the Lah numbers. More precisely, $\sum_{n\geq 0} L_{n,k}\frac{z^n}{n!} = \frac{1}{k!}(\frac{z}{1-z})^k$, from which one easily obtains by the methods of the previous section the formula $L_{n,k} = \frac{n!}{k!}\binom{n-1}{k-1}$.

Example. We want to compute the Stirling series $\widehat{S}(z,k) = \sum_{n\geq 0} S_{n,k}\frac{z^n}{n!}$. Since we are interested only in k-partitions, we set $g(k) = 1$, $g(i) = 0$ for $i \neq k$, and $f(n) = 1$ for $n \geq 1$. Hence $\widehat{G}(z) = \frac{z^k}{k!}$, $\widehat{F}(z) = e^z - 1$, and so

$$\widehat{S}(z,k) = \frac{(e^z-1)^k}{k!}. \tag{4}$$

Summation over k yields the exponential generating function for the Bell numbers $\mathrm{Bell}(n) = \sum_{k=0}^{n} S_{n,k}$,

$$\sum_{n\geq 0} \mathrm{Bell}(n)\frac{z^n}{n!} = e^{e^z-1}. \tag{5}$$

More generally, we get

$$\begin{aligned}\sum_{n\geq 0}\left(\sum_{k\geq 0} S_{n,k}x^k\right)\frac{z^n}{n!} &= \sum_{k\geq 0}\left(\sum_{n\geq 0} S_{n,k}\frac{z^n}{n!}\right)x^k \\ &= \sum_{k\geq 0}\frac{\left(x(e^z-1)\right)^k}{k!} = e^{x(e^z-1)}.\end{aligned} \tag{6}$$

There is also a permutation version for the composition formula.

Theorem 3.5. *Let $f,g : \mathbb{N}_0 \to \mathbb{C}$ with $f(0) = 0$ and $h : \mathbb{N}_0 \to \mathbb{C}$ be defined by*

$$\begin{aligned} h(|X|) &= \sum_{\sigma\in S(X)} f(|C_1|)\cdots f(|C_k|)g(k) \quad (|X|>0), \\ h(0) &= g(0), \end{aligned} \tag{7}$$

where $C_1, \ldots, C_k$ are the cycles of σ. Then

$$\hat{H}(z) = \hat{G}\left(\sum_{n\geq 1} f(n)\frac{z^n}{n}\right).$$

Proof. There are $(j-1)!$ ways to make a j-set into a cyclic permutation. Hence we may write

$$h(|X|) = \sum_{\{B_1,\ldots,B_k\}\in\prod(X)} (|B_1|-1)!f(|B_1|)\cdots(|B_k|-1)!f(|B_k|)g(k),$$

and obtain from the composition formula

$$\hat{H}(z) = \hat{G}\left(\sum_{n\geq 1}(n-1)!f(n)\frac{z^n}{n!}\right) = \hat{G}\left(\sum_{n\geq 1} f(n)\frac{z^n}{n}\right). \qquad \square$$

Corollary 3.6. *Let $f : \mathbb{N}_0 \to \mathbb{C}$ and $h : \mathbb{N}_0 \to \mathbb{C}$ be defined through*

$$h(|X|) = \sum_{\sigma\in S(X)} f(|C_1|)\cdots f(|C_k|) \quad (|X|>0),$$
$$h(0) = 1,$$

where $C_1, \ldots, C_k$ are the cycles of σ. Then

$$\hat{H}(z) = e^{\sum_{n\geq 1} f(n)\frac{z^n}{n}}.$$

Example. If we set $g(n) = 1$ for all n, and $f(1) = f(2) = 1$, $f(n) = 0$ for $n \geq 3$, then we are counting permutations with no cycles of length greater than or equal to 3, in other words involutions. Our theorem immediately yields the familiar result

$$\sum_{n\geq 0} i_n \frac{z^n}{n!} = e^{z+\frac{z^2}{2}}.$$

Example. As a particularly interesting application we can now verify that $\log(1+z) = \sum_{n\geq 1}(-1)^{n-1}\frac{z^n}{n}$ and $e^z - 1$ are compositional inverses. Set $g(k) = 1$ for all $k \geq 1$, that is, $\hat{G}(z) = e^z - 1$, and $f(n) = (-1)^{n-1}$. Then $\hat{H}(z) = e^{\log(1+z)} - 1$, where by (7),

$$h(n) = \sum_{k=0}^{n}(-1)^{n-k}s_{n,k} \;\; (n \geq 1), \quad h(0) = 0.$$

Now we know that $x^{\underline{n}} = \sum_{k=0}^{n}(-1)^{n-k}s_{n,k}x^k$, and for $x = 1$, this equals 1 for $n = 1$, and 0 otherwise. It follows that $\widehat{H}(z) = z$ as claimed.

Number of Trees.
One of the most beautiful applications of the exponential formula concerns Cayley's formula for the number of labeled trees with n vertices. Let T_n be this number, and take $\{1,\ldots,n\}$ as vertex-set.

Example. $T_3 = 3$:

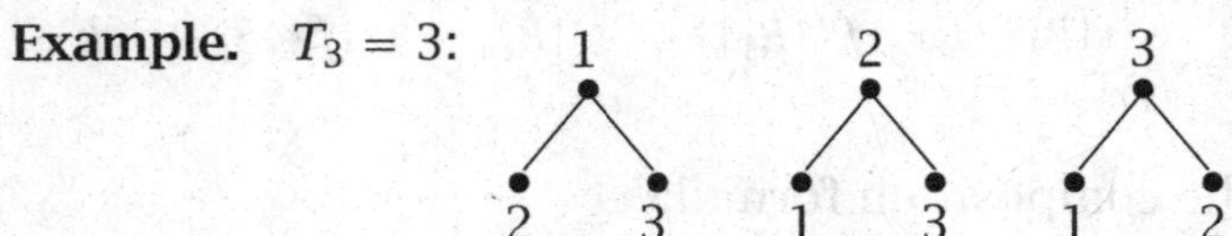

Instead of ordinary trees we consider *rooted* trees, which are better suited for applying the exponential formula. In a rooted tree we designate one vertex as the root. Each tree gives rise to n rooted trees; thus $t_n = nT_n$, where t_n is the number of rooted trees. For example, the first tree in our example produces the three rooted trees

More generally, we speak of *rooted forests* when every component tree is rooted. Let f_n be the number of rooted forests on n vertices, with $f_0 = 1$ by definition. The following picture shows $T_{n+1} = f_n$, and hence

$$t_{n+1} = (n+1)f_n\,. \tag{8}$$

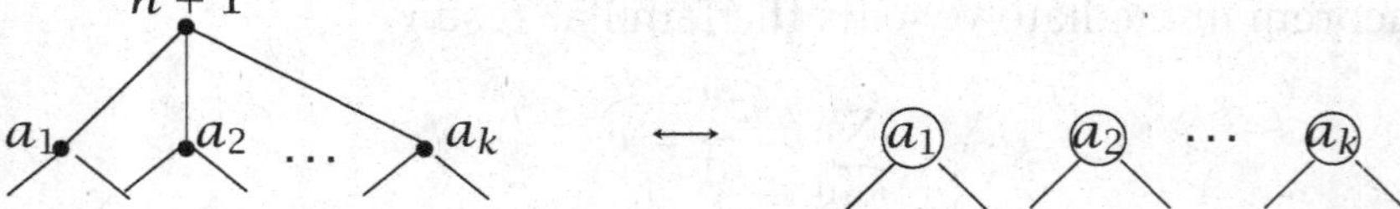

tree on $n+1$ vertices rooted forest on n vertices

Set $y = \widehat{T}(z) = \sum_{n\geq 1} t_n \frac{z^n}{n!}$, $\widehat{F}(z) = \sum_{n\geq 0} f_n \frac{z^n}{n!}$. The exponential formula tells us that $\widehat{F}(z) = e^y$. On the other hand, by (8),

$$z\widehat{F}(z) = \sum_{n\geq 0} f_n \frac{z^{n+1}}{n!} = \sum_{n\geq 0} t_{n+1}\frac{z^{n+1}}{(n+1)!} = \widehat{T}(z) = y\,.$$

In other words, $ze^y = y$, $z = ye^{-y}$, and so $\hat{T}(ye^{-y}) = y$. Substituting this into $\hat{T}(z)$ gives

$$\begin{aligned}\sum_{k\geq 1} t_k \frac{y^k}{k!} e^{-ky} &= \sum_{k\geq 1} t_k \frac{y^k}{k!} \sum_{\ell \geq 0} \frac{(-ky)^\ell}{\ell!} \\ &= \sum_{n\geq 0} \left(\sum_{k\geq 1} \binom{n}{k} t_k (-k)^{n-k} \right) \frac{y^n}{n!} = y\,.\end{aligned}$$

Hence (t_n) is the unique sequence with

$$\sum_{k=1}^{n} \binom{n}{k} t_k (-k)^{n-k} = \begin{cases} 1 & n = 1, \\ 0 & n \neq 1. \end{cases} \tag{9}$$

Let us compute the first values:

$n = 1: \quad t_1 = 1,$
$n = 2: \quad -2t_1 + t_2 = 0 \Longrightarrow t_2 = 2,$
$n = 3: \quad 3t_1 - 6t_2 + t_3 = 0 \Longrightarrow t_3 = 9,$
$n = 4: \quad -4t_1 + 24t_2 - 12t_3 + t_4 = 0 \Longrightarrow t_4 = 64.$

This should be enough to conjecture $t_k = k^{k-1}$; that is, we want to show that

$$\sum_{k=1}^{n} (-1)^{n-k} \binom{n}{k} k^{n-1} = 0 \text{ for } n \geq 2\,.$$

This does not look very promising, but it is simple! Take the polynomial $f(x) = x^{n-1}$. Then the Newton formula (9) in Section 2.4 does the job. With $T_n = \frac{t_n}{n}$ we have thus proved Cayley's formula.

Theorem 3.7. *There are precisely n^{n-2} trees on n vertices.*

Such a beautiful and simple result calls for equally simple proofs, and indeed there are plenty of them. You may look up four more proofs in Aigner–Ziegler.

Lagrange Inversion Formula.
Not only is this approach to counting trees nice, it also leads to a combinatorial proof of the famous inversion formula of Lagrange.

Theorem 3.8. *Suppose $F(z) = zG(F(z))$, $G(0) \neq 0$. Then*

$$[z^n]F(z) = \frac{1}{n}[z^{n-1}]G(z)^n \quad (n \geq 1)\,. \tag{10}$$

Proof. We write $F(z)$ and $G(z)$ in exponential form, $\hat{F}(z) = \sum_{n\geq 1} f(n)\frac{z^n}{n!}$, $\hat{G}(z) = \sum_{n\geq 0} g(n)\frac{z^n}{n!}$. For a rooted tree T on $\{1,\ldots,n\}$ let

$$g^T := g(0)^{r_0} g(1)^{r_1} g(2)^{r_2} \cdots,$$

where r_i is the number of vertices in T with out-degree i (edges pointing away from the root). The sequence $(r_0, r_1, r_2, \ldots)$ is called the *type* of T. Since T has $n-1$ edges, we have

$$\sum_{i\geq 0} r_i = n, \quad \sum_{i\geq 0} i r_i = n - 1. \tag{11}$$

Example.

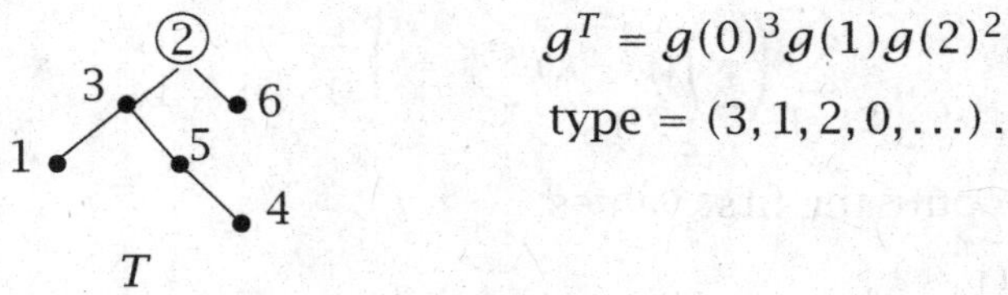

Let $f(n) = \sum_T g^T$ over all rooted trees on $\{1,\ldots,n\}$. For example, $f(1) = g(0)$, $f(2) = 2g(0)g(1)$, $f(3) = 6g(0)g(1)^2 + 3g(0)^2 g(2)$.

Claim 1. *$\hat{F}(z) = \sum_{n\geq 1} f(n)\frac{z^n}{n!}$ is the solution of the functional equation $\hat{F}(z) = z\hat{G}(\hat{F}(z))$.*

The claim is easily proved using the composition formula. Consider a rooted tree T on $\{1,\ldots,n,n+1\}$ whose root r has out-degree k:

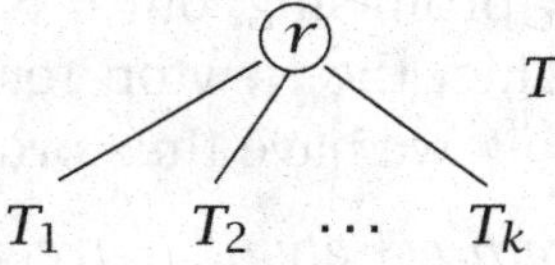

The vertices of the T_i's form a partition of $\{1,\ldots,n+1\} \setminus r$. Let $h(n)$ be defined as in (3) for the blocks $T_1,\ldots,T_k$; then $\hat{H}(z) = \hat{G}(\hat{F}(z))$. Furthermore, we see that $f(n+1) = (n+1)h(n)$, since the term $g(k)$ in (3) takes care of the out-degree of the root, and there are $n+1$ ways to choose the root. It follows that

$$\hat{F}(z) = \sum_{n\geq 0} f(n+1)\frac{z^{n+1}}{(n+1)!} = z\sum_{n\geq 0} h(n)\frac{z^n}{n!} = z\hat{H}(z) = z\hat{G}(\hat{F}(z)).$$

Since $[z^n]F(z) = \frac{f(n)}{n!}$, it remains to show that

$$f(n) = (n-1)![z^{n-1}]G(z)^n. \tag{12}$$

Claim 2. *There are precisely* $\binom{n-1}{d_1 d_2 \dots d_n}$ *rooted trees on* $\{1,\dots,n\}$ *in which vertex* i *has out-degree* d_i, $\sum_{i=1}^{n} d_i = n-1$.

Since $\binom{n-1}{d_1 \dots d_n}$ is the number of sequences of length $n-1$ in which i appears exactly d_i times, the claim will follow if we can find a bijection between the trees and these sequences. This bijection is provided by the famous *Prüfer code*, associating to every tree T a word $a_1 a_2 \dots a_{n-1}$.

Example. Consider the following tree

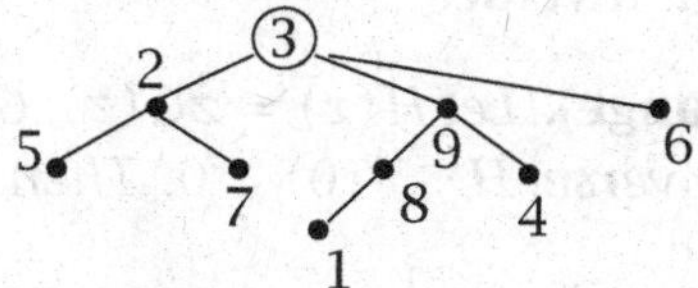

The word is constructed as follows:

1. Take the leaf with the smallest number; a_1 is then the number of the unique neighbor.
2. Delete the leaf and the incident edge, and go to 1. Iterate $n-1$ times.

In our example, we obtain the word 89232393. You are asked in the exercises to prove that this is indeed a bijection.

The proof of (12) is now readily established. By Claim 2, there are precisely

$$\binom{n}{r_0\ r_1 \dots} \frac{(n-1)!}{0!^{r_0} 1!^{r_1} \dots}$$

rooted trees on $\{1,\dots,n\}$ with type $(r_0, r_1, \dots)$. Hence

$$f(n) = \sum_{(r_0,r_1,\dots)} \binom{n}{r_0\ r_1 \dots r_n} \frac{(n-1)!}{0!^{r_0} 1!^{r_1} \dots} g(0)^{r_0} g(1)^{r_1} \cdots, \tag{13}$$

where the sum ranges over all sequences with $\sum r_i = n$, $\sum i r_i = n-1$. On the other hand,

$$G(z)^n = \left(g(0) + g(1)\frac{z}{1!} + g(2)\frac{z^2}{2!} + \cdots\right)^n,$$

and thus

$$[z^{n-1}]G(z)^n = \sum_{(r_0,r_1,\dots)} \binom{n}{r_0\ r_1 \dots r_n} \frac{g(0)^{r_0} g(1)^{r_1} \cdots}{0!^{r_0} 1!^{r_1} \cdots}. \tag{14}$$

Comparison of (13) and (14) finishes the proof. □

Example. We can now give a quick proof of Cayley's result $t_n = n^{n-1}$ for the number of rooted trees. We have seen that $\widehat{T}(z) = ze^{\widehat{T}(z)}$, and hence $\widehat{G}(z) = e^z$. Formula (10) yields

$$\frac{t_n}{n!} = \frac{1}{n}[z^{n-1}]e^{nz} = \frac{1}{n}\frac{n^{n-1}}{(n-1)!},$$

and thus $t_n = n^{n-1}$.

The Lagrange inversion formula gives in principle the coefficients of the compositional inverse.

Corollary 3.9 (Lagrange). *Let $H(z) = zG(z)$, $G(0) \neq 0$, and $H^{\langle -1\rangle}(z)$ the compositional inverse, $H^{\langle -1\rangle}(0) = 0$. Then*

$$[z^n]H^{\langle -1\rangle}(z) = \frac{1}{n}[z^{n-1}]G^{-n}(z) \ (n \geq 1).$$

Proof. Note that $F(z) = H^{\langle -1\rangle}(z)$ satisfies the equation

$$F(z) = zG^{-1}(F(z)),$$

since

$$H(F(z)) = F(z)G(F(z)) = zG^{-1}(F(z))G(F(z)) = z.$$

By (10),

$$[z^n]F(z) = \frac{1}{n}[z^{n-1}]G^{-n}(z),$$

and we are finished. □

Example. What is $H^{\langle -1\rangle}(z)$ for $H(z) = z(1-z)$? We have $G(z) = 1-z$, $G^{-n}(z) = \frac{1}{(1-z)^n}$, and thus

$$[z^n]H^{\langle -1\rangle}(z) = \frac{1}{n}[z^{n-1}]\frac{1}{(1-z)^n} = \frac{1}{n}\binom{2n-2}{n-1} = C_{n-1} \quad \text{(Catalan)}.$$

The answer is therefore $H^{\langle -1\rangle}(z) = zC(z)$.

Once we know the answer, we see that it also follows from the defining equation $C(z) = zC^2(z) + 1$ of the Catalan series. Indeed,

$$\begin{aligned} H(H^{\langle -1\rangle}(z)) &= zC(z)(1 - zC(z)) = zC(z) - z^2C^2(z) \\ &= z^2C^2(z) + z - z^2C^2(z) = z. \end{aligned}$$

Similarly, with $H(z) = z(1+z)$ we obtain $H^{\langle -1 \rangle}(z) = zC(-z)$.

Inversion Relations.

The Lagrange formula can also be used to establish the kind of inversion relations studied in Section 2.4. Suppose $a_n = \sum_{k=0}^n c_{n,k} b_k$ for all n, and thus

$$\sum_{n \ge 0} a_n z^n = \sum_{n \ge 0} \left(\sum_{k=0}^n c_{n,k} b_k \right) z^n = \sum_{k \ge 0} b_k \left(\sum_{n \ge 0} c_{n,k} z^n \right).$$

Assume that we succeed in expressing the inner sum in the form $H(z)^k$; then $A(z) = B(H(z))$. Setting $y = H(z)$, $z = H^{\langle -1 \rangle}(y)$, this gives $B(y) = A(H^{\langle -1 \rangle}(y))$, and now we work backward, interchanging the summation to obtain $b_n = \sum_{k=0}^n d_{n,k} a_k$.

Example. Suppose $a_n = \sum_{k=0}^n \left[\binom{n+1-k}{k} + \binom{n-k}{k-1} \right] b_{n-k}$. With $k \mapsto n-k$ this gives

$$\begin{aligned}
\sum_{n \ge 0} a_n z^n &= \sum_{n \ge 0} \left(\sum_{k=0}^n \left[\binom{k+1}{n-k} + \binom{k}{n-k-1} \right] b_k \right) z^n \\
&= \sum_{k \ge 0} b_k z^k \sum_{n \ge 0} \binom{k+1}{n-k} z^{n-k} \\
&\quad + \sum_{k \ge 0} b_k z^{k+1} \sum_{n \ge 0} \binom{k}{n-k-1} z^{n-k-1} \\
&= \sum_{k \ge 0} b_k z^k (1+z)^{k+1} + \sum_{k \ge 0} b_k z^{k+1} (1+z)^k \\
&= \sum_{k \ge 0} b_k (z(1+z))^k \cdot (1+2z);
\end{aligned}$$

hence $A(z) = (1+2z)B(z(1+z))$. With $y = H(z) = z(1+z)$ we have $z = H^{\langle -1 \rangle}(y) = yC(-y)$ by the previous example. Now $1 + 2yC(-y) = 1 - (1 - \sqrt{1+4y}) = \sqrt{1+4y}$, and we get

$$B(y) = \frac{A(yC(-y))}{\sqrt{1+4y}} = \sum_{k \ge 0} a_k \frac{(yC(-y))^k}{\sqrt{1+4y}}.$$

Now Exercise 3.25 comes to our help and we can work backward:

$$\sum_{k\geq 0} a_k \frac{(yC(-y))^k}{\sqrt{1+4y}} = \sum_{k\geq 0} a_k y^k \sum_{n\geq 0} \binom{2n-k}{n-k}(-y)^{n-k}$$
$$= \sum_{n\geq 0}\left[\sum_{k=0}^{n}(-1)^{n-k}\binom{2n-k}{n-k}a_k\right]y^n .$$

With the substitution $k \mapsto n-k$ we arrive at an inversion formula of Chebyshev type:

$$a_n = \sum_{k=0}^{n}\left[\binom{n+1-k}{k}+\binom{n-k}{k-1}\right]b_{n-k} \iff$$
$$b_n = \sum_{k=0}^{n}(-1)^k\binom{n+k}{k}a_{n-k} .$$

Exercises

3.44 Let h_n be the number of ordered set partitions of $\{1,\ldots,n\}$. Compute $\sum_{n\geq 0} h_n \frac{z^n}{n!}$.

3.45 Prove the formula $\mathrm{Bell}(n) = \frac{1}{e}\sum_{k\geq 0}\frac{k^n}{k!}$.

▷ **3.46** Let k_n be the number of ordered set-partitions of $\{1,\ldots,n\}$ where in addition, every block is linearly ordered. Compute k_n with the compositional formula and give a direct proof. Example: $k_2 = 4$ with $12, 21, 1|2, 2|1$.

3.47 Let m_n be the number of $n\times n$-matrices over $\{0,1\}$ with each row and column sum equal to 2.
a. Prove $m_n = n!\sum_\sigma 2^{-c(\sigma)}$, $c(\sigma) = \#$ cycles of σ, where σ ranges over all fixed-point-free permutations in $S(n)$. Hint: Use the matrix to define a set of fixed-point-free permutations on the rows.
b. Prove that $\sum_{n\geq 0} m_n \frac{z^n}{(n!)^2} = e^{-\frac{z}{2}}(1-z)^{-\frac{1}{2}}$.

▷ **3.48** Use Corollary 3.9 and $\log(1+z) = (e^z-1)^{\langle -1\rangle}$ to find a formula involving the Bernoulli numbers.

3.49 Verify the bijection given by the Prüfer code that was used in the proof of the Lagrange inversion formula.

3.50 Set $p_n(x) = \sum_{k=0}^{n} S_{n,k}x^k$, $S_{n,k}$ = Stirling numbers. Use exponential generating functions to re-prove $p_n(x+y) = \sum_{k=0}^{n}\binom{n}{k}p_k(x)p_{n-k}(y)$ (see Exercise 1.33).

3.51 Let $h_k(n)$ be the number of permutations in $S(n)$ all of whose cycle lengths are divisible by k. Compute $h_k(n)$.

* * *

3.52 There are n parking spaces $1, 2, \ldots, n$ available for n drivers. Each driver has a favorite space, driver i space $f(i)$, $1 \le f(i) \le n$. The drivers arrive one by one. When driver i arrives he tries to park his car in space $f(i)$. If it is taken he moves down the line to take the first free space greater than $f(i)$, if any. Example: $n = 4$, $f = 3221$; then driver $1 \to 3, 2 \to 2, 3 \to 4, 4 \to 1$, but for $f = 2332$ we have $1 \to 2, 2 \to 3, 3 \to 4, 2 \to ?$ Let $p(n)$ be the number of sequences f that allow each driver to park his car; f is then called a *parking* sequence. Prove: a. f is a parking sequence if and only if $\#\{i : f(i) \le k\} \ge k$. b. $p(n) = (n+1)^{n-1}$. This looks like Cayley's formula for the number of trees, can you find a bijection?

▷ **3.53** Let a_n be the number of permutations in $S(n)$ in which every cycle has odd length, and b_n the number of permutations in which all cycles have even length. Prove $\widehat{A}(z) = \sqrt{\frac{1+z}{1-z}}$, $\widehat{B}(z) = \frac{1}{\sqrt{1-z^2}}$. What relation follows between a_n and b_n?

3.54 Find a closed form for the sum $\sum_k (-1)^k S_{n,k} 2^{n-k} k!$.
Hint: Consider the Bernoulli numbers and Exercise 3.34.

▷ **3.55** Let $c_{n,i,k}$ be the number of graphs on n vertices, with i edges and k components. Use the exponential formula to prove

$$\sum_{n,i,k \ge 0} c_{n,i,k} \alpha^i \beta^k \frac{z^n}{n!} = \left[\sum_{n \ge 0} (1+\alpha)^{\binom{n}{2}} \frac{z^n}{n!} \right]^{\beta}.$$

3.56 Mimic the proof of the Lagrange formula (using rooted forests) to derive the general formula (notation the same as in the theorem): $[z^n]F(z)^k = \frac{k}{n}[z^{n-k}]G(z)^n$. Derive the Corollary $[z^n]H^{\langle -1 \rangle}(z)^k = \frac{k}{n}[z^{n-k}]G^{-n}(z)$, where $H(z) = zG(z)$.

3.57 Use the previous exercise to prove the general formula $[z^n]A(F(z)) = \frac{1}{n}[z^{n-1}]A'(z)G(z)^n$.

▷ **3.58** Show that there are $\binom{n-1}{k-1} n^{n-k}$ rooted forests on $\{1, \ldots, n\}$ with k components.

3.59 Prove $[z^n]C(z)^k = \frac{k}{2n+k}\binom{2n+k}{n}$, $C(z)$ = Catalan series, using Exercise 3.56 or alternatively $C(z) = zC^2(z) + 1$ and induction.

3.60 Prove the inversion formula

$$a_n = \sum_k \binom{n-k}{k} b_{n-k} \Leftrightarrow b_n = \sum_k (-1)^k \frac{n-k}{n+k} \binom{n+k}{k} a_{n-k}\,.$$

▷ **3.61** A tree on $\{0, 1, \ldots, n\}$ is called *alternating* if for every vertex i all neighbors are either greater than i or all are smaller than i. Let h_n be their number. Prove that $\hat{H}(z) = \sum h_n \frac{z^n}{n!}$ satisfies the equation $\hat{H}(z) = e^{\frac{z}{2}(\hat{H}(z)+1)}$. Deduce from this $h_n = \frac{1}{2^n} \sum_{k=0}^{n} \binom{n}{k} (k+1)^{n-1}$.

Example: $h_2 = 2$, 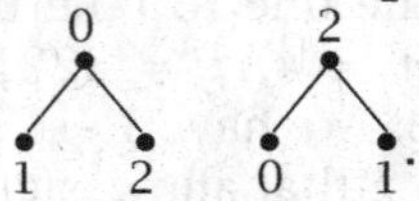

. Hint: Consider alternating forests.

3.62 Use Exercise 3.57 to find the sum of the first n terms in the expansion $(1 - \frac{1}{2})^{-n} = 1 + \frac{1}{2}n + \frac{1}{4}\binom{n+1}{2} + \frac{1}{8}\binom{n+2}{3} + \cdots$. Example: For $n = 2$ we get $1 + \frac{2}{2} = 2$, for $n = 3$, $1 + \frac{3}{2} + \frac{6}{4} = 4$, and for $n = 4$ we get 8. Conjecture? You may also use a summation formula (see Exercise 3.31).

▷ **3.63** Show the Abel identity $\sum_{k=0}^{n} \binom{n}{k} (k+1)^{k-1} (n-k+1)^{n-k} = (n+2)^n$. Hint: Use $y = \hat{T}(z) = \sum_{n \geq 1} t_n \frac{z^n}{n!}$, t_n = number of rooted trees and $y = ze^y$, and take the derivative.

3.4 Number-Partitions and Infinite Products

The study of number-partitions by means of generating functions goes back to Euler, and was in fact the starting point and first highlight for this method.

Consider the infinite product of series

$$(1 + z + z^2 + \cdots)(1 + z^2 + z^4 + \cdots) \cdots (1 + z^k + z^{2k} + \cdots) \cdots$$

Expanding the product we recognize that the coefficient of z^n is the number of ways to write $n = a_1 \cdot 1 + a_2 \cdot 2 + \ldots + a_n \cdot n$. But this is just the partition $n = \underbrace{1 \ldots 1}_{a_1} \underbrace{2 \ldots 2}_{a_2} \ldots$, that is $[z^n] = p(n)$, and we have the identity

$$\sum_{n \geq 0} p(n) z^n = \prod_{i \geq 1} \frac{1}{1 - z^i}\,. \tag{1}$$

More precisely, we see that the first factor $\frac{1}{1-z} = 1 + z + z^2 + \cdots$ accounts for the 1's in the partition, $\frac{1}{1-z^2}$ for the 2's, and so on. Thus if we restrict ourselves to partitions with largest part less than or equal to k, or equivalently with at most k summands, then

$$\sum_{n\geq 0} p(n;\leq k)z^n = \frac{1}{(1-z)(1-z^2)\cdots(1-z^k)} \tag{2}$$

results. With $p(n;k) = p(n;\leq k) - p(n;\leq k-1)$ this implies

$$\sum_{n\geq 0} p(n;k)z^n = \frac{z^k}{(1-z)\cdots(1-z^k)}\,. \tag{3}$$

Comparing (2) and (3) we obtain our old result

$$p(n;k) = p(n-k;\leq k)\,.$$

Partitions with unequal parts are clearly counted by

$$\sum_{n\geq 0} p_d(n)z^n = (1+z)(1+z^2)(1+z^3)\cdots\,. \tag{4}$$

Now recall equation (7) in Section 2.3:

$$\prod_{i\geq 1}(1+z^i) = \prod_{i\geq 1}\frac{1}{(1-z^{2i-1})}\,,$$

which gives the unexpected result

$$p_d(n) = p_o(n)\,, \tag{5}$$

where $p_o(n)$ is the number of partitions of n into odd parts. Such a simple identity calls for a bijection proof; we present two of them in the exercises, and two more in Chapter 5.

Example. For $n = 7$, $p_d(7) = p_o(7) = 5$ with the partitions $7, 61, 52, 43, 421$ and $7, 511, 331, 31111, 1111111$, respectively.

Partition Functions.
If we want to enumerate partitions according to the number of parts and highest summand, we use generating functions in two variables q and z. Expanding the left-hand side, we obtain

$$(1+qz+q^2z^2+\cdots)(1+qz^2+q^2z^4+\cdots)\cdots = \sum_{k\geq 0}\sum_{n\geq 0} p(n;k)z^n q^k;$$

thus

$$\sum_{k\geq 0}\sum_{n\geq 0} p(n;k)z^n q^k = \frac{1}{\prod_{i\geq 1}(1-qz^i)}\,. \tag{6}$$

More precisely, setting $p_m(z,k) = \sum_{n\geq 0} p(n;k;\leq m)z^n$ we have

$$P_m(z,q) = \sum_{k\geq 0} p_m(z,k)q^k = \frac{1}{\prod_{i=1}^{m}(1-qz^i)}\,. \tag{7}$$

From (7) we immediately infer

$$(1-qz)P_m(z,q) = (1-qz^{m+1})P_m(z,qz)\,,$$

which translates for the coefficient series $p_m(z,k)$ into the identity

$$p_m(z,k) - zp_m(z,k-1) = z^k p_m(z,k) - z^{m+k}p_m(z,k-1)\,,$$

or

$$p_m(z,k) = z\frac{1-z^{m+k-1}}{1-z^k}p_m(z,k-1)\,.$$

Iteration down to $k=0$ gives

$$\sum_{n\geq 0} p(n;k;\leq m)z^n = z^k\frac{(1-z^m)\cdots(1-z^{m+k-1})}{(1-z)(1-z^2)\cdots(1-z^k)}\,. \tag{8}$$

Furthermore, with $p(n;k;m) = p(n;k;\leq m) - p(n;k;\leq m-1)$,

$$\sum_{n\geq 0} p(n;k;m)z^n = z^{m+k-1}\frac{(1-z^m)(1-z^{m+1})\cdots(1-z^{m+k-2})}{(1-z)(1-z^2)\cdots(1-z^{k-1})}\,. \tag{9}$$

The two products in (8) and (9) should look familar to you; they are Gaussian coefficients multiplied by an appropriate power of z. In fact, the formulas follow directly from Proposition 1.1 in Section 1.6, where we proved via lattice paths that

$$\sum_{n\geq 0} p(n;\leq k;\leq m)z^n = \begin{bmatrix} m+k \\ k \end{bmatrix}_z = \frac{(1-z^{m+1})\cdots(1-z^{m+k})}{(1-z)\cdots(1-z^k)}\,.$$

The identities (8) and (9) are then obtained from the recurrence of the Gaussian coefficients.

Now let us enumerate the partitions with unequal summands by the number of parts. This is analogously given by

$$\sum_{n,k\geq 0} p_d(n;k)z^n q^k = (1+qz)(1+qz^2)(1+qz^3)\cdots.$$

To express the right-hand side we could proceed as before, but here is a quicker way. The q-binomial theorem in Section 1.6 states with the substitutions $x \mapsto q$, $q \mapsto z$,

$$(1+qz)(1+qz^2)\cdots(1+qz^n) = \sum_{i=0}^{n} z^{\binom{i+1}{2}} \begin{bmatrix} n \\ i \end{bmatrix}_z q^i.$$

Thus letting $n \to \infty$, we immediately obtain

$$\sum_{n,k\geq 0} p_d(n;k)z^n q^k = \sum_{i\geq 0} \frac{z^{\binom{i+1}{2}}}{(1-z)\cdots(1-z^i)} q^i, \tag{10}$$

since for fixed i the Gaussian coefficient $\left[{n \atop i}\right]_z = \sum_{j\geq 0} p(j;\leq i;\leq n-i)z^j$ goes to $\sum_{j\geq 0} p(j;\leq i)z^j = \frac{1}{(1-z)\cdots(1-z^i)}$ according to (2). Note that the limiting process is formally justified as explained in Section 2.3. Several other examples of "proof by going to infinity" are contained in the exercises.

Example. Let us enumerate the number $p_{sc}(n)$ of self-conjugate partitions of n. Exercise 1.53 says that their number is the same as that of partitions with all parts odd and distinct. Hence

$$\sum_{n\geq 0} p_{sc}(n)z^n = \prod_{i\geq 1}(1+z^{2i-1}). \tag{11}$$

But there is another way to look at self-conjugate partitions $\lambda = \lambda^*$. Take the Ferrers diagram and suppose that the largest square in the upper left-hand corner is of the form $k \times k$; this is called the *Durfee square* of λ. In the example we have $k = 3$:

$\lambda = 7543211$

Since $\lambda = \lambda^*$, the partitions (of $\frac{n-k^2}{2}$) on the right and below the square are the same, and the number of parts is at most k. In our example, those "small" partitions are 421. For fixed k, the contribution to $p_{sc}(n)$ is therefore $[z^{\frac{n-k^2}{2}}]\frac{1}{(1-z)\cdots(1-z^k)}$. With the substitution $z \mapsto z^2$ this is $[z^{n-k^2}]\frac{1}{(1-z^2)\cdots(1-z^{2k})} = [z^n]\frac{z^{k^2}}{(1-z^2)\cdots(1-z^{2k})}$. Summing over k, we obtain

$$\sum_{n\geq 0} p_{sc}(n)z^n = \sum_{k\geq 0} \frac{z^{k^2}}{(1-z^2)\cdots(1-z^{2k})},$$

which proves the astounding identity

$$\prod_{i\geq 1}(1+z^{2i-1}) = \sum_{k\geq 0} \frac{z^{k^2}}{\prod_{i=1}^{k}(1-z^{2i})}. \tag{12}$$

Euler's Pentagonal Theorem.

Let us return to our first identity

$$\sum_{n\geq 0} p(n)z^n = \prod_{i\geq 1}\frac{1}{1-z^i},$$

and let us study the inverse $\prod_{i\geq 1}(1-z^i)$ of the partition function. The first terms in the expansion are

$$\prod_{i\geq 1}(1-z^i) = 1 - z - z^2 + z^5 + z^7 - z^{12} - z^{15} + z^{22} + z^{26} \mp \cdots.$$

It appears that all coefficients are $0, 1$, and -1, and that the nonzero coefficients after the first come in pairs, with exponents $1,2; 5,7; 12,15; 22,26; \ldots$. A little experimentation reveals the pattern: The exponents seem to come in pairs $\left(\frac{3j^2-j}{2}, \frac{3j^2+j}{2}\right)$ or shorter $\left(\frac{3j^2+j}{2}: j \in \mathbb{Z}\right)$ with sign $(-1)^j$, and this is what we want to show.

Set $\prod_{i\geq 1}(1-z^i) = \sum_{n\geq 0} a(n)z^n$. Then

$$\sum_{n\geq 0} a(n)z^n \cdot \sum_{n\geq 0} p(n)z^n = 1,$$

that is,

$$\sum_{k=0}^{n} a(k)p(n-k) = 0 \quad \text{for } n \geq 1. \tag{13}$$

Our conjecture thus reads

$$a(k)=\begin{cases}1 & k=\frac{3j^2+j}{2}, \quad j\in\mathbb{Z}\text{ even},\\ -1 & k=\frac{3j^2+j}{2}, \quad j\in\mathbb{Z}\text{ odd},\\ 0 & \text{otherwise.}\end{cases}$$

If we set $b(j)=\frac{3j^2+j}{2}$ $(j\in\mathbb{Z})$, and substitute $b(j)$ into (13), the conjecture takes on the form

$$\sum_{j\text{ even}} p(n-b(j)) = \sum_{j\text{ odd}} p(n-b(j)) \quad (j\in\mathbb{Z}),$$

and this, of course, calls for a bijection

$$\bigcup_{j\text{ even}} \mathrm{Par}(n-b(j)) \longrightarrow \bigcup_{j\text{ odd}} \mathrm{Par}(n-b(j)).$$

The following marvelous bijection ϕ was found by Bressoud and Zeilberger (in fact, ϕ is an involution). Let $\lambda=\lambda_1\lambda_2\dots\lambda_t\in\mathrm{Par}(n-b(j))$. Then

$$\phi\lambda=\begin{cases} t+3j-1,(\lambda_1-1),\dots,(\lambda_t-1) & \text{if } t+3j\ge\lambda_1,\\ (\lambda_2+1),\dots,(\lambda_t+1),\underbrace{1,\dots,1}_{\lambda_1-3j-t-1} & \text{if } t+3j<\lambda_1,\end{cases} \tag{14}$$

where we omit possible 0's at the end.

Once you know ϕ, the verification is straightforward (see Exercise 3.70).

Example. Take $n=12$, $j=-2$; thus $b(-2)=5$, and consider $\lambda=3211\in\mathrm{Par}(7)=\mathrm{Par}(12-b(-2))$. The second case applies, and $\phi\lambda=3221111$ since $\lambda_1-3j-t-1=3+6-4-1=4$. Thus $\phi\lambda\in\mathrm{Par}(11)=\mathrm{Par}(12-1)=\mathrm{Par}(12-b(-1))$, $b(-1)=1$. Applying ϕ to $\mu=3221111\in\mathrm{Par}(12-b(-1))$, the first case holds, and we return to the original partition $\phi\mu=3211$.

In summary, we have proved Euler's famous "pentagonal" theorem.

Theorem 3.10. *We have*

$$\prod_{i\ge1}(1-z^i)=1+\sum_{j\ge1}(-1)^j\left(z^{\frac{3j^2-j}{2}}+z^{\frac{3j^2+j}{2}}\right). \tag{15}$$

Here is the reason for the name "pentagonal." The number $\frac{3j^2-j}{2}$ counts the number of dots in nested pentagons up to side length j (check it!).

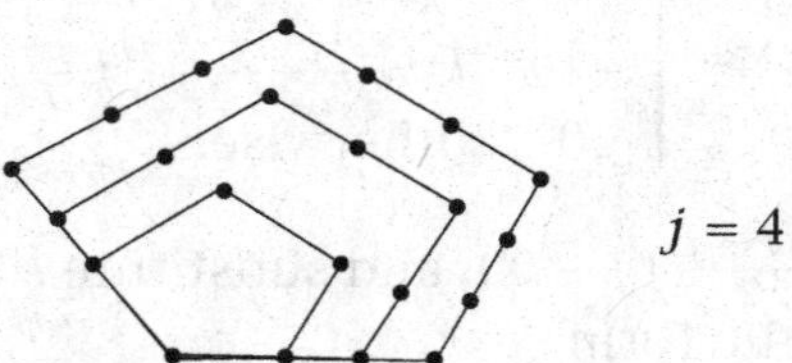

Not only is Euler's theorem beautiful, it also yields an efficient method for calculating the partition numbers $p(n)$ for small n. Looking at (13), we obtain

$$\begin{aligned} p(n) = p(n-1) + p(n-2) - p(n-5) - p(n-7) \\ + p(n-12) + p(n-15) \mp \cdots . \end{aligned}$$

As an example,

$$\begin{aligned} p(7) &= p(6) + p(5) - p(2) - p(0) = 11 + 7 - 2 - 1 = 15, \\ p(8) &= p(7) + p(6) - p(3) - p(1) = 15 + 11 - 3 - 1 = 22. \end{aligned}$$

Jacobi's Triple Product Theorem.
A far-reaching generalization of Euler's theorem is provided by the equally famous triple product theorem of Jacobi.

Theorem 3.11. *We have the identity*

$$\prod_{k\geq 1}(1+zq^k)(1+z^{-1}q^{k-1})(1-q^k) = \sum_{n=-\infty}^{\infty} q^{\frac{n(n+1)}{2}} z^n . \tag{16}$$

Proof. Set

$$F(z) = \prod_{k\geq 1}(1+zq^k)(1+z^{-1}q^{k-1}) = \sum_{n=-\infty}^{\infty} a_n(q)z^n .$$

Then the theorem claims that

$$F(z) = \frac{1}{\prod\limits_{k\geq 1}(1-q^k)} \sum_{n=-\infty}^{\infty} q^{\frac{n(n+1)}{2}} z^n , \tag{17}$$

where the factor in front is just the partition function in the variable q. The series $F(z)$ is a so-called *Laurent series*, where the index-set ranges over $\mathbb{Z}$ instead of the familiar set $\mathbb{N}_0$. We could argue analytically that $F(z)$ converges for $|z| < 1$. But in the spirit of the book we proceed in purely formal terms. Write

$$F(z) = (1 + zq)(1 + zq^2) \cdots (1 + z^{-1})(1 + z^{-1}q) \cdots .$$

What is the coefficient of $z^n q^m$, $n \in \mathbb{Z}$, $m \in \mathbb{N}_0$? If $n \geq 0$, then we have to take $n + i$ factors from the first product and i from the second, so that the q-exponents add to m, for all $i \geq 0$. In other words,

$$[z^n q^m]F(z) = \# \text{ pairs } \lambda_1\lambda_2 \ldots \lambda_{n+i},\ \mu_1\mu_2 \ldots \mu_i ,$$

where

$$\lambda_1 > \lambda_2 > \cdots > \lambda_{n+i} \geq 1, \quad \mu_1 > \mu_2 > \cdots > \mu_i \geq 0,$$
$$\sum \lambda_j + \sum \mu_j = m .$$

So we have to consider all pairs of these partitions λ, μ whose parts add up to m, and of these there are only *finitely* many. Similarly, when $n < 0$, then we have to take i factors out of the first product and $-n + i$ out of the second. Again this reduces to pairs of partitions, and so $[z^n q^m]F(z)$ is well-defined. The coefficient $a_n(q)$ in $F(z)$ is thus a bona fide generating function for every $n \in \mathbb{Z}$.

The definition of $F(z)$ implies the functional equation

$$\begin{aligned} F(qz) &= \prod_{k \geq 1} (1 + zq^{k+1})(1 + z^{-1}q^{k-2}) \\ &= F(z)\frac{1 + z^{-1}q^{-1}}{1 + zq} = z^{-1}q^{-1}F(z); \end{aligned}$$

thus

$$\sum_{n=-\infty}^{\infty} a_n(q)q^n z^n = \sum_{n=-\infty}^{\infty} a_n(q)q^{-1}z^{n-1} .$$

Comparing coefficients for z^{n-1} we get $a_n(q) = q^n a_{n-1}(q)$. For $n \geq 0$, iteration gives $a_n(q) = q^{\frac{n(n+1)}{2}} a_0(q)$, and for negative indices $-n$, $a_{-n}(q) = q^{n-1}a_{-(n-1)}(q)$ implies $a_{-n}(q) = q^{\frac{n(n-1)}{2}} a_0(q)$, and hence $a_n(q) = q^{\frac{n(n+1)}{2}} a_0(q)$ for all $n \in \mathbb{Z}$.

In summary, $F(z) = a_0(q)\sum_{n=-\infty}^{\infty} q^{\frac{n(n+1)}{2}} z^n$, and it remains to compute the series $a_0(q)$. Set $a_0(q) = \sum_{m\geq 0} b_m q^m$. By the same argument as above,

$$\begin{aligned} b_m &= \#\lambda_1,\dots,\lambda_i,\mu_1,\dots,\mu_i \text{ with} \\ \lambda_1 &> \cdots > \lambda_i \geq 1,\ \mu_1 > \cdots > \mu_i \geq 0, \quad \sum \lambda_j + \sum \mu_j = m\,. \end{aligned} \tag{18}$$

Claim. $b_m = p(m)$.

You were already asked to show this in Exercise 1.63. The bijection $\phi : \mathrm{Par}(m) \to (\lambda, \mu)$ in the following figure does it.

Example. $m = 26 = 7 + 6 + 4 + 4 + 3 + 1 + 1$

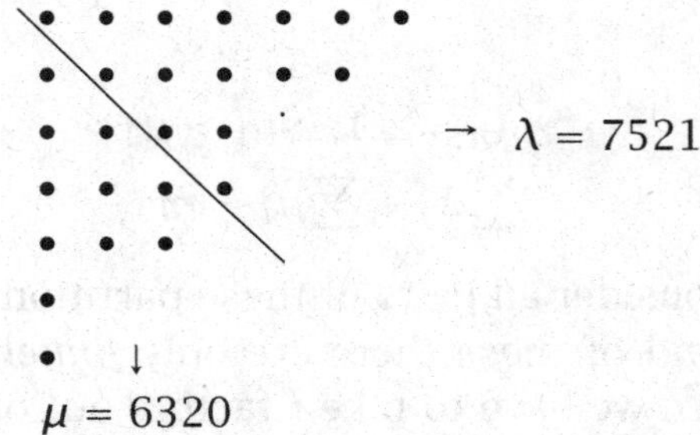

Hence $a_0(q) = \prod_{k\geq 1} \frac{1}{1-q^k}$, and the theorem follows. □

Specialization of z and q in Jacobi's theorem now leads to a plethora of identities relating infinite products and sums. For example, if we set $q = q^3$, $z = -q^{-1}$ in the triple product theorem, then

$$\prod_{k\geq 1}(1-q^{3k-1})(1-q^{3k-2})(1-q^{3k}) = \sum_{n=-\infty}^{\infty}(-1)^n q^{\frac{3n^2+n}{2}},$$

and this is precisely Euler's Theorem 3.10.

Exercises

3.64 Let $c_{n,k}$ be the number of ordered k-partitions of n, $n \geq k \geq 1$. Show that $\sum_{n\geq 1} c_{n,k} z^n = (z + z^2 + z^3 + \cdots)^k$ and deduce our old result $c_{n,k} = \binom{n-1}{k-1}$.

▷ **3.65** Prove $p_o(n) = p_d(n)$ by the following bijection $\phi : \mathrm{Par}_o(n) \to \mathrm{Par}_d(n)$, due to Glaisher.
Let $\lambda \in \mathrm{Par}_o(n)$, $n = \underbrace{\lambda_1 + \cdots + \lambda_1}_{n_1} + \underbrace{\lambda_2 + \cdots + \lambda_2}_{n_2} + \cdots$.
Write $n_1 = 2^{m_1} + 2^{m_2} + \cdots + 2^{m_k}$ in binary form, and similarly the other n_i. Then $\phi\lambda = 2^{m_1}\lambda_1 + \cdots + 2^{m_k}\lambda_1 + \cdots$. Example: $n = 25$, $\lambda = 555331111$; then $\phi\lambda = (2+1)5 + (2)3 + (4)1 = 10 + 5 + 6 + 4$.

3.66 Generalizing $p_d(n) = p_o(n)$, show that the number of partitions of n in which every part appears at most $k - 1$ times equals the number of partitions in which every part is not divisible by k, for every $k \geq 2$.

3.67 Show that the following two sets of partitions have the same size:
a. the partitions of n in which the even summands appear at most once,
b. the partitions of n for which every summand appears at most three times.

▷ **3.68** Prove a result of I. Schur: The number of partitions of n into parts congruent to 1 or 5 (mod 6) equals the number of partitions of n into distinct parts all congruent to 1 or 2 (mod 3).

3.69 Prove the following identities:
a. $\prod_{i\geq 1} \frac{1}{1-qz^i} = \sum_{k\geq 0} \frac{q^k z^k}{(1-z)\cdots(1-z^k)} = \sum_{k\geq 0} \frac{q^k z^{k^2}}{(1-z)\cdots(1-z^k)(1-qz)\cdots(1-qz^k)}$
b. $\prod_{i\geq 1}(1 + qz^{2i-1}) = \sum_{k\geq 0} \frac{q^k z^{k^2}}{(1-z^2)(1-z^4)\cdots(1-z^{2k})}$.
Hint: For the second equality in (a) consider the Durfee square.

3.70 Verify the Bressoud-Zeilberger bijection (14).

▷ **3.71** Show that $\prod_{i\geq 1}(1 - z^i) = \sum_{n\geq 0}(p_{d,e}(n) - p_{d,o}(n))z^n$, where $p_{d,e}(n)$ and $p_{d,o}(n)$ count the partitions in $\mathrm{Par}_d(n)$ with an even and odd number of parts, respectively. Hence by Euler's theorem the difference $p_{d,e}(n) - p_{d,o}(n)$ is always $0, \pm 1$. Hint: Use $\prod_{i\geq 1}(1 + qz^i)$ and set $q = -1$.

3.72 Prove the identity

$$\prod_{k\geq 1}(1 - q^{4k-3})(1 - q^{4k-1})(1 - q^{4k}) = \sum_{n=-\infty}^{\infty}(-1)^n q^{2n^2+n}.$$

* * *

3.73 Give another proof of $p_o(n) = p_d(n)$ using Sylvester's bijection, modifying the Ferrers diagram as in the figure. Example: $21 = 7 + 5 + 3 + 3 + 1 + 1 + 1$.

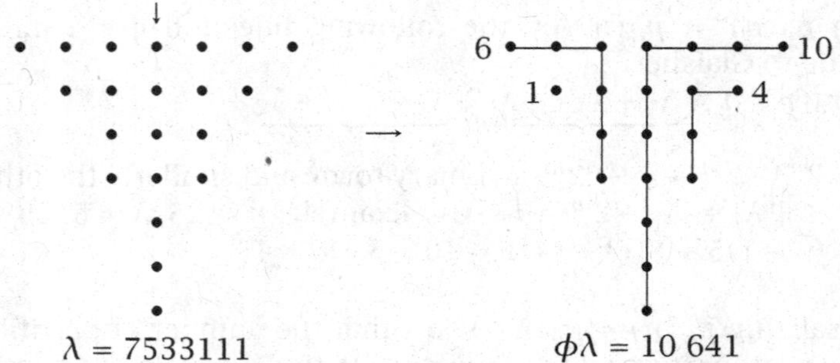

3.74 Show that Sylvester's bijection proves the following: The number of *distinct* odd parts that appear in $\lambda \in \text{Par}_o(n)$ is the same as the number of runs in $\phi\lambda \in \text{Par}_d(n)$, where a run is a maximal sequence of consecutive integers. Example: 86521 has the three runs $\{8\}, \{6,5\}, \{2,1\}$.

▷ **3.75** Use generating functions to re-prove Exercise 1.60.

3.76 Let $r(n)$ be the number of partitions of n whose parts differ by at least 2, and $s(n)$ the number of those in which, in addition, the summand 1 does not appear. Prove

a. $\sum_{n\geq 0} r(n)z^n = \sum_{k\geq 0} \frac{z^{k^2}}{\prod_{i=1}^{k}(1-z^i)}$, b. $\sum_{n\geq 0} s(n)z^n = \sum_{k\geq 0} \frac{z^{k^2+k}}{\prod_{i=1}^{k}(1-z^i)}$.

Hint: Use the fact that $k^2 = 1+3+5+\cdots+(2k-1)$. The famous Rogers–Ramanujan identities will show to which infinite products these sums are equal (see the highlight at the end of Chapter 10).

▷ **3.77** By taking $\frac{P'(z)}{P(z)}$ for the partition function $P(z) = \sum_{n\geq 0} p(n)z^n$ prove the recurrence $p(n) = \frac{1}{n}\sum_{i=1}^{n} \sigma(i)p(n-i)$, where $\sigma(i)$ is the sum of divisors of i. Give also a combinatorial proof. Hint: Exercise 2.31.

3.78 Given a set $S \subseteq \mathbb{N}$, let $p_S(n)$ and $p_{d,S}(n)$ denote the number of partitions in $\text{Par}(n)$ respectively $\text{Par}_d(n)$ that use only parts from S. Call (S,T) an *Euler pair* if $p_S(n) = p_{d,T}(n)$ for all n. For example, S = odd integers and $T = \mathbb{N}$ is an Euler pair. Prove that (S,T) is an Euler pair if and only if $2T \subseteq T$ and $S = T \setminus 2T$, where $2T = \{2t : t \in T\}$. Consider the example $S = \{1\}$, $T = \{1, 2, 2^2, 2^3, \ldots\}$. What does it imply?

3.79 Prove that $p(n)$ equals the number of partitions into distinct parts whose odd-indexed parts sum to n. Example: $p(5) = 7$, with the partitions $5, 5+1, 5+2, 5+3, 5+4, 4+3+1$, and $4+2+1$.
Hint: $p(n;k) = \#\lambda \in \text{Par}_d$ with $\lambda_1 + \lambda_3 + \lambda_5 + \cdots = n$, $|\lambda| = 2n-k$.

3.80 Show that $p(n; \leq 3)$ is the integer nearest to $\frac{(n+3)^2}{12}$.

▷ **3.81** Show that $\prod_{k\geq 1}(1-q^k)^3 = \sum_{n\in\mathbb{Z}}(-1)^n n q^{\frac{n(n+1)}{2}} = \sum_{n\geq 0}(-1)^n(2n+1)q^{\frac{n(n+1)}{2}}$. Hint: Use the triple product theorem, make the substitution $z = y - 1$, and consider both sides at $y = 0$.

▷ **3.82** Prove the following finite version of Jacobi's triple product theorem:

$$\prod_{k=1}^{N}(1+zq^k)(1+z^{-1}q^{k-1}) = \sum_{n=-N}^{N} q^{\frac{n(n+1)}{2}} \begin{bmatrix} 2N \\ N+n \end{bmatrix}_q z^n,$$

and deduce Jacobi's theorem by letting N go to ∞.
Hint: Use the q-binomial theorem with $n = 2N$, and make some clever substitutions.

3.83 Let r, m be integers with $1 \le r < \frac{m}{2}$. Apply Jacobi's theorem to prove

$$\prod_{k\ge 1}(1-q^{(k-1)m+r})(1-q^{km-r})(1-q^{km}) = \sum_{n=-\infty}^{\infty}(-1)^n q^{\frac{mn^2+(m-2r)n}{2}}.$$

For $m = 5$, $r = 1$ or 2, we get the formulas

$$\prod_{k\ge 1}(1-q^{5k-4})(1-q^{5k-1})(1-q^{5k}) = \sum_{n=-\infty}^{\infty}(-1)^n q^{\frac{n(5n+3)}{2}},$$

$$\prod_{k\ge 1}(1-q^{5k-3})(1-q^{5k-2})(1-q^{5k}) = \sum_{n=-\infty}^{\infty}(-1)^n q^{\frac{n(5n+1)}{2}}.$$

Highlight: Ramanujan's Most Beautiful Formula

In the section on partition identities we have seen several formulas involving products of the form $\prod(1-aq^k)$. To shorten these expressions we employ the following concise notation familiar to all partition theorists:

$$(a;q) := (1-a)(1-aq)(1-aq^2)\cdots = \prod_{i\geq 0}(1-aq^i).$$

When there are several infinite products we write

$$(a_1,\ldots,a_s;q) = (a_1;q)\cdots(a_s;q),$$

and even more generally,

$$\left(\begin{matrix} a_1,\ldots,a_s \\ b_1,\ldots,b_t \end{matrix};q\right) = \frac{\prod\limits_{i=1}^{s}(a_i;q)}{\prod\limits_{j=1}^{t}(b_j;q)}.$$

As an example, the ordinary partition function may be abbreviated as

$$\sum_{n\geq 0} p(n)q^n = \frac{1}{(q;q)}. \tag{1}$$

Jacobi's triple product Theorem 3.11 translates into

$$(-zq,-z^{-1},q;q) = \sum_{n\in\mathbb{Z}} q^{\frac{n(n+1)}{2}} z^n,$$

or with the substitution $z \mapsto -z$ into

$$(zq,z^{-1},q;q) = \sum_{n\in\mathbb{Z}} (-1)^n q^{\frac{n(n+1)}{2}} z^n. \tag{2}$$

Our goal is to prove the following formula due to Ramanujan, which his mentor and collaborator Hardy regarded as his most beautiful.

Theorem. *We have*

$$\sum_{n\geq 0} p(5n+4)q^n = 5\frac{(q^5;q^5)^5}{(q;q)^6}.$$

In particular, $p(5n+4)$ *is a multiple of* 5 *for every* n.

The following elegant approach was suggested by Michael Hirschhorn.

Paving the Way.
Note first that by classifying the factors in $(q;q) = \prod_{k\geq 1}(1-q^k)$ according to the exponents modulo 5, we have

$$(q;q) = (q, q^2, q^3, q^4, q^5; q^5),$$

and therefore

$$(q;q) = \frac{(q, q^4, q^5; q^5)(q^2, q^3, q^5; q^5)}{(q^5; q^5)}. \tag{3}$$

Each of the factors in the numerator contains three terms, and this calls, of course, for an application of Jacobi's theorem.
Making in (2) the substitutions $z \mapsto q^{-4}, q \mapsto q^5$ respectively $z \mapsto q^{-3}, q \mapsto q^5$, we see that (3) becomes

$$\begin{aligned}(q;q) &= \frac{1}{(q^5;q^5)} \sum_{r\in\mathbb{Z}} (-1)^r q^{\frac{5r^2-3r}{2}} \cdot \sum_{s\in\mathbb{Z}} (-1)^s q^{\frac{5s^2-s}{2}} \\ &= \frac{1}{(q^5;q^5)} \left[\sum_{r,s\in\mathbb{Z}} (-1)^{r+s} q^{r+2s} q^{\frac{5r^2+5s^2-5r-5s}{2}} \right]. \end{aligned} \tag{4}$$

Next we split the sum in (4) into five sums $\sum_i$, according to the residue of $r+2s$ modulo 5, that is, $r+2s = 5n+i$, $i = 0,1,2,3,4$. Let us just compute $\sum_0$; the other sums are dealt with by analogous arguments.

Suppose $r+2s = 5n$, and thus $r \equiv n \pmod 2$. We set $r = n-2t$, and hence $s = 2n+t$, and replace the running indices r and s by n and t. We have $(-1)^r = (-1)^n$, $(-1)^s = (-1)^t$, and

$$\frac{5r^2+5s^2-5r-5s}{2} = \frac{25n^2+25t^2-15n+5t}{2}.$$

Moving the terms involving n to the front we get

$$\Sigma_0 = \sum_{n\in\mathbb{Z}} (-1)^n q^{\frac{25n^2-5n}{2}} \cdot \sum_{t\in\mathbb{Z}} (-1)^t q^{\frac{25t^2+5t}{2}}. \tag{5}$$

The substitutions $z \mapsto q^{-15}, q \mapsto q^{25}$ in (2) for the first sum in (5) respectively $z \mapsto q^{-10}, q \mapsto q^{25}$ for the second sum yield (check it!)

$$\Sigma_0 = (q^{10}, q^{15}, q^{25}; q^{25}) \cdot (q^{10}, q^{15}, q^{25}; q^{25}),$$

and thus as contribution to (4),

$$\frac{(q^{10}, q^{15}, q^{25}; q^{25})(q^{10}, q^{15}, q^{25}; q^{25})}{(q^5; q^5)} = (q^{25}; q^{25}) \frac{(q^{10}, q^{15}; q^{25})}{(q^5, q^{20}; q^{25})}. \tag{6}$$

For the other sums we obtain the contributions

$$\begin{aligned} \frac{1}{(q^5; q^5)} \Sigma_1 &= -q(q^{25}; q^{25}), \\ \frac{1}{(q^5; q^5)} \Sigma_2 &= -q^2 (q^{25}; q^{25}) \frac{(q^5, q^{20}; q^{25})}{(q^{10}, q^{15}; q^{25})}, \end{aligned} \tag{7}$$

while Σ_3 and Σ_4 are both equal to 0.

With the abbreviation

$$R(q) = \left(\begin{matrix} q^2, q^3 \\ q, q^4 \end{matrix} ; q^5 \right)$$

we have thus proved another formula of Ramanujan:

$$(q; q) = (q^{25}; q^{25})[R(q^5) - q - q^2 R(q^5)^{-1}]. \tag{8}$$

Since $(q; q) = (q^5; q^5)(q, q^2, q^3, q^4; q^5)$ we may write (8) also as

$$(q^5; q^5)(q, q^2, q^3, q^4; q^5) = (q^{25}; q^{25})[R(q^5) - q - q^2 R(q^5)^{-1}]. \tag{9}$$

Using Roots of Unity.
Denote by Ω the set of fifth roots of unity, that is, of the roots of the polynomial $x^5 - 1$. Thus $x^5 - 1 = \prod_{\omega \in \Omega} (x - \omega)$. Note that for any a,

$$\prod_{\omega \in \Omega} (1 - a\omega) = a^5 \prod_{\omega} (a^{-1} - \omega) = a^5 (a^{-5} - 1) = 1 - a^5. \tag{10}$$

With the substitution $q \mapsto \omega q$, equation (9) becomes

$$\begin{aligned} &(q^5; q^5)(\omega q, \omega^2 q^2, \omega^3 q^3, \omega^4 q^4; q^5) = \\ &\qquad (q^{25}; q^{25})[R(q^5) - \omega q - \omega^2 q^2 R(q^5)^{-1}]. \end{aligned} \tag{11}$$

Now we take the product of the left-hand side over all $\omega \in \Omega$. Since $\{\omega^i : \omega \in \Omega\} = \Omega$ for $i = 1, 2, 3, 4$, we have for fixed i, using (10),

$$\prod_{\omega}(\omega^i q^i;q^5) = \prod_{\omega}(\omega q^i;q^5) = \prod_{k\geq 0}\prod_{\omega}(1-\omega q^{5k+i})$$
$$= \prod_{k\geq 0}(1-q^{25k+5i}) = (q^{5i};q^{25}).$$

Taking the product over all ω, the left-hand side of (11) therefore becomes

$$(q^5;q^5)^5(q^5,q^{10},q^{15},q^{20};q^{25}) = \frac{(q^5;q^5)^6}{(q^{25};q^{25})}. \qquad (12)$$

Next we look at the right-hand side. Set $R(q^5) = c$, and regard the second factor in (11) as the polynomial $p(\omega) = c - \omega q - \omega^2 q^2 c^{-1}$ in ω. It is easily seen that

$$p(\omega) = c\left(1 - \frac{q}{c}\tau\omega\right)\left(1 - \frac{q}{c}\hat{\tau}\omega\right),$$

where $\tau = \frac{1+\sqrt{5}}{2}$, $\hat{\tau} = \frac{1-\sqrt{5}}{2}$. By (10),

$$\prod_{\omega} p(\omega) = c^5\left(1 - \frac{q^5}{c^5}\tau^5\right)\left(1 - \frac{q^5}{c^5}\hat{\tau}^5\right)$$
$$= c^5 - q^5(\tau^5 + \hat{\tau}^5) + \frac{q^{10}}{c^5}(\tau\hat{\tau})^5.$$

Now $\tau\hat{\tau} = -1$, and it is easily verified that $\tau^5 + \hat{\tau}^5 = 11$. So the product over ω on the right-hand side of (11) equals

$$(q^{25};q^{25})^5[R(q^5)^5 - 11q^5 - q^{10}R(q^5)^{-5}], \qquad (13)$$

and so, equating (12) and (13),

$$1 = \frac{(q^{25};q^{25})^6}{(q^5,q^5)^6}[R(q^5)^5 - 11q^5 - q^{10}R(q^5)^{-5}]. \qquad (14)$$

The Final Step.
Look at the equations (14) and (8) and divide (14) by (8). This gives

$$\sum_{n\geq 0} p(n)q^n = \frac{1}{(q;q)} = \frac{(q^{25};q^{25})^5}{(q^5;q^5)^6} \cdot \frac{R(q^5)^5 - 11q^5 - q^{10}R(q^5)^{-5}}{R(q^5) - q - q^2R(q^5)^{-1}}.$$

Setting again $c = R(q^5)$, the second factor is a rational function in q, and polynomial division readily yields the polynomial

$$f(q) = c^{-4}q^8 - c^{-3}q^7 + 2c^{-2}q^6 - 3c^{-1}q^5 \\ + 5q^4 + 3cq^3 + 2c^2q^2 + c^3q + c^4, \quad (15)$$

and thus

$$\sum_{n\geq 0} p(n)q^n = \frac{(q^{25};q^{25})^5}{(q^5;q^5)^6} f(q).$$

Now look at the coefficients of powers q^{5n+4}. In $(q^{25};q^{25})$ and $(q^5;q^5)$ all exponents are multiples of 5, and this is also true for $R(q^5)$ and $R(q^5)^{-1}$. So the exponents in the expression (15) of $f(q)$ are all incongruent to 4 (modulo 5) except for the term $5q^4$, and we conclude that

$$\sum_{n\geq 0} p(5n+4)q^{5n+4} = \frac{(q^{25};q^{25})^5}{(q^5;q^5)^6} 5q^4 ,$$

that is,

$$\sum_{n\geq 0} p(5n+4)q^{5n} = 5\frac{(q^{25};q^{25})^5}{(q^5;q^5)^6} .$$

Now replace q by $q^{1/5}$, and out comes Ramanujan's formula

$$\sum_{n\geq 0} p(5n+4)q^{n} = 5\frac{(q^{5};q^{5})^5}{(q;q)^6} . \quad (16)$$

We may express (16) also as follows. Since

$$\begin{aligned} \frac{(q^5;q^5)}{(q;q)} &= \frac{1}{(q,q^2,q^3,q^4;q^5)} \\ &= \prod_{k\geq 1} \frac{1}{(1-q^{5k-1})(1-q^{5k-2})(1-q^{5k-3})(1-q^{5k-4})}, \end{aligned}$$

we obtain

$$\sum_{n\geq 0} p(5n+4)q^n = 5\left(\sum_{n\geq 0} p(n)q^n\right)\left(\sum_{n\geq 0} s(n)q^n\right)^5 ,$$

where $s(n)$ is the number of partitions of n with no part divisible by 5.

Notes and References

Solving recurrences and computing sums via generating functions belongs again to the classical part of enumerative combinatorics. The books by Goulden–Jackson and Wilf contain a wealth of examples. The composition formula has many origins; we follow here the approach taken by Foata and Schützenberger. For an extensive treatment see Chapter 5 in Stanley's book. A more general setup, called the theory of species, was proposed by Joyal. Cayley's formula on the number of trees was anticipated by Borchardt, and was known even earlier to Sylvester. See the book by Aigner and Ziegler for four more proofs and historical references. Andrews is the authoritative source for the theory of number-partitions. There you find references to all the important results such as Euler's pentagonal theorem or Jacobi's triple product theorem. For connections to additive number theory and the many contributions of Ramanujan the book of Hardy–Wright is recommended. The proof of the highlight follows the paper by Hirschhorn.

1. M. Aigner and G. Ziegler (2003): *Proofs from THE BOOK*, 3rd edition. Springer, Berlin.
2. G.E. Andrews (1976): *The Theory of Partitions.* Addison-Wesley, Reading.
3. D. Foata and M. Schützenberger (1970): *Théorie Géometrique des Polynômes Euleriens.* Lecture Notes Math. 138. Springer, Berlin.
4. I.P. Goulden and D.M. Jackson (1983): *Combinatorial Enumeration.* Wiley, New York.
5. G.H. Hardy and E.M. Wright (1954): *An Introduction to the Theory of Numbers*, 3rd edition. Oxford Univ. Press, London.
6. M.D. Hirschhorn (2000): An identity of Ramanujan, and applications. In: *q-Series From a Contemporary Perspective.* Contemporary Mathematics, vol. 254, 229–234. American Math. Society.
7. A. Joyal (1981): Une théorie combinatoire des séries formelles. *Advances Math.* 42, 1–82.
8. R.P. Stanley (1999): *Enumerative Combinatorics*, vol. 2. Cambridge Univ. Press, Cambridge/New York.
9. H.S. Wilf (1994): *generatigfunctionology*, 2nd edition. Academic Press, San Diego.

4 Hypergeometric Summation

We have seen many sums of the form $s(n) = \sum_k a_{n,k}$ such as $\sum_k \binom{r}{k}\binom{s}{n-k} = \binom{r+s}{n}$ or $\sum_k \binom{n+k}{m+2k}\binom{2k}{k}\frac{(-1)^k}{k+1} = \binom{n-1}{m-1}$. In this chapter we want to review some techniques that permit an almost automatic evaluation of sums like these, and we also study when such a sum can be presented in closed form at all. It is clear that when the sums get more complicated, doing them by hand becomes out of the question. So the emphasis lies on algorithms that can be and have been implemented in computer packages.

4.1 Summation by Elimination

Suppose we are given numbers $F(n,k)$ with $n,k \in \mathbb{Z}$, and we want to compute the sum $s(n) = \sum_k F(n,k)$. Of course, this makes sense only when there is only a finite number of nonzero parts $F(n,k)$ involved, that is, the sum extends over a finite range $\sum_{k=-m}^{M} F(n,k)$ with $m, M \geq 0$. We then say that $F(n,k)$ becomes *eventually* 0 as $k \to \pm\infty$, and denote this fact by $\lim_{k\to\pm\infty} F(n,k) = 0$. As an example, for an integer n, $\lim_{k\to\pm\infty} \binom{n}{k} = 0$ holds precisely when n is nonnegative.

Using Operators.
The key idea in this section is to rewrite recurrence relations in operator notation. Take, for example, the recurrence $a(n+2) = 3a(n+1) - (n-1)a(n)$. With the operator $N\colon a(n) \to a(n+1)$ this translates into the equation $(N^2 - 3N + (n-1)I)a(n) = 0$, where I is the identity operator. Now we introduce operators for both parameters n and k.

Let N and K be the linear shift operators $N : F(n,k) \to F(n+1,k)$ and $K : F(n,k) \to F(n,k+1)$, and define N^i, K^i $(i \in \mathbb{Z})$ as shifts by i. We extend this definition by linearity to all (finite) expressions of the form

$$\sum_{i,j\in\mathbb{Z}} p_{i,j}(n,k)N^iK^j\,, \tag{1}$$

where all $p_{i,j}(n,k)$ are *polynomials* in n and k.

Example. We have $(N^2K - 2nNK^2)F(n,k) = F(n+2,k+1) - 2nF(n+1,k+2)$.

The operators in (1) thus form a (noncommutative) ring generated by N, K, n, k, observing the relations

$$NK = KN,\quad kN = Nk,\quad nK = Kn,\quad Nn = (n+1)N,$$
$$Kk = (k+1)K,\quad nk = kn\,.$$

As in our example above, operators $P(N,K,n,k)$ that satisfy $P(N,K,n,k)F(n,k) = 0$ for all n and k give rise to recurrences for $F(n,k)$, and we call them accordingly *recurrence operators.*

Example. The Pascal recurrence $\binom{n+1}{k+1} = \binom{n}{k} + \binom{n}{k+1}$ for $F(n,k) = \binom{n}{k}$ corresponds to the operator equation

$$NK\binom{n}{k} = (I+K)\binom{n}{k};$$

hence $P = NK - I - K$ is a recurrence operator for $\binom{n}{k}$.

If P is a recurrence operator, then so is every *left* multiple AP, since $AP(F(n,k)) = A(PF(n,k)) = 0$. Here is the main idea. If $P_1,\ldots,P_t$ are recurrence operators, then so is every combination $A_1P_1 + \cdots + A_tP_t$. Suppose we succeed by an appropriate choice of the A_i to eliminate k, that is,

$$R(N,K,n) = \sum_{i=1}^{t} A_i(N,K,n,k)P_i(N,K,n,k)$$

no longer depends on k. Setting $S(N,n) = R(N,I,n)$, we obtain by polynomial division $K^j - I = (K-I)(K^{j-1} + \cdots + I)$,

$$R(N,K,n) - S(N,n) = (K-I)\overline{R}(N,K,n)\,.$$

With $C = -\overline{R}$ this gives

$$S(N,n) = \sum_{i=1}^{t} A_iP_i + (K-I)C\,. \tag{2}$$

Claim. *$S(N,n)$ is a recurrence operator for the sum $s(n) = \sum_k F(n,k)$.* To prove this we set $G(n,k) = C(N,K,n)F(n,k)$ and infer from (2),

$$S(N,n)F(n,k) = (K-I)G(n,k) = G(n,k+1) - G(n,k),$$

and hence by linearity

$$S(N,n)\sum_k F(n,k) = \sum_k (G(n,k+1) - G(n,k)) = 0,$$

provided that $\lim_{k\to\pm\infty} G(n,k) = 0$.

Notice that the same conclusion holds when in (2) the operator $C(N,K,n,k)$ also depends on k. But keep in mind that we multiply the recurrence operators P_i from the *left*, and $K-I$ by C on the *right*. Let us state this as our first result.

Proposition 4.1. *Let P_i be recurrence operators for $F(n,k)$. If*

$$S(N,n) = \sum_{i=1}^{t} A_i(N,K,n,k)P_i(N,K,n,k) + (K-I)C(N,K,n,k)$$

does not depend on K, k, then $S(N,n)$ is a recurrence operator for the sum $s(n) = \sum_k F(n,k)$, whenever $\lim_{k\to\pm\infty} C(N,K,n,k)F(n,k) = 0$.

Example. The simplest case arises when there is a single recurrence operator P that does not depend on k. For the binomial coefficients $F(n,k) = \binom{n}{k}$, we have the recurrence operator from above

$$P = NK - I - K = (K-I)(N-I) + N - 2I;$$

hence $N - 2I$ is a recurrence operator for $s(n) = \sum_k \binom{n}{k}$. In other words, $s(n+1) = 2s(n)$, and we conclude that $s(n) = 2^n$ since $s(0) = 1$. Note that in this case, $G(n,k) = (I-N)\binom{n}{k} = \binom{n}{k} - \binom{n+1}{k} = -\binom{n}{k-1}$, which becomes eventually 0 for $n \geq 0$, so the result $s(n) = 2^n$ is justified for $n \geq 0$.

In most cases a single recurrence operator will not do, so we try to find two such operators P and Q (and under certain technical conditions two will suffice), eliminate k in $AP + BQ$, calculate "mod $K-I$," and read off the recurrence for $s(n)$.

Example. Consider $\sum_k \binom{n}{k}\binom{m}{k}$, $m \in \mathbb{Z}$ fixed.
With $F(n,k) = \binom{n}{k}\binom{m}{k}$, we have

$$\frac{F(n+1,k)}{F(n,k)} = \frac{n+1}{n-k+1}, \quad \frac{F(n,k+1)}{F(n,k)} = \frac{(n-k)(m-k)}{(k+1)^2}.$$

By cross-multiplication we obtain

$$(n-k+1)F(n+1,k) - (n+1)F(n,k) = 0,$$
$$(k+1)^2F(n,k+1) - (n-k)(m-k)F(n,k) = 0,$$

and thus the recurrence operators (we omit I)

$$P = (n-k+1)N - (n+1), \; Q = (k+1)^2K - (n-k)(m-k).$$

Ordered according to powers of k this gives

$$P = -kN + (n+1)(N-I), \; Q = (K-I)k^2 + k(m+n) - mn.$$

We can now eliminate k by multiplying P on the left by $m+n+1$, and Q by N. Mod $K-I$ this gives

$$S(N,n) = (m+n+1)(n+1)(N-I) - m(n+1)N.$$

The sum $s(n)$ therefore satisfies the recurrence (dividing by $n+1$)

$$(n+1)s(n+1) = (m+n+1)s(n),$$

that is,

$$s(n) = \frac{m+n}{n}s(n-1).$$

With $s(0) = 1$ this yields

$$s(n) = \binom{m+n}{n} = \sum_k \binom{n}{k}\binom{m}{k},$$

our old Vandermonde convolution.

Example. This is all familiar and easy, but here is a mechanical proof of one of the most famous formulas involving binomial coefficients which no longer looks so easy, *Dixon's identity*:

$$\sum_k (-1)^k \binom{n+m}{n+k}\binom{m+p}{m+k}\binom{p+n}{p+k} = \frac{(n+m+p)!}{n!m!p!},$$

$$m, p \in \mathbb{N}_0 \text{ fixed.} \quad (3)$$

We proceed as before with $F(n,k)$ being the summand.

Cross-multiplication gives

$$\frac{F(n+1,k)}{F(n,k)} = \frac{n+1+m}{n+1+k} \cdot \frac{n+1+p}{n+1-k},$$
$$\frac{F(n,k+1)}{F(n,k)} = -\frac{m-k}{n+k+1} \cdot \frac{p-k}{m+k+1} \cdot \frac{n-k}{p+k+1},$$

and hence the recurrence operators

$$\begin{aligned} P &= N(n^2-k^2) - (n+1+m)(n+1+p) \\ &= -k^2N + Nn^2 - (n+1+m)(n+1+p), \end{aligned}$$

$$\begin{aligned} Q &= K(n+k)(m+k)(p+k) + (m-k)(n-k)(p-k) \\ &= 2k^2(n+m+p) + 2mnp \pmod{K-I}, \end{aligned}$$

or dividing by 2,

$$Q = k^2(n+m+p) + mnp \pmod{K-I}.$$

To eliminate k we multiply P on the left by $n+m+p+1$, Q by N, and obtain with a short computation

$$S(N,n) = (n+1)N - (n+m+p+1).$$

Hence $(n+1)s(n+1) = (n+m+p+1)s(n)$, or

$$s(n) = \frac{n+m+p}{n}s(n-1),$$

and this yields with $s(0) = \binom{m+p}{m}$,

$$s(n) = \binom{n+m+p}{n}\binom{m+p}{m} = \frac{(n+m+p)!}{n!m!p!}.$$

Exercises

4.1 Compute $\sum_k \binom{n-k}{k}$ by elimination, $n \geq 0$. Notice that for fixed n this is not a finite sum, so modify the definition of the summand $F(n,k)$.

4.2 Translate the recurrence for the Stirling numbers $s_{n,k}$ into an operator equation, and verify $\sum_{k=0}^{n} s_{n,k} = n!$.

▷ **4.3** Generalize the previous exercise to the following situation: Suppose we have the recurrence $F(n+1,k+1) = f(n)F(n,k) + g(n)F(n,k+1)$ with $F(n,k) = 0$ for $n < k$ and for $k < 0$. Then

$$s(n) = \sum_{k=0}^{n} F(n,k) = a\left[\prod_{i=0}^{n-1}(f(i)+g(i))\right], \quad a = F(0,0).$$

* * *

4.4 Compute $\sum_k (-1)^k \binom{n+m}{n+k}\binom{n+m}{m+k}$, $m \in \mathbb{N}_0$ fixed.

▷ **4.5** Show that $\sum_k \frac{(-1)^k}{m+k}\binom{n}{k} = \frac{1}{m}\binom{m+n}{n}^{-1}$, $m \geq 1$ fixed.

4.6 Evaluate $\sum_{k=0}^{n} k\binom{m-k-1}{m-n-1}$, $m > n \geq 0$, by elimination.

4.7 Find $\sum_{k=0}^{n}(-4)^k\binom{n+k}{2k}$, $n \in \mathbb{N}_0$.

▷ **4.8** Compute $s(n) = \sum_k (-1)^k \binom{2n}{k}^3$, which gives another identity of Dixon.

4.2 Indefinite Sums and Closed Forms

We now address the question of what we mean that a sequence $(a(n))$ can be "nicely" presented. The simplest case arises when $\frac{a(n)}{a(n-1)}$ is a constant c for all n. Such sequences are called *geometric* since $a(n) = c^n a(0)$.

Definition. A sequence $(a(n))_{n \geq 0}$ is said to be *hypergeometric* or a *closed form* (*CF*) if $\frac{a(n)}{a(n-1)} = \frac{p(n)}{q(n)}$ is a rational function in n.

Hence $(a(n))$ is a *CF* if $q(n)a(n) - p(n)a(n-1) = 0$, or in the language of the previous section, when $q(n)N - p(n)$ is a recurrence operator of first order.

To every sequence $(a(n))$ there corresponds an *indefinite* sum $S(n)$ with $a(n) = S(n+1) - S(n)$. Gosper's remarkable algorithm completely answers the following question.

Question. *Given a CF $(a(n))$, when is the indefinite sum sequence $(S(n))$ also a closed form?*

Remark. If $(S(n))$ is a *CF*, then also $(a(n))$, since with $a(n) = S(n+1) - S(n)$ we obtain

$$\frac{a(n)}{a(n-1)} = \frac{S(n+1) - S(n)}{S(n) - S(n-1)} = \frac{\frac{S(n+1)}{S(n)} - 1}{1 - \frac{S(n-1)}{S(n)}} = \text{rational function.}$$

To get a feeling for the main idea, let us work backward. The simplest *CF* that is not geometric is $S(n) = n!$ with $\frac{S(n)}{S(n-1)} = n$. Here $a(n) = S(n+1) - S(n) = n \cdot n!$. Now we have

$$\frac{a(n)}{a(n-1)} = \frac{n}{n-1} \cdot \frac{n}{1} = \frac{p(n)}{p(n-1)} \cdot \frac{q(n)}{r(n)}\,.$$

The first factor is called the "polynomial" part, and the second $\frac{q(n)}{r(n)}$ the "factorial" part. Note that

$$S(n+1) = (n+1)! = \frac{a(n)(n+1)}{n} = \frac{a(n)q(n+1)}{p(n)}\,.$$

Let's make it a bit more complicated. Consider $S(n) = (n+2)n!$, $a(n) = S(n+1) - S(n) = (n+3)(n+1)! - (n+2)n! = (n^2+3n+1)n!$, so $p(n) = n^2 + 3n + 1$,

$$\frac{a(n)}{a(n-1)} = \frac{p(n)}{p(n-1)} \cdot \frac{n}{1}\,;$$

hence $q(n) = n$, $r(n) = 1$. For $S(n+1)$ this gives

$$S(n+1) = (n+3)(n+1)! = n!(n+1)(n+3) = \frac{a(n)q(n+1)}{p(n)} f(n)\,,$$

where $f(n) = n + 3$.

Gosper's Algorithm.
Gosper's amazing discovery states that if $(S(n))$ is a *CF*, then it can always be written in this form, where $p(n), q(n), f(n)$ are polynomials determined by the sequence $(a(n))$. The equality then also *computes* the indefinite sum $S(n+1)$.

Let us carefully go through the stages of the algorithm. Let $(a(n))$ be a closed form, $a(n) = S(n+1) - S(n)$.

Step 1. Set

$$\frac{a(n)}{a(n-1)} = \frac{p(n)}{p(n-1)} \cdot \frac{q(n)}{r(n)}, \quad \gcd(q(n), r(n)) = 1, \tag{1}$$

where $p(n), q(n), r(n)$ are polynomials in n, and $p(n)$ is a polynomial of *maximal* degree such that $\frac{a(n)}{a(n-1)}$ can be written in the form (1).

Lemma 4.2. *The maximality of the degree of* $p(n)$ *implies* $\gcd(q(n), r(n+j)) = 1$ *for all* $j \geq 0$.

Proof. Suppose to the contrary that $g(n) = \gcd(q(n), r(n+j)) \neq 1$ for some $j > 0$. Then $g(n-j) \mid r(n)$ and

$$\frac{q(n)}{r(n)} = \frac{g(n)}{g(n-1)} \cdot \frac{g(n-1)}{g(n-2)} \cdots \frac{g(n-j+1)}{g(n-j)} \frac{q^*(n)}{r^*(n)}$$

with $q(n) = g(n)q^*(n), r(n) = g(n-j)r^*(n)$. Hence we may move $g(n)g(n-1)\cdots g(n-j+1)$ into the polynomial part, contradicting the maximality of the degree. □

Note that the lemma gives a recipe for how to find $p(n)$ of maximal degree. Start with $p(n) = 1$ or some obvious $p(n)$ as in the examples. If the second factor contains $q(n)$ and $r(n)$ with $\gcd(q(n), r(n+j)) \neq 1$ for some $j \geq 0$, apply the procedure of the lemma and continue.

Step 2. Set

$$S(n+1) = \frac{a(n)q(n+1)}{p(n)} f(n). \tag{2}$$

Of course, we can always do this, since everything apart from $f(n)$ is known.

Lemma 4.3. $(S(n))$ *is a CF if and only if* $f(n)$ *is a rational function.*

Proof. If $f(n)$ is a quotient of polynomials, then $(S(n))$ is clearly a closed form since $(a(n))$ is. Suppose, conversely, that $(S(n))$ is a *CF.* Then

$$f(n) = \frac{p(n)S(n+1)}{q(n+1)a(n)} = \frac{p(n)S(n+1)}{q(n+1)(S(n+1)-S(n))}$$
$$= \frac{p(n)}{q(n+1)(1-\frac{S(n)}{S(n+1)})},$$

thus $f(n)$ is rational. □

Step 3. *$f(n)$ satisfies the functional equation*

$$q(n+1)f(n) - r(n)f(n-1) = p(n). \tag{3}$$

Indeed, we have

$$\begin{aligned} a(n) &= S(n+1) - S(n) \\ &= \frac{a(n)q(n+1)}{p(n)}f(n) - \frac{a(n-1)q(n)}{p(n-1)}f(n-1) \\ &\overset{(1)}{=} a(n)\left(\frac{q(n+1)f(n)}{p(n)} - \frac{p(n-1)r(n)q(n)}{p(n-1)q(n)p(n)}f(n-1)\right) \\ &= a(n)\left(\frac{q(n+1)f(n)}{p(n)} - \frac{r(n)f(n-1)}{p(n)}\right), \end{aligned}$$

and (3) follows. □

What makes the algorithm work is the surprising fact that not only is f rational, it is a *polynomial* in n.

Lemma 4.4. *A rational function $f(n)$ that satisfies (3) (under the hypothesis (1)) is a polynomial.*

Proof. Let $f(n) = \frac{c(n)}{d(n)}$, gcd $(c(n), d(n)) = 1$. In order to prove $d(n) = 1$, we will show more generally that gcd $(d(n), d(n+k)) = 1$ for *all* $k \geq 0$; the case $k = 0$ will then prove $d(n) = 1$. Suppose this is false, and let $j \geq 0$ be maximal with

$$\gcd(d(n), d(n+j)) = g(n) \neq 1. \tag{4}$$

If $d(n) \neq 1$, then such an index j exists. Indeed, if α is a root of $g(n)$, then α and $\beta = \alpha + j$ are roots of $d(n)$. Hence, if $j > \max(\beta - \alpha : \alpha, \beta$ roots of $d(n))$, then (4) cannot be satisfied. It follows from the definition of j that

$$\gcd(d(n), d(n+j+1)) = 1. \tag{5}$$

With $f(n) = \frac{c(n)}{d(n)}$ equation (3) can be written as

$$q(n+1)c(n)d(n-1) - r(n)c(n-1)d(n) = p(n)d(n)d(n-1)\,. \quad (6)$$

We derive the desired contradiction by showing that

A. $g(n-1) \mid q(n)$,
B. $g(n-1) \mid r(n+j)$.

Lemma 4.2 implies then $g(n) = 1$, and we are through.

Proof of A. Let $h(n) = \gcd(g(n), d(n-1))$. Then we have $h(n) \mid g(n)$, hence $h(n+1) \mid g(n+1) \mid d(n+j+1)$ and $h(n+1) \mid d(n)$. Thus $h(n) = 1$ by (5). Looking at equation (6), we infer from $g(n) \mid d(n)$ that $g(n) \mid q(n+1)$ holds (since $g(n)$ is relatively prime to $d(n-1)$ and $c(n)$), and therefore $g(n-1) \mid q(n)$.

Proof of B. Let $k(n) = \gcd(g(n-j-1), d(n))$. Then we have $k(n+j+1) \mid g(n) \mid d(n)$ and also $k(n+j+1) \mid d(n+j+1)$, and so $k(n) = 1$ by (5). Furthermore, $g(n) \mid d(n+j)$ implies $g(n-j-1) \mid d(n-1)$. Looking again at equation (6), this gives $g(n-j-1) \mid r(n)c(n-1)d(n)$. Since $g(n-j-1)$ and $d(n)$ are relatively prime, we have $g(n-j-1) \mid r(n)c(n-1)$. Since $g(n-j-1) \mid d(n-1)$, and $d(n-1), c(n-1)$ are relatively prime, we have, in fact, $g(n-j-1) \mid r(n)$ or $g(n-1) \mid r(n+j)$. This proves B, and thus the lemma. □

Now we are all set for the algorithm of Gosper.

Input. A closed form $(a(n))$.

(I) Set $\frac{a(n)}{a(n-1)} = \frac{p(n)}{p(n-1)}\frac{q(n)}{r(n)}$, $p(n)$ of maximal degree.

(II) Write $f(n) = \sum_{i=0}^{L} f_i n^i$ in indeterminates f_i, and solve the equation

$$q(n+1)f(n) - r(n)f(n-1) = p(n)\,.$$

If no solution exists, then $(S(n))$ is not a closed form.

(III) Otherwise,

$$S(n+1) = \frac{a(n)q(n+1)}{p(n)} f(n)$$

is the desired indefinite sum with $a(n) = S(n+1) - S(n)$. For the definite sum we thus obtain

$$\sum_{i=0}^{n} a(i) = S(n+1) - S(1) + a(0)$$

or

$$\sum_{i=r}^{n} a(i) = S(n+1) - S(r+1) + a(r)$$

for some starting index r.

Two remarks are in order:

1. In the proof of Lemma 4.4 we need that $q(n)$ and $r(n+j)$ are relatively prime for all $j \geq 0$. One way to check this is to use resultants, known from linear algebra.
2. How large is the degree L of $f(n)$? Equation (3) implies $L = \deg p(n) - \max(\deg q(n), \deg r(n))$, except when the highest coefficients of $q(n+1)$ and $r(n)$ cancel. Then a larger L may exist.

Now it's time for examples.

Example. $a(n) = (n^2+n-1)(n-1)!$, $n \geq 1$.

(I) $\frac{a(n)}{a(n-1)} = \frac{n^2+n-1}{(n-1)^2+(n-1)-1} \cdot \frac{n-1}{1}$, $q(n) = n-1$, $r(n) = 1$.

(II) $nf(n) - f(n-1) = n^2+n-1$; hence $\deg f(n) = 1$, $f(n) = f_0 + f_1 n$. Comparing coefficients gives the system of equations

$$\begin{aligned} n^0 &: -f_0 + f_1 = -1, \\ n^1 &: f_0 - f_1 = 1, \\ n^2 &: \phantom{-f_0 + {}} f_1 = 1, \end{aligned}$$

with solution $f_0 = 2$, $f_1 = 1$.

(III) $S(n+1) = (n-1)!n(n+2) = (n+2)n!$, and hence

$$\sum_{i=1}^{n} a(i) = (n+2)n! - 2.$$

Example. Is the sum $\sum_{i=0}^{n} i!$ a closed form? Here we have $a(n) = n!$, and thus

$$\frac{a(n)}{a(n-1)} = n,$$

$p(n) = 1$, $q(n) = n$, $r(n) = 1$. The functional equation is

$$(n+1)f(n) - f(n-1) = 1,$$

and this clearly has no solution. Hence $(S(n))$ is not a closed form.

Example. Let us now look at the partial sums of binomial coefficients. We have already raised the question in Section 1.2 whether the following partial sums are closed forms:

$$\sum_{i=0}^{n}(-1)^i\binom{m}{i} \quad \text{and} \quad \sum_{i=0}^{n}\binom{m}{i}, \quad m \geq 0 .$$

In the first case we find for $a(n) = (-1)^n\binom{m}{n}$

$$\frac{a(n)}{a(n-1)} = -\frac{m-n+1}{n} = \frac{n-m-1}{n},$$

$p(n) = 1$, $q(n) = n - m - 1$, $r(n) = n$. Note that $n - m - 1$ and $n + j$ are relatively prime for all $j \geq 0$. The equation is

$$(n-m)f(n) - nf(n-1) = 1 .$$

Here we encounter the special case that the highest coefficients of $q(n+1) = n-m$ and $r(n) = n$ cancel. Setting $f(n) = f_0$, we obtain as solution $f_0 = -\frac{1}{m}$, and thus

$$S(n+1) = (-1)^n\binom{m}{n}(n-m)\left(-\frac{1}{m}\right) = (-1)^n\binom{m-1}{n},$$

and with $S(1) = a(0) = 1$,

$$\sum_{i=0}^{n}(-1)^i\binom{m}{i} = (-1)^n\binom{m-1}{n} .$$

In the second case, $a(n) = \binom{m}{n}$,

$$\frac{a(n)}{a(n-1)} = \frac{m-n+1}{n},$$

$p(n) = 1$, $q(n) = m - n + 1$, $r(n) = n$,

$$(m-n)f(n) - nf(n-1) = 1 .$$

Since the highest coefficients of $q(n+1)$ and $r(n)$ do not cancel out, there is no solution, and $(S(n))$ is not a closed form.

Exercises

4.9 Is the sum $\sum_{i=0}^{n} C_i$ of the Catalan numbers a closed form?

4.10 Can the harmonic numbers $H_n = \sum_{i=1}^{n} \frac{1}{i}$ be written as a closed form?

4.11 Prove $\sum_{i=0}^{n} \binom{m+i}{i} = \binom{m+n+1}{n}$ using Gosper's algorithm.

▷ **4.12** Use Gosper's method to find $\sum_{i=2}^{n} \frac{1}{i^2-1}$. Verify your answer using $\frac{1}{i^2-1} = \frac{1}{2}(\frac{1}{i-1} - \frac{1}{i+1})$. What about $\sum_{i=2}^{n} \frac{i}{i^2-1}$? Is it still a closed form?

4.13 For which values of z is $\sum_{k=0}^{n} \binom{m}{k} z^k$ a closed form?

4.14 Prove that $\sum_{k=0}^{n} \frac{2^k}{k+1}$ is not a closed form.

* * *

▷ **4.15** Show that $a(n,k) = \sum_k \frac{\binom{n}{k}}{2^n}$ is not Gosper-summable in k, but that $a(n+1,k) - a(n,k)$ is.

4.16 Consider $\sum_{k=0}^{n} \binom{-m}{k}$, $m \geq 1$. What is the polynomial $p(n)$ in Gosper's algorithm? Evaluate the sum for $m \leq 4$.

4.17 Find a constant α for which $\sum_{i=0}^{n} \binom{m}{i}(i+\alpha)$ is a closed form, and evaluate the sum.

▷ **4.18** Find the unique value $a \in \mathbb{Q}$ such that $\sum_{k=0}^{n} \binom{2k}{k} a^k$ is a closed form, and compute it for this value a.

4.19 Evaluate the sum $\sum_{k=0}^{n} k^m 2^k$ for $m = 1, 2, 3$.

▷ **4.20** Evaluate $\sum_{k=1}^{n-1} \frac{k^2-2k-1}{k^2(k+1)^2} 2^k$, using Gosper's algorithm.

4.3 Recurrences for Hypergeometric Sums

Let us return to our original problem. Given $F(n,k)$, we want to compute the sum $s(n) = \sum_k F(n,k)$, or at least find a recurrence for $s(n)$. Gosper's algorithm finds the partial sums when $F(n,k)$ is a hypergeometric term in k, but in general, we want to sum over *all* k. We know for example that $\sum \binom{n}{k}$ is not indefinitely summable in the sense of Gosper, but of course $\sum_k \binom{n}{k} = 2^n$.

Zeilberger's Algorithm.
Zeilberger ingeniously extended Gosper's algorithm to handle the definite summation for a vast variety of cases. Suppose $F(n,k)$

is a hypergeometric term with respect to both n and k. We know from Section 4.1 that with $G(n,k) = C(N,K,n,k)F(n,k)$ for some $C(N,K,n,k)$,

$$S(N,n)F(n,k) = G(n,k+1) - G(n,k)\,, \tag{1}$$

where $S(N,n)$ is a recurrence operator for the sum $s(n)$. Suppose that $G(n,k)$ is a hypergeometric term in k. Then $G(n,k)$ is just the indefinite sum of $S(N,n)F(n,k)$, and can thus be computed via Gosper's algorithm. For precise conditions, when this is the case we refer to the book by Petovšek–Wilf–Zeilberger.

We proceed as follows. Set $H(n,k) = S(N,n)F(n,k)$, where the operator is written $S(N,n) = \sum_{j=0}^{M} s_j(n)N^j$, with $s_j(n)$ polynomials in n. We "guess" the degree M of $S(N,n)$, or try $M = 1, 2, \ldots$. Since $G(n,k)$ is a closed form, some degree must eventually work.

The algorithm works as follows:

(I) Set $H(n,k) = S(N,n)F(n,k)$.

(II) Compute $p(k), q(k), r(k)$, where these are polynomials in k with polynomial coefficients in n.

(III) From $q(k+1)f(k) - r(k)f(k-1) = p(k)$, $f(k) = \sum_{i=0}^{L} f_i(n)k^i$ there results a system of equations for $f_0(n), \ldots, f_L(n)$ and $s_0(n), \ldots, s_M(n)$.

A solution of this system gives the recurrence operator $S(N,n)$ and at the same time the "certificate" $G(n,k)$ with

$$S(N,n)F(n,k) = G(n,k+1) - G(n,k)\,.$$

Example. Consider $s(n) = \sum_k \binom{n}{k}\binom{n+k}{k}(-1)^k$. Let us try all methods we have seen so far.

1. Elimination. $F(n,k) = \binom{n}{k}\binom{n+k}{k}(-1)^k$ gives

$$\frac{F(n+1,k)}{F(n,k)} = \frac{n+1+k}{n-k+1}, \text{ hence } P = N(n-k) - (n+1+k) = -(N+I)k + Nn - (n+1);$$

$$\frac{F(n,k+1)}{F(n,k)} = -\frac{(n+1+k)(n-k)}{(k+1)^2},$$

$$\text{hence } Q = Kk^2 + (n-k)(n+1+k) = -k + n^2 + n \pmod{K-I}\,.$$

With

$$S(N,n) = -P + (N+I)Q = -Nn + (n+1) + (N+I)(n^2+n)$$
$$= (n+1)^2N + (n+1)^2 \,,$$

we obtain the recurrence

$$s(n+1) = -s(n), \text{ and thus } s(n) = (-1)^n \,.$$

2. Gosper–Zeilberger. With $S(N,n) = s_0(n) + s_1(n)N$ we have

$$H(n,k) = s_0(n)\binom{n}{k}\binom{n+k}{k}(-1)^k$$
$$+ s_1(n)\binom{n+1}{k}\binom{n+1+k}{k}(-1)^k$$
$$= \frac{(n+k)!(-1)^k}{k!^2(n+1-k)!}\big(\underbrace{k(s_1-s_0)+(n+1)(s_0+s_1)}_{p(k)}\big)\,,$$

$$\frac{H(n,k)}{H(n,k-1)} = \frac{p(k)}{p(k-1)}\frac{(k+n)(k-2-n)}{k^2}\,,$$

$$q(k) = (k+n)(k-2-n), \quad r(k) = k^2 \,,$$

$$(k^2-(n+1)^2)f(k) - k^2f(k-1) = k(s_1-s_0) + (n+1)(s_0+s_1)\,.$$

This equation is satisfied with $s_0(n) = s_1(n) = n+1$, $f(k) = -2$, and we obtain

$$S(N,n) = (n+1) + (n+1)N\,,$$

and thus again $s(n+1) = -s(n)$, $s(n) = (-1)^n$.

The Gosper–Zeilberger algorithm seems to be more complicated, and it is a little distressing that Vandermonde does it in one line:

$$\sum_k \binom{n}{k}\binom{n+k}{k}(-1)^k = \sum_k \binom{n}{k}\binom{-n-1}{k} = \binom{-1}{n} = (-1)^n\,.$$

But consider now the following innocuous looking sum, where we just delete $(-1)^k$.

Example. $\sum_k \binom{n}{k}\binom{n+k}{k}$. The elimination method is messy, and Vandermonde does not help either. Let's try Gosper–Zeilberger with $S(N,n) = s_0 + s_1N + s_2N^2$. Here

$$\begin{aligned} H(n,k) &= s_0\binom{n}{k}\binom{n+k}{k} + s_1\binom{n+1}{k}\binom{n+1+k}{k} \\ &\quad + s_2\binom{n+2}{k}\binom{n+2+k}{k} \\ &= \frac{(n+k)!}{k!^2(n-k+2)!}p(k) \end{aligned}$$

with $p(k) = s_0(n-k+1)(n-k+2) + s_1(n+1+k)(n-k+2) + s_2(n+1+k)(n+2+k)$. Now we have

$$\frac{H(n,k)}{H(n,k-1)} = \frac{p(k)}{p(k-1)}\frac{(n+k)(n-k+3)}{k^2},$$

$$q(k) = (n+k)(n-k+3), \quad r(k) = k^2 .$$

The functional equation

$$(n+k+1)(n-k+2)f(k) - k^2 f(k-1) = p(k)$$

leads therefore to $\deg f = 0$, that is $f = f(n)$, and thus to the following system of equations:

$$\begin{aligned} k^0 : (n+1)(n+2)f &= (n+1)(n+2)(s_0+s_1+s_2) \\ &\Longrightarrow f = s_0+s_1+s_2, \\ k^1 : \qquad\qquad f &= s_0(-2n-3) + s_1 + s_2(2n+3) \\ &= (2n+3)(s_2-s_0) + s_1, \\ k^2 : \qquad\quad -2f &= s_0 - s_1 + s_2 . \end{aligned}$$

This is easily solved with $s_0 = n+1$, $s_1(n) = -6n-9$, $s_2(n) = n+2$, $f(n) = -4n-6$, and we obtain the recurrence operator

$$S(N,n) = (n+2)N^2 - (6n+9)N + (n+1) .$$

The sum $s(n)$ therefore satisfies the recurrence

$$ns(n) = (6n-3)s(n-1) - (n-1)s(n-2) = 0 \qquad (2)$$

with initial values $s(0) = 1$, $s(1) = 3$.

This recurrence relation should look familiar. The sum $\sum_k \binom{n}{k}\binom{n+k}{k}$ is just the central Delannoy number $D_{n,n}$, whose recurrence (2) was derived in Section 3.2 using generating functions. The advantage of the present method is that it is completely mechanical. It amounts to solving a system of linear equations, and we can safely leave this to a computer. And what's more, it tells us that no recurrence of first order exists for $D_{n,n}$ (since Gosper–Zeilberger returns "no solution"). The algorithm will find a recurrence of some order if one exists at all, and in most cases it has minimal order.

Example. The most involved sum that we will explicitly compute is the famous *identity of Pfaff–Saalschütz:*

$$\sum_k \binom{m-r+s}{k}\binom{n+r-s}{n-k}\binom{r+k}{m+n} = \binom{r}{m}\binom{s}{n} \qquad (m,n \in \mathbb{N}_0)\,. \tag{3}$$

We try $S(N,n) = s_0(n) + s_1(n)N$, and set

$$\begin{aligned} H(n,k) &= S(N,n)F(n,k) \\ &= s_0\binom{m-r+s}{k}\binom{n+r-s}{n-k}\binom{r+k}{m+n} \\ &\quad + s_1\binom{m-r+s}{k}\binom{n+1+r-s}{n+1-k}\binom{r+k}{m+n+1}. \end{aligned}$$

The usual method leads to (check it!)

$$\begin{aligned} p(k) &= s_0(n+1-k)(m+n+1) \\ &\qquad + s_1(n+1+r-s)(r+k-m-n)\,, \\ q(k) &= (n+2-k)(m-r+s-k+1)(r+k)\,, \\ r(k) &= k(r-s+k)(r+k-m-n)\,. \end{aligned}$$

The functional equation $q(k+1)f(k) - r(k)f(k-1) = p(k)$ has the solution $f(k) = 1$, $s_0(n) = s-n$, $s_1(n) = -(n+1)$. This gives the recurrence operator $S(N,n) = (n+1)N - (s-n)$, and thus

$$s(n) = \frac{s-n+1}{n}s(n-1)$$

for the desired sum. Hence $s(n) = \binom{s}{n}s(0)$, and with $s(0) = \binom{r}{m}$ we obtain the result $s(n) = \binom{r}{m}\binom{s}{n}$, as claimed.

Finite Summation.

The Gosper–Zeilberger algorithm can also be used to sum over a specified range, as in the following example.

Example. We want to compute $s_n(z) = \sum_{k=0}^{n} \binom{n+k}{k} z^k$. Elimination will not work, since $\binom{n+k}{k}$ does not become eventually 0. Let us try our new approach with $S(N, n) = t_0(n) + t_1(n)N$. We obtain

$$\begin{aligned} H(n,k) &= t_0 \binom{n+k}{k} z^k + t_1 \binom{n+1+k}{k} z^k \\ &= \binom{n+k}{k} \frac{z^k}{n+1} \underbrace{\left((n+1)t_0 + (n+1+k)t_1\right)}_{p(k)}, \end{aligned}$$

$$\frac{H(n,k)}{H(n,k-1)} = \frac{p(k)}{p(k-1)} \frac{n+k}{k} z, \; q(k) = (n+k)z, \; r(k) = k,$$

$$(n+k+1)zf(k) - kf(k-1) = (n+1)t_0 + (n+1+k)t_1 .$$

The functional equation clearly has the solution $f(k) = 1$, $t_0(n) = 1$, $t_1(n) = z - 1$. So, $S(N, n) = (z-1)N + I$ is a recurrence operator, leading to $S_n(z) = \frac{1}{1-z} S_{n-1}(z)$, and thus with $S_0(z) = \frac{1}{1-z}$ to $S_n(z) = \frac{1}{(1-z)^{n+1}}$, when we sum over *all* k. Of course, we have known this expression since Chapter 2, but now we want to compute the *finite* sum $s_n(z)$. Gosper–Zeilberger tells us that

$$\begin{aligned} H(n,k) &= S(N,n)\binom{n+k}{k} z^k = \binom{n+k}{k} z^k + (z-1)\binom{n+1+k}{k} z^k \\ &= G(n,k+1) - G(n,k), \end{aligned} \tag{4}$$

so let us compute the certificate $G(n, k+1)$. Proceeding as usual, we obtain

$$\begin{aligned} G(n,k+1) &= \frac{H(n,k)q(k+1)}{p(k)} f(k) = \binom{n+k}{k} z^{k+1} \frac{n+k+1}{n+1} \\ &= \binom{n+k+1}{k} z^{k+1} . \end{aligned}$$

Summing equation (4) from $k = 0$ to $k = n + 1$, we get

$$\begin{aligned} s_n(z) + \binom{2n+1}{n+1} z^{n+1} + (z-1)s_{n+1}(z) &= G(n, n+2) - G(n, 0) \\ &= \binom{2n+2}{n+1} z^{n+2}, \end{aligned}$$

hence

$$s_{n+1}(z) = \frac{1}{1-z}\left[s_n(z) + (1-2z)\binom{2n+1}{n}z^{n+1}\right],$$

or

$$s_n(z) = \frac{1}{1-z}\left[s_{n-1}(z) + (1-2z)\binom{2n-1}{n}z^{n}\right], \quad s_0(z) = 1. \quad (5)$$

We immediately see that $s_n(\frac{1}{2}) = 2s_{n-1}(\frac{1}{2})$, which produces the nice sum $\sum_{k=0}^{n}\binom{n+k}{k}2^{-k} = 2^n$. And we get a bonus out of (5). Multiplying (5) by $(1-z)^n$, we obtain

$$(1-z)^n s_n(z) = (1-z)^{n-1}s_{n-1}(z) + (1-z)^{n-1}(1-2z)\binom{2n-1}{n-1}z^n.$$

Unwrapping this recurrence for $(1-z)^n s_n(z)$, we obtain with some easy manipulations the even nicer identity

$$(1-z)^n\sum_{k=0}^{n}\binom{n+k}{k}z^k = 1 + \frac{1-2z}{2-2z}\sum_{k=1}^{n}\binom{2k}{k}(z(1-z))^k.$$

Exercises

4.21 Determine a recurrence for $\sum_{k=0}^{n}\frac{1}{k!(n-k)!}$ and compute the sum.

4.22 Do the same for $\sum_{k=0}^{n}\binom{n}{k}^2$.

▷ **4.23** Compute $\sum_k(-1)^k\binom{n-k}{k}2^{n-2k}$, i.e., the value of the Chebyshev polynomial at $x = 2$.

4.24 What recurrence does the Gosper-Zeilberger algorithm produce for $s(n) = \sum_k\binom{n}{2k}$?

* * *

▷ **4.25** Prove the identity

$$\sum_{k=0}^{2n}\binom{k}{n}z^k = (\frac{z}{1-z})^n\left[1 + z(1-2z)\sum_{j=0}^{n-1}\binom{2j+1}{j}(z(1-z))^j\right]$$

and derive $\sum_{k=0}^{2n}\binom{k}{n}(\frac{1}{2})^k = 1$ for all n. Hint: Use $F(n,k) = \binom{k}{n}z^k$.

4.26 Find a recurrence satisfied by $s(n) = \sum_k\frac{n!}{k!2^k(n-2k)!}$. The recurrence should look familiar; what does $s(n)$ count?

4.27 Re-prove Dixon's identity using the Gosper–Zeilberger method.

▷ **4.28** Prove that $\sum_k \binom{n-1}{k} n^{-k}(k+1)! = n$, and give a combinatorial argument. Hint: Multiply by n^{n-1} and consider mappings from $\{1,\dots,n\}$ into itself.

4.29 We have looked at the sum $\sum_{k=0}^{n} \binom{n}{k}^{-1}$ in Exercise 1.20. Re-prove the recurrence given there by Gosper–Zeilberger. Hint: Prove on the way $\frac{n+2}{\binom{n}{k}} - \frac{2n+2}{\binom{n+1}{k}} = \frac{n-k}{\binom{n}{k+1}} - \frac{n+1-k}{\binom{n}{k}}$ and sum over k.

4.30 Let $s(n) = \sum_{k\geq 0} \binom{2n-2k}{n-k}\binom{2k}{k}^2$. Derive the recurrence $n^3 s(n) = 8(2n-1)(2n^2-2n+1)s(n-1) - 256(n-1)^3 s(n-2)$.

4.4 Hypergeometric Series

We have assembled a large repertoire of summation methods, and they work almost mechanically, especially when binomial coefficients are involved. And some formulas, like the Vandermonde identity, seem to crop up every other time, allowing us to simplify a complicated expression. We discuss now a general framework in which all summations with rational quotients of consecutive terms, i.e., hypergeometric sequences, will find their place.

Definition. A general *hypergeometric series* is a power series in the variable z with $m+n$ parameters $a_1,\dots;a_m, b_1,\dots,b_n$,

$$F\begin{pmatrix} a_1,\dots,a_m \\ b_1,\dots,b_n \end{pmatrix} ; z\Big) = \sum_{k\geq 0} \frac{a_1^{\overline{k}}\cdots a_m^{\overline{k}}}{b_1^{\overline{k}}\cdots b_n^{\overline{k}}} \frac{z^k}{k!}, \tag{1}$$

provided that the denominator is nonzero for all k.

Examples.

a. When $m = n = 0$, then $F(\,;z) = \sum_{k\geq 0} \frac{z^k}{k!} = e^z$. We could, of course, also write $F({1 \atop 1};z) = e^z$.

b. $m=1$, $n=0$. $F({1 \atop };z) = F({1,1 \atop 1};z) = \sum_{k\geq 0} z^k = \frac{1}{1-z}$, and in general $F({a,1 \atop 1};z) = \sum_{k\geq 0} a^{\overline{k}} \frac{z^k}{k!} = \sum_{k\geq 0}\binom{a+k-1}{k} z^k = \frac{1}{(1-z)^a}$.

c. $m=0$, $n=1$. $F({ \atop 1};z) = F({1 \atop 1,1};z) = \sum_{k\geq 0}\frac{z^k}{k!^2}$.

d. $m=2$, $n=1$. This is the most famous hypergeometric series studied since Euler's time: $F({a,b \atop c};z) = \sum_{k\geq 0}\frac{a^{\overline{k}} b^{\overline{k}}}{c^{\overline{k}}}\frac{z^k}{k!}$.

As an example,

$$zF\left(\begin{matrix}1,1\\2\end{matrix};-z\right) = z\sum_{k\geq 0}\frac{(-z)^k}{k+1} = \sum_{k\geq 1}\frac{(-1)^{k-1}}{k}z^k = \log(1+z).$$

The Hypergeometric Setup.
Why do we call them hypergeometric series? Let t_k be the k-th summand of $F(z)$; then

$$\begin{aligned}\frac{t_{k+1}}{t_k} &= \frac{a_1^{\overline{k+1}}\cdots a_m^{\overline{k+1}}}{a_1^{\overline{k}}\cdots a_m^{\overline{k}}}\cdot\frac{b_1^{\overline{k}}\cdots b_n^{\overline{k}}}{b_1^{\overline{k+1}}\cdots b_n^{\overline{k+1}}}\frac{k!}{(k+1)!}\frac{z^{k+1}}{z^k}\\ &= \frac{(k+a_1)(k+a_2)\cdots(k+a_m)}{(k+b_1)(k+b_2)\cdots(k+b_n)}\frac{z}{k+1}.\end{aligned}$$

The quotient t_{k+1}/t_k is thus a *rational* function in k. Now, every rational function $p(k)/q(k)$ over $\mathbb{C}$ can be decomposed into linear factors in the numerator and denominator. The $-a_i$'s are the roots of $p(k)$, and the $-b_j$'s the roots of $q(k)$. If $k+1$ does not appear in the denominator, then we add it to both the numerator and denominator. There remains a constant, which shall be z. So the expression (1) captures all series with hypergeometric terms in k.

Example. Let $\frac{t_{k+1}}{t_k} = \frac{k^2+7k+10}{4k^2-1}$. We have

$$\frac{t_{k+1}}{t_k} = \frac{(k+2)(k+5)(k+1)}{(k+\frac{1}{2})(k-\frac{1}{2})(k+1)}\left(\frac{1}{4}\right),$$

and obtain

$$\sum_{k\geq 0} t_k = t_0 F\left(\begin{matrix}2,5,1\\\frac{1}{2},-\frac{1}{2}\end{matrix};\frac{1}{4}\right).$$

A note of caution: We must make sure that the denominators are all nonzero. However, if this is not the case, one can often proceed with some limiting process.

Example. Vandermonde convolution:

$$\sum_{k\geq 0}\binom{r}{k}\binom{s}{n-k} = \binom{r+s}{n} \quad (n\in\mathbb{N}_0). \tag{2}$$

With $t_k = \binom{r}{k}\binom{s}{n-k}$ we obtain

$$\frac{t_{k+1}}{t_k} = \frac{r-k}{k+1} \cdot \frac{n-k}{s-n+k+1} = \frac{(k-r)(k-n)}{(k+s-n+1)(k+1)}, \quad t_0 = \binom{s}{n}.$$

Interpreted as a hypergeometric series (1) this becomes

$$\binom{s}{n} F\left(\begin{matrix}-r,-n\\ s-n+1\end{matrix};1\right) = \binom{r+s}{n} \quad (n \in \mathbb{N}_0).$$

With $a = -r$, $c = s - n + 1$ we can rewrite this as

$$F\left(\begin{matrix}a,-n\\ c\end{matrix};1\right) = \frac{\binom{-a+c+n-1}{n}}{\binom{c+n-1}{n}} = \frac{\binom{a-c}{n}}{\binom{-c}{n}},$$

obtaining the formula

$$F\left(\begin{matrix}a,-n\\ c\end{matrix};1\right) = \frac{(a-c)^{\underline{n}}}{(-c)^{\underline{n}}} = \frac{(c-a)^{\overline{n}}}{c^{\overline{n}}} \quad (n \in \mathbb{N}_0). \tag{3}$$

Specializations of (3) lead to a multitude of summation formulas that have their root in the Vandermonde identity. All we have to make sure is that the second parameter on top is a nonpositive integer.

Examples. Consider $\sum_k \binom{n}{k}/\binom{m}{k}$, $m \geq n \geq 0$. We have

$$\frac{t_{k+1}}{t_k} = \frac{n-k}{k+1} \cdot \frac{k+1}{m-k} = \frac{(k-n)(k+1)}{(k-m)(k+1)}, \quad t_0 = 1;$$

hence

$$\sum_{k \geq 0} \binom{n}{k} \Big/ \binom{m}{k} = F\left(\begin{matrix}1,-n\\ -m\end{matrix};1\right) = \frac{(m+1)^{\underline{n}}}{m^{\underline{n}}} = \frac{m+1}{m-n+1}.$$

Let us next look at the sum $\sum_k \binom{n+k}{2k}\binom{2k}{k}\frac{(-1)^k}{k+1}$ $(m, n \in \mathbb{N}_0)$, which we have computed in Section 3.2 by a rather lengthy argument via generating functions. We quickly obtain

$$\begin{aligned}\frac{t_{k+1}}{t_k} &= \frac{(n+k+1)(n-k)}{(2k+2)(2k+1)} \cdot \frac{(2k+2)(2k+1)}{(k+1)(k+1)} \cdot \frac{k+1}{k+2}(-1)\\ &= \frac{(k+n+1)(k-n)}{(k+2)(k+1)}, \quad t_0 = 1;\end{aligned}$$

hence

$$\sum_{k\geq 0}\binom{n+k}{2k}\binom{2k}{k}\frac{(-1)^k}{k+1} = F\left({n+1,-n \atop 2};1\right) = \frac{(1-n)^{\overline{n}}}{2^{\overline{n}}}$$
$$= \begin{cases} 1 & n=0, \\ 0 & n>0. \end{cases} \tag{4}$$

Although the sum looks as if the Catalan numbers C_k were involved, it is really Vandermonde in disguise.

The General Program.

The general approach is now clear: Take any (hypergeometric) summation formula, write it as a hypergeometric series, and specialize.

One particularly interesting candidate is the Pfaff-Saalschütz identity proved in the previous section:

$$\sum_k \binom{m-r+s}{k}\binom{n+r-s}{n-k}\binom{r+k}{m+n} = \binom{r}{m}\binom{s}{n} \qquad (m,n\in\mathbb{N}_0).$$

In hypergeometric terms this becomes

$$\frac{t_{k+1}}{t_k} = \frac{m-r+s-k}{k+1}\cdot\frac{n-k}{r-s+k+1}\cdot\frac{r+k+1}{r+k-m-n+1}$$
$$= \frac{(k-m+r-s)(k-n)(k+r+1)}{(k+r-s+1)(k+r-m-n+1)(k+1)},$$
$$t_0 = \binom{n+r-s}{n}\binom{r}{m+n}.$$

Hence

$$F\left({r-m-s,\ r+1,-n \atop r+1-s,\ r-m-n+1};1\right) = \frac{\binom{r}{m}\binom{s}{n}}{\binom{r}{m+n}\binom{n+r-s}{n}} = \frac{s^{\underline{n}}\binom{m+n}{n}}{(r-m)^{\underline{n}}\binom{r-s+n}{n}}$$
$$= \frac{s^{\underline{n}}}{(r-m)^{\underline{n}}}\cdot\frac{(-m-1)^{\underline{n}}}{(s-r-1)^{\underline{n}}}.$$

If we set $a=r-m-s$, $b=r+1$, $c=r+1-s$, then we get on the right-hand side $s=b-c$, $-m-1=a-c$, $r-m=a+b-c$, $s-r-1=-c$. This gives the Pfaff-Saalschütz identity in the hypergeometric setting:

$$F\left({a,b,-n \atop c,a+b-c-n+1};1\right) = \frac{(a-c)^{\underline{n}}(b-c)^{\underline{n}}}{(-c)^{\underline{n}}(a+b-c)^{\underline{n}}}$$
$$= \frac{(c-a)^{\overline{n}}(c-b)^{\overline{n}}}{c^{\overline{n}}(c-a-b)^{\overline{n}}} \qquad (n\in\mathbb{N}_0). \tag{5}$$

The formula thus applies to any $a_1, a_2, a_3 = -n, b_1, b_2$:

$$F\left({a_1,a_2,-n \atop b_1,b_2};1\right) = \frac{(a_1-b_1)^{\underline{n}}(a_2-b_1)^{\underline{n}}}{(-b_1)^{\underline{n}}(a_1+a_2-b_1)^{\underline{n}}} = \frac{(b_1-a_1)^{\overline{n}}(b_1-a_2)^{\overline{n}}}{b_1^{\overline{n}}(b_1-a_1-a_2)^{\overline{n}}}, \tag{6}$$

where $n \in \mathbb{N}_0$, $b_1 + b_2 = a_1 + a_2 + a_3 + 1$.

Example. Let us compute $\sum_{k\geq 0}\binom{n+k}{2k}\binom{2k}{k}\frac{(-1)^k}{k+1+m}$ $(m, n \in \mathbb{N}_0)$. We obtain

$$\frac{t_{k+1}}{t_k} = -\frac{(n+k+1)(n-k)}{(2k+2)(2k+1)}\cdot\frac{(2k+2)(2k+1)}{(k+1)(k+1)}\cdot\frac{k+1+m}{k+2+m} = \frac{(k+n+1)(k+1+m)(k-n)}{(k+2+m)(k+1)(k+1)}, \quad t_0 = \frac{1}{m+1},$$

and thus by (5),

$$\sum_{k\geq 0}\binom{n+k}{2k}\binom{2k}{k}\frac{(-1)^k}{k+1+m} = \frac{1}{m+1}F\left({n+1,m+1,-n \atop 1,m+2};1\right) = \frac{1}{m+1}\frac{n^{\underline{n}}m^{\underline{n}}}{(-1)^{\underline{n}}(n+m+1)^{\underline{n}}} = (-1)^n\frac{m^{\underline{n}}}{(m+n+1)^{\underline{n+1}}},$$

generalizing formula (4) from above.

Example. Now let us try something different. Obviously, the sum $\sum_k(-1)^k\binom{n}{k}^2$ is 0 for n odd, so let us assume $n = 2m$. The usual procedure gives

$$\frac{t_{k+1}}{t_k} = \frac{(n-k)^2}{(k+1)^2}(-1), \; t_0 = 1,$$

and thus

$$\sum_k(-1)^k\binom{2m}{k}^2 = F\left({-2m,-2m \atop 1};-1\right). \tag{7}$$

But so far we don't know how to evaluate hypergeometric series at $z = -1$. The following simple computation comes to our help. The identity $(1-z)^r(1+z)^r = (1-z^2)^r$ gives by convolution

$$\sum_{k=0}^{n} (-1)^k \binom{r}{k}\binom{r}{n-k} = (-1)^{n/2}\binom{r}{n/2}[n \text{ even}];$$

hence

$$\sum_{k=0}^{2m} (-1)^k \binom{r}{k}\binom{r}{2m-k} = (-1)^m \binom{r}{m}.$$

For the left-hand side we obtain

$$\frac{t_{k+1}}{t_k} = -\frac{r-k}{k+1}\,\frac{2m-k}{k+r-2m+1}, \quad t_0 = \binom{r}{2m}.$$

This gives

$$\sum_{k=0}^{2m} (-1)^k \binom{r}{k}\binom{r}{2m-k} = \binom{r}{2m} F\left(\begin{matrix}-r,-2m\\ r-2m+1\end{matrix}; -1\right) = (-1)^m \binom{r}{m},$$

and thus *Kummer's identity*

$$F\left(\begin{matrix}-r,-2m\\ r-2m+1\end{matrix}; -1\right) = (-1)^m \frac{\binom{r}{m}}{\binom{r}{2m}}, \quad m \in \mathbb{N}_0. \tag{8}$$

If we set $r = 2m$, then our sum becomes

$$\sum_{k=0}^{n} (-1)^k \binom{n}{k}^2 = \begin{cases} (-1)^{n/2}\binom{n}{n/2} & n \text{ even},\\ 0 & n \text{ odd}. \end{cases}$$

A Transformation Formula.
We could go on like this for a long time; the database for hypergeometric identities is enormous and still growing. To finish let us discuss the following useful reflection law:

$$\frac{1}{(1-z)^a} F\left(\begin{matrix}a,b\\ c\end{matrix}; \frac{-z}{1-z}\right) = F\left(\begin{matrix}a,c-b\\ c\end{matrix}; z\right), \quad c > 0. \tag{9}$$

For the left-hand side we get directly from the definition

$$\sum_{k\geq 0} \frac{a^{\overline{k}} b^{\overline{k}}}{c^{\overline{k}}} \frac{(-z)^k}{k!} \frac{1}{(1-z)^{k+a}} = \sum_{k\geq 0} \frac{a^{\overline{k}} b^{\overline{k}}}{c^{\overline{k}}} \frac{(-z)^k}{k!} \sum_{\ell\geq 0} \binom{a+k+\ell-1}{\ell} z^\ell$$
$$= \sum_{n\geq 0} z^n \sum_{k\geq 0} \frac{a^{\overline{k}} b^{\overline{k}}}{c^{\overline{k}}} \frac{(-1)^k}{k!} \binom{n+a-1}{n-k}$$
$$= \sum_{n\geq 0} \frac{a^{\overline{n}} z^n}{n!} \left(\sum_{k\geq 0} \frac{b^{\overline{k}} (-n)^{\overline{k}}}{c^{\overline{k}} k!} \right)$$
$$= \sum_{n\geq 0} F\left({b,-n \atop c};1\right) \frac{a^{\overline{n}} z^n}{n!},$$

and the result follows from the Vandermonde identity (3).

Example. If we set on the right-hand side of (9), $z=-1$, $a=-r$, $b=r+1$, $c=r-2n+1$, then (8) and (9) yield

$$(-1)^n \frac{\binom{r}{n}}{\binom{r}{2n}} = F\left({-r,-2n \atop r-2n+1};-1\right) = 2^r F\left({-r,r+1 \atop r-2n+1};\frac{1}{2}\right),$$

and thus the identity

$$\sum_{k\geq 0} \frac{(-r)^{\overline{k}} (r+1)^{\overline{k}}}{(r-2n+1)^{\overline{k}}} \frac{1}{2^k k!} = \frac{(-1)^n \binom{r}{n}}{2^r \binom{r}{2n}},$$

or

$$\sum_{k\geq 0} (-\frac{1}{2})^k \frac{\binom{r}{k}\binom{r+k}{k}}{\binom{r+k-2n}{k}} = \frac{(-1)^n \binom{r}{n}}{2^r \binom{r}{2n}} \quad \text{for } r \geq 2n.$$

Taking $n=1$, we obtain the special sum

$$\sum_{k=0}^{r} (-\frac{1}{2})^k \frac{\binom{r}{k}\binom{r+k}{k}}{\binom{r+k-2}{k}} = \frac{1}{2^{r-1}(1-r)} \quad (r \geq 2).$$

Exercises

4.31 Show that the following identities all come from Vandermonde:

a. $\sum_{k\geq 0} \binom{n-k}{m}\binom{r}{k-s}(-1)^k = (-1)^{m+n}\binom{r-m-1}{n-m-s}$, $m,n,s \in \mathbb{N}_0$,

b. $\sum_{k\geq 0} \binom{m}{n+k}\binom{r+k}{s}(-1)^k = (-1)^{m+n}\binom{r-n}{s-m}$, $m,n,s \in \mathbb{N}_0$,

c. $\sum_{k=0}^{n}\binom{r+k}{k}=\binom{r+n+1}{n}$,

d. $\sum_{k=0}^{n}\binom{k+r}{k}\binom{n-k+s}{n-k}\frac{r}{k+r}=\binom{r+s+n}{n}$ (r, s fixed).

4.32 Compute $\sum_{k=0}^{n}\binom{m}{k}\binom{n}{k}k$, m fixed. Since $t_0=0$, you must make an index shift.

▷ **4.33** Compute $\sum_{k=0}^{n}k\binom{m-k}{m-n}$, $m \geq n$ fixed.

4.34 Look at Exercise 4.4 and its solution, and translate it into a hypergeometric series equation. What do you get?

▷ **4.35** Write Dixon's identity (3) of Section 4.1 as a formula involving hypergeometric series.

4.36 When you differentiate a general hypergeometric series $F\left(\begin{matrix}a_1,\dots,a_m\\b_1,\dots,b_n\end{matrix};z\right)$ you get another one, what is it?

▷ **4.37** Use the usual procedure to evaluate $\sum_{k=0}^{n}(-1)^k\binom{r}{k}\binom{r+n-k}{n-k}\frac{s}{s+n-k}$, $s \geq 0$.

4.38 Prove the identity $F\left(\begin{matrix}a,b\\c\end{matrix};z\right)=(1-z)^{c-a-b}F\left(\begin{matrix}c-a,c-b\\c\end{matrix};z\right)$ by applying the reflection law (9) twice.

* * *

4.39 Prove $\sum_{k=0}^{n}\binom{2n+1}{2k+1}\binom{m+k}{2n}=\binom{2m}{2n}$, $m \in \mathbb{N}$ fixed, using the identity of Pfaff-Saalschütz.

4.40 Verify $F\left(\begin{matrix}2a,2b\\a+b+\frac{1}{2}\end{matrix};z\right)=F\left(\begin{matrix}a,b\\a+b+\frac{1}{2}\end{matrix};4z(1-z)\right)$ by showing that both functions satisfy the same differential equation

$$z(1-z)F''(z)+(a+b+\tfrac{1}{2})(1-2z)F'(z)-4abF(z)=0\,.$$

Derive from this the identity $F\left(\begin{matrix}2a,2b\\a+b+\frac{1}{2}\end{matrix};\frac{1}{2}\right)=F\left(\begin{matrix}a,b\\a+b+\frac{1}{2}\end{matrix};1\right)$, due to Gauss.

▷ **4.41** We have noted that $\sum_{k>n}\binom{n+k}{k}2^{-k}=2^n$ in the previous section. Translate the equation into hypergeometric series. How is it connected to the previous exercise?

4.42 Prove the polynomial identity

$$\sum_{k=0}^{m}\binom{m+r}{k}x^{m-k}=\sum_{k=0}^{m}(-1)^k\binom{-r}{k}(1+x)^{m-k}\quad(m\in\mathbb{N}_0)\,.$$

Hint: Considered as polynomials in r both sides have degree m; find $m+1$ values where they agree.

▷ **4.43** Use the previous exercise to show that

$$\sum_{k=0}^{m} \binom{m+r}{k}\binom{m-k}{n} x^{m-n-k} = \sum_{k=0}^{m} (-1)^k \binom{-r}{k}\binom{m-k}{n}(1+x)^{m-n-k}$$

and deduce from this that

$$F\left(\begin{matrix} a, -n \\ c \end{matrix}; 1-z\right) = \frac{(a-c)^{\overline{n}}}{(-c)^{\overline{n}}} F\left(\begin{matrix} a, -n \\ a-n-c+1 \end{matrix}; z\right).$$

4.44 Show that $\sum_{k\geq 0}(-2)^k \frac{\binom{2n}{k}^2}{\binom{4n}{k}} = (-1)^n \frac{\binom{2n}{n}}{\binom{4n}{2n}}$.
Hint: Use (8), and the previous exercise.

▷ **4.45** Prove the following variants of Vandermonde's relation (3), where $N, n \in \mathbb{N}_0$:

a. $F\left(\begin{matrix} -\frac{N}{2} - \frac{1}{2}, -\frac{N}{2} \\ n + \frac{1}{2} \end{matrix}; 1\right) = 2^N \frac{(2n)!(n+N)!}{n!(2n+N)!}$

b. $F\left(\begin{matrix} -\frac{N}{2} + \frac{1}{2}, -\frac{N}{2} \\ n+1 \end{matrix}; 1\right) = 2^{-N} \frac{(2n+2N)!n!}{(2n+N)!(n+N)!}$,

c. $F\left(\begin{matrix} -\frac{N}{2} + \frac{1}{2}, -\frac{N}{2} \\ -n-N+\frac{1}{2} \end{matrix}; 1\right) = 2^N \frac{(n+N)!(2n+N)!}{(2n+2N)!n!}$.

4.46 Use the previous exercise to prove the following formulas, $m, n, \ell \in \mathbb{N}_0$:

a. $\sum_k \binom{m}{2k}\binom{k}{n} = 2^{m-2n-1}\left[\binom{m-n}{n} + \binom{m-n-1}{n-1}\right]$

b. $\sum_k \frac{1}{(m-n-2k)!(n+k)!(k-\ell)!4^k} = 2^{n-m} \frac{(2m-2\ell)!}{(m-n-2\ell)!(m-\ell)!(n+m)!}$

c. $\sum_k (-1)^k \binom{m-1}{k}\binom{2m-2k-1}{m+n-1} = 2^{m-n-1} \frac{m+n}{m}\binom{m}{n}$.

Highlight: New Identities from Old

Hypergeometric functions have occupied a prominent place in analysis and algebra at least since Euler's time. Many of the great names made an imprint, and the theory abounds with impressive identities, some of them too long for a single line.

Among the recent developments, one idea stands out in its simplicity and elegance—the method of WZ pairs, named after their inventors Herb Wilf and Doron Zeilberger.

WZ Pairs.
Suppose we know or conjecture an identity of the form

$$\sum_{k\geq 0} a(n,k) = s(n)\,.$$

Assuming $s(n) \neq 0$ and setting $F(n,k) = \frac{a(n,k)}{s(n)}$ we rewrite this as

$$\sum_{k\geq 0} F(n,k) = 1\,.$$

In other words, our task is to prove that $\sum_k F(n,k)$ is a constant (in this case = 1) *independent* of n.

Here is the idea: Suppose we can find another function $G(n,k)$ such that for all nonnegative integers n and k,

$$F(n+1,k) - F(n,k) = G(n,k+1) - G(n,k) \tag{1}$$

holds. Then we are done, provided that for all n,

$$G(n,0) = 0 \quad \text{and} \quad \lim_{k\to\infty} G(n,k) = 0\,. \tag{2}$$

Indeed, summing (1) over all k we get

$$\sum_{k\geq 0} F(n+1,k) - \sum_{k\geq 0} F(n,k) = 0\,,$$

and thus $\sum_{k\geq 0} F(n,k) =$ constant for all n. It then remains to check the sum for *one* value of n.

A pair (F,G) satisfying (1) is called a WZ *pair*, and G a *mate* for F. We also know how to find $G(n,k)$. Input $F(n+1,k) - F(n,k)$ as a function of k into Gosper's algorithm, and output $G(n, k+1)$ as an indefinite sum. Note that when both $F(n+1,k)/F(n,k)$ and $G(n,k+1)/G(n,k)$ are rational functions, then so is $G(n,k)/F(n,k)$, since

$$\frac{G(n,k)}{F(n,k)} = \frac{F(n+1,k)/F(n,k)-1}{G(n,k+1)/G(n,k)-1}.$$

Hence we get the mate in the form $G(n,k) = F(n,k)R(n,k)$, $R(n,k)$ rational.

Example. Let us look at Vandermonde's identity

$$\sum_{k\geq 0}\binom{n}{k}\binom{m}{k} = \binom{n+m}{n}, \quad m \in \mathbb{N}_0,$$

that is,

$$F(n,k) = \frac{\binom{n}{k}\binom{m}{k}}{\binom{n+m}{n}}.$$

The input

$$a(k) = F(n+1,k) - F(n,k) = \frac{\binom{n}{k}\binom{m}{k}}{\binom{n+m}{n}}\,\frac{k(n+m+1)-m(n+1)}{(n-k+1)(n+m+1)}$$

returns the mate (check it!)

$$G(n,k) = \frac{\binom{n}{k-1}\binom{m}{k-1}}{\binom{n+m+1}{n}}\,\frac{k-1-m}{m+1}. \tag{3}$$

Condition (2) is clearly satisfied, and with the single evaluation $\sum_{k\geq 0} F(0,k) = 1$ the proof is complete.

This is nice, and the method of WZ pairs can be used to furnish quick verifications of practically all standard identities. The real beauty, however, lies in the fact that it produces new identities from old. We follow the informative (and entertaining) book A = B by Petovšek, Wilf, and Zeilberger, and some clever variants by Ira Gessel.

The Companion Identity.
Suppose (F, G) is a WZ pair that satisfies (2). Is there also a natural identity associated with $G(n,k)$? There is, and again it is surprisingly simple. Suppose that in addition to (1) and (2), the following condition holds:

For every $k \geq 0$, *the limit* $f_k = \lim_{n\to\infty} F(n,k)$ *exists, and is finite.* (4)

Summing (1) from $n = 0$ to $n = N$ gives

$$F(N+1, i) - F(0, i) = \sum_{n=0}^{N} G(n, i+1) - \sum_{n=0}^{N} G(n, i),$$

and with $N \to \infty$,

$$f_i - F(0, i) = \sum_{n\geq 0} G(n, i+1) - \sum_{n\geq 0} G(n, i).$$

Now we sum the last equation from $i = 0$ to $i = k - 1$, and obtain the *companion identity*

$$\sum_{i=0}^{k-1} (f_i - F(0, i)) = \sum_{n\geq 0} G(n, k) \qquad (k \geq 1), \tag{5}$$

since $G(n, 0) = 0$ according to (2).

Example. Let us look at the Vandermonde pair (F, G) from above. Clearly, $F(0, i) = \delta_{0,i}$, and

$$f_i = \lim_{n\to\infty} \frac{\binom{n}{i}\binom{m}{i}}{\binom{n+m}{m}} = 0 \text{ for } m > i,$$

since the leading power in the numerator is n^i, while it is n^m in the denominator. Replacing k by $k+1$ in (3), we get the companion identity

$$\sum_{n\geq 0} \frac{\binom{n}{k}}{\binom{n+m+1}{n}} = \frac{m+1}{(k+1)\binom{m}{k+1}} \text{ for } m > k \geq 0,$$

which according to Wilf–Zeilberger is not immediately deducible from the known database of hypergeometric functions.

As illustration, $m = k+1$ yields the infinite family of identities

$$\sum_{n\geq 0} \frac{\binom{n}{k}}{\binom{n+k+2}{n}} = \frac{k+2}{k+1} \qquad (k \geq 0)\,.$$

Typically, the original sum for the $F(n,k)$'s is finite, while the companion identity has infinitely many terms.

The Gamma Function.
The real power of the method is displayed by the following clever idea. In the definition (1) of WZ pairs there is no reason why n and k shouldn't be arbitrary complex numbers. So let us call (F,G) a WZ pair if (1) is satisfied for $n,k \in \mathbb{C}$. We can now manufacture new WZ pairs out of a given pair (F,G). It is easily seen that with $(F(n,k),G(n,k))$ the following are also WZ pairs:

a. $(F(n+\alpha,k+\beta),G(n+\alpha,k+\beta))$, $\alpha,\beta \in \mathbb{C}$,
b. $(\gamma F(n,k),\gamma G(n,k))$, $\gamma \in \mathbb{C}$,
c. $(F(-n,k),-G(-n-1,k))$,
d. $(F(n,-k),-G(n,-k+1))$.

We will illustrate with an example how this approach opens the door to a plethora of new hypergeometric identities. To do this we need the classical *gamma function*, which extends the factorial $n!$ to complex numbers. The function

$$\Gamma(z) = \int_0^\infty t^{z-1}e^{-t}dt$$

is defined (by analytical continuation) for all complex numbers except the negative integers. Using partial integration, one obtains

$$\Gamma(z+1) = z\Gamma(z)\,, \tag{6}$$

and so in particular, with $\Gamma(1) = 1$,

$$\Gamma(n+1) = n! \;\text{ for } n \in \mathbb{N}_0\,.$$

The identity (6) implies for the rising factorials

$$a^{\overline{k}} = \frac{\Gamma(a+k)}{\Gamma(a)} \qquad (k \in \mathbb{N}_0)\,, \tag{7}$$

and further, for $k \in \mathbb{N}_0$,

$$a^{\overline{-k}} = \frac{\Gamma(a-k)}{\Gamma(a)} = \frac{1}{(a-1)\cdots(a-k)} = (-1)^k \frac{1}{(1-a)^{\overline{k}}}. \tag{8}$$

Abbreviating

$$\Gamma\binom{a_1,\dots,a_s}{b_1,\dots,b_t} = \frac{\prod_{i=1}^{s} \Gamma(a_i)}{\prod_{j=1}^{t} \Gamma(b_j)},$$

a typical hypergeometric term can thus be expressed with the help of the gamma function as

$$\frac{a_1^{\overline{k}} \cdots a_s^{\overline{k}} z^k}{b_1^{\overline{k}} \cdots b_t^{\overline{k}} k!} = \Gamma\binom{a_1+k,\dots,a_s+k,b_1,\dots,b_t}{a_1,\dots,a_s,b_1+k,\dots,b_t+k,k+1} z^k.$$

The following formula, called the *Chu-Vandermonde identity*, generalizes the terminating sum (3) of the last section from $b = -n$ to arbitrary $b \in \mathbb{C}$:

$$F\left(\begin{matrix}a,b\\c\end{matrix};1\right) = \Gamma\binom{c,c-a-b}{c-a,c-b}. \tag{9}$$

Indeed, for $b = -n$ the right-hand side becomes by (7),

$$\Gamma\binom{c,c-a+n}{c-a,c+n} = \frac{(c-a)^{\overline{n}}}{c^{\overline{n}}}.$$

Using WZ pairs.

Let us start with the identity of Exercise 4.40,

$$F\left(\begin{matrix}2a,2b\\a+b+\frac{1}{2}\end{matrix};\frac{1}{2}\right) = F\left(\begin{matrix}a,b\\a+b+\frac{1}{2}\end{matrix};1\right),$$

which we can rewrite by (9) as

$$\sum_{k\geq 0} \frac{(2a)^{\overline{k}}(2b)^{\overline{k}}}{(a+b+\frac{1}{2})^{\overline{k}}k!}\left(\frac{1}{2}\right)^k \Gamma\binom{a+\frac{1}{2},b+\frac{1}{2}}{a+b+\frac{1}{2},\frac{1}{2}} = 1.$$

So far, the running variable n does not appear. To get a candidate for a WZ pair we make the substitution $a \mapsto a-n$, and obtain as summand

$$\frac{(2a-2n)^{\overline{k}}(2b)^{\overline{k}}}{(a-n+b+\frac{1}{2})^{\overline{k}}k!}\left(\frac{1}{2}\right)^k \Gamma\binom{a-n+\frac{1}{2}, b+\frac{1}{2}}{a-n+b+\frac{1}{2}, \frac{1}{2}}.$$

Using (8), the Γ-term can be written as $(\frac{1}{2}-a-b)^{\overline{n}}/(\frac{1}{2}-a)^{\overline{n}}$, times a factor that depends only on a and b and can therefore be deleted. Our candidate is thus

$$F(n,k) = \frac{(2a-2n)^{\overline{k}}(2b)^{\overline{k}}}{(a-n+b+\frac{1}{2})^{\overline{k}}k!}(\frac{1}{2})^k \frac{(\frac{1}{2}-a-b)^{\overline{n}}}{(\frac{1}{2}-a)^{\overline{n}}}. \tag{10}$$

Application of the Gosper algorithm gives the WZ mate

$$G(n,k) = F(n,k)\frac{2k(a-n+b+k-\frac{1}{2})}{(2a-n+k-1)(2a-n+k-2)}. \tag{11}$$

Condition (2) is satisfied when $2a-2n$ is a negative integer. For example, the choice $a=-\frac{1}{2}$ yields

$$F(n,k) = \frac{(-2n-1)^{\overline{k}}(2b)^{\overline{k}}}{(b-n)^{\overline{k}}k!}(\frac{1}{2})^k \frac{(1-b)^{\overline{n}}}{n!},$$

and $\sum_{k\geq 0} F(n,k)$ is independent of n. The case $n=0$ yields

$$\sum_{k\geq 0} F(0,k) = 1 - \frac{2b}{b}\frac{1}{2} = 0,$$

and thus the identity

$$F\left(\begin{matrix}2b, -2n-1\\ b-n\end{matrix}; \frac{1}{2}\right) = 0 \qquad (n \in \mathbb{N}_0).$$

Next consider the substitution $k \mapsto k+c$ in (10). This gives, apart from a constant factor,

$$\frac{(2a-2n+c)^{\overline{k}}(2b+c)^{\overline{k}}}{(a-n+b+\frac{1}{2}+c)^{\overline{k}}(1+c)^{\overline{k}}}\left(\frac{1}{2}\right)^k \frac{(\frac{1}{2}-a-b)^{\overline{n}}}{(\frac{1}{2}-a)^{\overline{n}}}$$
$$\times \Gamma\binom{2a-2n+c, a-n+b+\frac{1}{2}}{a-n+b+\frac{1}{2}+c, 2a-2n}.$$

Now replace k by $-k$, n by $-n$, and consider the specializations $b=a-\frac{1}{2}$, $c=1-2a$. This gives the new term apart from a constant (using (8) again),

$$\frac{(2a-1)^{\overline{k}}(-n)^{\overline{k}}}{(-2n)^{\overline{k}}k!}\,2^k\,\frac{(\frac{1}{2}+a)^{\overline{n}}}{(2a)^{\overline{2n}}}\,\frac{(2n)!}{n!}\,.$$

The product of the last two factors is $(\frac{1}{2})^{\overline{n}}/a^{\overline{n}}$, which finally produces the candidate

$$F(n,k)=\frac{(2a-1)^{\overline{k}}(-n)^{\overline{k}}}{(-2n)^{\overline{k}}k!}\,2^k\,\frac{(\frac{1}{2})^{\overline{n}}}{a^{\overline{n}}}\,.$$

Gosper's algorithm returns the mate

$$G(n,k)=-\frac{1}{4}\,\frac{k(2n-k+1)}{(n+a)(n-k+1)}\,F(n,k)\,.$$

Now, $F(n,k)=0$ for $n+1\le k\le 2n$, and $G(n,k)=0$ for $n+2\le k\le 2n+1$; hence

$$\begin{aligned}\sum_{k=0}^{n+1}F(n+1,k)-\sum_{k=0}^{n+1}F(n,k)&=\sum_{k=0}^{n+1}(F(n+1,k)-F(n,k))\\&=\sum_{k=0}^{n+1}(G(n,k+1)-G(n,k))\\&=G(n,n+2)-G(n,0)=0.\end{aligned}$$

We conclude that $\sum_{k=0}^{n}F(n,k)$ is independent of n for $n\ge 1$. Checking in addition $n=0$ gives the formula

$$F\left(\begin{matrix}2a-1,-n\\-2n\end{matrix};2\right)=\frac{a^{\overline{n}}}{(\frac{1}{2})^{\overline{n}}}\qquad(n\in\mathbb{N}_0).$$

Let us close with an identity (one of more than 50 others) from Gessel's paper. It is derived from Dixon's identity and should convincingly demonstrate the power of WZ pairs: For $n\ge 3$,

$$F\left(\begin{matrix}-\frac{3}{2},1-\frac{6n}{11},\frac{1-n}{2},-\frac{n}{2},4-2n\\-\frac{6n}{11},\frac{2-2n}{3},\frac{3-2n}{3},\frac{4-2n}{3}\end{matrix};\frac{16}{27}\right)=\frac{1}{6}\,\frac{(-\frac{3}{2})^{\overline{n}}}{(-\frac{1}{2})^{\overline{n}}}\,.$$

Notes and References

Hypergeometric functions have been the subject of intensive study since Euler and Gauss. An excellent survey is the book by Andrews, Askey, and Roy. The organization of this chapter and, in particular, the treatment of more recent topics has greatly benefited from lecture notes of Zeilberger and the very readable book by Petovšek, Wilf, and Zeilberger. The references list the original papers on Gosper's algorithm and the extension due to Zeilberger. See also Chapter 5 in the book by Graham, Knuth, and Patashnik for further examples and ramifications. The presentation of WZ pairs in the highlight follows the articles by Wilf, Zeilberger, and Gessel.

1. G.E. Andrews, R. Askey, and R. Roy (1999): *Special Functions.* Cambridge Univ. Press, Cambridge.
2. I. Gessel (1995): Finding identities with the WZ method. *J. Symbolic Computation* 20, 537–566.
3. R.W. Gosper, Jr. (1975): Decision procedure for indefinite hypergeometric summation. *Proc. Natl. Acad. Sci. USA* 75, 40–42.
4. R.Ł. Graham, D.E. Knuth, and O. Patashnik (1994): *Concrete Mathematics*, 2nd edition. Addison-Wesley, Reading.
5. M. Petovšek, H.S. Wilf, and D. Zeilberger (1996): *A=B.* AK Peters, Wellesley.
6. H.S. Wilf and D. Zeilberger (1990): Rational functions certify combinatorial identities. *J. Amer. Math. Soc.* 3, 77–83.
7. D. Zeilberger (1990): *Three Recitations on Holonomic Sequences and Hypergeometric Series.* Lecture Notes.
8. D. Zeilberger (1991): The method of creative telescoping. *J. Symbolic Computation* 11, 195–204.

5 Sieve Methods

Sieve methods are used in essentially two variants. In order to determine the size of a set we overcount the set, subtract from this count, add again, subtract again, until finally the exact number of elements is determined. The classical example is the *principle of inclusion–exclusion*. In the second variant we sieve out the unwanted elements through a suitable weighting. This is the fundamental idea behind the *involution principle*.

5.1 Inclusion–Exclusion

Consider a finite set X and three subsets A, B, C. To obtain $|A \cup B \cup C|$ we take the sum $|A| + |B| + |C|$. Unless A, B, C are pairwise disjoint, we have an overcount, since the elements of $A \cap B$, $A \cap C$, $B \cap C$ have been counted twice. So we subtract $|A \cap B| + |A \cap C| + |B \cap C|$. Now the count is correct except for the elements in $A \cap B \cap C$ which have been added three times, but also subtracted three times. The answer is therefore

$$|A \cup B \cup C| = |A| + |B| + |C| - |A \cap B| - |A \cap C| - |B \cap C| + |A \cap B \cap C|,$$

or equivalently,

$$\begin{aligned}|X \setminus (A \cup B \cup C)| = |X| - |A| - |B| - |C| + |A \cap B| \\ + |A \cap C| + |B \cap C| - |A \cap B \cap C|.\end{aligned}$$

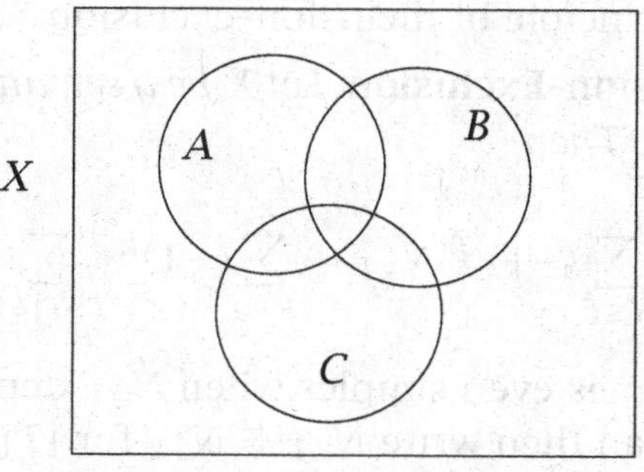

The following formula addresses the general case.

Let $A_1, A_2, \ldots, A_n$ be subsets of X. Then

$$|X \setminus \bigcup_{i=1}^{n} A_i| = |X| - \sum_{i=1}^{n} |A_i| + \sum_{i<j} |A_i \cap A_j| \mp \cdots + (-1)^n |A_1 \cap \cdots \cap A_n| . \quad (1)$$

For the proof we check how often an element $x \in X$ is counted on both sides. If $x \notin \bigcup_{i=1}^{n} A_i$, then it is counted once on either side. Suppose $x \in \bigcup_{i=1}^{n} A_i$, and more precisely, that x is in exactly m of the sets A_i. The count on the left-hand side is 0, and on the right-hand side we obtain

$$1 - \binom{m}{1} + \binom{m}{2} - \binom{m}{3} \pm \cdots + (-1)^m \binom{m}{m} = 0,$$

since $m \geq 1$, and (1) follows.

The standard interpretation leads to the principle of inclusion–exclusion. Suppose we are given a set X, called the *universe*, and a set $E = \{e_1, \ldots, e_n\}$ of properties that the elements of X may or may not possess. Let A_i be the subset of elements that enjoy property e_i (and possibly others). Then $|X \setminus \bigcup_{i=1}^{n} A_i|$ is the number of elements that possess *none* of the properties. Now consider a term on the right-hand side of (1). Clearly, $A_{i_1} \cap \cdots \cap A_{i_t}$ is the set of elements that possess the properties $e_{i_1}, \ldots, e_{i_t}$ (and maybe others). Using the notation

$$\begin{aligned} N_{\supseteq T} &:= \#\{x \in X : x \text{ possesses } \textit{at least} \text{ the properties in } T\}, \\ N_{=T} &:= \#\{x \in X : x \text{ possesses } \textit{precisely} \text{ the properties in } T\}, \end{aligned} \quad (2)$$

we arrive at the principle of inclusion–exclusion.

Principle of Inclusion–Exclusion. *Let X be a set, and $E = \{e_1, \ldots, e_n\}$ a set of properties. Then*

$$N_{=\emptyset} = \sum_{T \subseteq E} (-1)^{|T|} N_{\supseteq T} = \sum_{k=0}^{n} (-1)^k \sum_{T:|T|=k} N_{\supseteq T} . \quad (3)$$

The formula becomes even simpler when $N_{\supseteq T}$ depends only on the size $|T| = k$. We can then write $N_{\supseteq T} = N_{\geq k}$ for $|T| = k$, and call E a *homogeneous* set of properties. We will see in a moment that in this case $N_{=T} = N_{=k}$ also depends only on the cardinality of T. Hence for homogeneous properties, we have

$$N_{=0} = \sum_{k=0}^{n} (-1)^k \binom{n}{k} N_{\geq k}. \tag{4}$$

Example. One of the earliest examples concerns the computation of Euler's function $\varphi(n) = \#\{d : 1 \leq d \leq n,\ \gcd(d,n) = 1\}$. Suppose $n = p_1^{a_1} \cdots p_t^{a_t}$ is the prime decomposition of n. Let $X = \{1, 2, \ldots, n\}$, and e_i the property that p_i divides d. Clearly, $N_{=\emptyset} = \varphi(n)$, whereas $N_{\supseteq T}$ counts the integers $\leq n$ that are multiples of $\prod_{e_i \in T} p_i$. Hence $N_{\supseteq T} = \frac{n}{\prod_{e_i \in T} p_i}$, and (3) gives

$$\begin{aligned}\varphi(n) &= n - \sum_{i=1}^{t} \frac{n}{p_i} + \sum_{i<j} \frac{n}{p_i p_j} \mp \cdots + (-1)^t \frac{n}{p_1 \cdots p_t} \\ &= n\left(1 - \sum_{i=1}^{t} \frac{1}{p_i} + \sum_{i<j} \frac{1}{p_i p_j} \mp \cdots + (-1)^t \frac{1}{p_1 \cdots p_t}\right),\end{aligned}$$

or

$$\varphi(n) = n \prod_{i=1}^{t} \left(1 - \frac{1}{p_i}\right). \tag{5}$$

As an example, for $n = 20 = 2^2 \cdot 5$, we get

$$\varphi(20) = 20 \cdot \frac{1}{2} \cdot \frac{4}{5} = 8.$$

The numbers less than or equal to 20 relatively prime to 20 are $1, 3, 7, 9, 11, 13, 17, 19$. Note that (5) immediately implies $\varphi(mn) = \varphi(m)\varphi(n)$ when m and n are relatively prime.

Example. We know that the number of non-negative integer solutions of $x_1 + x_2 + \cdots + x_n = k$ is given by $\binom{n+k-1}{k} = \binom{n+k-1}{n-1}$. How many of the solutions satisfy the additional restriction $x_i < s$ for all i? No problem with inclusion-exclusion. The universe X is the set of all solutions, and property i means that $x_i \geq s$. Clearly, this set of properties is homogeneous, with $N_{=0}$ the desired number. Take $T = \{1, \ldots, j\}$; then $N_{\supseteq T}$ is the number of solutions with

$x_1 \geq s, \ldots, x_j \geq s$. Setting $y_i = x_i - s$ $(i \leq j)$, $y_i = x_i$ $(i > j)$, this is the same as the number of solutions of the system

$$y_1 + \cdots + y_n = k - js\,,$$

and the answer is therefore

$$N_{=0} = \sum_{j=0}^{n} (-1)^j \binom{n}{j} \binom{n+k-js-1}{n-1}. \tag{6}$$

In particular, for $s = 1$, all x_i must be 0, and (6) yields the binomial identity

$$\sum_{j=0}^{n} (-1)^j \binom{n}{j} \binom{n+k-j-1}{n-1} = \begin{cases} 1 & k = 0\,, \\ 0 & k > 0\,. \end{cases}$$

Inclusion–exclusion leads to alternating sums, and whenever a sum of the form $\sum_{k=0}^{n} (-1)^k \binom{n}{k} c_k$ is given it may be advantageous to look for a proof via inclusion–exclusion, by interpreting c_k as $N_{\geq k}$ for some set of homogeneous properties. Consider the sum

$$\sum_{k=0}^{m} (-1)^k \binom{m}{k} \binom{n-k}{r} \qquad (m \leq r \leq n)\,.$$

Since $c_0 = |X| = \binom{n}{r}$, we take as universe the family of all r-subsets R of an n-set. A little reflection shows that the following set of properties will work. Let $M = \{a_1, \ldots, a_m\}$ be a fixed m-set, and $E = \{e_1, \ldots, e_m\}$, where property e_i means that element a_i is *not* contained in R. Then $N_{\geq k} = \binom{n-k}{r}$, $N_{=0}$ counts all r-subsets R with $M \subseteq R$, and so

$$\sum_{k=0}^{m} (-1)^k \binom{m}{k} \binom{n-k}{r} = \binom{n-m}{r-m}.$$

Example. Here is another classical problem that is a bit more complicated. At a long dinner table with seats numbered $1, 2, \ldots, 2n$, n couples take their places. In how many ways can they be seated so that no couple sit next to each other? Let X be the set of all seatings; thus $|X| = (2n)!$, and e_i the property that the i-th couple sit side by side. Clearly, this is again a homogeneous set of properties, and we are looking for $N_{=0}$. Let $|T| = k$. In order to compute $N_{\supseteq T}$, consider the initial positions $a_1 < \cdots < a_k$, where these k couples sit next to each other. The a_i's are between 1 and $2n-1$ and a_i, a_{i+1} differ by at least 2 (since the next seat $a_i + 1$ is occupied by the partner). According to Exercise 1.8 there are $\binom{2n-k}{k}$ possible choices for the a_i's. Since the couples may be permuted in $k!$ ways, the couples interchanged, and the remaining seats taken arbitrarily, we obtain

$$N_{\geq k} = \binom{2n-k}{k} k! 2^k (2n-2k)! = (2n-k)! 2^k . \tag{7}$$

Inclusion-exclusion yields as the number of possible arrangements

$$N_{=0} = \sum_{k=0}^{n} (-2)^k \binom{n}{k} (2n-k)! . \tag{8}$$

Example. Recall the beautiful result that the number $p_o(n)$ of partitions of n into odd summands equals the number $p_d(n)$ of partitions of n into distinct parts. Here is a new proof that uses inclusion-exclusion in an unexpected way. Consider $p_o(n)$ first. Take as universe $X = \text{Par}(n)$, and let e_i be the property that the *even* summand i appears ($i \leq n$). How many partitions contain 2? Clearly $p(n-2)$, since we may add 2 to any $\lambda \in \text{Par}(n-2)$. Similarly, $p(n-6)$ counts the number of partitions that contain 2 and 4, and so on. Inclusion-exclusion yields therefore

$$\begin{aligned} p_o(n) = p(n) & \\ & - p(n-2) - p(n-4) - p(n-6) - \cdots \\ & + p(n-2-4) + p(n-2-6) + p(n-2-8) + \cdots \\ & - p(n-2-4-6) - \cdots . \end{aligned}$$

To compute $p_d(n)$, we consider the properties e_i, where e_i means that i appears at least twice. This gives

$$\begin{aligned} p_d(n) = p(n) & \\ & - p(n-1-1) - p(n-2-2) - p(n-3-3) - \cdots \\ & + p(n-1-1-2-2) + p(n-1-1-3-3) + \cdots . \end{aligned}$$

The computations agree in every row, and $p_o(n) = p_d(n)$ follows.

A Generalization of Inclusion–Exclusion.
So far we have expressed $N_{=\emptyset}$ in terms of $N_{\supseteq T}$ as in (3). It is natural to ask for a general formula expressing $N_{=A}$ in an analogous way. From the definition (2) we immediately infer

$$N_{\supseteq A} = \sum_{T \supseteq A} N_{=T}. \tag{9}$$

Now we are going to prove that, conversely,

$$N_{=A} = \sum_{T \supseteq A} (-1)^{|T|-|A|} N_{\supseteq T}, \tag{10}$$

which reduces to (3) for $A = \emptyset$. Note that this implies $N_{=A} = N_{=B}$ for $|A| = |B|$ when E is a homogeneous set of properties, as announced earlier. We prove the inversion relations (9), (10) for arbitrary set functions.

Theorem 5.1. *Let E be a finite set, and $f, g : 2^E \to K$ functions into a field K of characteristic* 0. *Then*

$$f(A) = \sum_{T \supseteq A} g(T)\ (\forall A) \iff g(A) = \sum_{T \supseteq A} (-1)^{|T|-|A|} f(T)\ (\forall A). \tag{11}$$

Proof. Assume the equality on the left-hand side. Then

$$\begin{aligned}\sum_{T \supseteq A} (-1)^{|T|-|A|} f(T) &= \sum_{T \supseteq A} (-1)^{|T|-|A|} \sum_{U \supseteq T} g(U) \\ &= \sum_{U \supseteq A} \left(\sum_{U \supseteq T \supseteq A} (-1)^{|T|-|A|} \right) g(U).\end{aligned}$$

If $|U \setminus A| = m$, then the inner summand is $\sum_{k=0}^{m} (-1)^k \binom{m}{k} = \delta_{m,0}$, whence $g(A)$ results. The other direction is proved in an analogous fashion. □

Formula (10) is now an immediate consequence by considering the functions $g(A) = N_{=A}$, $f(A) = N_{\supseteq A}$.

Corollary 5.2. *Let X be a universe, $E = \{e_1, \ldots, e_n\}$ a set of properties, and N_p the number of elements in X that possess precisely p properties. Then*

$$N_p = \sum_{k=p}^{n} (-1)^{k-p} \binom{k}{p} \sum_{T:|T|=k} N_{\supseteq T}. \tag{12}$$

In particular, if E is homogeneous, then

$$N_p = \binom{n}{p} \sum_{k=p}^{n} (-1)^{k-p} \binom{n-p}{k-p} N_{\geq k} . \tag{13}$$

Proof. We have by (10),

$$\begin{aligned} N_p &= \sum_{A:|A|=p} N_{=A} = \sum_{A:|A|=p} \sum_{T \supseteq A} (-1)^{|T|-|A|} N_{\supseteq T} \\ &= \sum_{T:|T| \geq p} (-1)^{|T|-p} \sum_{A:|A|=p, T \supseteq A} N_{\supseteq T} \\ &= \sum_{k=p}^{n} (-1)^{k-p} \binom{k}{p} \sum_{|T|=k} N_{\supseteq T} , \end{aligned}$$

which in the homogeneous case is

$$\sum_{k=p}^{n} (-1)^{k-p} \binom{k}{p} \binom{n}{k} N_{\geq k} = \binom{n}{p} \sum_{k=p}^{n} (-1)^{k-p} \binom{n-p}{k-p} N_{\geq k} . \qquad \square$$

Example. How many permutations in $S(n)$ have precisely p fixed points? We set $X = S(n)$, and let e_i be the property that i is a fixed point. Then E is clearly homogeneous with $N_{\geq k} = (n-k)!$, and so by (13),

$$\begin{aligned} N_p &= \binom{n}{p} \sum_{k=p}^{n} (-1)^{k-p} \binom{n-p}{k-p} (n-k)! = \binom{n}{p} (n-p)! \sum_{k=0}^{n-p} \frac{(-1)^k}{k!} \\ &= \binom{n}{p} D_{n-p} , \end{aligned}$$

which is also immediately clear from the definition of the derangement numbers.

The equivalence in Theorem 5.1 uses inversion "from above," since we consider all sets T containing a given A. By taking complements, the following inversion "from below" is easily seen.

Theorem 5.3. *Let E be a finite set, and $f, g : 2^E \to K$ two functions. Then*

$$f(A) = \sum_{T \subseteq A} g(T) \ (\forall A) \iff g(A) = \sum_{T \subseteq A} (-1)^{|A|-|T|} f(T) \ (\forall A) . \tag{14}$$

Note that in the homogeneous case $f(A) = u_k$, $g(A) = v_k$ for $|A| = k$, (14) reduces to the binomial inversion formula considered in Section 2.4.

An important generalization of inclusion-exclusion arises when the elements of X are weighted. Suppose $w : X \to K$ is a weight-function, which we extend to 2^X, setting $w(A) = \sum_{x\in A} w(x)$ with $w(\emptyset) = 0$. Now, for a set $E = \{e_1, \ldots, e_n\}$ of properties, let

$$\begin{aligned} W_{\supseteq T} &:= \sum\{w(x) : x \text{ possesses at least the properties of } T\}, \\ W_{=T} &:= \sum\{w(x) : x \text{ possesses precisely the properties in } T\}. \end{aligned} \tag{15}$$

Thus for the constant weighting $w = 1$ we get $W_{\supseteq T} = N_{\supseteq T}$ and $W_{=T} = N_{=T}$. Since $W_{\supseteq A} = \sum_{T\supseteq A} W_{=T}$ clearly holds, we obtain the following general principle of inclusion-exclusion.

General Principle of Inclusion-Exclusion. *Let X be a set, $w : X \to K$ a weighting, and $E = \{e_1, \ldots, e_n\}$ a set of properties. Then*

$$W_{=\emptyset} = \sum_{T\subseteq E} (-1)^{|T|} W_{\supseteq T} = \sum_{k=0}^{n} (-1)^k \sum_{T:|T|=k} W_{\supseteq T}, \tag{16}$$

and in general

$$W_p = \sum_{k=p}^{n} (-1)^{k-p} \binom{k}{p} \sum_{T:|T|=k} W_{\supseteq T}, \tag{17}$$

where $W_p = \sum w(x)$ over all x that possess precisely p properties.

Example. Let $M = (m_{ij})$ be an $n \times n$-matrix. The *permanent* of M is defined as $\operatorname{per}(M) = \sum_{\sigma\in S(n)} m_{1\sigma(1)} \cdots m_{n\sigma(n)}$; thus it uses the same summands as the determinant, but without signs. To compute $\operatorname{per}(M)$, we take as universe X the set of all mappings ρ from $\{1, 2, \ldots, n\}$ into itself with the weighting $w(\rho) = m_{1\rho(1)} \cdots m_{n\rho(n)}$. Let $E = \{e_1, e_2, \ldots, e_n\}$, where $e_i \in E$ means that $i \notin \operatorname{im}(\rho)$. Clearly, $W_{=\emptyset} = \operatorname{per}(M)$, and it remains to find an expression for $W_{\supseteq T}$. Now,

$$W_{\supseteq T} = \sum_{\rho} m_{1\rho(1)} \cdots m_{n\rho(n)}$$

over all mappings ρ with $\mathrm{im}(\rho) \subseteq \{1,\dots,n\} \setminus T$. Let us denote by $M|C$ the $n \times |C|$-submatrix with column-set C, and by $P(M|C)$ the product of its row-sums. Then

$$W_{\supseteq T} = P(M|\{1,\dots,n\} \setminus T),$$

as is easily seen by expanding the right-hand side. According to (16) we obtain

$$\mathrm{per}(M) = \sum_{k=1}^{n} (-1)^{n-k} \sum_{T:|T|=k} P(M|T). \tag{18}$$

As examples, consider the all-1's matrix J_n, and $J_n - I_n$, where I_n is the identity matrix. Clearly, $\mathrm{per}(J_n) = n!$, $\mathrm{per}(J_n - I_n) = D_n$ (derangement number), which yields the identities

$$n! = \sum_{k=1}^{n} (-1)^{n-k} \binom{n}{k} k^n,$$

$$D_n = \sum_{k=1}^{n} (-1)^{n-k} \binom{n}{k} (k-1)^k k^{n-k}.$$

Probability Theory.

A natural setting for inclusion-exclusion is probability theory. This just means that we divide all terms by $|X|$, and consider all elements of X equally likely. For example, the probability that x possesses all properties in T is $\frac{N_{\supseteq T}}{|X|}$. Let X and $E = \{e_1,\dots,e_n\}$ be as before, and consider the random variable $Z : X \to \{0,1,\dots,n\}$ counting the number of properties. If we set $\sigma_k = \frac{1}{|X|} \sum_{T:|T|=k} N_{\supseteq T}$, (12) becomes

$$\mathrm{Prob}(Z = p) = \sum_{k=p}^{n} (-1)^{k-p} \binom{k}{p} \sigma_k. \tag{19}$$

For the expectation $\mathbb{E}(Z) = \sum_{p=0}^{n} p\,\mathrm{Prob}(Z = p)$ we thus compute

$$\begin{aligned}
\mathbb{E}(Z) &= \sum_{p=0}^{n} p \sum_{k=p}^{n} (-1)^{k-p} \binom{k}{p} \sigma_k = \sum_{k=0}^{n} \sum_{p=1}^{k} (-1)^{k-p} p \binom{k}{p} \sigma_k \\
&= \sum_{k=1}^{n} (-1)^{k-1} k \sigma_k \sum_{p=1}^{k} (-1)^{p-1} \binom{k-1}{p-1} \\
&= \sum_{k=1}^{n} (-1)^{k-1} k \sigma_k \sum_{p=0}^{k-1} (-1)^{p} \binom{k-1}{p}.
\end{aligned}$$

The inner sum is 0 for $k \geq 2$, and 1 for $k = 1$, and we obtain

$$\mathbb{E}(Z) = \sigma_1 . \tag{20}$$

Of course, this can also be seen by considering the indicator variable $Z_i(x) = 1$ if x possesses property e_i, and 0 otherwise. Then $\mathbb{E}(Z_i) = \text{Prob}(x \text{ has } i) = \frac{N_{\supseteq\{i\}}}{|X|}$, and linearity of expectation implies

$$\mathbb{E}(Z) = \sum_{i=1}^{n} \mathbb{E}(Z_i) = \sigma_1 .$$

Similarly, the variance $\text{Var}(Z) = \mathbb{E}(Z^2) - \mathbb{E}(Z)^2$ is quickly computed to

$$\text{Var}(Z) = \sigma_1 + 2\sigma_2 - \sigma_1^2 . \tag{21}$$

Example. With n couples sitting down at a table with $2n$ seats, what is the expected number of couples that sit next to each other? According to (7), $\sigma_k = \binom{n}{k}\frac{(2n-k)!2^k}{(2n)!}$, and thus

$$\mathbb{E}(Z) = \sigma_1 = n\frac{(2n-1)!2}{(2n)!} = 1 ,$$

which may come as a mild surprise.

Inclusion–exclusion has, however, its drawbacks. If we want to compute $\text{Prob}(Z = 0) = \frac{N_{=0}}{|X|}$, we have to consider in principle 2^n terms. Number theorists have developed ingenious methods to obtain good bounds, considering only a small number of terms in (19). Let us just quote one result from probability theory, Chebyshev's inequality: For $a > 0$,

$$\text{Prob}(|Z - \mathbb{E}(Z)| \geq a) \leq \frac{\text{Var} Z}{a^2} .$$

In particular, in our case

$$\text{Prob}(Z = 0) \leq \text{Prob}(|Z - \mathbb{E}(Z)| \geq \mathbb{E}(Z)) \leq \frac{\text{Var} Z}{\mathbb{E}(Z)^2} ,$$

which by (20) and (21) becomes

$$\text{Prob}(Z = 0) \leq \frac{\sigma_1 + 2\sigma_2}{\sigma_1^2} - 1 .$$

Exercises

5.1 How many numbers less than one million are not of the form x^2 or x^3 or x^5?

5.2 Derive the formula for the Stirling numbers

$$S_{n,r} = \frac{1}{r!}\sum_{k=0}^{r}(-1)^{r-k}\binom{r}{k}k^n$$

using inclusion-exclusion.

▷ **5.3** Obtain a summation formula for the number of permutations in $S(n)$ with no cycles of length ℓ. Verify your answer, using the composition formula of Section 3.3.

5.4 Evaluate each sum by an inclusion-exclusion argument and check it using generating functions:

a. $\sum_{k=1}^{n}(-1)^k k\binom{n}{k}$,
b. $\sum_{k=0}^{n}(-1)^k\binom{n}{k}2^{n-k}$,
c. $\sum_{k=0}^{n}(-1)^k\binom{n}{k}\binom{n-k}{m-k}$,
d. $\sum_{k=0}^{n}(-1)^k\binom{n}{k}\binom{n+k+r-1}{r}$.

5.5 Prove the following identities:

a. $\sum_{k=p}^{n}(-1)^{k-p}\binom{2n-k}{n}\binom{n-p}{k-p} = \binom{n}{p}$,
b. $\sum_{k=p}^{n}(-1)^{k-p}\binom{k}{p}\binom{n}{k}2^{n-k} = \binom{n}{p}$.

▷ **5.6** Show that the number of permutations in $S(n)$ with an even number of fixed points is always greater than those with an odd number, for $n \geq 4$.

5.7 How many integer solutions $x_1 + x_2 + x_3 + x_4 = 30$ exist with the restriction $-10 \leq x_i \leq 20$?

5.8 Each of n men checks a hat and an umbrella when entering a restaurant. When they leave they are given a hat and an umbrella at random. What is the probability that no one gets his own hat or umbrella?

▷ **5.9** With the notation as in (12), prove

$$\sum_{p=0}^{n} N_p x^p = \sum_{k=0}^{n}(x-1)^k S_k, \text{ where } S_k = \sum_{T:|T|=k} N_{\supseteq T}\,.$$

5.10 What is the probability that $\sigma \in S(n)$ has exactly p fixed points when $n \to \infty$?

* * *

▷ **5.11** Color the integers 1 to $2n$ red or blue in such a way that if i is red then $i-1$ is red too. Use this to prove that

$$\sum_{k=0}^{n}(-1)^k\binom{2n-k}{k}2^{2n-2k}=2n+1,$$

and verify it using generating functions. Use $m+1$ colors to derive the identity

$$\sum_{k\ge 0}(-1)^k\binom{n-k}{k}m^k(m+1)^{n-2k}=\frac{m^{n+1}-1}{m-1}\qquad(m\ge 2).$$

5.12 Set $S_k=\sum_{T:|T|=k}N_{\supseteq T}$ in formula (3); thus $N_{=\emptyset}=\sum_{k=0}^{n}(-1)^kS_k$. Prove the Bonferroni inequalities:
a. $N_{=\emptyset}-\sum_{k=0}^{r}(-1)^kS_k\ge 0$ for r odd,
b. $N_{=\emptyset}-\sum_{k=0}^{r}(-1)^kS_k\le 0$ for r even.
Hint: Show first that $S_k=\sum_{p=k}^{n}\binom{p}{k}N_p$.

5.13 Prove $kS_k\le(n-k+1)S_{k-1}$ with S_k as in the previous exercise, and sharpen this to $kS_k\le S_{k-1}$ if $S_{k+1}=0$.

▷ **5.14** Determine the number of seatings of n couples at a table of length $2n$ as in (8), with no couple sitting side by side, but with the additional restriction that men and women take alternate seats.

5.15 The ménage problem of Lucas asks for the number of seatings as in the previous problem, but with the $2n$ persons sitting around a circular table. Determine this number.

5.16 Consider an $n\times n$-chessboard colored white and black as usual. The number of ways that n non-attacking rooks can be placed is clearly $n!$. Determine the number of ways of placing n non-attacking rooks on the board such that k of them are placed on a white square and $n-k$ on a black square. Hint: Interpret the condition "white" or "black" as a restriction on the associated permutation.

5.17 Galileo problem. Let $G(n,s)$ be the number of different outcomes after throwing n distinguishable dice, summing to s. Show that

$$G(n,s)=\sum_{k=0}^{r}(-1)^k\binom{n}{k}\binom{s-6k-1}{n-1},\quad r=\left\lfloor\frac{s-n}{6}\right\rfloor.$$

▷ **5.18** Let $C(n,k,s)$ be the number of k-subsets of $\{1,2,\dots,n\}$ that contain no run of s consecutive integers. Show that

$$C(n,k,s)=\sum_{i=0}^{\lfloor k/s\rfloor}(-1)^i\binom{n-k+1}{i}\binom{n-is}{n-k}.$$

Hint: Use formula (6).

5.19 Let $D(n,k,s)$ be the number of k-subsets of $\{1,\dots,n\}$ in which any two numbers are at least $s+1$ apart. Prove $D(n,k,s) = \binom{n-(k-1)s}{k}$, and show further that when $1,\dots,n$ are arranged around a circle, this number is $\frac{n}{n-ks}\binom{n-ks}{k}$.

5.20 Brun's sieve: Let $f : \{1,\dots,n\} \to \mathbb{N}_0$ be any function, and set $\mathcal{T} = \{T \subseteq \{1,\dots,n\} : |T \cap \{1,\dots,k\}| \le 2f(k) \text{ for all } k\}$. Prove that $N_{=\emptyset} \le \sum_{T\in\mathcal{T}} (-1)^{|T|} N_{\supseteq T}$.

▷ **5.21** Coupon collector's problem. A company produces s series of n coupons, and puts one coupon in each of sn packages. A customer collects r packages. What is the probability that he gets a complete set of n different coupons? What is the expected number of different coupons?

5.22 Let $\varphi(m)$ be Euler's φ-function, and n an integer that is greater than $m - \varphi(m)$. Prove that $\sum_{k=1}^{n} (-1)^k \binom{n}{k} k^m$ is divisible by m. Hint: Interpret the sum.

▷ **5.23** Let the natural number n have t distinct prime divisors $p_1,\dots,p_t$. Prove that $\sum k^2 = \frac{n^2}{3}\varphi(n) + \frac{(-1)^t}{6} p_1 \cdots p_t \varphi(n)$, where the sum extends over all k, $1 \le k \le n$, that are relatively prime to n. Find a similar formula for $\sum k^3$. Hint: Use a weighting.

5.24 A λ-coloring of a graph $G = (V,E)$ is a mapping $c : V \to \{1,2,\dots,\lambda\}$ such that $c(u) \ne c(v)$ if $\{u,v\} \in E$. Use inclusion-exclusion to prove that the number $\chi(\lambda)$ of λ-colorings is given by $\chi(\lambda) = \sum_{k=0}^{|V|} a_k \lambda^k$, where $a_k = \sum_{i=0}^{|E|} (-1)^i N(i,k)$, and $N(i,k)$ is the number of spanning subgraphs with i edges and k connected components. What is $a_{|V|}, a_{|V|-1}, a_1$ for simple graphs?

5.2 Möbius Inversion

A far-reaching generalization of inclusion–exclusion is provided by inversion on posets. Let $P_<$ be a poset. We call $P_<$ *locally finite* if every *interval* $[a,b] := \{x \in P : a \le x \le b\}$ is finite. Examples are $\mathbb{N}$ or $\mathbb{Z}$ with the natural order, the poset $\mathbb{B}(S)$ of all *finite* subsets of a set S ordered by inclusion, or the *divisor lattice* $\mathbb{D}$ on $\mathbb{N}$ with divisibility $m \mid n$ as ordering.

A convenient way to represent a poset P graphically is to draw P as a graph with the relation $<$ going upward, and where we include only the *covering relations* $a <\!\cdot\, b$ as edges. We say that b *covers* a if $[a, b] = \{a, b\}$. The other relations are then induced by transitivity.

Example. The picture shows the interval $[1, 60]$ in $\mathbb{D}$.

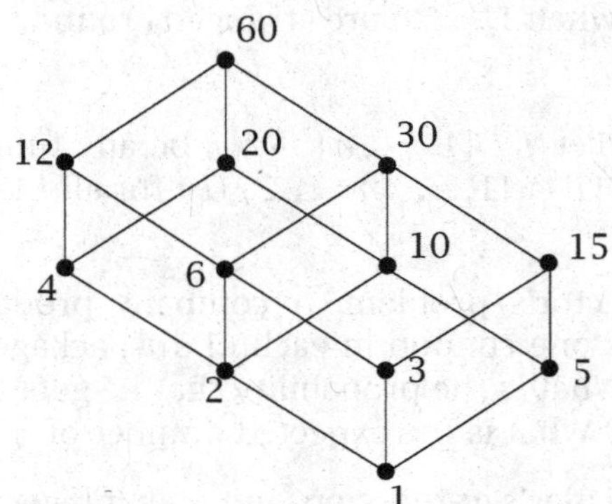

For a locally finite poset P we define the *incidence algebra* of P over a field K of characteristic 0 as

$$\mathbb{A}(P) = \{f : P^2 \to K : x \nleq y \Rightarrow f(x, y) = 0\}.$$

With the usual addition and scalar multiplication, $\mathbb{A}(P)$ becomes a vector space. Next we define the *convolution product* $f * g$ by

$$(f * g)(a, b) = \sum_{a \le x \le b} f(a, x)g(x, b) \quad \text{for } a \le b, \tag{1}$$

and $(f * g)(a, b) = 0$ if $a \nleq b$. By the assumption of local finiteness, the product in (1) is well-defined.

It is a routine matter to verify that the convolution product is associative, and has the Kronecker delta

$$\delta(x, y) = \begin{cases} 1 & x = y, \\ 0 & x \neq y, \end{cases}$$

as two-sided identity. The question when $f \in \mathbb{A}(P)$ has a multiplicative inverse is easily answered.

An element $f \in \mathbb{A}(P)$ has a (unique) two-sided inverse f^{-1} if and only if $f(x, x) \neq 0$ for all $x \in P$.

If f has an inverse f^{-1}, then $f(x, x)f^{-1}(x, x) = \delta(x, x) = 1$, and thus $f(x, x) \neq 0$ for all x. Conversely, let $f(x, x) \neq 0$ for all $x \in P$. We define the left inverse inductively by

$$f^{-1}(x,x) = \frac{1}{f(x,x)},$$

$$f^{-1}(x,y) = \frac{1}{f(y,y)}\left(-\sum_{x\le z<y} f^{-1}(x,z)f(z,y)\right).$$

In the same way the existence of a right inverse is proved, and that the two inverses are the same follows from associativity.

Now consider the function $\zeta \in \mathbb{A}(P)$ defined by

$$\zeta(x,y) = \begin{cases} 1 & \text{if } x \le y, \\ 0 & \text{if } x \not\le y. \end{cases}$$

Then ζ is called the *zeta-function* of P; ζ is obviously invertible.

Definition. The inverse $\mu = \zeta^{-1} \in A(P)$ is called the *Möbius function* of P.

The origin of the names zeta and Möbius function will be shortly explained. The following proposition is immediate fom (1), and could be alternatively used as definition for the Möbius function.

Proposition 5.4. *Let P be a locally finite poset. Then the Möbius function satisfies*

$$\begin{aligned} &\mu(a,a) = 1 \quad (a \in P), \\ &\mu(a,b) = -\sum_{a\le z<b} \mu(a,z) = -\sum_{a<z\le b} \mu(z,b) \quad (a<b). \end{aligned} \tag{2}$$

Example. In the following poset the numbers next to the elements a denote $\mu(0,a)$, where 0 is the minimal element

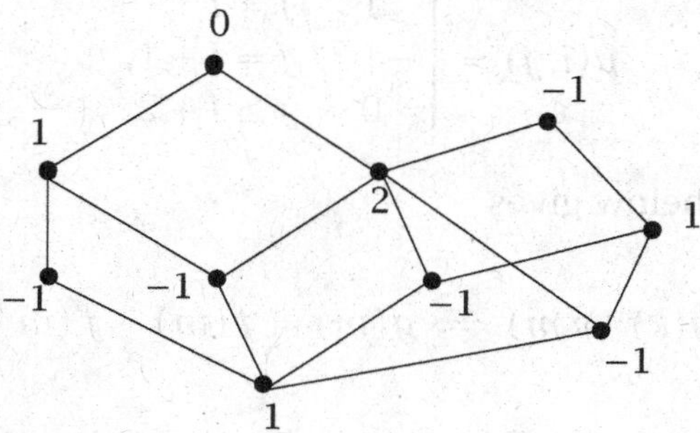

The main result of this section generalizes Theorems 5.1 and 5.3 of the previous section to arbitrary posets.

Theorem 5.5 (Möbius Inversion). *Let P be a locally finite poset, and $f, g : P \to K$. Then*

(i) *Inversion from below:*

$$f(a) = \sum_{x \le a} g(x) \ (\forall a \in P) \iff$$

$$g(a) = \sum_{x \le a} f(x)\mu(x,a) \ (\forall a \in P).$$

(ii) *Inversion from above:* (3)

$$f(a) = \sum_{x \ge a} g(x) \ (\forall a \in P) \iff$$

$$g(a) = \sum_{x \ge a} f(x)\mu(a,x) \ (\forall a \in P).$$

In (i) *we assume the existence of a minimum element* 0 *in* P, *and in* (ii) *of a maximum element* 1.

Proof. Set $\overline{f}(0,a) = f(a)$, and $\overline{f}(x,y) = 0$ for $x \ne 0$, and similarly $\overline{g}(0,a) = g(a)$, $\overline{g}(x,y) = 0$ for $x \ne 0$. The left-hand side in (i) is then equivalent to $\overline{f} = \overline{g} * \zeta$. This, in turn, is equivalent to $\overline{g} = \overline{f} * \mu$, which is precisely the right-hand side. The proof of (ii) is analogous, setting $\overline{f}(a,1) = f(a)$, $\overline{g}(a,1) = g(a)$. □

Difference Calculus.
Consider a chain of length n, $\mathbb{C}_n = \{0 < 1 < \cdots < n\}$. Formula (2) immediately implies for the Möbius function of $\mathbb{C}_n$,

$$\mu(i,j) = \begin{cases} 1 & j = i, \\ -1 & j = i+1, \\ 0 & j \ge i+2. \end{cases} \tag{4}$$

Inversion from below gives

$$f(m) = \sum_{k=0}^{m} g(k) \ (\forall m) \iff g(m) = f(m) - f(m-1) \ (\forall m),$$

which is, of course, just telescoping. Let us look at (4) from an operator standpoint. It is convenient to end the summation at $n-1$, that is,

$$f(n) = \sum_{k=0}^{n-1} g(k) \quad (\forall n) \iff g(n) = f(n+1) - f(n) \quad (\forall n)\,. \quad (5)$$

To $f(x)$ we associate the *(forward) difference operator* $\Delta f(x) = f(x+1) - f(x)$, and set

$$\sum g(x) = f(x) \iff \Delta f(x) = g(x) \quad (6)$$

if f, g are related by (5).

Hence we may regard Σ and Δ as the discrete analogues of *indefinite integration* and *derivative*. As in calculus, we can now compute *definite* sums, since

$$\sum_{k=a}^{b} g(k) = \sum_{k=a}^{b} (f(k+1) - f(k)) = f(b+1) - f(a)\,.$$

To calculate a sum we may therefore proceed as follows. Take an indefinite sum, and compute

$$\sum_{k=a}^{b} g(k) = \sum_{a}^{b+1} g(x) = f(x)\Big|_a^{b+1} = f(b+1) - f(a)\,. \quad (7)$$

Beware that the upper limit is $b+1$.

To make use of this idea we need a list of indefinite sums. For example,

$$\begin{aligned}\Delta x^{\underline{n}} &= (x+1)^{\underline{n}} - x^{\underline{n}} = (x+1)x^{\underline{n-1}} - x^{\underline{n-1}}(x-n+1)\\ &= nx^{\underline{n-1}},\end{aligned}$$

hence

$$\sum x^{\underline{n}} = \frac{x^{\underline{n+1}}}{n+1} \quad (n \ge 0)\,. \quad (8)$$

The falling factorials thus play the role of x^n in our difference calculus. We would like to extend (8) to negative n. From $\frac{x^{\underline{3}}}{x^{\underline{2}}} = x - 2$, $\frac{x^{\underline{2}}}{x^{\underline{1}}} = x - 1$, $\frac{x^{\underline{1}}}{x^{\underline{0}}} = x$, it is plausible to set $\frac{x^{\underline{0}}}{x^{\underline{-1}}} = \frac{1}{x^{\underline{-1}}} = x + 1$, and hence to define $x^{\underline{-1}} = \frac{1}{x+1}$, and in general

$$x^{\underline{-n}} = \frac{1}{(x+1)\cdots(x+n)} \quad (n > 0)\,. \quad (9)$$

It is an easy matter to verify that with (9), we have $\Delta x^{\underline{n}} = nx^{\underline{n-1}}$ for all $n \in \mathbb{Z}$, and thus

$$\sum x^{\underline{n}} = \frac{x^{\underline{n+1}}}{n+1} \quad (n \neq -1). \tag{10}$$

And what is $\sum x^{\underline{-1}}$? From $x^{\underline{-1}} = \frac{1}{x+1} = f(x+1) - f(x)$, we obtain $f(x) = 1 + \frac{1}{2} + \cdots + \frac{1}{x} = H_x$, or

$$\sum x^{\underline{-1}} = H_x. \tag{11}$$

The harmonic function H_x is thus the discrete analogue to the logarithm, and this is one reason why H_n appears in many summation formulas.

Example. Suppose we want to compute $\sum_{k=0}^{n} k^3$.
From $x^3 = \sum_{k=0}^{3} S_{3,k} x^{\underline{k}} = x^{\underline{3}} + 3x^{\underline{2}} + x^{\underline{1}}$ results

$$\begin{aligned}
\sum_{k=0}^{n} k^3 &= \sum_{0}^{n+1} x^3 = \sum_{0}^{n+1} x^{\underline{3}} + 3\sum_{0}^{n+1} x^{\underline{2}} + \sum_{0}^{n+1} x^{\underline{1}} \\
&= \frac{x^{\underline{4}}}{4}\Big|_0^{n+1} + 3\frac{x^{\underline{3}}}{3}\Big|_0^{n+1} + \frac{x^{\underline{2}}}{2}\Big|_0^{n+1} \\
&= \frac{(n+1)n(n-1)(n-2)}{4} + (n+1)n(n-1) + \frac{(n+1)n}{2} \\
&= \left(\frac{(n+1)n}{2}\right)^2.
\end{aligned}$$

There is also a rule for partial summation. Writing

$$\begin{aligned}
\Delta(u(x)v(x)) &= u(x+1)v(x+1) - u(x)v(x) \\
&= u(x+1)v(x+1) - u(x)v(x+1) \\
&\qquad + u(x)v(x+1) - u(x)v(x) \\
&= (\Delta u(x))v(x+1) + u(x)(\Delta v(x)),
\end{aligned}$$

we obtain

$$\sum u\Delta v = uv - \sum (\Delta u)Tv, \tag{12}$$

where $Tv(x) = v(x+1)$ is the translation operator.

Example. We have proved in Section 3.2 the formula $\sum_{k=1}^{n} H_k = (n+1)(H_{n+1} - 1)$. Let us evaluate the general sum $\sum_{k=1}^{n} \binom{k}{m} H_k$. From the Pascal recurrence $\binom{x+1}{m+1} = \binom{x}{m} + \binom{x}{m+1}$ follows $\Delta\binom{x}{m+1} = \binom{x}{m}$ or $\sum \binom{x}{m} = \binom{x}{m+1}$. If we set $u(x) = H_x$, $v(x) = \binom{x}{m}$, then partial summation yields

$$\begin{aligned}\sum_{k=1}^{n} \binom{k}{m} H_k = \sum_{1}^{n+1} \binom{x}{m} H_x &= \binom{x}{m+1} H_x \Big|_1^{n+1} - \sum_{1}^{n+1} \frac{1}{x+1} \binom{x+1}{m+1} \\ &= \binom{x}{m+1} H_x \Big|_1^{n+1} - \frac{1}{m+1} \sum_{1}^{n+1} \binom{x}{m} \\ &= \binom{x}{m+1} H_x \Big|_1^{n+1} - \frac{1}{m+1} \binom{x}{m+1} \Big|_1^{n+1} \\ &= \binom{n+1}{m+1} \left(H_{n+1} - \frac{1}{m+1} \right) \quad \text{for } m \geq 0 .\end{aligned}$$

Möbius Function.

Let us return to the general case. In order to apply Möbius inversion on the poset P we must be able to compute the Möbius function efficiently. There are many deep results; let us just collect some basic facts. First it is clear that $\mu(a, b) = \mu(c, d)$ if the intervals $[a, b]$ and $[c, d]$ are isomorphic posets. Consider next a product $P = P_1 \times P_2$ of two posets. This means that

$$(a, b) \underset{P}{\leq} (c, d) \iff a \underset{P_1}{\leq} c \text{ and } b \underset{P_2}{\leq} d .$$

Claim. $\mu((a, b), (c, d)) = \mu_1(a, c)\mu_2(b, d)$, where μ_i is the Möbius function of P_i.

If $a = c$, $b = d$, there is nothing to prove, so assume without loss of generality $b < d$. By (2) and induction, we have

$$\begin{aligned}
\mu((a,b),(c,d)) &= -\sum_{(a,b)\le(x,y)<(c,d)} \mu((a,b),(x,y)) \\
&= -\sum_{b\le y<d} \mu((a,b),(c,y)) \\
&\qquad - \sum_{a\le x<c}\sum_{b\le y\le d} \mu((a,b),(x,y)) \\
&= \mu_1(a,c)\left(-\sum_{b\le y<d} \mu_2(b,y)\right) \\
&\qquad - \sum_{a\le x<c} \mu_1(a,x)\left(\sum_{b\le y\le d} \mu_2(x,y)\right) \\
&= \mu_1(a,c)\mu_2(b,d)\,,
\end{aligned}$$

since the last sum is 0 because of $b < d$.

Suppose P and Q are finite posets with minimum and maximum elements $0_P, 1_P$, and $0_Q, 1_Q$, respectively. Then we abbreviate $\mu(P) = \mu_P(0_P, 1_P)$, and similarly $\mu(Q) = \mu(0_Q, 1_Q)$. By what we just proved,

$$\mu(P\times Q) = \mu(P)\mu(Q)\,. \tag{13}$$

Examples. Suppose $C = \mathbb{C}_{k_1}\times\cdots\times\mathbb{C}_{k_t}$ is a product of chains of lengths $k_1,\ldots,k_t \ge 1$. Then by (4) and (13),

$$\mu(C) = \begin{cases} (-1)^t & \text{if } k_1=\cdots=k_t=1\,, \\ 0 & \text{if } k_i\ge 2 \text{ for some } i\,. \end{cases} \tag{14}$$

Consider the Boolean poset $\mathbb{B}(n)$ of all subsets of an n-set S, ordered by inclusion. Then $\mathbb{B}(n) = \mathbb{C}_1\times\cdots\times\mathbb{C}_1$, as is easily seen via the characteristic vector, and thus $\mu(\mathbb{B}(n)) = (-1)^n$. More precisely, for $A\subseteq T\subseteq S$ we have $[A,T] = \{A\subseteq U\subseteq T\}\cong\mathbb{B}(t)$, where $t = |T\smallsetminus A|$, and hence

$$\mu(A,T) = (-1)^{|T|-|A|}\,.$$

Möbius inversion from above in the Boolean poset reduces thus precisely to Theorem 5.1, and similarly, inversion from below yields Theorem 5.3.

Let us consider next the divisor lattice $\mathbb{D}$, and suppose $\ell \mid m$ with $\frac{m}{\ell} = p_1^{k_1}\cdots p_t^{k_t}$ being the decomposition into prime powers. A moment's thought shows that $[\ell,m]\cong\mathbb{C}_{k_1}\times\cdots\times\mathbb{C}_{k_t}$. In the example $[1,60]$ from above we have $60 = 2^2 \cdot 3 \cdot 5$, and thus $[1,60] \cong \mathbb{C}_2 \times \mathbb{C}_1 \times \mathbb{C}_1$. In general, equation (14) gives

$$\mu(\ell,m) = \begin{cases} 1 & \text{if } \ell = m, \\ (-1)^t & \text{if } \frac{m}{\ell} = p_1 \cdots p_t,\ p_i \text{ distinct primes,} \\ 0 & \text{otherwise.} \end{cases} \tag{15}$$

At this point it is convenient to introduce the *Möbius function* $\overline{\mu}(n)$ from number theory:

$$\overline{\mu}(n) = \begin{cases} 1 & n = 1, \\ (-1)^t & \text{if } n = p_1 \cdots p_t,\ p_i \text{ distinct primes,} \\ 0 & \text{otherwise.} \end{cases} \tag{16}$$

Hence we have $\mu(\ell,m) = \overline{\mu}(\frac{m}{\ell})$ for $\ell \mid m$.

Remark. This connection is the origin of the name Möbius function, and the term zeta-function is also borrowed from number theory. We may relate a certain subalgebra of the incidence algebra of $\mathbb{D}$ to so-called *Dirichlet series*, mapping $f \in \mathbb{A}(\mathbb{D})$ to $\sum_{n\geq 1} \frac{\overline{f}(n)}{n^s}$, where $\overline{f}(n) = f(\ell,m)$, $\frac{m}{\ell} = n$, and s is a complex variable. Thus $\zeta \in \mathbb{A}(\mathbb{D})$ is associated with the famous Riemann ζ-function $\zeta(s) = \sum_{n\geq 1} \frac{1}{n^s}$ and μ with $\sum_{n\geq 1} \frac{\overline{\mu}(n)}{n^s}$. The relation $\mu = \zeta^{-1}$ then translates into $\zeta(s)^{-1} = \sum_{n\geq 1} \frac{\overline{\mu}(n)}{n^s}$, as can be formally proved by comparing coefficients.

Example. Consider a finite field $K = GF(q)$ with q elements. A basic result of algebra states that

$$x^{q^n} - x = \prod f(x),$$

where $f(x)$ runs through all irreducible polynomials over $GF(q)$ with leading coefficient 1, and degree d, $d \mid n$. Let U_d be the number of these irreducible polynomials of degree d. Then

$$q^n = \sum_{d\mid n} dU_d\,,$$

since every polynomial of degree d contributes d roots of $x^{q^n} - x$. Möbius inversion over $[1,n]$ thus yields

$$U_n = \frac{1}{n} \sum_{d\mid n} \overline{\mu}(d) q^{\frac{n}{d}}\,.$$

For $q = 2$, $n = 3$, we obtain

$$U_3 = \frac{1}{3}[2^3 - 2] = 2\,,$$

with the polynomials $x^3 + x^2 + 1$ and $x^3 + x + 1$.

Möbius inversion sometimes provides a convenient way to evaluate determinants. Let $P = \{a_1, \dots, a_n\}$ be a lower semilattice. That is, P is a poset, and to every pair a_i, a_j there exists a unique maximal element $a_i \wedge a_j \le a_i, a_j$. Let $f, g : P \to K$ be functions with $f(a) = \sum_{x \le a} g(x)$ for all $a \in P$,and denote by $F = (f_{ij})$ the $n \times n$-matrix with $f_{ij} = f(a_i \wedge a_j)$. Then

$$\det F = g(a_1)g(a_2)\cdots g(a_n)\,. \tag{17}$$

For the proof let $G = (g_{ij})$ be the diagonal matrix with $g_{ii} = g(a_i)$, and $Z = (z_{ij})$ with

$$z_{ij} = \begin{cases} 1 & a_i \le a_j, \\ 0 & a_i \not\le a_j. \end{cases}\,.$$

We have $Z^T G Z = F$, since

$$\begin{aligned}(Z^T G Z)_{i,j} &= \sum_{k=1}^{n} \sum_{\ell=1}^{n} z_{ki} g_{k\ell} z_{\ell j} = \sum_{k=1}^{n} z_{ki} g_{kk} z_{kj} \\ &= \sum_{\substack{a_k \le a_i \\ a_k \le a_j}} g(a_k) = \sum_{a_k \le a_i \wedge a_j} g(a_k) = f(a_i \wedge a_j) = f_{ij}\,.\end{aligned}$$

Now, $\det Z = 1$ (clear?), and it follows that

$$\det F = \det G = g(a_1)\cdots g(a_n)\,.$$

Hence whenever we are able to invert $f(a) = \sum_{x \le a} g(x)$, we can read off $\det F$ from (17).

Exercises

5.25 Prove associativity of the convolution product $*$.

5.26 Let $\eta \in \mathbb{A}(P)$ be defined by $\eta = \zeta - \delta$. Show that $\eta^k(a, b)$ counts the numbers of chains from a to b of length k. Determine η^k for the chain $\mathbb{C}_n$ and the Boolean poset $\mathbb{B}(n)$.

▷ **5.27** Show that $\mu = \sum_{k\ge 0}(-1)^k\eta^k$ in $\mathbb{A}(P)$, and apply this to the posets $\mathbb{C}_n$ and $\mathbb{B}(n)$.

5.28 Interpret $\zeta^k(a,b)$ for a poset P, and compute ζ^k for $\mathbb{C}_n$ and $\mathbb{B}(n)$.

5.29 Show that the incidence algebra $\mathbb{A}(P)$ is commutative if and only if P is an antichain, that is, there are no relations in P.

5.30 Compute the Möbius function for the poset

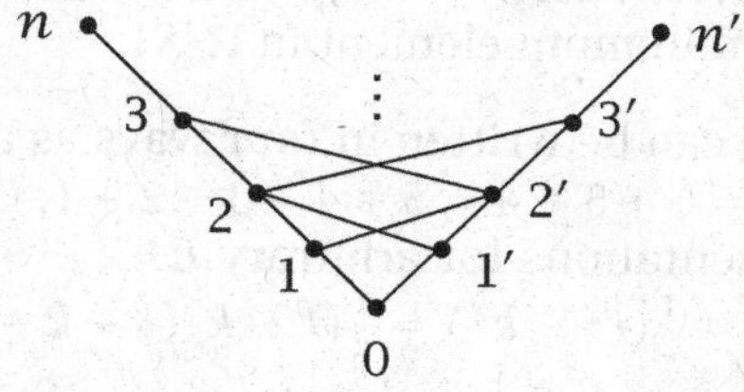

that is, $i < j$, $i < j'$, $i' < j$, $i' < j'$, if $i < j$ as natural numbers.

▷ **5.31** Interpret $\varphi(n) = n - \sum_i \frac{n}{p_i} + \sum_{i<j}\frac{n}{p_ip_j} \mp \cdots$ in the poset $\mathbb{D}$, and derive $n = \sum_{d|n}\varphi(d)$ with Möbius inversion.

5.32 Let F_x be the Fibonacci function. Show that $\Delta F_x = F_{x-1}$, and re-prove $\sum_{k=0}^{n} F_k = F_{n+2} - 1$.

5.33 Show that $\sum c^x = \frac{c^x}{c-1}$ $(c \ne 1)$, and use this to evaluate $\sum_{k=0}^{n} k2^k$ and $\sum_{k=0}^{n} k^2 2^k$.

▷ **5.34** Evaluate $\sum_{k=1}^{n-1}\frac{H_k}{(k+1)(k+2)}$ with partial summation.

5.35 Show that $\sum_{k\ge 0}\eta^k(a,b) = \frac{1}{2\delta-\zeta}(a,b)$, $a \le b$, where η is defined as in Exercise 5.26, and apply this to the posets $\mathbb{C}_n$ and $\mathbb{B}(n)$.

* * *

▷ **5.36** Prove $\mu^r = \sum_{k\ge 0}(-1)^k\binom{r+k-1}{k}\eta^k$ in $\mathbb{A}(P)$, and deduce from this the following identities:
a. $\sum_k (-1)^{n-k}\binom{r+k-1}{k}k!S_{n,k} = r^n$, b. $\sum_k (-1)^k\binom{r+k-1}{k}\binom{n-1}{k-1} = (-1)^n\binom{r}{n}$.

5.37 Let $\mathbb{L}(n,q)$ be the poset of all subspaces of an n-dimensional vector space over the finite field $GF(q)$. Prove $\mu(\mathbb{L}(n,q)) = (-1)^n q^{\binom{n}{2}}$. Hint: Use Exercise 1.74.

▷ **5.38** Use the previous exercise to compute the number of surjective linear transformations from an n-dimensional vector space to an r-dimensional vector space, both over $GF(q)$.

5.39 Prove that the number of $n \times m$-matrices over $GF(q)$ with rank r is given by

$$\begin{bmatrix} m \\ r \end{bmatrix}_q \sum_{k=0}^{r} (-1)^{r-k} \begin{bmatrix} r \\ k \end{bmatrix}_q q^{nk+\binom{r-k}{2}} .$$

▷ **5.40** Let S be an n-set. The set $\Pi(S)$ of all partitions becomes a poset with the *refinement relation*. That is, $\pi \leq \sigma$ if any block of π is contained in a block of σ. Compute $\mu(\Pi(S))$. Hint: For a map $h : S \to X$ let $\ker(h)$ be the partition of S induced by the pre-images. Define $g : \Pi(S) \longrightarrow \mathbb{R}[x]$ by $g(\pi) = \#\{h \in \mathrm{Map}(S, X) : \ker(h) = \pi\}$, where X is an x-set, and consider $g(0)$, where 0 is the minimum element in $\Pi(S)$.

5.41 The number 15 can be written in four ways as a sum of consecutive integers: $15 = 8 + 7 = 6 + 5 + 4 = 5 + 4 + 3 + 2 + 1$. Compute the number $f(n)$ of these representations for arbitrary n.
Hint: Consider $\sum_k^{\ell} x = \frac{1}{2}(\ell^{\underline{2}} - k^{\underline{2}}) = \frac{1}{2}(\ell - k)(\ell + k - 1)$.

5.42 Compute $\sum_{k=0}^{n}(-1)^k/\binom{n}{k}$ with the difference calculus.

5.43 Determine $\sum_{k=1}^{n}(-1)^k\binom{n}{k}H_k$ with partial summation. The result is very simple. Deduce with binomial inversion a new formula for H_n.

▷ **5.44** The sorting algorithm Quicksort has an average running time Q_n observing the recurrence $Q_n = (n-1) + \frac{1}{n}\sum_{k=1}^{n}(Q_{k-1} + Q_{n-k})$, $Q_0 = 0$. Use the difference calculus to compute Q_n. Hint: Write the recurrence as $nQ_n = n(n-1) + 2\sum_{k=0}^{n-1} Q_k$ $(n \geq 1)$.

5.45 Determine $\sum_{n \leq x} \overline{\mu}(n) \lfloor \frac{x}{n} \rfloor$. Hint: $\lfloor \frac{x}{n} \rfloor = \sum_{n|k, k \leq x} 1$.

5.46 Evaluate the determinant of the $n \times n$-matrix $M = (m_{ij})$, where $m_{ij} = \gcd(i, j)$.

▷ **5.47** Suppose $\mathcal{D}$ is a down-set of 2^X, that is, $A \in \mathcal{D}$, $B \subseteq A$ implies $B \in \mathcal{D}$. List the members of $\mathcal{D}$ as $U_1, U_2, \ldots$, and let $M = (m_{ij})$ be the matrix with

$$m_{ij} = \begin{cases} 1 & U_i \cap U_j = \emptyset \\ 0 & \text{otherwise} \end{cases} .$$

Prove $\det M = (-1)^{\ell}$, $\ell = \#$ (odd-sized sets in $\mathcal{D}$).

5.3 The Involution Principle

We have seen many examples in which identities or counting formulas were found by producing suitable bijections. The *involution principle* generalizes this idea.

Let $S = S^+ \cup S^-$ be a partition of a finite set S into two parts, the *positive* part S^+ and the *negative* part S^-. We then call S a *signed* set. What we are interested in is the difference $|S^+| - |S^-|$. Let us call an involution $\varphi : S \to S$ *alternating* (or *sign-reversing*) if x and φx are in different parts of S whenever $\varphi x \neq x$. It follows for the set of fixed points that

$$|S^+| - |S^-| = |\mathrm{Fix}\, S^+| - |\mathrm{Fix}\, S^-|\,. \tag{1}$$

If, in particular, $\mathrm{Fix}\, S^- = \emptyset$, that is, all of S^- is carried over into S^+ by φ, then

$$|S^+| - |S^-| = |\mathrm{Fix}\, S^+|\,. \tag{2}$$

Our method thus proceeds as follows: In order to enumerate a set X,

1. embed X into a signed set $S = S^+ \cup S^-$ with $X \subseteq S^+$,
2. find an alternating involution φ with $\mathrm{Fix}\, S^- = \emptyset$, $\mathrm{Fix}\, S^+ = X$.

Then $|X| = |S^+| - |S^-|$.

Example. Catalan Paths. We want to count all lattice paths from $(0,0)$ to (n,n) with the usual $(1,0)$- and $(0,1)$-steps that never go beyond the diagonal $y = x$. Let A_n be their number.

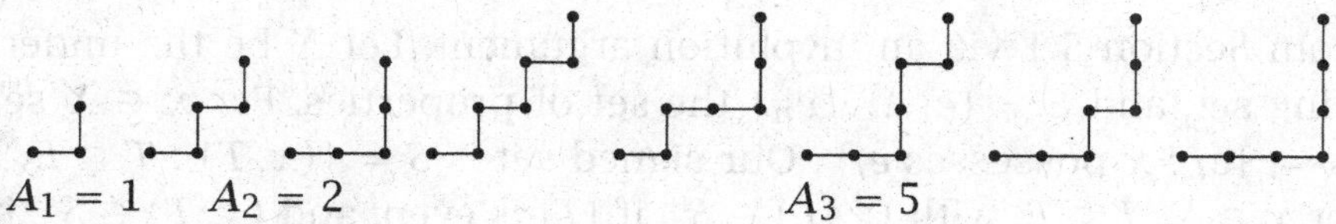

Let S^+ be the set of *all* lattice paths from $(1,0)$ to $(n+1,n)$, and S^- all paths from $(0,1)$ to $(n+1,n)$. By our old result from Section 1.2, $|S^+| = \binom{2n}{n}$, $|S^-| = \binom{2n}{n-1}$. For $W \in S^+ \cup S^-$, let φW be the path in which the part from $(1,0)$ or $(0,1)$ until the *first* arrival at the diagonal is reflected (and the rest kept unchanged). If W never hits the diagonal, then $\varphi W = W$. As an example,

(4,3)
(1,0)
W

(4,3)
(0,1)
(0,0)
φW

It is clear that φ is an alternating involution. Furthermore, $\operatorname{Fix} S^- = \emptyset$ because any path in S^- must eventually cross the diagonal. The paths in $\operatorname{Fix} S^+$ correspond precisely to the Catalan paths by moving them one step to the left, and we obtain

$$A_n = |\operatorname{Fix} S^+| = |S^+| - |S^-| = \binom{2n}{n} - \binom{2n}{n-1} = \frac{1}{n+1}\binom{2n}{n}.$$

Hence the number of these paths is precisely the Catalan number C_n.

It is convenient to represent Catalan paths (by a rotation and reflection) as paths from $(0,0)$ to $(2n,0)$ with diagonal steps up $(1,1)$ and down $(1,-1)$ that never fall below the x-axis.

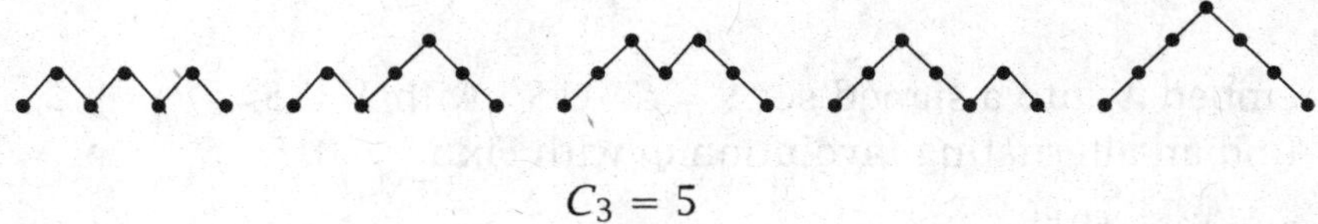

$C_3 = 5$

Example. Inclusion–Exclusion. Let us re-prove the formula

$$N_{=\emptyset} = \sum_{T \subseteq E} (-1)^{|T|} N_{\supseteq T}$$

from Section 5.1 via an involution argument. Let X be the underlying set, and $E = \{e_1, \dots, e_n\}$ the set of properties. For $x \in X$ set $E_x = \{e_i : x \text{ possesses } e_i\}$. Our signed set is $S = \{(x,T) : T \subseteq E_x\}$ for $x \in X$, $T \subseteq E$, with $(x,T) \in S^+$ if $|T|$ is even, and $(x,T) \in S^-$ if $|T|$ is odd. For $x \in X$ let $m(x) = \max\{i : e_i \in E_x\}$ whenever $E_x \neq \emptyset$. Now define $\varphi : S \to S$ by

$$\varphi(x,T) = \begin{cases} (x, T \smallsetminus e_{m(x)}) & \text{if } e_{m(x)} \in T, \\ (x, T \cup e_{m(x)}) & \text{if } e_{m(x)} \notin T, \end{cases}$$

whenever E_x is nonempty, and $\varphi(x,\emptyset) = (x,\emptyset)$ for $E_x = \emptyset$.

The mapping φ is clearly an alternating involution, and the only fixed points are $(x,\emptyset)$ with $E_x = \emptyset$. Since $|\emptyset| = 0$ is even, we obtain $|\operatorname{Fix} S| = |\operatorname{Fix} S^+| = N_{=\emptyset}$. Furthermore,

$$N_{\supseteq T} = \Big| \bigcup_{x : T \subseteq E_x} (x,T) \Big|,$$

and the involution principle gives

$$N_{=\emptyset} = |\mathrm{Fix}\, S^+| = |S^+| - |S^-| = \sum_{T\subseteq E} (-1)^{|T|} N_{\supseteq T}\,.$$

Just as for inclusion–exclusion we can readily extend the involution principle to the weighted case. Let $S = S^+ \cup S^-$ be a signed set, and suppose the elements x of S have weights $w(x)$. We call $\varphi : S \to S$ an alternating *weight-preserving* involution if $w(\varphi x) = w(x)$ for all $x \in S$. It then follows that

$$w(S^+) - w(S^-) = w(\mathrm{Fix}\, S^+) - w(\mathrm{Fix}\, S^-)\,. \tag{3}$$

Example. Vandermonde Determinant. We want to prove

$$\det \begin{pmatrix} x_1^{n-1} & x_2^{n-1} & \dots & x_n^{n-1} \\ x_1^{n-2} & x_2^{n-2} & \dots & x_n^{n-2} \\ & & \dots & \\ x_1 & x_2 & \dots & x_n \\ 1 & 1 & \dots & 1 \end{pmatrix} = \prod_{1\le i<j\le n} (x_i - x_j)\,. \tag{4}$$

This formula is of eminent importance in linear algebra, and will play a prominent role in Chapter 8, on symmetric functions. The following proof is certainly not the quickest, but it very nicely illustrates the applicability of the involution principle.

The left-hand side in (4) contains $n!$ summands, and the right-hand side $2^{\binom{n}{2}}$; hence we must sieve out $2^{\binom{n}{2}} - n!$ summands. The determinant equals

$$\det = \sum_{\sigma\in S(n)} (\mathrm{sign}\,\sigma) x_{\sigma(1)}^{n-1} x_{\sigma(2)}^{n-2} \cdots x_{\sigma(n)}^{0}\,, \tag{5}$$

whereas the right-hand side is the sum of all expressions

$$(-1)^m x_1^{a_1} x_2^{a_2} \cdots x_n^{a_n} \tag{6}$$

with $\sum_{i=1}^n a_i = \binom{n}{2}$, where $m = \#\{j : x_j \text{ is taken from } x_i - x_j\}$.

Our set consists of the *tournaments* on $\{1,2,\dots,n\}$, that is, all directed graphs on the vertex-set $\{1,2,\dots,n\}$ with precisely one directed edge $i \to j$ or $j \to i$ for any pair $i \neq j$. Hence there are $2^{\binom{n}{2}}$ tournaments, and we now want to associate to each tournament a unique term as in (6).

Let T be a tournament, and $e : k \to \ell$ an edge in T. We call k the *winner*, and define the *weight* $w(e)$ by

$$w(e) = x_k \text{ with sign } e = \begin{cases} 1 & k < \ell. \\ -1 & k > \ell, \end{cases}$$

If we define the weight $w(T)$ of the tournament T as

$$w(T) = \prod_{e \in T} w(e), \quad \text{sign } T = \prod_{e \in T} \text{sign } e,$$

then it is clear that we obtain precisely the expression (6),

$$(-1)^m x_1^{a_1} \cdots x_n^{a_n}, \quad w(T) = x_1^{a_1} \cdots x_n^{a_n}, \quad (-1)^m = \text{sign } T,$$

where a_i is the out-degree $d^-(i)$, or in our language the number of wins of i.

In sum, we get

$$\prod_{1 \le i < j \le n} (x_i - x_j) = \sum_T (\text{sign } T) w(T). \tag{7}$$

Example. Consider the tournament

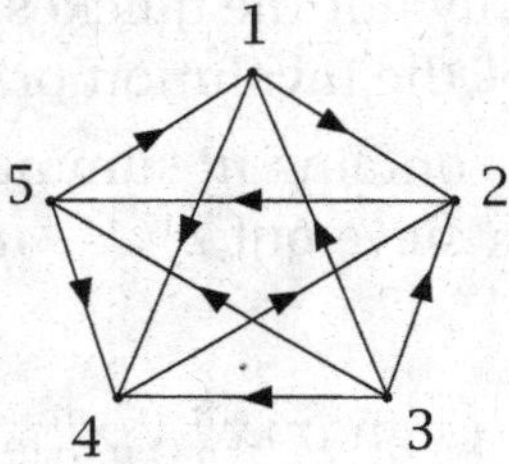

We pick the winners from the product

$$(\underline{x_1} - x_2)(x_1 - \underline{x_3})(\underline{x_1} - x_4)(x_1 - \underline{x_5})(x_2 - \underline{x_3}) \\ \times (x_2 - \underline{x_4})(\underline{x_2} - x_5)(\underline{x_3} - x_4)(\underline{x_3} - x_5)(x_4 - \underline{x_5})$$

and obtain as weight $x_1^2 x_2 x_3^4 x_4 x_5^2$ with sign $= -1$.

Among the tournaments there is a special type, the *transitive* tournaments, which satisfy that $i \to j, j \to k$ implies $i \to k$. Every transitive tournament can be thought of as a (unique) linear order or permutation $\sigma(1)\sigma(2)\ldots\sigma(n)$, where $\sigma(1)$ is the overall winner,

$\sigma(2)$ the second best, and so on. Hence there are $n!$ transitive tournaments.

Example. The permutation 35142 corresponds to the transitive tournament

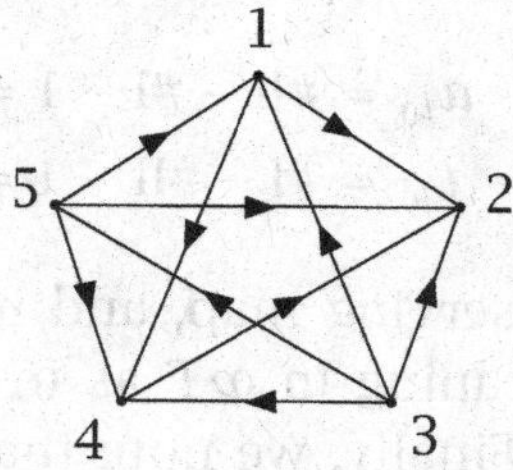

For a transitive tournament T_σ we obtain

$$w(T_\sigma) = x_{\sigma(1)}^{n-1} x_{\sigma(2)}^{n-2} \cdots x_{\sigma(n)}^{0}, \quad \text{sign } T_\sigma = (-1)^{\text{inv}(\sigma)} = \text{sign } \sigma,$$

and so

$$\det = \sum_{\sigma \in S(n)} (\text{sign } \sigma) w(T_\sigma). \tag{8}$$

Comparing (7) and (8) we see that we must sieve out the non-transitive tournaments. Let S be the set of non-transitive tournaments, with $S^+ = \{T \in S : \text{sign } T = 1\}$, $S^- = \{T \in S : \text{sign } T = -1\}$. All that remains to find is an alternating weight-preserving involution on S without fixed points.

A little reflection shows that any non-transitive tournament contains two vertices i and j with $d^-(i) = d^-(j)$. Consider $T \in S$, and $d^-(i) = a_i$. Among all double occurrences $a_i = a_j$ choose i_0 to be the minimal index, and among all $a_j = a_{i_0}$, let j_0 be minimal. We may assume $i_0 \to j_0$, the other case being analogous. For $k \neq i_0, j_0$ we have the following four possibilities:

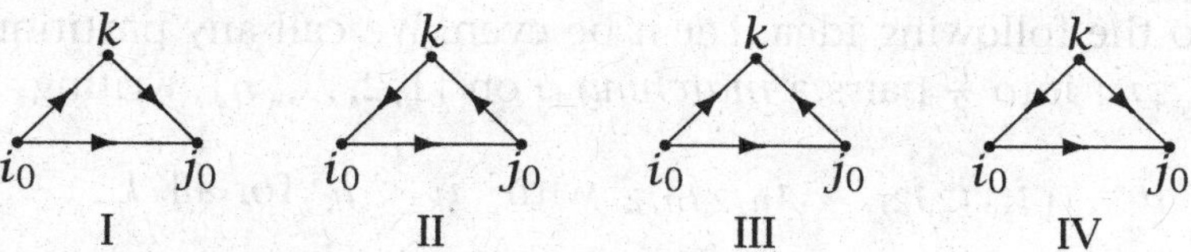

Since $d^-(i_0) = d^-(j_0)$, there must be at least one triangle of type II, more precisely,

$$\#\text{II} = \#\text{I} + 1.$$

Now we define the tournament φT by reversing $i_0 \to j_0$, and the other two directions in all triangles of types I and II, keeping the rest unchanged. Set $d^-(k) = b_k$ in φT, then clearly $b_k = a_k$ for all $k \neq i_0, j_0$, and for i_0 and j_0 we obtain

$$\begin{aligned} b_{i_0} - a_{i_0} &= \#\text{II} - \#\text{I} - 1 = 0\,, \\ b_{j_0} - a_{j_0} &= \#\text{I} - \#\text{II} + 1 = 0\,. \end{aligned}$$

Thus φ is a weight-preserving map, and it is an involution, since i_0, j_0 have the same meaning in φT as in T, returning in $\varphi(\varphi T)$ the original directions. Finally, we note that φ reverses two edges in every triangle of type I or II plus the single edge $i_0 \to j_0$, hence altogether an odd number. It follows that sign $\varphi T = -$sign T, and we are finished.

Example. The Pfaffian. As our final example we discuss an important theorem from linear algebra, where the involution principle is applied in an unexpected way. Let $A = (a_{ij})$ be a real skew-symmetric $n \times n$-matrix, that is, $A^T = -A$. If n is odd, then

$$\det A = \det A^T = (-1)^n \det A = -\det A\,,$$

and hence $\det A = 0$.

But what happens when n is even? For $n = 4$ one easily computes

$$\det \begin{pmatrix} 0 & a_{12} & a_{13} & a_{14} \\ -a_{12} & 0 & a_{23} & a_{24} \\ -a_{13} & -a_{23} & 0 & a_{34} \\ -a_{14} & -a_{24} & -a_{34} & 0 \end{pmatrix} = (a_{12}a_{34} - a_{13}a_{24} + a_{14}a_{23})^2\,. \quad (9)$$

We see that the terms on the right-hand side are of the form $a_{i_1 j_1} a_{i_2 j_2}$ with $\{i_1, j_1, i_2, j_2\} = \{1, 2, 3, 4\}$, with different signs. This leads to the following idea. Let n be even. We call any partition of $\{1, 2, \ldots, n\}$ into $\frac{n}{2}$ pairs a *matching* μ on $\{1, 2, \ldots, n\}$, writing

$$\mu = i_1 j_1, i_2 j_2, \ldots, i_{n/2} j_{n/2} \quad \text{with} \quad i_k < j_k \text{ for all } k\,.$$

There are clearly $(n-1)(n-3)\cdots 3 \cdot 1$ different matchings altogether. Given the skew-symmetric matrix A, we use the notation

$$a_\mu = a_{i_1 j_1} a_{i_2 j_2} \cdots a_{i_{n/2} j_{n/2}}\,.$$

To define the sign of μ we list $1, 2, \ldots$ as in the diagram; $\#\mu$ denotes the number of *crossings*, and the sign of μ is sign $\mu = (-1)^{\#\mu}$.

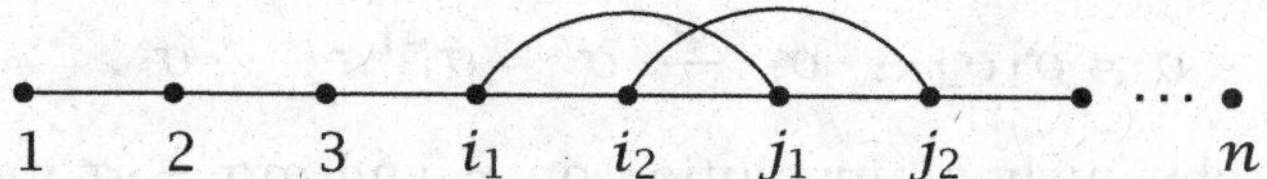

Example. For $n = 4$ we have three matchings:

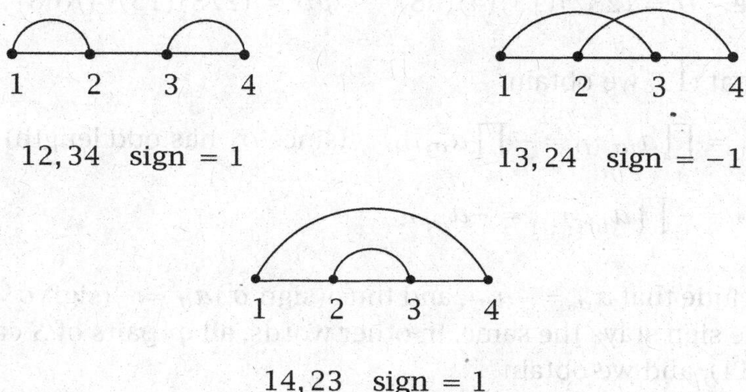

Note that the signs are precisely the same as in (9), and this is no coincidence, as we are now going to prove.

Definition. Let A be a real skew-symmetric $n \times n$-matrix, n even. The *Pfaffian* $\text{Pf}(A)$ is defined as

$$\text{Pf}(A) = \sum_{\mu} (\text{sign } \mu) a_{\mu} .$$

Theorem 5.6 (Cayley). *Let A be a real skew-symmetric $n \times n$-matrix, n even. Then*

$$\det A = [\text{Pf}(A)]^2 . \tag{10}$$

Proof. We start with the usual expression for the determinant,

$$\det A = \sum_{\sigma \in S(n)} (\text{sign } \sigma) a_{\sigma}, \quad \text{where} \quad a_{\sigma} = a_{1\sigma(1)} a_{2\sigma(2)} \cdots a_{n\sigma(n)} . \tag{11}$$

Let S be the subset of those permutations that possess at least one cycle of odd length. We define $\varphi : S \to S$ as follows. For $\sigma \in S$, let

$\sigma = \sigma_1\sigma_2 \cdots \sigma_t$ be the cycle decomposition, where σ_1 shall be the cycle of odd length that contains the smallest element among all cycles of odd length, and define

$$\sigma = \sigma_1\sigma_2 \cdots \sigma_t \xrightarrow{\varphi} \sigma' = \sigma_1^{-1}\sigma_2 \cdots \sigma_t \,. \tag{12}$$

Then φ is obviously an involution on S, and $\varphi\sigma = \sigma$ means that $\sigma_1 = (k)$ is a cycle of length 1. In this case, k is a fixed point of σ, which implies $a_\sigma = 0$, since $a_{kk} = 0$.

Example. $\sigma = (237)(15)(4)(68) \longrightarrow \varphi\sigma = (273)(15)(4)(68)\,.$

Looking at (12) we obtain

$$\begin{aligned} a_{\sigma_1} &= \prod_i a_{i\sigma_1(i)} = -\prod_i a_{\sigma_1(i)i} \quad \text{(since } \sigma_1 \text{ has odd length)} \\ &= -\prod_i a_{i\sigma_1^{-1}(i)} = -a_{\sigma_1^{-1}}\,. \end{aligned}$$

We conclude that $a_\sigma = -a_{\sigma'}$, and thus $(\text{sign}\ \sigma)a_\sigma = -(\text{sign}\ \sigma')a_{\sigma'}$, since the sign stays the same. In other words, all φ-pairs of S cancel out in (11), and we obtain

$$\det A = \sum_{\sigma\in E} (\text{sign}\ \sigma)a_\sigma\,, \tag{13}$$

where $E \subseteq S(n)$ is the set of permutations all of whose cycles have *even* length.

To prove $[\text{Pf}(A)]^2 = \det A$, it suffices therefore to find a bijection $(\mu_1, \mu_2) \xrightarrow{\phi} \sigma \in E$ with

$$(\text{sign}\ \mu_1)a_{\mu_1} \cdot (\text{sign}\ \mu_2)a_{\mu_2} = (\text{sign}\ \sigma)a_\sigma\,. \tag{14}$$

The mapping ϕ is explained in the following picture.

Example. $\mu_1 = 14, 28, 35, 67$, $\mu_2 = 15, 26, 34, 78$.

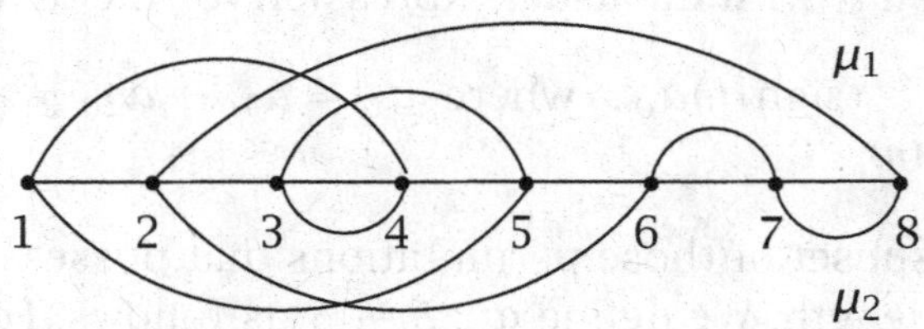

The pair (μ_1, μ_2) decomposes $\{1, 2, \ldots, n\}$ into disjoint directed cycles of even length (these are the cycles of σ), where we choose the orientation in a cycle, starting at the smallest element in the direction of μ_1.

In our example, this gives

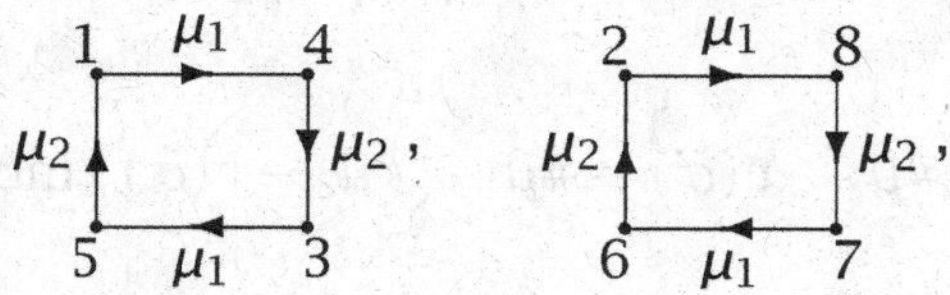

thus $\sigma = (1\,4\,3\,5)(2\,8\,7\,6)$.

The reverse construction is now clear.

Example. Given $\sigma = (1\,3)(2\,6\,5\,8)(4\,10\,9\,7)$ in E, then we have pictorially

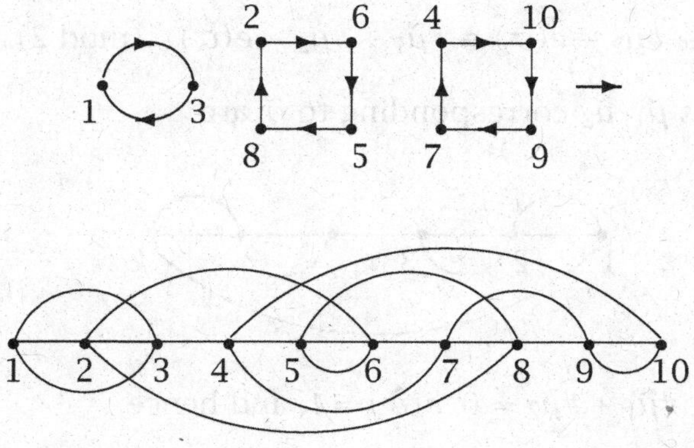

and obtain $\mu_1 = 13, 26, 4\,10, 58, 79$, $\mu_2 = 13, 28, 47, 56, 9\,10$.

It remains to check (14) for $\sigma = \phi(\mu_1, \mu_2)$. We clearly have

$$a_{\mu_1} a_{\mu_2} = (-1)^{e(\sigma)} a_\sigma, \quad e_\sigma = \#\{i : \sigma(i) < i\}. \qquad (15)$$

Let $\sigma = \sigma_1 \sigma_2 \cdots \sigma_t$ be the cycle decomposition; then sign $\sigma = (-1)^t$, since all σ_i have even length, and thus negative sign. Equation (14) is therefore equivalent to

$$\text{sign } \mu_1 \cdot \text{sign } \mu_2 = (-1)^{e(\sigma)+t},$$

or

$$\#\mu_1 + \#\mu_2 - e(\sigma) \equiv t \pmod 2 . \tag{16}$$

For the final step you are asked in the exercises to prove the following: Let $\sigma \in E$, $\sigma = \phi(\mu_1, \mu_2)$, and $\hat{\sigma} \in E$ the permutation that arises from σ by interchanging the positions of i and $i+1$ in the cycle decomposition. Then the parity of $\#\mu_1 + \#\mu_2 - e(\sigma)$ stays invariant, that is,

$$\#\mu_1 + \#\mu_2 - e(\sigma) \equiv \#\hat{\mu}_1 + \#\hat{\mu}_2 - e(\hat{\sigma}) \pmod 2 , \tag{17}$$

where $\phi(\hat{\mu}_1, \hat{\mu}_2) = \hat{\sigma}$.

We may transform $\sigma = \sigma_1\sigma_2 \cdots \sigma_t$ by a sequence of such exchanges into the natural order from 1 to n (clear?),

$$\tilde{\sigma} = \underbrace{(1 \dots k)}_{\sigma_1}\underbrace{(k+1 \dots)}_{\sigma_2} \dots \underbrace{(\dots n)}_{\sigma_t} ,$$

where by the exercise,

$$\#\mu_1 + \#\mu_2 - e(\sigma) \equiv \#\tilde{\mu}_1 + \#\tilde{\mu}_2 - e(\tilde{\sigma}) \pmod 2 .$$

The matchings $\tilde{\mu}_1, \tilde{\mu}_2$ corresponding to $\tilde{\sigma}$ are

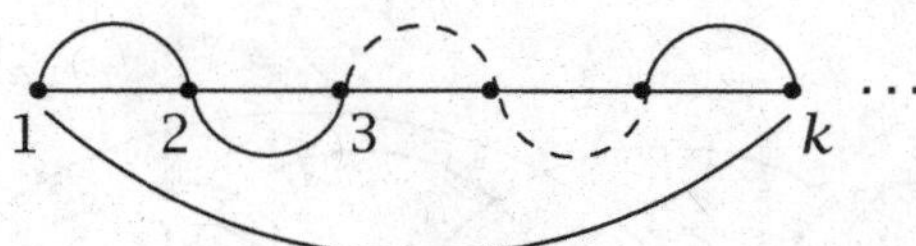

It follows that $\#\tilde{\mu}_1 + \#\tilde{\mu}_2 = 0$, $e(\tilde{\sigma}) = t$, and hence

$$\#\mu_1 + \#\mu_2 - e(\sigma) \equiv t \pmod 2 .$$

This proves (16), and thus the theorem. □

General Involution Principle.

The *general involution principle*, introduced by Garsia and Milne, carries these ideas to two signed sets S and T. One of their motivations was to find concrete combinatorial bijections between two sets X and Y, where equality $|X| = |Y|$ had been proven by, say, generating functions. Recall Section 3.4, which contains a number of these proofs via infinite products, relating two sets of partitions.

The general setup is as follows. Consider two signed sets $S = S^+ \cup S^-$, $T = T^+ \cup T^-$, and suppose $f : S \to T$ is a sign-preserving bijection. This, of course, implies $|S^+| = |T^+|$, $|S^-| = |T^-|$, and $|S^+| - |S^-| = |T^+| - |T^-|$.

Theorem 5.7 (Involution Principle). *Let $f : S \to T$ be a sign-preserving bijection between two signed sets S and T, and let φ and ψ be alternating involutions on S and T, respectively, with $\mathrm{Fix}_\varphi S \subseteq S^+$, $\mathrm{Fix}_\psi T \subseteq T^+$. Then $|\mathrm{Fix}_\varphi S| = |\mathrm{Fix}_\psi T|$, and for each $x \in \mathrm{Fix}_\varphi S$ there is a least integer $m(x)$ such that*

$$f((\varphi f^{-1} \psi f)^{m(x)})(x) \in \mathrm{Fix}_\psi T. \qquad (18)$$

The map $\beta : x \mapsto f((\varphi f^{-1} \psi f)^{m(x)})(x)$ is a bijection from $\mathrm{Fix}_\varphi S$ to $\mathrm{Fix}_\psi T$.

Proof. Since $|\mathrm{Fix}_\varphi S| = |S^+| - |S^-| = |T^+| - |T^-| = |\mathrm{Fix}_\psi T|$, it remains to verify (18). We have the two involutions φ and ψ, and we may regard f and f^{-1} together as a third fixed-point-free involution on $S \cup T$. We now construct the following graph G on the vertex-set $S \cup T$. The edges are $\{s,t\}$ with $t = f(s)$, and all pairs $\{a,b\}$, $a \neq b \in S$ with $b = \varphi a$, and all pairs $\{c,d\}$, $c \neq d \in T$, with $d = \psi c$. The picture should make this clear:

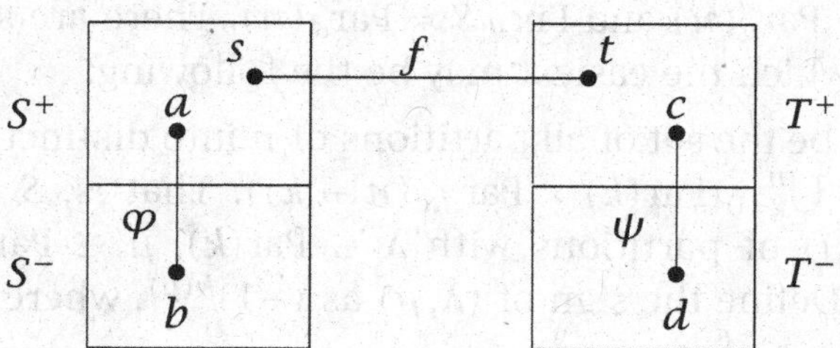

Every vertex is incident to an f-edge and at most one edge produced by φ and ψ. Hence G decomposes into disjoint circuits and paths, and it is clear that the vertices of degree 1 are precisely the points in $\mathrm{Fix}_\varphi S \cup \mathrm{Fix}_\psi T$. In other words, the endpoints of the paths pair off the elements in $\mathrm{Fix}_\varphi S \cup \mathrm{Fix}_\psi T$, and it remains to show that if one end is in $\mathrm{Fix}_\varphi S$, the other is in $\mathrm{Fix}_\psi T$. But this is easy. Let $x \in \mathrm{Fix}_\varphi S$ and look at the path in G

$$\underset{x \in \mathrm{Fix}_\varphi S}{\bullet} \xrightarrow{\ f\ } \underset{T^+}{\bullet} \xrightarrow{\ \psi\ } \underset{T^-}{\bullet} \xrightarrow{\ f^{-1}\ } \underset{S^-}{\bullet} \xrightarrow{\ \varphi\ } \underset{S^+}{\bullet} \cdots$$

If the path were to end with an f^{-1}-edge terminating at y, then $y \in S^-$, which is not a fixed point. Hence it must end in $\mathrm{Fix}_\psi T$, and (18) gives the desired bijection. □

In practice, the method then goes as follows. To find a bijection between X and Y, embed X into a signed set S, and Y into a signed set T with a bijection $f : S \to T$. Find alternating involutions with $X = \mathrm{Fix}_\varphi S^+$, $\mathrm{Fix}_\varphi S^- = \emptyset$, $Y = \mathrm{Fix}_\psi T^+$, $\mathrm{Fix}_\psi T^- = \emptyset$, and construct $\beta : X \to Y$ as in (18). Frequently, the sets X and Y may be very small compared to S and T, and of course, the number of iterations in (18) may be very large. Garsia and Milne invented their principle to give the first bijective proof of the famous Rogers–Ramanujan identities, which we will prove as a highlight in Chapter 10 by totally different methods.

Example. The simplest case arises when $S = T$ and f is the identity. If φ and ψ are two alternating involutions on the signed set S with $\mathrm{Fix}_\varphi S \subseteq S^+$, $\mathrm{Fix}_\psi S \subseteq S^+$, then $|\mathrm{Fix}_\varphi S| = |\mathrm{Fix}_\psi S|$, and the involution principle will produce an explicit bijection.

As an illustration let us re-prove $|\mathrm{Par}_o(n)| = |\mathrm{Par}_d(n)|$. To do this we have to find a signed set S and two involutions φ and ψ such that $\mathrm{Fix}_\varphi S = \mathrm{Par}_o(n)$ and $\mathrm{Fix}_\psi S = \mathrm{Par}_d(n)$. There are several possibilities, of which the easiest may be the following:

Let $\mathrm{Par}_{d,e}(n)$ be the set of all partitions of n into distinct even parts, and set $S = \bigcup_{k=0}^{n}(\mathrm{Par}(k) \times \mathrm{Par}_{d,e}(n-k))$. That is, S consists of all pairs (λ, μ) of partitions with $\lambda \in \mathrm{Par}(k)$, $\mu \in \mathrm{Par}_{d,e}(n-k)$, $k = 0, \dots, n$. Define the sign of (λ, μ) as $(-1)^{b(\mu)}$, where $b(\mu)$ is the number of parts of μ.

For $(\lambda, \mu) \in S$, $\varphi(\lambda, \mu)$ is declared as follows: Take the smallest *even* part e of λ or μ. If e is in μ, move e to λ; otherwise, move e from λ to μ. Clearly, $\mathrm{Fix}_\varphi S = \{(\lambda, \emptyset) : \lambda \in \mathrm{Par}_o(n)\} \subseteq S^+$ (since $\emptyset$ has 0 parts), and φ is alternating.

To define ψ we proceed accordingly. For $(\lambda, \mu) \in S$, let i be the smallest repeated number in λ, and $2j$ the smallest part in μ. If $i < j$ or $\mu = \emptyset$, ψ moves ii to μ, creating a summand $2i$. If $i \geq j$, ψ moves $2j$ from μ to λ, creating jj. Again it is immediate that

$\text{Fix}_\psi S = \{(\lambda, \emptyset) : \lambda \in \text{Par}_d(n)\} \subseteq S^+$, and ψ is alternating. The involution principle thus implies $|\text{Par}_o(n)| = |\text{Par}_d(n)|$.

As an example consider $n = 6$. We have $\text{Par}_o(6) = \{51, 33, 3111, 111111\}$, $\text{Par}_d(6) = \{6, 51, 42, 321\}$. Running the algorithm in (18) we obtain

$$51 \times \emptyset \xrightarrow{f} 51 \times \emptyset \in \text{Fix}_\psi,$$

$$33 \times \emptyset \xrightarrow{f} 33 \times \emptyset \xrightarrow{\psi} \emptyset \times 6 \xrightarrow{f} \emptyset \times 6 \xrightarrow{\varphi} 6 \times \emptyset \xrightarrow{f} 6 \times \emptyset \in \text{Fix}_\psi,$$

$$3111 \times \emptyset \xrightarrow{f} 3111 \times \emptyset \xrightarrow{\psi} 31 \times 2 \xrightarrow{f} 31 \times 2 \xrightarrow{\varphi} 321 \times \emptyset \xrightarrow{f} 321 \times \emptyset \in \text{Fix}_\psi,$$

$$111111 \times \emptyset \xrightarrow{\psi} 1111 \times 2 \xrightarrow{\varphi} 21111 \times \emptyset \xrightarrow{\psi} 211 \times 2 \xrightarrow{\varphi} 2211 \times \emptyset \xrightarrow{\psi} 22 \times 2$$
$$\xrightarrow{\varphi} 222 \times \emptyset \xrightarrow{\psi} 2 \times 4 \xrightarrow{\varphi} \emptyset \times 42 \xrightarrow{\psi} 11 \times 4 \xrightarrow{\varphi} 411 \times \emptyset \xrightarrow{\psi} 4 \times 2 \xrightarrow{\varphi} 42 \times \emptyset \in \text{Fix}_\psi.$$

In a sense, the involutions are modeled after the "forbidden" configurations as in the inclusion-exclusion proof in Section 5.1. But the method does give an explicit bijection. And what is this bijection? You are asked in the exercises to show that it is, surprisingly, precisely Glaisher's correspondence discussed in Exercise 3.65.

Exercises

5.48 Determine $\sum_{k=0}^{n}(-1)^k\binom{n}{k}^2$ with an alternating involution on a signed set.

5.49 Let A_n be the number of Catalan paths. Verify $A_n = C_n$ directly by proving the recurrence $A_n = \sum_{k=0}^{n-1} A_k A_{n-1-k}$, $A_0 = 1$.

5.50 Generalize the result about Catalan paths, counting the number of lattice paths from $(0,0)$ to (m,n), $m \geq n$, that never cross the diagonal $y = x$.

▷ **5.51** Let S be the family of k-subsets of $\{1, 2, \ldots, n\}$, n even. For $A \in S$ let $w(A) = \sum_{i \in A} i$, and set $S^+ = \{A \in S : w(A) \text{ even}\}$, $S^- = \{A \in S : w(A) \text{ odd}\}$. Find an alternating involution to show that

$$|S^+| - |S^-| = \begin{cases} 0, & k \text{ odd}, \\ (-1)^{k/2}\binom{n/2}{k/2}, & k \text{ even}. \end{cases}$$

5.52 Find suitable involutions to prove the following identities:

a. $\sum_{k=p}^{n}\binom{n}{k}\binom{k}{p}(-1)^k = \delta_{n,p}(-1)^p$,

b. $\sum_{k=0}^{m}\binom{n}{k}\binom{n}{m-k}(-1)^k = \begin{cases} 0 & m \text{ odd}, \\ (-1)^{m/2}\binom{n}{m/2} & m \text{ even}, \end{cases}$

c. $\sum_{k=0}^{m}\binom{n+k-1}{k}\binom{n}{m-k}(-1)^{m-k} = \delta_{m,0}$.

▷ **5.53** Define a signed weighted set of partitions and an alternating involution to prove the obvious identity

$$\frac{(1-z^{k+1})(1-z^{k+2})\cdots}{(1-z)(1-z^2)(1-z^3)\cdots} = \frac{1}{(1-z)(1-z^2)\cdots(1-z^k)}.$$

5.54 Consider the set S of all graphs G on $\{1,\ldots,n\}$ with m edges. A graph G is called *even* (or Eulerian) if every vertex has even degree. Define an alternating involution on the set of pairs (G,A), $A \subseteq \{1,\ldots,n\}$ for which the fixed points are precisely the pairs (G,A), G even.

* * *

▷ **5.55** Write $\sigma = \sigma(1)\sigma(2)\ldots\sigma(n) \in S(n)$ in word form, and define $D(\sigma) = \{i : i \geq \sigma(i)\}$, $P(\sigma) = \frac{1}{|D(\sigma)|}\sum_{i\in D(\sigma)}(i+\sigma(i))$. Prove that $\sum_{\sigma\in S(n)} P(\sigma) = (n+1)!$. Hint: Look at σ^* with $\sigma^*(i) = j$ if $\sigma(n+1-j) = n+1-i$, and consider $P(\sigma) + P(\sigma^*)$.

5.56 Re-prove

$$\sum_{k=0}^{n}(-1)^k\binom{n}{k}^3 = \begin{cases} (-1)^{n/2}(3n/2)!/(n/2)!^3 & n \text{ even}, \\ 0 & n \text{ odd} \end{cases}$$

(see Exercise 4.8).

5.57 Use the involution principle to re-prove the result of Exercise 3.68: The number of partitions of n into parts congruent to 1 or 5 (mod 6) equals the number of partitions of n into distinct parts congruent to 1 or 2 (mod 3). Hint: Model your set and the involutions according to the forbidden parts.

5.58 Use involutions to prove $\sum_{k=m}^{n} S_{n,k}(-1)^{k-m}s_{k,m} = \delta_{m,n}$, where $S_{n,k}$ and $s_{k,m}$ are the Stirling numbers.

5.59 Find a proper weighting to prove $x^n = \sum_{k=0}^{n}(-1)^{n-k}S_{n,k}x^{\overline{k}}$.

▷ **5.60** Prove assertion (17) of the text.

5.61 Show that the bijection $\phi : \mathrm{Par}_o(n) \longrightarrow \mathrm{Par}_d(n)$ produced in the text is precisely Glaisher's correspondence (Exercise 3.65).

▷ **5.62** Consider the length $h(n)$ of the algorithm to produce the mate of $11\ldots1 \in \mathrm{Par}_o(n)$, counting only the occurrences $\psi,\varphi,\psi,\varphi,\ldots,\varphi$. For example: $h(2) = h(3) = 2$, $h(4) = h(5) = 6$, $h(6) = h(7) = 12$. Does $h(2n+1) = h(2n)$ always hold? What is the mate in $\mathrm{Par}_d(n)$?

5.63 Extend the involution principle to the following general situation: Let $S = S^+ \cup S^-$, $T = T^+ \cup T^-$ be two signed sets, and set $(S \cup T)^+ = S^+ \cup T^-$, $(S \cup T)^- = S^- \cup T^+$. We call f a *signed bijection* between S and T if f is an alternating involution on $S \cup T$ without fixed points. Note that this implies $|S^+| - |S^-| = |T^+| - |T^-|$. Suppose φ and ψ are alternating involutions on S and T. Then there exists a signed bijection between $\mathrm{Fix}_{\varphi} S$ and $\mathrm{Fix}_{\psi} T$.

5.4 The Lemma of Gessel–Viennot

We come to a most elegant theorem, one that reveals via involutions an astounding connection between lattice paths and determinants. It was originally proved by Lindström, but its combinatorial significance was seen first and expounded by Gessel and Viennot, and the name has stuck ever since.

Let us start with the usual permutation description of the determinant of an $n \times n$-matrix $M = (m_{ij})$:

$$\det M = \sum_{\sigma \in S(n)} (\operatorname{sign} \sigma) m_{1\sigma(1)} \cdots m_{n\sigma(n)} . \tag{1}$$

We may represent M as a directed weighted bipartite graph in an obvious way, letting the rows correspond to vertices $A_1, \ldots, A_n$, and the columns to $B_1, \ldots, B_n$:

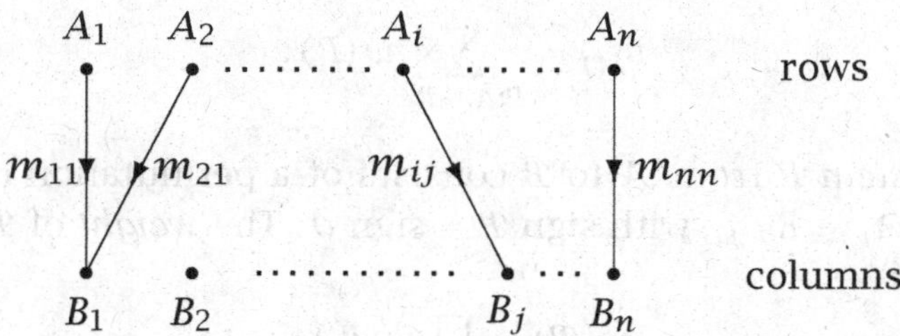

We can now give (1) the following interpretation. The *left-hand* side of (1) is the determinant of the *path matrix*, where the (i, j)-entry is the weight m_{ij} of the unique directed path from A_i to B_j. The *right-hand* side is the weighted signed sum over all *vertex-disjoint path systems* from $\mathcal{A} = \{A_1, \ldots, A_n\}$ to $\mathcal{B} = \{B_1, \ldots, B_n\}$. Such a system $\mathcal{P}_\sigma$ is determined by

$$P_1 : A_1 \to B_{\sigma(1)}, \ldots, P_n : A_n \to B_{\sigma(n)} ,$$

and the *weight* of $\mathcal{P}_\sigma$ is the product of the weights of the individual paths,

$$w(\mathcal{P}_\sigma) = w(P_1) \cdots w(P_n).$$

In this interpretation, (1) becomes

$$\det M = \sum_\sigma (\operatorname{sign} \sigma) w(\mathcal{P}_\sigma).$$

Recall that we used a similar idea in the computation of the Vandermonde matrix in the last section.

The lemma of Gessel–Viennot generalizes this idea from bipartite to arbitrary graphs. Let $G = (V, E)$ be a finite directed graph without directed circuits. We call such a graph *acyclic*. In particular, in an acyclic graph there are only finitely many directed paths between any two vertices, where we include the trivial paths $A \to A$ of length 0. Every edge e has a weight $w(e)$. Let $P : A \to B$ be a directed path from A to B. Then the *weight* of P is

$$w(P) = \prod_{e \in P} w(e), \tag{2}$$

with $w(P) = 1$ if P has length 0.

Suppose $\mathcal{A} = \{A_1, \ldots, A_n\}$ and $\mathcal{B} = \{B_1, \ldots, B_n\}$ are two n-sets of vertices, which need not be disjoint. To $\mathcal{A}$ and $\mathcal{B}$ we associate the *path matrix* $M = (m_{ij})$, where

$$m_{ij} = \sum_{P: A_i \to B_j} w(P). \tag{3}$$

A *path system* $\mathcal{P}$ from $\mathcal{A}$ to $\mathcal{B}$ consists of a permutation σ, and n paths $P_i : A_i \to B_{\sigma(i)}$, with sign $\mathcal{P}$ = sign σ. The *weight* of $\mathcal{P}$ is

$$w(\mathcal{P}) = \prod_{i=1}^{n} w(P_i). \tag{4}$$

Finally, we call the path system $\mathcal{P}$ *vertex-disjoint* if no two paths have a vertex in common. Let VD be the family of vertex-disjoint path systems.

Lemma 5.8 (Gessel–Viennot–Lindström). *Let $G = (V, E)$ be a directed acyclic graph with weights on E, $\mathcal{A} = \{A_1, \ldots, A_n\}$, $\mathcal{B} = \{B_1, \ldots, B_n\}$, and M the path-matrix from $\mathcal{A}$ to $\mathcal{B}$. Then*

$$\det M = \sum_{\mathcal{P}\in VD} (\operatorname{sign}\mathcal{P})w(\mathcal{P}). \tag{5}$$

Proof. A typical summand of M is

$$\begin{aligned}(\operatorname{sign}\sigma)&m_{1\sigma(1)}\cdots m_{n\sigma(n)}\\ &= (\operatorname{sign}\sigma)\left(\sum_{P_1:A_1\to B_{\sigma(1)}} w(P_1)\right)\cdots\left(\sum_{P_n:A_n\to B_{\sigma(n)}} w(P_n)\right).\end{aligned}$$

Summation over σ immediately yields

$$\det M = \sum_{\mathcal{P}} (\operatorname{sign}\mathcal{P})w(\mathcal{P}),$$

where $\mathcal{P}$ runs through *all* path systems from $\mathcal{A}$ to $\mathcal{B}$. To prove (5), all we need to show is that

$$\sum_{\mathcal{P}\in N} (\operatorname{sign}\mathcal{P})w(\mathcal{P}) = 0, \tag{6}$$

where N is the family of all path systems that are *not* vertex-disjoint. And this is where involutions come into the game. We define a fixed-point-free involution φ on N with $w(\varphi\mathcal{P}) = w(\mathcal{P})$ and $\operatorname{sign}(\varphi\mathcal{P}) = -\operatorname{sign}\mathcal{P}$. This will prove (6), and thus the lemma.

The involution φ is declared in the most natural way. Suppose $\mathcal{P} \in N$ with paths $P_i : A_i \to B_{\sigma(i)}$. Among the crossing paths we define

i_0 = smallest index such that P_{i_0} crosses with some P_j,

X = first common vertex on P_{i_0},

j_0 = smallest index such that $X \in P_{i_0} \cap P_{j_0}$ $(j_0 > i_0)$.

The situation is thus as in the figure:

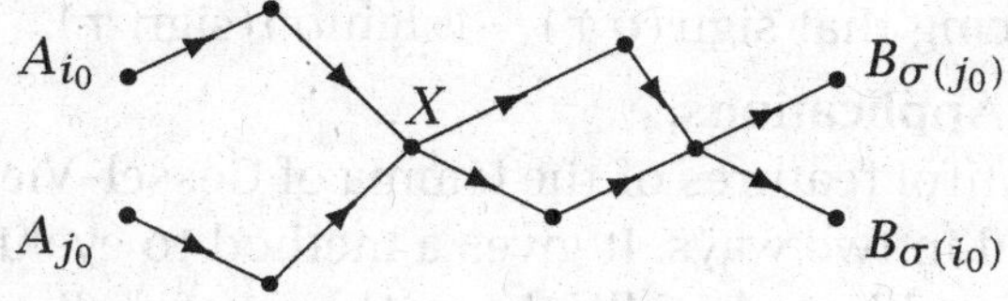

Now we construct $\varphi\mathcal{P} = \{P_1',\dots,P_n'\}$ as follows: For $k \neq i_0, j_0$ set $P_k' = P_k$. The new path P_{i_0}' goes from A_{i_0} to X along P_{i_0}, and then

along P_{j_0} from X to $B_{\sigma(j_0)}$. Similarly, P'_{j_0} goes along P_{j_0} from A_{j_0} to X, and then proceeds along P_{i_0} to $B_{\sigma(i_0)}$. Clearly, $\varphi(\varphi\mathcal{P}) = \mathcal{P}$, since i_0, X, j_0 retain their significance in $\varphi\mathcal{P}$. Since $\mathcal{P}$ and $\varphi\mathcal{P}$ use the same edges, we have $w(\varphi\mathcal{P}) = w(\mathcal{P})$, and finally, sign $(\varphi\mathcal{P}) = -$sign $\mathcal{P}$, since σ' equals σ times the transposition (i_0, j_0). □

Matrix Theorems.
Before discussing a number of applications, let us see how the lemma proves some theorems on determinants "at a glance."

Example. $\det M^T = \det M$. Just look at the bipartite graph corresponding to M from bottom to top, that is, the path systems from $\mathcal{B}$ to $\mathcal{A}$. Since sign σ^{-1} = sign σ, the result follows.

Example. $\det(MM') = (\det M)(\det M')$. We concatenate the two bipartite graphs as in the figure:

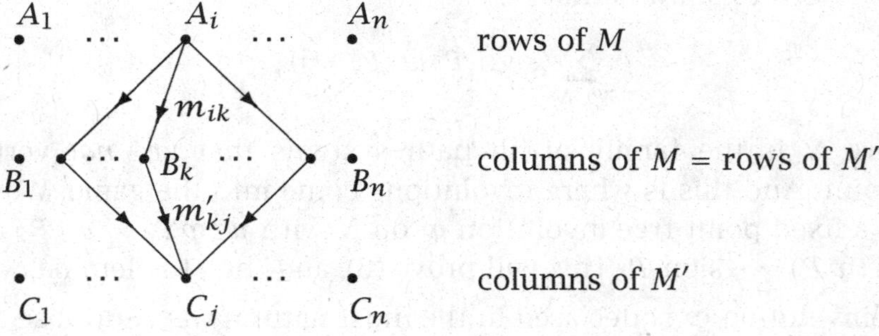

Consider the paths from $\mathcal{A}$ to C. The path matrix is

$$\sum_{P:A_i \to C_j} w(P) = \sum_{k=1}^{n} m_{ik} m'_{kj},$$

that is, MM'. The vertex-disjoint path systems from $\mathcal{A}$ to C correspond to pairs of systems from $\mathcal{A}$ to $\mathcal{B}$ and $\mathcal{B}$ to C, and the result follows on noticing that sign $(\sigma\tau)$ = (sign σ)(sign τ).

Combinatorial Applications.
One of the beautiful features of the lemma of Gessel–Viennot is that it can be applied in two ways. It gives a method to evaluate a determinant if we are able to handle the path systems $\mathcal{P}$ on the right-hand side of (5). But it can also lead to an effective enumeration of the path systems if we succeed in computing the determinant. The following examples should make this clear.

A word on the notation: We write a matrix $(m_{ij})_{i,j=1}^{n}$ with the understanding that i is the row-index and j the column-index. For example, the Vandermonde matrix is written (x_j^{n-i}).

Example. Binomial Determinants. Let $0 \le a_1 < \cdots < a_n$ and $0 \le b_1 < \cdots < b_n$ be two sets of natural numbers. Our task is to compute $\det M$ with $m_{ij} = \binom{a_i}{b_j}$, the binomial coefficient. The method consists in finding a suitable directed graph G and vertex sets $\mathcal{A}$ and $\mathcal{B}$ such that M is precisely the path matrix.

We take the lattice graph with steps up and to the right, and with all weights equal to 1. Let $A_i = (0, -a_i)$, $B_j = (b_j, -b_j)$ as in the figure:

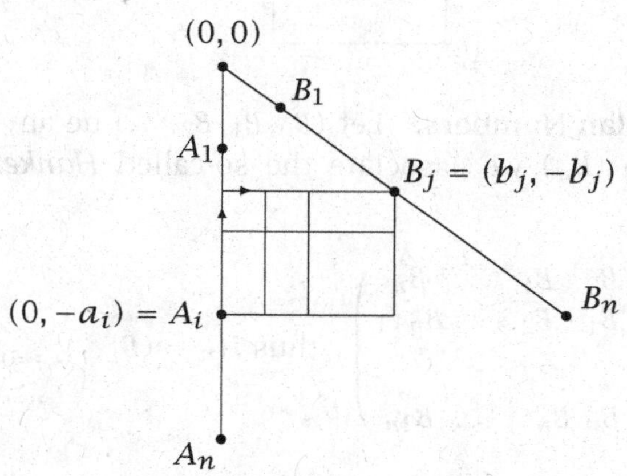

Then $m_{ij} = \binom{b_j+(a_i-b_j)}{b_j} = \binom{a_i}{b_j}$. It is clear that all vertex-disjoint path systems must correspond to the identity permutation, and we obtain

$$\det\left(\binom{a_i}{b_j}\right)_{i,j=1}^{n} = \#\text{vertex-disjoint path systems}$$
$$P_1 : A_1 \to B_1, \ldots, P_n : A_n \to B_n\,.$$

As an example, suppose $a_i = m + i - 1$, $b_j = j - 1$; then

$$M = \begin{pmatrix} \binom{m}{0} & \binom{m}{1} & \cdots & \binom{m}{n-1} \\ \binom{m+1}{0} & \binom{m+1}{1} & \cdots & \binom{m+1}{n-1} \\ \vdots & & & \\ \binom{m+n-1}{0} & \binom{m+n-1}{1} & \cdots & \binom{m+n-1}{n-1} \end{pmatrix}.$$

In this case the graph looks as follows, and it is plain that there is only the one vertex-disjoint path system depicted in the figure. It follows that $\det M = 1$.

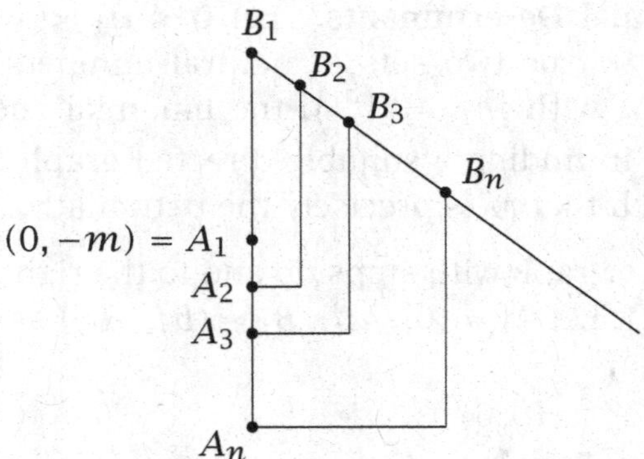

Example. Catalan Numbers. Let $(B_0, B_1, B_2, \ldots)$ be any sequence of numbers. To (B_n) we associate the so-called *Hankel matrices* $H_n, H_n^{(1)}$:

$$H_n = \begin{pmatrix} B_0 & B_1 & \ldots & B_n \\ B_1 & B_2 & \ldots & B_{n+1} \\ \vdots & & & \\ B_n & B_{n+1} & \ldots & B_{2n} \end{pmatrix}, \text{ thus } H_n = (B_{i+j})_{i,j=0}^n,$$

$$H_n^{(1)} = \begin{pmatrix} B_1 & B_2 & \ldots & B_{n+1} \\ B_2 & B_3 & \ldots & B_{n+2} \\ \vdots & & & \\ B_{n+1} & B_{n+2} & \ldots & B_{2n+1} \end{pmatrix}, \; H_n^{(1)} = (B_{i+j+1})_{i,j=0}^n .$$

Every sequence (B_n) uniquely determines the sequence $\det H_0$, $\det H_0^{(1)}, \det H_1, \det H_1^{(1)}, \ldots$. If, conversely, this latter sequence is given, and if all $\det H_n \neq 0$, $\det H_n^{(1)} \neq 0$, then (B_n) can be uniquely recovered. Indeed, $\det H_0 = B_0$, $\det H_0^{(1)} = B_1$, $\det H_1 = \det \binom{B_0 B_1}{B_1 B_2} = B_0 B_2 - B_1^2$, which gives B_2 since $B_0 \neq 0$, and so on.

The following result gives an unexpected and beautiful characterization of the Catalan numbers.

Claim. *The Catalan numbers C_n are the unique sequence of real numbers with* $\det H_n = \det H_n^{(1)} = 1$ *for all n.*

For the proof we consider the lattice graph above the x-axis, this time with diagonal steps $(1,1)$ or $(1,-1)$, and all weights 1. Let $A_i = (-2i, 0)$, $B_j = (2j, 0)$, $i, j = 0, 1, \ldots, n$. For $n = 3$ we obtain

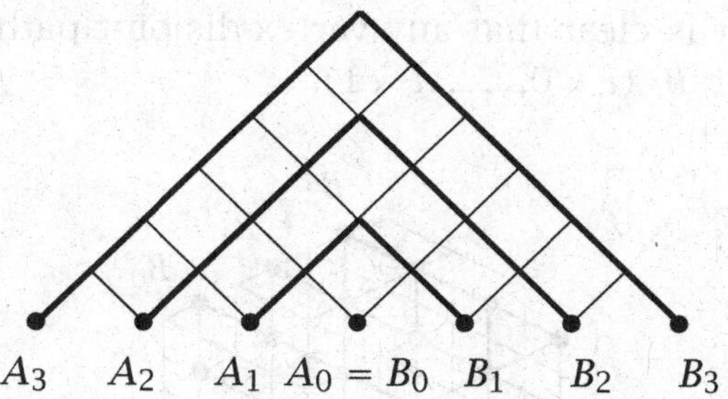

We know from the previous section that C_k counts the number of these paths of length $2k$. In other words, the path matrix $M = (m_{ij})$ satisfies $m_{ij} = C_{i+j}$, and thus $M = H_n$. Since there is again only one vertex-disjoint path system from $\mathcal{A}$ to $\mathcal{B}$, shown by the bold edges in the figure, we conclude that $\det H_n = 1$. You are asked in the exercises to prove similarly $\det H_n^{(1)} = 1$. Hankel matrices play an important role in general recurrences; we will return to this topic in Chapter 7.

Now we want to discuss two famous examples that will shed light on the second feature of the lemma.

Example. Rhombic Tilings. We are given a hexagon of side length n that is triangulated. A rhombus consists of two triangles with a common side. What we want to determine is the number $h(n)$ of decompositions of the hexagon into rhombi.

The figure on the right shows a rhombic tiling for $n = 3$.

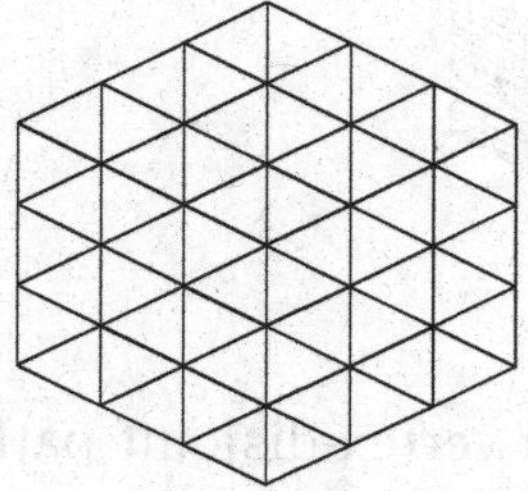

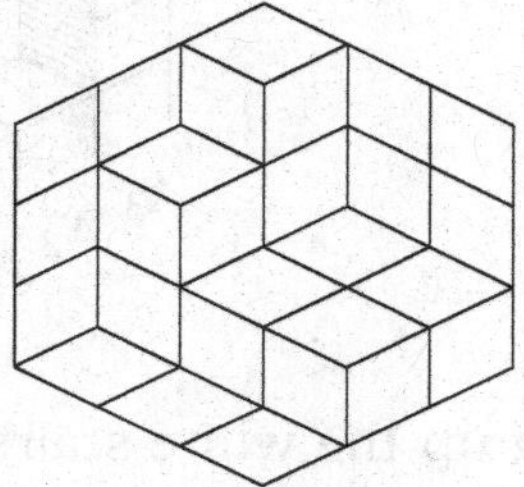

Clearly, $h(1) = 2$, and $h(2) = 20$ can still be done by hand. But already $n = 3$ appears rather hopeless (in fact, $h(3) = 980$).

We associate to the hexagon a directed graph as in the figure with edges directed upward and to the right (all weights equal to 1). Let $\mathcal{A} = \{A_0, A_1, \ldots, A_{n-1}\}$ be the vertices on the left bottom line, and $\mathcal{B} = \{B_0, B_1, \ldots, B_{n-1}\}$ those on the right top line, as indicated in the figure. Again it is clear that any vertex-disjoint path system must connect A_i with B_i $(i = 0, \ldots, n-1)$.

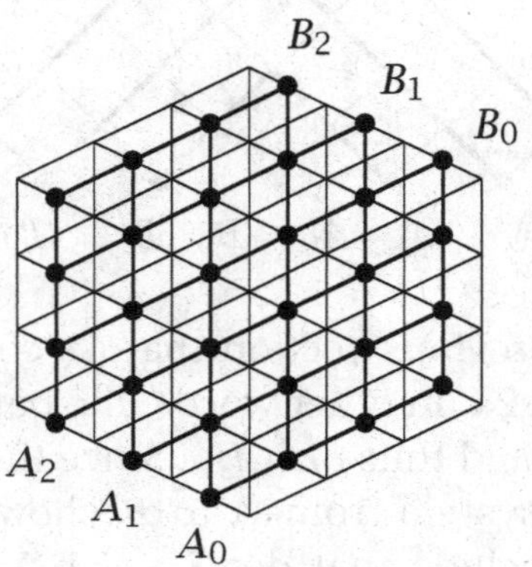

Now comes an elegant observation: The vertex-disjoint path systems $\mathcal{P} : \mathcal{A} \to \mathcal{B}$ correspond bijectively to the rhombic tilings. To see this, look at such a tiling in a "3-dimensional" fashion. We shade all rhombi of the form and keep the others white, as in the figure:

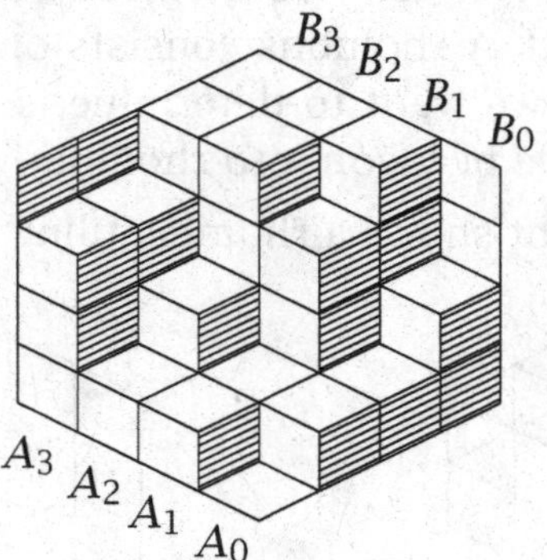

Climbing up the white stairs, we obtain a vertex-disjoint path system $P_i : A_i \to B_i$. Conversely, if such a system is given, then we keep the rhombi along the paths white, and fill in the others in a unique way. Hence we have

$$h(n) = \#\text{ vertex-disjoint path systems},$$

and this is by Gessel–Viennot equal to the determinant of the path-matrix M. Now the paths $P : A_i \to B_j$ correspond to ordinary lattice paths from $A_i = (-i, i)$ to $B_j = (n - j, n + j)$, which gives

$$m_{ij} = \binom{n - j + i + n + j - i}{n + i - j} = \binom{2n}{n + i - j},$$

and thus

$$h(n) = \det\left(\binom{2n}{n + i - j}\right)_{i,j=0}^{n-1}. \tag{7}$$

This determinant is easily computed (you are asked in Exercise 5.75 to do it), and out comes a remarkable formula due to MacMahon:

$$h(n) = \prod_{i=0}^{n-1} \frac{\binom{2n+i}{n}}{\binom{n+i}{n}}. \tag{8}$$

For $n = 3$, this gives

$$h(3) = \frac{\binom{6}{3}\binom{7}{3}\binom{8}{3}}{\binom{3}{3}\binom{4}{3}\binom{5}{3}} = \frac{20 \cdot 35 \cdot 56}{1 \cdot 4 \cdot 10} = 980,$$

and further, $h(4) = 232848$.

Example. Plane Partitions. One of the great successes of the lemma of Gessel–Viennot was its application to the enumeration of plane partitions. A *plane partition* of n is an array of integers $\lambda_{ij} \geq 1$,

$$\begin{array}{l} \lambda_{11}\ \lambda_{12} \ldots \lambda_{1s} \\ \lambda_{21}\ \lambda_{22} \ldots \\ \vdots \\ \lambda_{r1}\ \lambda_{r2} \ldots \end{array} \qquad \text{with} \quad \sum \lambda_{ij} = n,$$

such that every row $\lambda_{i1} \geq \lambda_{i2} \geq \cdots$ and every column $\lambda_{1j} \geq \lambda_{2j} \geq \cdots$ is weakly monotone. It follows that

$$\lambda_{ik} \geq \lambda_{j\ell} \quad \text{whenever} \quad i \leq j,\ k \leq \ell. \tag{9}$$

Example. $n = 15$:

$$\begin{array}{l} 3\ 2\ 1 \\ 2\ 2\ 1 \\ 2\ 1 \\ 1 \end{array}$$

Every plane partition consists of r ordinary number partitions (of the rows) and s partitions (of the columns). Let $pp(r,s,t)$ be the number of plane partitions with at most r rows and at most s columns, and in which the highest summand is at most t, and $pp(n;r,s,t)$ the number of those for which $|\lambda| = n$, with $pp(0;r,s,t) = 1$.

In other words, if we represent the numbers λ_{ij} by stacks of λ_{ij} unit cubes, we ask for the number of arrangements of stacks that fit into an $r \times s \times t$-box. For example, the partition from above corresponds to

It is convenient to consider $r \times s$-arrays with $\lambda_{ij} \leq t$ by filling up the remaining cells with 0's.

Example. $pp(2,2,2) = 20$:

$$\underbrace{\begin{matrix}00\\00\end{matrix}}_{n=0}\quad \underbrace{\begin{matrix}10\\00\end{matrix}}_{n=1}\quad \underbrace{\begin{matrix}11 & 10 & 20\\00 & 10 & 00\end{matrix}}_{n=2}\quad \underbrace{\begin{matrix}11 & 21 & 20\\10 & 00 & 10\end{matrix}}_{n=3}$$

$$\underbrace{\begin{matrix}11 & 22 & 20 & 21\\11 & 00 & 20 & 10\end{matrix}}_{n=4}\quad \underbrace{\begin{matrix}21 & 22 & 21\\11 & 10 & 20\end{matrix}}_{n=5}\quad \underbrace{\begin{matrix}22 & 21 & 22\\11 & 21 & 20\end{matrix}}_{n=6}\quad \underbrace{\begin{matrix}22\\21\end{matrix}}_{n=7}\quad \underbrace{\begin{matrix}22\\22\end{matrix}}_{n=8}$$

$$\sum_{n\geq 0} pp(n;2,2,2)q^n = 1 + q + 3q^2 + 3q^3 + 4q^4 + 3q^5 + 3q^6 + q^7 + q^8\,.$$

Every plane partition λ gives by reflection a conjugate partition λ^*, which implies

$$pp(r,s,t) = pp(s,r,t)\,. \tag{10}$$

To determine $pp(r,s,t)$ we consider the usual lattice graph G with steps up and to the right, and the sets $\mathcal{A} = \{A_1,\ldots,A_r\}$, $\mathcal{B} = \{B_1,\ldots,B_r\}$, where $A_i = (-i,i)$, $B_j = (t-j,s+j)$, $i,j = 1,\ldots,r$.

Example. $r = 5, s = 3, t = 4$:

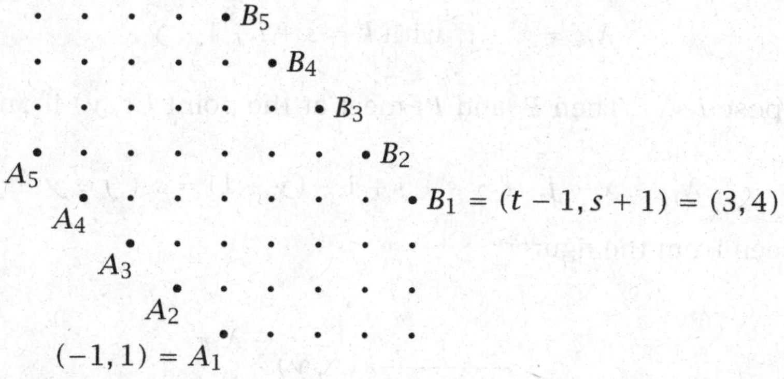

B_5, B_4, B_3, B_2, $B_1 = (t-1, s+1) = (3,4)$

A_5, A_4, A_3, A_2, $(-1,1) = A_1$

The path matrix $M = (m_{ij})$ is therefore given by

$$m_{ij} = \binom{t-j+i+s+j-i}{s+j-i} = \binom{t+s}{s+j-i}.$$

Consider now the vertex-disjoint path systems from $\mathcal{A}$ to $\mathcal{B}$. Again only $\sigma = \text{id}$ is possible. Let $\mathcal{P} = \{P_1, \ldots, P_r\}$ be such a system, $P_i : A_i \to B_i$. We know from Section 1.6 that every such path corresponds bijectively to a partition $\lambda_{i1}\lambda_{i2}\ldots\lambda_{ik}$ with $\lambda_{i1} \le t$, $k \le s$, by looking at the Ferrers diagram above the path. If $k < s$, then we fill up with 0's. Hence every such path corresponds uniquely to a partition $\lambda_{i1}\lambda_{i2}\ldots\lambda_{is}$ with $\lambda_{i1} \le t$.

Claim. *Let $\lambda_1, \ldots, \lambda_r$ be the partitions corresponding to the paths $P_i : A_i \to B_i$. Then $\lambda_1 \ldots \lambda_r$ form a plane partition, (that is, (9) is satisfied) if and only if $\mathcal{P} = \{P_1, \ldots, P_r\}$ is a vertex-disjoint path system.*

For the proof consider a path $P_i : A_i \to B_i$ with partition $\lambda_{i1}\lambda_{i2}\ldots\lambda_{is}$, and mark the points at the end of an upward step. These are precisely the marks where the points in the Ferrers diagram of this row end.

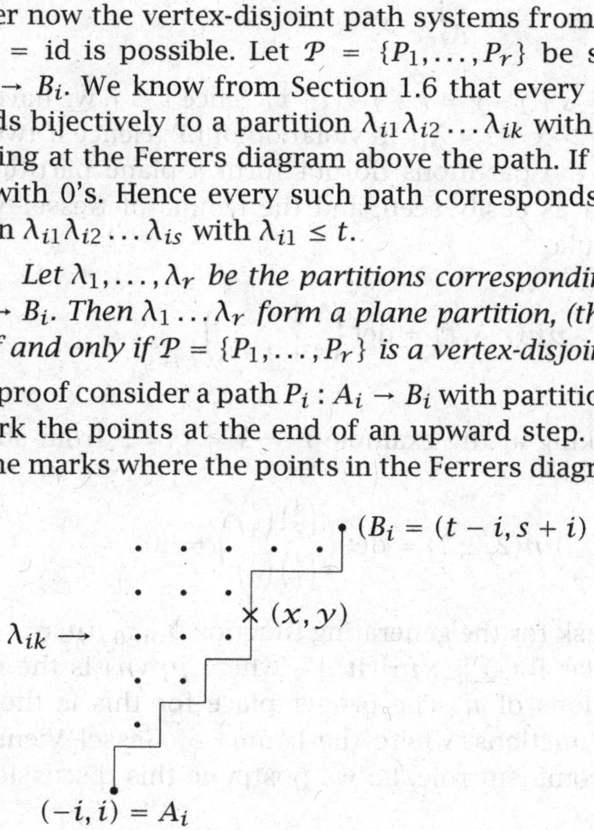

$(B_i = (t-i, s+i)$

$\times\ (x,y)$

$\lambda_{ik} \longrightarrow$

$(-i,i) = A_i$

Looking at the diagram we see that

$$\lambda_{ik} = x + i, \text{ with } k = s + i + 1 - y\,. \tag{11}$$

Suppose $i < j$. Then P_i and P_j meet at the point (x, y) if and only if

$$\lambda_{j\ell} \geq x + j, \quad \ell = s + j + 1 - (y + 1) = s + j - y \quad .$$

as seen from the figure:

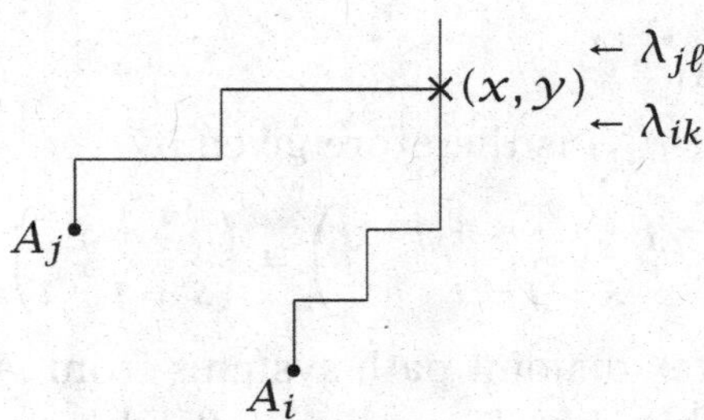

This implies $\ell = s + j - y = k + j - (i + 1)$. Since $i < j$, we have $k \leq \ell$, and $\lambda_{j\ell} \geq x + j > x + i = \lambda_{ik}$ in violation of (9). Hence if two paths intersect, then the partitions do not form a plane partition. The converse is just as easily seen, and the lemma of Gessel–Viennot yields the formula

$$pp(r, s, t) = \det \left(\binom{t + s}{s + j - i} \right)_{i,j=1}^{r} . \tag{12}$$

Example. Looking at the example $r = s = t = 2$ from above we again obtain

$$pp(2, 2, 2) = \det \begin{pmatrix} \binom{4}{2} & \binom{4}{3} \\ \binom{4}{1} & \binom{4}{2} \end{pmatrix} = 20\,.$$

We could now ask for the generating function $\sum_{n \geq 0} pp(n; r, s, t) q^n$ and, in particular for $\sum_{n \geq 0} pp(n) q^n$, where $pp(n)$ is the number of plane partitions of n. The proper place for this is the theory of symmetric functions, where the lemma of Gessel–Viennot will again play a prominent role. So we postpone this discussion until Chapter 8.

Exercises

5.64 Show that the involution used in the proof of the lemma of Gessel-Viennot does not work if we just choose i_0 minimal, then the minimal j_0 such that the paths P_{i_0}, P_{j_0} intersect, and then the first common point X on P_{i_0}. What could go wrong?

▷ **5.65** Show "graphically" that $\det M = 0$ if the rows of M are linearly dependent.

5.66 Generalize the product theorem for determinants to the Binet-Cauchy formula: Let M be an $n \times p$-matrix and M' an $p \times n$-matrix, with $n \le p$. Then $\det(MM') = \sum_{|R|=n} (\det M_R)(\det M'_R)$, where M_R is the $n\times n$-submatrix of M with columns in R, and M'_R the corresponding submatrix of M' with rows in R.

5.67 Prove the usual formula, developing $\det M$ according to the i-th row or the j-th column.

▷ **5.68** The *Motzkin number* is the number of lattice paths from $(0,0)$ to $(n,0)$ with horizontal steps and diagonal steps up or down, staying above $y = 0$. For example $M_3 = 4$ with . Show that $\det H_n = 1$, where H_n is the Hankel matrix of the sequence (M_n).

5.69 Show that $\det H_n^{(1)} = 1$ for $H_n^{(1)} = (C_{i+j+1})_{i,j=0}^{n}$, C_n Catalan.

5.70 Let $M = \left(\binom{a_i}{b_j}\right)_{i,j=1}^{n}$, and suppose $a_k = b_k$. Prove that

$$\det M = \det\left(\binom{a_i}{b_j}\right)_{i,j=1}^{k-1} \cdot \det\left(\binom{a_i}{b_j}\right)_{i,j=k+1}^{n}.$$

* * *

▷ **5.71** Let $\lambda_1\lambda_2\ldots\lambda_k$ be a partition with Ferrers diagram D. Given a cell s of D, let ℓ be the lowest cell below s, and r the cell farthest to the right of s. Insert in s the number $h(s)$ of paths starting in ℓ, ending in r, with all steps up and to the right. Example: $\lambda = 43311$

6	3	1	1
3	2	1	
1	1	1	
1			
1			

Let M be the Durfee square of λ (largest square contained in D). Prove $\det M = 1$.

5.72 Deduce from the recurrence $S_{n,k} = S_{n-1,k-1} + kS_{n-1,k}$ ($S_{n,k}$ Stirling number) the formula $\det(S_{m+i,j})_{i,j=1}^{n} = (n!)^m$, $m \geq 0, n \geq 1$. Hint: Use a lattice graph that mirrors the recurrence.

▷ **5.73** Generalize the previous exercise to recurrences $a_{n,k} = a_{n-1,k-1} + c_k a_{n-1,k}$, $a_{0,0} = 1$, where $(c_0, c_1, c_2, \ldots)$ is a fixed sequence with $c_i \neq 0$ for $i \geq 1$. Prove $\det(a_{m+i,j})_{i,j=1}^{n} = (c_1 c_2 \cdots c_n)^m \cdot \sum_{0 \leq i_n \leq \cdots \leq i_1 \leq m} s_1^{i_1} s_2^{i_2} \cdots s_n^{i_n}$, where $s_i = \frac{c_{i-1}}{c_i}$. Look at the special case $c_i = q^{-i}$. What do you get for the right-hand side?

5.74 Compute $\det\left(\binom{m+i-1}{k+j-1}\right)_{i,j=1}^{n}$ directly (without using Gessel–Viennot). For $k = 1$ you get

$$\det\binom{m+i-1}{j}_{i,j=1}^{n} = \binom{m+n-1}{n},$$

which also counts the number of sequences $m - 1 \geq a_1 \geq a_2 \geq \cdots \geq a_n \geq 0$. Give a bijective proof of this last formula using Gessel–Viennot.

5.75 Complete the analysis of the rhombic tilings by showing that

$$\det\left(\binom{2n}{n+i-j}\right)_{i,j=0}^{n-1} = \prod_{i=0}^{n-1} \frac{(2n+i)^{\underline{n}}}{(n+i)^{\underline{n}}}.$$

▷ **5.76** Let $S = \{(a_i, b_i) : i = 1, \ldots, n\}$ be a set of lattice points with $0 \leq a_1 \leq \cdots \leq a_n \leq r$, $0 \leq b_1 \leq \cdots \leq b_n \leq s$. Count the number of lattice paths from $(0,0)$ to (r,s) with steps up and to the right that avoid the set S. Do the same thing using inclusion–exclusion. What identity do you get?

5.77 Choose $r = s = n + 1$ and $a_i = b_i = i$ $(i = 1, \ldots, n)$ in the previous exercise. Derive a determinant formula for the Catalan numbers C_n, and prove further that $\sum_{k=0}^{n} \binom{2k}{k} C_{n-k} = \binom{2n+1}{n}$.

5.78 Let M be an $n \times n$-matrix, and define $\overline{M}$ to be the usual cofactor matrix, $\overline{m}_{ij} = (-1)^{i+j} \det M_{j,i}$, where $M_{j,i}$ is the matrix with row j and column i deleted. Use Gessel–Viennot to prove $M\overline{M} = (\det M) I_n$.

5.79 Re-prove the Vandermonde result $\det(x_j^{n-i})_{i,j=1}^{n} = \prod_{i<j}(x_i - x_j)$ using weighted lattice paths.

Highlight: Tutte's Matrix–Tree Theorem

A classical theorem of graph theory, due essentially to Kirchhoff, relates the number of spanning trees of a connected graph to the determinant of a certain matrix. We discuss an important generalization to directed graphs, first established by W. T. Tutte, and give a proof via the lemma of Gessel-Viennot. The proof was suggested by Jürgen Schütz.

Arborescences in Directed Graphs.
Suppose $T = (V, E)$ is a directed graph and u a vertex. We say that T is an *arborescence converging to* u if there is a directed path from any $v \neq u$ to u, and the out-degree $d_T^-(v)$ equals 1 for all $v \neq u$ (and $d_T^-(u) = 0$). This implies, of course, that there is exactly one such path from any v to u. Given a directed graph $G = (V, E)$, then a subgraph $T \subseteq G$ is called a *spanning arborescence converging to* u if T is an arborescence towards u, and $V(T) = V(G)$.

Example. The bold edges constitute a spanning arborescence converging to u.

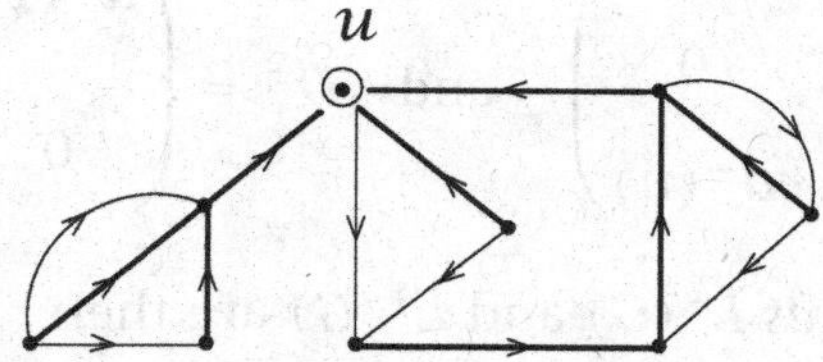

Similarly, $T \subseteq G$ is a *spanning arborescence diverging from* u if there is (precisely) one directed path in T from u to any vertex $v \neq u$ and in-degree $d_T^+(v) = 1$ for $v \neq u$, $d_T^+(u) = 0$.

Given $G = (V, E)$, our goal is to compute the numbers $t^-(G, u)$ and $t^+(G, u)$ of spanning arborescences in G converging to u and diverging from u, respectively.

Since loops clearly play no role, we may and will assume that G contains no loops, but there may be multiple edges between $v \neq w$ in either direction. Before we start our discussion of $t^-(G, u)$ and $t^+(G, u)$ let us note that these quantities yield also a formula for the number $t(G)$ of spanning trees of an undirected graph G.

Indeed, replace each edge $\{v, w\}$ of G by a pair of directed edges (v, w), (w, v) in each direction, and call the resulting graph G'.

Then we have, for any $u \in V$,

$$t(G) = t^-(G', u) = t^+(G', u)\,. \tag{1}$$

To see (1), note that each spanning arborescence in G' converging to u (or diverging from u) yields a spanning tree in G, by deleting the orientations. Conversely, let T be a spanning tree of G. Orienting the edges of T in the (unique) way toward u (respectively away from u) gives a spanning arborescence in G, and the correspondences are clearly bijections.

Laplace Matrix.
Let $G = (V, E)$ be a directed loopless graph with $V = \{1, \dots, n\}$, and denote by $A = (a_{ij})$ the $n \times n$-matrix with

$$a_{ij} = \text{ number of directed edges } i \to j\,.$$

Furthermore, let D^- and D^+ be the diagonal out-degree respectively in-degree matrices,

$$D^- = \begin{pmatrix} d^-(1) & & 0 \\ & \ddots & \\ 0 & & d^-(n) \end{pmatrix} \quad \text{and} \quad D^+ = \begin{pmatrix} d^+(1) & & 0 \\ & \ddots & \\ 0 & & d^+(n) \end{pmatrix}.$$

The *Laplace matrices* $L^-(G)$ and $L^+(G)$ are then

$$L^-(G) = D^- - A \quad \text{and} \quad L^+(G) = D^+ - A\,.$$

Finally, for any matrix $M = (m_{ij})$ we denote as usual by $M_{s,t}$ the $(n-1) \times (n-1)$-submatrix where row s and column t have been deleted.

Now we are all set for the theorem of Tutte.

Theorem. *Let $G = (V, E)$ be a loopless directed graph, $V = \{1, \dots, n\}$. Then for any $s, t \in V$ (where $s = t$ is possible),*

$$\begin{aligned} t^-(G, s) &= (-1)^{s+t} \det L^-(G)_{s,t}\,, \\ t^+(G, s) &= (-1)^{s+t} \det L^+(G)_{t,s}\,. \end{aligned} \tag{2}$$

Let us note first that the second formula in (2) follows from the first. Assume the formula for t^-, and let G' be the graph derived from G by reversing all directions. Then

$$t^+(G,s) = t^-(G',s),\ L^+(G) = L^-(G')^T,$$

and therefore for $s,t \in V$,

$$\begin{aligned} t^+(G,s) = t^-(G',s) &= (-1)^{s+t} \det L^-(G')_{s,t} \\ &= (-1)^{s+t} \det L^+(G)_{t,s}, \end{aligned}$$

as claimed. It thus suffices to prove (2) for converging arborescences.

Using the Lemma of Gessel–Viennot.
Let the directed graph $G = (V,E)$ be given. In order to prove (2) we construct a weighted directed graph H and two vertex sets $\mathcal{A}$ and $\mathcal{B}$ of H whose path matrix is precisely $L^-(G)$. It remains then to find a correspondence between the vertex-disjoint path systems of H and the arborescences in G.

The graph H is declared as follows. The vertex set is $(V \times \{0\}) \dot\cup (V \times \{1\}) \dot\cup E$, where we use the notation $A_k = (k,0)$ and $B_k = (k,1)$ for $k = 1,\dots,n$. To every edge $e = (k,\ell) \in E$ we have the H-edges $A_k \to e$, $e \to B_k$, each of weight 1, and further $e \to B_\ell$ of weight -1. The figure shows a small example, where the vertices of H corresponding to E are depicted by hollow circles.

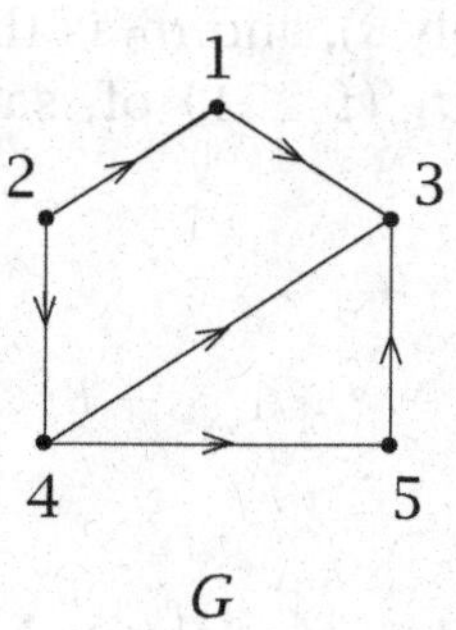

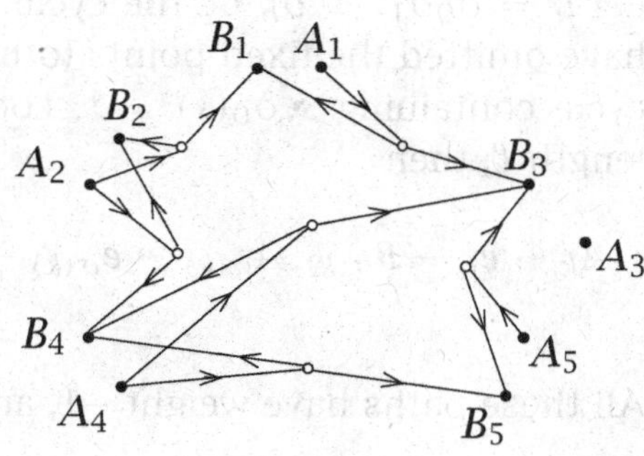

The following facts are immediate from the definition:

1. Directed paths from A_k to B_k are of the form $A_k \to e \to B_k$, where e is any edge leaving k in G. Each such path has weight 1, and hence the total weight of these paths is $d_G^-(k)$.
2. For $i \neq j$, the paths from A_i to B_j are $A_i \xrightarrow{1} e \xrightarrow{-1} B_j$, where e is an edge $e = (i,j)$ in G. The weight of the path is -1, and hence the total weight $-a_{ij}$. We conclude that $L^-(G)$ is precisely the path matrix with respect to $\mathcal{A} = \{A_1, \dots, A_n\}$ and $\mathcal{B} = \{B_1, \dots, B_n\}$.

Next we delete all edges in H incident with A_s and B_t, and insert a single edge $A_s \to B_t$ of weight 1 (or no edge if $s = t$). Call the new graph H'. The path matrix in H' from $\mathcal{A}$ to $\mathcal{B}$ is clearly $L^-(G)'_{s,t}$, where $L^-(G)'_{s,t}$ is the matrix that results by setting all elements in the s-th row and t-th column of $L(G^-)$ to 0, except the entry (s,t), which is equal to 1. In particular,

$$\det L^-(G)'_{s,t} = (-1)^{s+t} \det L^-(G)_{s,t}\,, \tag{3}$$

which is the right-hand side of (2).

Now we look at the vertex-disjoint path systems in H'. Let $\mathcal{P} = \{P_1, \dots, P_n\}$ be such a system, $P_i : A_i \to B_{\sigma(i)}$. The paths are of the form

$$\begin{aligned} &P_s : A_s \to B_{\sigma(s)} = B_t, \;\; w(P_s) = 1\,, \\ &P_k : A_k \to e_k \to B_{\sigma(k)}\,, \;\; k \neq s\,, \end{aligned}$$

where k is the initial vertex of $e_k \in E$, and either $\sigma(k)$ is also the initial vertex of e_k, in which case $\sigma(k) = k$ with $w(P_k) = 1$, or $\sigma(k)$ is the terminal vertex of $e_k = (k, \sigma(k))$ with $w(P_k) = -1$.

Let $\sigma = \sigma_0\sigma_1 \cdots \sigma_h$ be the cycle decomposition of σ in which we have omitted the fixed points (other than possibly s), and σ_0 is the cycle containing $s, \sigma_0(s) = t$. Look at a cycle σ_i $(i \geq 1)$ of, say, length ℓ; then

$$\begin{aligned} A_k \to e_k \to B_{\sigma(k)}, A_{\sigma(k)} \to e_{\sigma(k)} \to B_{\sigma^2(k)}, \\ \dots, A_{\sigma^{\ell-1}(k)} \to e_{\sigma^{\ell-1}(k)} \to B_k\,. \end{aligned}$$

All these paths have weight -1, and so

$$(\operatorname{sign}\sigma_i)(w(P_k)w(P_{\sigma(k)}) \cdots w(P_{\sigma^{\ell-1}(k)})) = (-1)^{\ell-1}(-1)^{\ell} = -1\,.$$

Similarly for σ_0, we obtain

$$(\operatorname{sign}\sigma_0)(w(P_s)w(P_{\sigma(s)})\cdots w(P_{\sigma^{\ell-1}(s)}))=1,$$

since $w(P_s)=1$, while the other paths have weight -1.

In summary, we see that

$$(\operatorname{sign}\mathcal{P})w(\mathcal{P})=(-1)^h, \tag{4}$$

with h as defined above, $\sigma=\sigma_0\sigma_1\cdots\sigma_h$.

For the final step we associate to each vertex-disjoint path system $\mathcal{P}$ the subgraph $T_{\mathcal{P}}=(V,\{e_k : k\neq s, k=1,\ldots,n\})$ of G. Since the paths in $\mathcal{P}$ are vertex-disjoint we obtain

$$d^-_{T_{\mathcal{P}}}(k)=\begin{cases}1 & k\neq s,\\ 0 & k=s.\end{cases} \tag{5}$$

The following two assertions will finish the proof.

A. If T is a spanning arborescence in G converging to s, then there is precisely one vertex-disjoint path system in H' with $T_{\mathcal{P}}=T$. Furthermore, $(\operatorname{sign}\mathcal{P})w(\mathcal{P})=1$.

B. On the set of all vertex-disjoint path systems $\mathcal{P}$ for which $T_{\mathcal{P}}$ is *not* an arborescence converging to u, there is an involution φ with

$$(\operatorname{sign}\varphi\mathcal{P})w(\varphi\mathcal{P})=-(\operatorname{sign}\mathcal{P})w(\mathcal{P}).$$

The lemma of Gessel–Viennot will then give

$$\begin{aligned}(-1)^{s+t}\det L^-(G)_{s,t} &= \sum_{\mathcal{P}\in VD}(\operatorname{sign}\mathcal{P})w(\mathcal{P})\\ &= \sum_{T_{\mathcal{P}}}(\operatorname{sign}\mathcal{P})w(\mathcal{P}),\ T_{\mathcal{P}}\text{ arborescence to } s,\\ &= t^-(G,s).\end{aligned}$$

Proof of A. Let T be a spanning arborescence in G converging to s, and denote by e_k the unique edge leaving $k\neq s$ in T. Suppose $Q: t=k_1\to\cdots\to k_\ell\to s$ is the path from t to s. The system $\mathcal{P}$ in H',

$$\begin{aligned}&P_s : A_s\to B_t,\\ &P_{k_i} : A_{k_i}\to e_{k_i}\to B_{k_{i+1}}\quad (i=1,\ldots,\ell),\\ &P_k : A_k\to e_k\to B_k\quad (k\text{ not on }Q),\end{aligned}$$

is clearly vertex-disjoint with $T_{\mathcal{P}} = T$ and $(\operatorname{sign}\mathcal{P})w(\mathcal{P}) = 1$ by (4), since all cycles not containing s are fixed points. Uniqueness is easily established by induction on the number of vertices.

Proof of B. If $T_{\mathcal{P}}$ is not an arborescence toward s, then by (5) it must contain directed circuits. Let k be minimal such that k is on a circuit C in $T_{\mathcal{P}}$, $C = \{i_1 \xrightarrow{e_{i_1}} i_2 \xrightarrow{e_{i_2}} \cdots \longrightarrow i_\ell \xrightarrow{e_{i_\ell}} i_1\}$. Note that $k \notin \sigma_0$. For the paths P_{i_j} in $\mathcal{P}$ there are two possibilities:

$$P_{i_j} : A_{i_j} \to e_{i_j} \to B_{i_j} \quad \text{for all } j = 1,\ldots,\ell, \quad \text{or}$$
$$P_{i_j} : A_{i_j} \to e_{i_j} \to B_{i_{j+1}} \quad \text{for all } j = 1,\ldots,\ell.$$

Denote by $\varphi(\mathcal{P})$ the system in which the paths P_{i_j} follow the other possibility. Since $T_{\varphi(\mathcal{P})} = T_{\mathcal{P}}$, φ is an involution. Now we observe that φ changes the number h in $\sigma = \sigma_0\sigma_1\cdots\sigma_h$ precisely by 1, and we conclude from (4) that

$$(\operatorname{sign}\varphi\mathcal{P})w(\varphi\mathcal{P}) = -(\operatorname{sign}\mathcal{P})w(\mathcal{P}),$$

and have thus proved the theorem.

We get a bonus out of the theorem. Suppose $G = (V,E)$ is an *Eulerian* directed graph, that is, $d^-(i) = d^+(i)$ for all $i \in V$, hence $L^-(G) = L^+(G)$. Then for $s,t \in V$, we conclude from (2) that

$$\begin{aligned} t^-(G,s) &= (-1)^{s+t}\det L^-(G)_{s,t} = (-1)^{s+t}\det L^+(G)_{s,t} \\ &= t^+(G,t). \end{aligned}$$

Hence all quantities $t^-(G,i)$ and $t^+(G,i)$ are the same, and we obtain the following beautiful result:

Corollary. *In an Eulerian directed graph G, there is the same number of arborescences converging to a vertex u (or diverging from u), for all $u \in V$.*

Finally, the bijection $G \to G'$ spelled out in (1) gives the corresponding result for undirected graphs, which is usually called Kirchhoff's theorem.

Corollary. *Let $G = (V,E)$ be a loopless undirected graph, A the adjacency matrix, and D the diagonal matrix of the vertex-degrees, $L(G) = D - A$. Then*

$$t(G) = (-1^{s+t}L(G)_{s,t}$$

for any $s,t \in V$.

Notes and References

The origin of the principle of inclusion–exclusion is unclear. The combinatorial form is sometimes attributed to Sylvester, but it may have been known to Euler when he evaluated the function $\varphi(n)$, now named after him. In probabilistic terms the principle can be traced back to de Moivre and Poincaré. The general theory of Möbius inversion on posets was advanced in the influential paper of Rota; for a detailed treatment see the books by Stanley and Aigner. A nice introduction to most of the topics of this chapter is given by Stanton and White. More on the Pfaffian and other algebraic questions can be found in the book by Godsil. The general involution principle was introduced by Garsia and Milne. The significance of the lemma of Gessel–Viennot, with many applications, was, logically enough, expounded by Gessel and Viennot. The original statement of the lemma appears in the paper of Lindström. For some important generalizations see also Stembridge. The matrix–tree theorem of the highlight appears (with a different proof) in Chapter 6 of the book by Tutte.

1. M. Aigner (1997): *Combinatorial Theory.* Reprint of the 1979 edition. Springer, Berlin.
2. A. Garsia and S. Milne (1981): Method for constructing bijections for classical partition identities. *Proc. Natl. Acad. Sci. USA* 78, 2026–2028.
3. I. Gessel and G. Viennot (1985): Binomial determinants, paths, and hook length formulae. *Advances Math.* 58, 300–321.
4. C.D. Godsil (1993): *Algebraic Combinatorics.* Chapman & Hall, New York.
5. B. Lindström (1973): On the vector representation of induced matroids. *Bull. London Math. Soc.* 5, 85–90.
6. G.-C. Rota (1964): On the foundations of combinatorial theory I. Theory of Möbius functions. *Z. Wahrscheinlichkeitstheorie* 2, 340–368.
7. R.P. Stanley (1997): *Enumerative Combinatorics*, vol. 1, 2nd edition. Cambridge Univ. Press, Cambridge.

8. D. Stanton and D. White (1986): *Constructive Combinatorics.* Springer, Berlin.

9. J. Stembridge (1990): Nonintersecting paths, Pfaffians, and plane partitions. *Advances Math.* 83, 96–131.

10. W.T. Tutte (1984): *Graph Theory.* Addison–Wesley, Reading.

6 Enumeration of Patterns

Many enumeration problems are of a different kind from those we have discussed so far. They are determined by *symmetries* on the underlying structure. How to approach problems of this type is the content of the present chapter.

6.1 Symmetries and Patterns

The following three examples should make clear what it is all about. We will solve all three problems as we go along.

Examples.

A. Consider a necklace with n beads colored black or white, lying on the table. How many different colored necklaces are there? First of all, what do we mean by "different"? We consider two necklaces as equal if one can be transformed into the other by some rotation. As an example, for $n = 4$ we obtain 6 different necklaces:

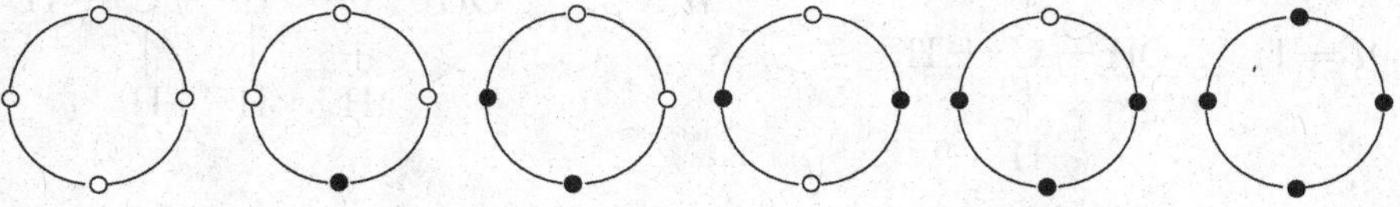

Suppose we color the four beads with three colors: W(hite), B(lack), and R(ed). Classifying the types with $W \geq B \geq R$ we get the following table:

W	B	R	number
4	0	0	1
3	1	0	1
2	2	0	2
2	1	1	3

In each type we may permute the colors, which gives altogether $3 \cdot 1 + 6 \cdot 1 + 3 \cdot 2 + 3 \cdot 3 = 24$ necklaces. Of course, what we really want to know is the general result for n beads and r colors.

B. Consider a cube whose faces are colored black or white. We will consider two colored cubes as equal if they can be transformed into each other by a rotation of the cube. The following table is readily verified:

color type		number
6	0	$2 \cdot 1$
5	1	$2 \cdot 1$
4	2	$2 \cdot 2$
3	3	$1 \cdot 2$
		10

Again we are interested in the general case in which r colors are available.

C. Let us now look at an example that is a bit more involved. An *alcohol* consists of one OH-group (1-valent), C-atoms (4-valent), and H-atoms (1-valent). We are interested in the number of alcohols with precisely n C-atoms. The figure shows the alcohols for $n \leq 3$:

```
n = 0:   OH—H

                   H
                   |
n = 1:   OH—  C  — H
                   |
                   H

                H    H
                |    |
n = 2:   OH— C — C — H
                |    |
                H    H

                H    H    H
                |    |    |
n = 3:   OH— C — C — C — H
                |    |    |
                H    H    H

                H         H
                |         |
         OH— C ——— C — H
                |         |
            H — C — H    H
                |
                H
```

In all examples "equality" is determined by a group of symmetries. In example A, it consists of the *cyclic* group C_n of all rotations through an angle $i\frac{360°}{n}$ $(i = 0, \ldots, n-1)$. In example B, it is the group of symmetries of the cube, and in the last example the symmetric group $S(3)$ of the three substructures attached to OH–C.

The Problem.

This leads to the following setup. We are given finite sets N and R. Any mapping $f : N \to R$ can be thought of as a coloring of N with colors in R. Next we are given a group $G \leq S(N)$ of permutations on N, where $S(N)$ denotes the full symmetric group on N.

In problem A, $N = \{1, \ldots, n\}$ and $G = C_n$ is the cyclic group of rotations. Colored necklaces correspond to mappings $f \in \mathrm{Map}(N, R)$. Clearly, two necklaces f, f' are "equal" if and only if $f' = f \circ g$ for some $g \in G$. Let us call $f, f' \in \mathrm{Map}(N, R)$ *equivalent*, $f \sim f'$, if $f' = f \circ g$ for some $g \in G$. The relation $\sim$ is an equivalence relation (since G is a group), and we call the equivalence classes the *patterns* of $\mathrm{Map}(N, R)$ under G. Thus we are interested in the number of patterns.

Example. Suppose $N = \{1, \ldots, n\}$. If $G = S(n)$, then the patterns correspond to n-multisets of R. The number of patterns is therefore $\binom{r+n-1}{n}$, with $|R| = r$. On the other hand, if $G = \{\mathrm{id}\}$, then the number of patterns is r^n.

Before we state the main problem we need an important generalization. Consider problem B, where G is the group of symmetries of the cube, and N the six faces. Instead of the faces we could just as well color the edges ($|N| = 12$), or the vertices ($|N| = 8$). Each time, the same group G *acts* as a group of permutations on the set N under consideration.

Hence we give the following definition. Let G be a finite group. We say that G *acts* on N if there is a homomorphism $\tau : G \to S(N)$, $g \mapsto \overline{g}$. We call $\overline{g} \in S(N)$ the permutation *induced by* g, and have

$$\overline{gh}(a) = \overline{g}(a)\overline{h}(a) \text{ for all } a \in N .$$

We set again

$$f' \sim f \iff f' = f \circ \overline{g} \text{ for some } g \in G ,$$

and call the resulting equivalence classes the *patterns* of $\mathrm{Map}(N, R)$ under the action of G.

In problem B, $\overline{g}$ is thus the actual permutation of the faces (or edges, or vertices) induced by the symmetry $g \in G$.

Problem. *Given N, R, and a group G that acts on N, count the number of patterns of* $\mathrm{Map}(N,R)$ *under G.*

Let us study this problem in full generality. Let G be a group that acts on the set X. As before, we set $y \sim x$ if $y = \overline{g}x$ for some $g \in G$, and call the equivalence classes the *patterns*. In group theory the patterns are called *orbits*, but we stick to the more expressive term "patterns". Thus in our previous setup, $X = \mathrm{Map}(N,R)$, and $f' = \overline{g}f$ if $f' = f \circ \overline{g}$.

The pattern containing x is therefore $M(x) = \{\overline{g}x : g \in G\}$. The following concepts pave the way to our first important result. The *stabilizer* of $x \in X$ is $G_x = \{g \in G : \overline{g}x = x\}$, which is clearly a subgroup of G. The *fixed-point set* of $g \in G$ is $X_g = \{x \in X : \overline{g}x = x\}$. Counting the pairs $\{(x,g) : \overline{g}x = x\}$ in two ways immediately gives

$$\sum_{x \in X} |G_x| = \sum_{g \in G} |X_g| \,. \tag{1}$$

Lemma 6.1. *Let G act on X. Then for any $x \in X$,*

$$|M(x)| = \frac{|G|}{|G_x|} \,. \tag{2}$$

Proof. We have $M(x) = \{\overline{g}x : g \in G\}$. Now,

$$\overline{g}x = \overline{h}x \iff \overline{g}^{-1}\overline{h}x = x \iff g^{-1}h \in G_x \iff h = ga, a \in G_x \,.$$

This means that for every $g \in G$, exactly $|G_x|$ group elements, namely all $h = ga$ with $a \in G_x$, give the *same* element $\overline{h}x = \overline{g}x \in M(x)$, and (2) follows immediately. □

Results (1) and (2) imply the following famous lemma of Burnside–Frobenius.

Lemma 6.2 (Burnside–Frobenius). *Let the group G act on X, and let $\mathcal{M}$ be the set of patterns. Then*

$$|\mathcal{M}| = \frac{1}{|G|} \sum_{g \in G} |X_g| \,. \tag{3}$$

Proof. Let $y \in M(x)$, or equivalently $M(y) = M(x)$. Then by (2), $|G_x| = |G_y|$, and therefore

$$|G| = |M(x)||G_y| = \sum_{y \in M(x)} |G_y| .$$

Summation over all patterns together with (1) gives

$$|\mathcal{M}||G| = \sum_{\text{patterns}\, M(x)} \sum_{y \in M(x)} |G_y| = \sum_{y \in X} |G_y| = \sum_{g \in G} |X_g| ,$$

and thus the result. □

The importance of the lemma rests on the fact that in many examples the sets X_g are easily enumerated, whereas the patterns $M(x)$ may be harder to count.

Example. Consider the necklace problem with $n = p$ a prime number and r colors. We set $X = \text{Map}(N, R)$ and want to determine $|X_{\rho_i}|$, where ρ_i is the rotation through the angle $i\frac{360°}{p}$. Obviously, $|X_{\rho_0}| = r^p$. Consider $f \in X_{\rho_i}$, that is, $f = f \circ \rho_i$ for $0 < i \le p - 1$. Since i and p are relatively prime, a moment's thought shows that all beads must be colored alike. This means that $|X_{\rho_i}| = r$, and the lemma implies

$$|\mathcal{M}| = \frac{1}{p}(r^p + (p-1)r) .$$

Since the right-hand side is an integer, we have proved Fermat's theorem

$$r^p \equiv r \pmod{p} ,$$

or

$$r^{p-1} \equiv 1 \pmod{p}$$

if r and p are relatively prime.

The following idea helps to greatly reduce the number of $g \in G$ that we have to consider in (3). Call two elements $g, g' \in G$ *conjugate* if $g' = h^{-1}gh$ for some $h \in G$. Clearly, conjugacy is an equivalence relation on G, whose equivalence classes are called *conjugacy classes* $C(g)$.

Claim. *If g and g' are conjugate, then $|X_g| = |X_{g'}|$.*

Suppose $g' = h^{-1}gh$, and consider the bijection $\overline{h} : X \to X, x \mapsto \overline{h}x$. We have

$$x \in X_{g'} \iff \overline{h}^{-1}\overline{g}\overline{h}x = x \iff \overline{g}(\overline{h}x) = \overline{h}x \iff \overline{h}x \in X_g ,$$

and thus $|X_{g'}| = |X_g|$.

It suffices therefore to pick a representative out of each conjugacy class; let us call such a set a *transversal* of the conjugacy classes.

Corollary 6.3. *Let G act on X, and let T be a transversal of the conjugacy classes of G. Then*

$$|\mathcal{M}| = \frac{1}{|G|} \sum_{g \in T} |C(g)||X_g| . \tag{4}$$

Example. If the group G is commutative, then $h^{-1}gh = g$ for any h, and the conjugacy classes consist of singletons. So in this case, for example for the cyclic group C_n, the corollary does not help.

But consider now the *dihedral group* D_n, $n \geq 3$. It consists of all rotations of a regular n-gon plus n reflections. For the necklace problem this means that we may take off the necklace and put it back on the other way around. Let $N = \{0, 1, \ldots, n-1\}$ with 0 on top as in the figure.

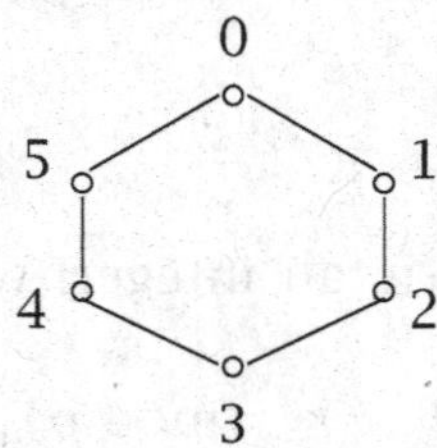

We write the group D_n in multiplicative form. Thus the subgroup of rotations is $\{\rho^0 = 1, \rho, \rho^2, \ldots, \rho^{n-1}\}$, where

$$\rho^k : i \to i + k \pmod{n} \quad (k = 0, \ldots, n-1) .$$

Let σ be the reflection along the vertical middle axis,

$$\sigma : i \to n - i \pmod{n} .$$

Then $D_n = \{\rho^0, \rho, \ldots, \rho^{n-1}, \sigma, \sigma\rho, \ldots, \sigma\rho^{n-1}\}$, with

$$\sigma\rho^k = \rho^{-k}\sigma .$$

It is easy to see that the rotations split into the conjugacy classes

$$\{\rho^0\}, \{\rho, \rho^{n-1}\}, \ldots, \{\rho^{\frac{n-1}{2}}, \rho^{\frac{n+1}{2}}\} \quad \text{for odd } n ,$$

and

$$\{\rho^0\}, \{\rho, \rho^{n-1}\}, \ldots, \{\rho^{\frac{n}{2}-1}, \rho^{\frac{n}{2}+1}\}, \{\rho^{\frac{n}{2}}\} \quad \text{for even } n.$$

For odd n, there is one class for the reflections, and for even n there are the two classes

$$\{\sigma, \sigma\rho^2, \sigma\rho^4, \ldots, \sigma\rho^{n-2}\}, \quad \{\sigma\rho, \sigma\rho^3, \sigma\rho^5, \ldots, \sigma\rho^{n-1}\}.$$

A Famous Example.

To demonstrate the full power of the lemma of Burnside–Frobenius let us look at the following famous chessboard problem. In how many ways can we place n non-attacking rooks on an $n \times n$-chessboard? These arrangements $\{(i, \pi_i) : i = 1, \ldots, n\}$ correspond bijectively to the $n!$ permutations $\pi \in S(n)$. But how many patterns are there? The symmetry group of the board is the dihedral group D_4, $|D_4| = 8$. Let us denote as before the rotations in clockwise fashion by $1, \rho, \rho^2, \rho^3$, and by σ the reflection along the vertical axis.

Here are the patterns up to $n = 4$:

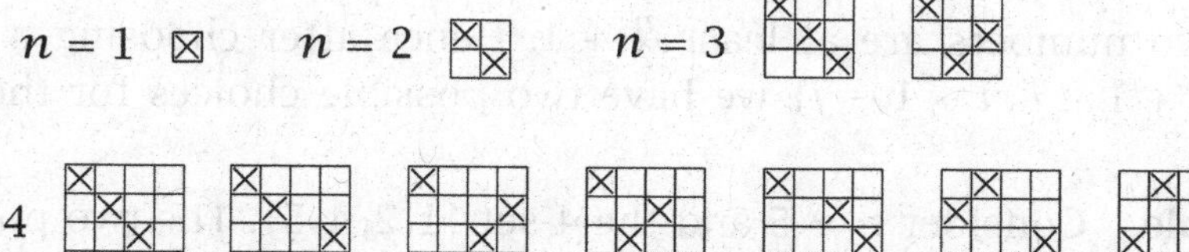

To obtain the general result, let $N^2 = \{(i, j) : 1 \le i, j \le n\}$ be the cells of the board, and X the set of non-attacking rook arrangements. We identify $\Pi = \{(i, \pi_i) : i = 1, \ldots, n\} \in X$ with the permutation $\pi \in S(n)$, and write $i \xrightarrow{\pi} j$ if $(i, j) \in \Pi$. In order to apply the lemma we must determine where a cell (i, j) is moved by the elements of D_4. The following table shows this (check it!):

$$\begin{aligned}
(i,j) &\xrightarrow{\rho^0} (i,j) & (i,j) &\xrightarrow{\sigma} (i, n+1-j) \\
&\xrightarrow{\rho} (j, n+1-i) & &\xrightarrow{\sigma\rho} (j, i) \\
&\xrightarrow{\rho^2} (n+1-i, n+1-j) & &\xrightarrow{\sigma\rho^2} (n+1-i, j) \\
&\xrightarrow{\rho^3} (n+1-j, i) & &\xrightarrow{\sigma\rho^3} (n+1-j, n+1-i).
\end{aligned} \tag{5}$$

Notice that any $h \in D_4$ carries $\Pi \in X$ again into an $h\Pi \in X$. It remains to determine the size of X_g for $g \in D_4$, where by the remarks on conjugacy we may restrict ourselves to $\rho^0, \rho, \rho^2, \sigma$, and $\sigma\rho$.

We just consider ρ; the other cases are settled in a similar fashion. Let $\Pi \in X_\rho$; we describe Π by looking at its corresponding permutation π. If $i \xrightarrow{\pi} j$, then $j \xrightarrow{\pi} n+1-i$ according to (5). This, in turn, implies $n+1-i \xrightarrow{\pi} n+1-j \xrightarrow{\pi} i$. Hence π decomposes into cycles of the form $(i, j, n+1-i, n+1-j)$, but of course the cycle may close before $n+1-j$.

Suppose $i \xrightarrow{\pi} i$. Then $i \xrightarrow{\pi} i \xrightarrow{\pi} n+1-i = i$, that is, $i = \frac{n+1}{2}$. So this can happen only for odd n, and it is clear that in this case the central cell $\left(\frac{n+1}{2}, \frac{n+1}{2}\right)$ is fixed. Now let $i \xrightarrow{\pi} j \neq i$. If (i,j) is a 2-cycle of π, then $i \xrightarrow{\pi} j \xrightarrow{\pi} n+1-i = i \xrightarrow{\pi} n+1-j = j$; thus $i = j = \frac{n+1}{2}$, which cannot be. Hence there are no 2-cycles, and similarly there are no 3-cycles. We therefore have the following result:

If n is odd, then $\pi \in X_\rho$ consists of one fixed point and $\frac{n-1}{4}$ 4-cycles; thus $n = 4t+1$. If n is even, then $\pi \in X_\rho$ consists of $\frac{n}{4}$ 4-cycles; thus $n = 4t$. Furthermore, we see that in every 4-cycle $(i, j, n+1-i, n+1-j)$, two numbers are less than or equal $\frac{n}{2}$, and two numbers are at least $\frac{n}{2}+1$. Hence after choosing a set $\{i, j, n+1-i, n+1-j\}$ we have two possible choices for the 4-cycle.

Example. Consider $n = 5$ and the 4-set $\{1, 2, 4, 5\}$. The two possible 4-cycles are $(1, 2, 5, 4)$ and $(1, 4, 5, 2)$, corresponding to the rook configurations $\{(1,2), (2,5), (3,3), (4,1), (5,4)\}$ and $\{(1,4), (2,1), (3,3), (4,5), (5,2)\}$:

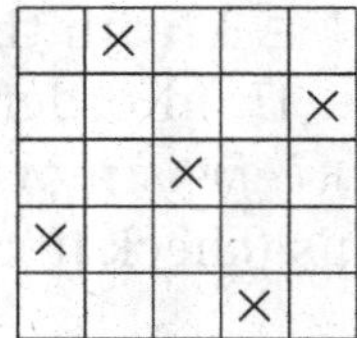

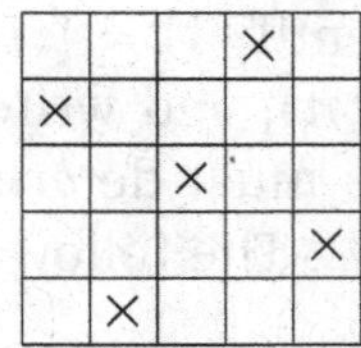

Both configurations are invariant under ρ, which they must be.

It remains to count in how many ways we can split the set N (or $N \smallsetminus \{\frac{n+1}{2}\}$) into 4-sets of the form $\{i, j, n+1-i, n+1-j\}$. But this is easy. We may pair 1 with any other number $j \leq \frac{n}{2}$; the remaining numbers n and $n+1-j$ are then determined. Now we choose the next element, and so on. Altogether this gives

$$\tfrac{n-3}{2}\cdot\tfrac{n-7}{2}\cdots 1 \quad \text{for } n = 4t+1\,,$$
$$\tfrac{n-2}{2}\cdot\tfrac{n-6}{2}\cdots 1 \quad \text{for } n = 4t$$

possibilities. Taking into account that every 4-set may be made into two 4-cycles we obtain

$$|X_\rho| = |X_{\rho^3}| = \begin{cases} (n-3)(n-7)\cdots 2 & n = 4t+1\,, \\ (n-2)(n-6)\cdots 2 & n = 4t\,. \end{cases} \tag{6}$$

The result for ρ^2 is

$$|X_{\rho^2}| = \begin{cases} (n-1)(n-3)\cdots 2 & n \ \text{ odd,} \\ n(n-2)\cdots 2 & n \ \text{ even,} \end{cases} \tag{7}$$

and, of course, $|X_{\rho^0}| = n!$.

The table (5) easily implies $|X_\sigma| = |X_{\sigma\rho^2}| = 0$, whereas for $\sigma\rho$ (and by conjugacy for $\sigma\rho^3$), $\pi \in X_{\sigma\rho}$ consists of 2-cycles and 1-cycles. Thus

$$|X_{\sigma\rho}| = |X_{\sigma\rho^3}| = i_n\,, \tag{8}$$

where i_n is the number of involutions in $S(n)$. In Exercise 1.39 we derived the recurrence $i_{n+1} = i_n + n i_{n-1}$ ($i_0 = 1$), which gives the small values

n	0	1	2	3	4	5	6	7	8
i_n	1	1	2	4	10	26	76	232	764

With some straightforward algebraic manipulations we arrive with (6), (7), (8), and the lemma of Burnside–Frobenius at the final result. Let $\mathcal{R}_n$ be the patterns of non-attacking rook configurations. Then for $n \ge 2$,

$$|\mathcal{R}_n| = \begin{cases} \frac{1}{8}\left(n! + 2i_n + (2t)!2^{2t} + (2t)!\frac{2}{t!}\right) & \text{for } n = 4t, 4t+1\,, \\ \frac{1}{8}\left(n! + 2i_n + (2t+1)!2^{2t+1}\right) & \text{for } n = 4t+2, 4t+3\,. \end{cases}$$

For $n = 4$ we compute $|\mathcal{R}_4| = \frac{1}{8}(24 + 20 + 2\cdot 4 + 2\cdot 2) = 7$, verifying our example from above, and further, $|\mathcal{R}_5| = 23$, $|\mathcal{R}_6| = 115$, $|\mathcal{R}_7| = 694$, $|\mathcal{R}_8| = 5282$.

Exercises

6.1 A flag is divided into n horizontal stripes. How many black–white coloring patterns exist in which the flag can be turned upside down and look the same?

6.2 How many patterns do we obtain when we color the 12 edges of a cube with 12 different colors?

▷ **6.3** Consider a necklace with 6 beads, colored black or white, and let C_6 act on the necklace (rotations only) or D_6 (also reflections). Compute the number of patterns. There are exactly two patterns that are different under C_6, but equal under D_6; which?

6.4 Use the lemma of Burnside–Frobenius to count the number of four-bead necklaces that can be formed with three colors such that no two consecutive beads receive the same color. Consider C_4 and D_4.

▷ **6.5** Count the number of lattice paths from $(0,0)$ to (n,n) with steps up and to the right, when two paths are considered equal if one can be moved on top of the other by a rotation or reflection. Example:

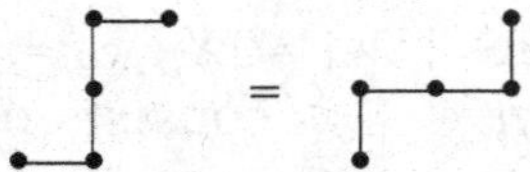

Hint: Represent a path as a sequence of 0's and 1's and determine the group that acts on these sequences.

6.6 Show that there are precisely three vertex-patterns of the octahedron in which three vertices are red, two are blue, and one is green. Draw them.

6.7 Complete the computation in the example of the non-attacking rook patterns by determining the sizes of the other fixed-point sets.

▷ **6.8** Consider a triangular array of $\binom{n+1}{2}$ balls on a table, for example

for $n = 3$. Find the number of patterns (under C_3) when three colors are available.

6.9 A rotating table has a pocket at each corner that may contain 1, 2, or 3 balls. Compute the number of different arrangements with a total number of 8 balls.

* * *

6.10 A company introduces quadratic 3×3-cards with two holes punched in them as identity cards, for example,

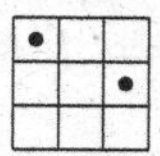

The two sides of the cards are not distinguishable. How many people can be employed?

▷ **6.11** Let N be an n-set, $G \le S(N)$. We call two subsets A and B G-equivalent if $B = \{g(a) : a \in A\}$ for some $g \in G$. Show for the number m of inequivalent subsets that $m = \frac{1}{|G|}\sum_{g\in G} 2^{c(g)}$, where $c(g)$ denotes the number of cycles of g.

6.12 Use the previous exercise to prove $\sum_{\sigma\in S(n)} 2^{c(\sigma)} = (n+1)!$.

▷ **6.13** Determine the number of non-isomorphic multiplication tables on two elements 0 and 1. There are $2^4 = 16$ tables. Two multiplications $x\cdot y$ and $x \circ y$ are *isomorphic* if there is a bijection φ with $\varphi(x\cdot y) = \varphi(x)\circ\varphi(y)$. Determine this number also for three elements.

6.14 A *caterpillar* is a tree in which deletion of end-vertices (vertices of degree 1) leaves a path. Example: . Prove that the number of non-isomorphic caterpillars on n vertices is $2^{n-4} + 2^{\lfloor n/2\rfloor -2}$ for $n \ge 3$.

6.15 Let p be a prime, and r, ℓ positive integers. Generalize Fermat's theorem to prove that $r^{p^\ell} - r^{p^{\ell-1}}$ is divisible by p^ℓ.

6.16 Consider an arbitrary group G. The group G acts on itself by $\overline{h} : G \to G$, $\overline{h}(g) = h^{-1}gh$, for $h \in G$. The patterns under this action are the conjugacy classes. Describe the stabilizer G_g under this action.

▷ **6.17** Consider the symmetric group $S(n)$. Show that if π and σ are conjugate, then they have the same type. Use the previous exercise and Exercises 1.45, 1.46 to conclude that two permutations are conjugate if and only if they have the same type.

6.18 Let p be a prime, and consider the (trivial) action of $S(p)$ on a single point $N = \{a\}$. Use (4) and the previous exercise to deduce Wilson's theorem from number theory: $(p-1)! \equiv -1 \pmod{p}$.

6.2 The Theorem of Pólya-Redfield

Let us return to our original problem in which we are given two sets N and R, and a group G that acts on N. The lemma of Burnside–

Frobenius gives the number of patterns of $\text{Map}(N,R)$, where the action on $\text{Map}(N,R)$ is

$$f' = \overline{g}f \iff f' = f \circ \overline{g}. \tag{1}$$

We see with (1) that we may more generally consider subsets $\mathcal{F} \subseteq \text{Map}(N,R)$ such that $f \circ \overline{g} \in \mathcal{F}$ whenever $f \in \mathcal{F}$ holds. Let us call these families $\mathcal{F}$ *closed under* G. Two obvious examples are the injective mappings $\text{Inj}(N,R)$ and the surjective mappings $\text{Surj}\,(N,R)$.

For many applications it is important to specify the patterns further. In the necklace problem A we could ask how many patterns exist with k white beads. In problem C it is of interest to determine the number of alcohols with exactly k C-atoms.

Let us treat this in full generality. We associate to $j \in R$ a variable x_j, and define the *weight* $w(f)$ of $f \in \mathcal{F}$ by

$$w(f) = \prod_{i \in N} x_{f(i)}. \tag{2}$$

Now if $f' \sim f$, $f' = f \circ \overline{g}$, then

$$w(f') = \prod_{i \in N} x_{f'(i)} = \prod_{i \in N} x_{f(\overline{g}i)} = \prod_{i \in N} x_{f(i)} = w(f),$$

since with $i \in N$, $\overline{g}i$ runs through all of N as well. We can therefore unambiguously define the weight of a pattern M by

$$w(M) = w(f) \quad (f \in M). \tag{3}$$

With this we come to the following general problem.

Problem. *Given sets N and R, $|N| = n$, $|R| = r$, a group G that acts on N, and variables x_j $(j \in R)$; let $\mathcal{M}_{\mathcal{F}}$ be the set of patterns of the G-closed family $\mathcal{F} \subseteq$* $\text{Map}(N,R)$. *Determine the weight enumerator*

$$w(\mathcal{F};G) = \sum_{M \in \mathcal{M}_{\mathcal{F}}} w(M). \tag{4}$$

In particular, if we set $x_j = 1$ for all $j \in R$, then we obtain the number $|\mathcal{M}_{\mathcal{F}}|$ of all patterns of $\mathcal{F}$.

Example. Consider the necklace problem with $n = 4$, $r = 2$, and the variables W and B. Then

$$w(\mathrm{Map}(N,R);C_4) = W^4 + W^3B + 2W^2B^2 + WB^3 + B^4.$$

If we color the faces of the cube white or black, then

$$w(\mathrm{Map}(N,R);G) = W^6+W^5B+2W^4B^2+2W^3B^3+2W^2B^4+WB^5+B^6.$$

Theorem 6.4. *Let N and R be sets, G a group that acts on N, $\mathcal{F}$ a G-closed class, and x_j $(j \in R)$ variables. Then*

$$w(\mathcal{F};G) = \sum_{M\in\mathcal{M}_{\mathcal{F}}} w(M) = \frac{1}{|G|}\sum_{g\in G}\ \sum_{f\in\mathcal{F},f\circ\overline{g}=f} w(f). \tag{5}$$

Proof. Let M be a pattern of $\mathcal{F}$ under G. If we apply G to M, then, of course, we obtain only one pattern, namely M. By the lemma of Burnside–Frobenius,

$$1 = \frac{1}{|G|}\sum_{g\in G}|M_g|, \tag{6}$$

where $M_g = \{f \in M : f \circ \overline{g} = f\}$. We know that all $f \in M$ have the same weight $w(f) = w(M)$. Multiplication of (6) by $w(M)$ therefore gives

$$w(M) = \frac{1}{|G|}\sum_{g\in G}|M_g|w(f) = \frac{1}{|G|}\sum_{g\in G}\ \sum_{f\in M:f\circ\overline{g}=f} w(f),$$

and summation over all patterns yields (5). □

Theorem of Pólya–Redfield.

To apply Theorem 6.4 we must be able to evaluate the inner sum $\sum_{f\in\mathcal{F}:f\circ\overline{g}=f} w(f)$. In general, this raises grave difficulties, but for the most important case $\mathcal{F} = \mathrm{Map}(N,R)$ there is an elegant solution due to Pólya and Redfield.

Consider $f \in \mathrm{Map}(N,R)$ with $f \circ \overline{g} = f$, and let

$$(a, \overline{g}a, \overline{g}^2a, \ldots, \overline{g}^{k-1}a)$$

be a cycle of $\overline{g}$. Then $f(a) = f(\overline{g}a) = f(\overline{g}^2a) = \cdots = f(\overline{g}^{k-1}a)$, that is, f is *constant* on all cycles of $\overline{g}$. It follows that $f \circ \overline{g} = f$ if and only if f is constant on all cycles of $\overline{g}$. Now suppose $\overline{g}$ has type $1^{c_1}2^{c_2}\ldots n^{c_n}$, and f maps the 1-cycles to $f_{11},\ldots,f_{1c_1}$, the 2-cycles to $f_{21},\ldots,f_{2c_2}$, and so on. This gives the weight of f as

$$w(f) = x_{f_{11}} \cdots x_{f_{1c_1}} x^2_{f_{21}} \cdots x^2_{f_{2c_2}} \cdots x^n_{f_{n1}} \cdots x^n_{f_{nc_n}} . \tag{7}$$

Since *all* mappings f appear that are constant on the cycles of $\overline{g}$, we conclude that

$$\sum_{f:f\circ\overline{g}=f} w(f) = \Big(\sum_{j\in R} x_j\Big)^{c_1} \Big(\sum_{j\in R} x_j^2\Big)^{c_2} \cdots \Big(\sum_{j\in R} x_j^n\Big)^{c_n} . \tag{8}$$

In other words, to evaluate an inner sum $\sum_{f:f\circ\overline{g}=f} w(f)$ in (5) only the cycle structure of $\overline{g}$ matters, and this suggests the following definition.

Definition. Let G be a group that acts on N, $|N| = n$. The *cycle index* of G is the polynomial in the variables $z_1, \ldots, z_n$ given by

$$Z(G; z_1, \ldots, z_n) = \frac{1}{|G|} \sum_{g\in G} z_1^{c_1(\overline{g})} \cdots z_n^{c_n(\overline{g})} ,$$

where $t(\overline{g}) = 1^{c_1(\overline{g})} \ldots n^{c_n(\overline{g})}$ is the type of $\overline{g}$.

With this definition we immediately infer from (5) and (8) the following fundamental theorem.

Theorem 6.5 (Pólya–Redfield). *Let N and R be sets, $|N| = n$, $|R| = r$, and G a group acting on N, x_j $(j \in R)$ variables. Then*

$$\begin{aligned} w(\mathrm{Map}(N,R); G) &= \sum_{M\in\mathcal{M}} w(M) \\ &= Z\Big(G; \sum_{j\in R} x_j, \sum_{j\in R} x_j^2, \ldots, \sum_{j\in R} x_j^n\Big) . \end{aligned} \tag{9}$$

In particular,

$$|\mathcal{M}| = Z(G; r, r, \ldots, r) . \tag{10}$$

The Three Problems.

With this theorem in hand we can easily solve our three problems. Each time, we have to compute the cycle index of the group in question.

Problem A. Let $N = \{0, 1, \ldots, n-1\}$, and $\rho^k : i \to i + k \pmod n$ as before. It is clear that ρ^k decomposes N into cycles of the same length d; hence $t(\rho^k) = d^{n/d}$ for some divisor d of n. Furthermore,

d is the smallest number such that after d rotations ρ^k we arrive again at 0. This implies that d is the smallest number with $n \mid dk$. It follows that $\frac{n}{\gcd(n,k)} \mid d$, and hence $d = \frac{n}{\gcd(n,k)}$ by the minimality of d. In the solution of Exercise 1.9 it was shown that there are exactly $\varphi(d)$ numbers k with $\gcd(n,k) = \frac{n}{d}$, where φ is Euler's function. In sum, we obtain the beautiful formula

$$Z(C_n; z_1, \dots, z_n) = \frac{1}{n} \sum_{d|n} \varphi(d) z_d^{n/d} . \tag{11}$$

By the theorem of Pólya–Redfield the number of n-necklaces with r colors is thus given by

$$|\mathcal{M}_{n,r}| = \frac{1}{n} \sum_{d|n} \varphi(d) r^{n/d} . \tag{12}$$

For $n = 4$ this gives

$$|\mathcal{M}_{4,r}| = \frac{1}{4}(r^4 + r^2 + 2r) .$$

For example, there are 70 such necklaces with 4 colors, and 616 with 7 colors.

Problem B. One learns in geometry that every symmetry of the cube possesses an axis. This axis passes through opposite faces (more precisely through the middle points), through diametrically opposite edges, or through diametrically opposite vertices. Hence there are 3 face axes, 6 edge axes, and 4 vertex axes. The figure shows that every face axis produces three symmetries (rotations) apart from the identity.

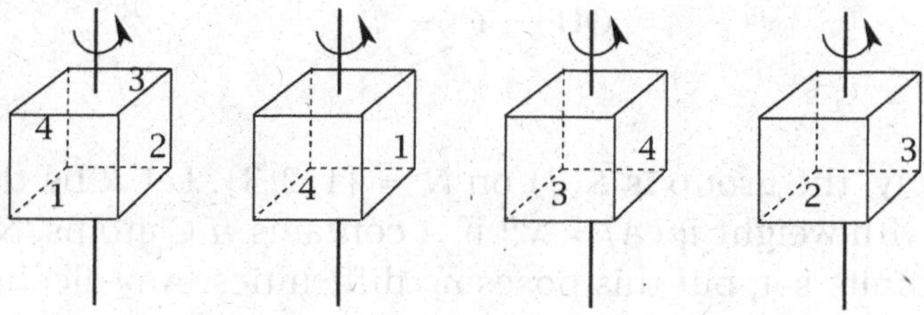

Altogether, G consists of the 24 symmetries

identity	1
face axes	$3 \cdot 3$
edge axes	$6 \cdot 1$
vertex axes	$4 \cdot 2$
	24

It is now an easy matter to determine the cycle index for the action on the faces:

identity	z_1^6,
face axis	$3(z_1^2 z_2^2 + 2z_1^2 z_4)$,
edge axis	$6z_2^3$,
vertex axis	$8z_3^2$,

and so

$$Z(G; z_1, \dots, z_6) = \frac{1}{24}(z_1^6 + 3z_1^2 z_2^2 + 6z_1^2 z_4 + 6z_2^3 + 8z_3^2). \tag{13}$$

By the theorem the number of face patterns with r colors is

$$\begin{aligned} |\mathcal{M}_r| &= \frac{1}{24}(r^6 + 3r^4 + 12r^3 + 8r^2) \\ &= \frac{1}{24} r^2 (r+1)(r^3 - r^2 + 4r + 8). \end{aligned} \tag{14}$$

For $r = 2$ and $r = 3$ we obtain $|\mathcal{M}_2| = 10$ and $|\mathcal{M}_3| = 57$.

Problem C. We want to determine the generating function $A(x) = \sum_{n \geq 0} a_n x^n$, where a_n is the number of alcohols with n C-atoms. We have $a_0 = 1$, and for $n \geq 1$ we call the C-atom attached to OH the *root* of the alcohol. There are three subalcohols attached to the root (the root now plays the role of OH), which may be permuted arbitrarily:

```
            1
           /
OH — C — 2
           \
            3
```

Accordingly, the group is $S(3)$ on $N = \{1, 2, 3\}$. Let R be the set of alcohols with weight $w(A) = x^n$ if A contains n C-atoms. Note that R is an infinite set, but this poses no difficulties. Any alcohol corresponds therefore to a map $f : N \to R$, and the different alcohols are precisely the patterns under $S(3)$. The cycle index of $S(3)$ is

$$Z(S_3; z_1, z_2, z_3) = \frac{1}{6}(z_1^3 + 3z_1z_2 + 2z_3);$$

furthermore, $\sum_{A\in R} w(A)^k = A(x^k)$. Taking the root into account we arrive with Pólya-Redfield at the functional equation

$$A(x) = 1 + \frac{x}{6}[A(x)^3 + 3A(x)A(x^2) + 2A(x^3)],$$

which gives by comparing coefficients

$$a_n = \frac{1}{6}\left[\sum_{i+j+k=n-1} a_ia_ja_k + 3\sum_{i+2j=n-1} a_ia_j + 2a_{\frac{n-1}{3}}\right] \quad (n \geq 1).$$

The first values are

n	0	1	2	3	4	5	6	7	8
a_n	1	1	1	2	4	8	17	39	89

The figure shows the 8 alcohols for $n = 5$, where only the OH-group and the C-atoms are given:

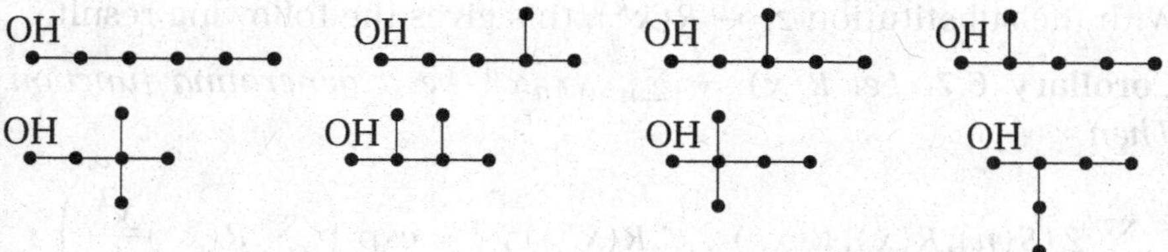

A Generalization.
The alcohol example can immediately be generalized to the following situation. Let R be a set (possibly countably infinite), and suppose every element of R has a weight x^n $(n \geq 0)$; $R(x) = \sum_{n\geq 0} r_nx^n$ is the generating function, where r_n is the number of elements of R of weight x^n (assumed to be finite). Now let G act on the n-set N, and consider as before the patterns of $\text{Map}(N,R)$ under G.

Corollary 6.6. *Let m_k be the number of patterns of weight x^k in* $\text{Map}(N,R)$ *under G, where all m_k are assumed to be finite. Then*

$$\begin{aligned} w(\text{Map}(N,R);G) &= \sum_{k\geq 0} m_kx^k \\ &= Z(G; R(x), R(x^2), \ldots, R(x^n)). \end{aligned} \tag{15}$$

Proof. As in problem C, $\sum_{a\in R} w(a)^k = R(x^k)$, and the result follows. □

When $G = S(n)$ we get from this a very interesting formula. Consider the permutation version of the exponential formula in Section 3.3 with the notation there, and let $f(n) = z_n$ $(n \geq 1)$. Then with

$$h(n) = \sum_{\sigma\in S(n)} z_1^{c_1(\sigma)} z_2^{c_2(\sigma)} \cdots z_n^{c_n(\sigma)},$$

we have

$$\hat{H}(y) = \sum_{n\geq 0} h(n)\frac{y^n}{n!} = \exp\left(\sum_{k\geq 1} z_k \frac{y^k}{k}\right).$$

But $\frac{h(n)}{n!} = \frac{1}{n!}\sum_{\sigma\in S(n)} z_1^{c_1(\sigma)} \cdots z_n^{c_n(\sigma)} = Z(S(n); z_1, \ldots, z_n)$, and so we obtain

$$\sum_{n\geq 0} Z(S(n); z_1, \ldots, z_n) y^n = \exp\left(\sum_{k\geq 1} z_k \frac{y^k}{k}\right). \tag{16}$$

With the substitution $z_k \mapsto R(x^k)$, this gives the following result.

Corollary 6.7. *Let $R(x) = \sum_{n\geq 0} r_n x^n$ be a generating function. Then*

$$\sum_{n\geq 0} Z(S(n); R(x), R(x^2), \ldots, R(x^n)) y^n = \exp\left(\sum_{k\geq 1} R(x^k)\frac{y^k}{k}\right). \tag{17}$$

Example. We want to determine the generating function $U(x) = \sum_{n\geq 1} u_n x^n$, where u_n is the number of isomorphism classes of rooted trees with n vertices. Two rooted trees (T, v) and (T', v') are isomorphic if $\varphi : T \to T'$ is an isomorphism that carries v into v'. Thus, for example, $u_1 = u_2 = 1$, $u_3 = 2$, $u_4 = 4$. The figure shows the trees:

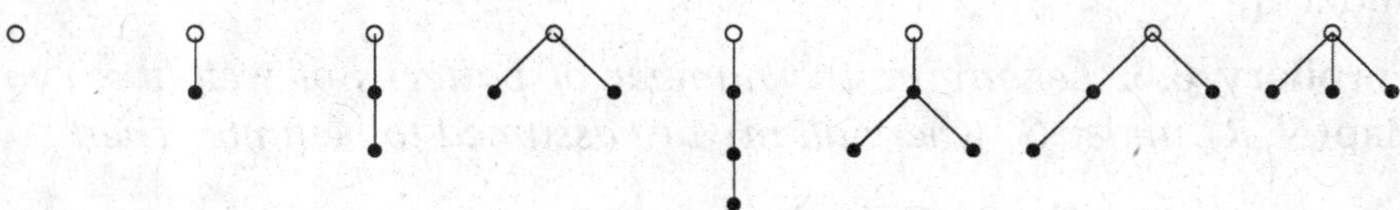

Suppose the root v has degree n. Let $N = \{1, \ldots, n\}$, and R the set of rooted trees, with $w(T) = x^n$ if T has n vertices:

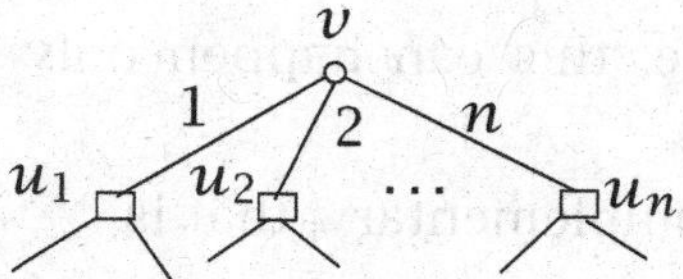

Considering the neighbors $u_1, \ldots, u_n$ of v as roots of the subtrees, we may permute these subtrees arbritrarily, and hence obtain from (15), taking the root into account,

$$V_n(x) = xZ(S(n); U(x), U(x^2), \ldots, U(x^n)),$$

where $V_n(x)$ is the generating function of the number of isomorphism classes when the root has degree $n \geq 0$. Summing over n, we get with (17) for $y = 1$ the functional equation

$$U(x) = xe^{\sum_{k\geq 1} \frac{U(x^k)}{k}} . \tag{18}$$

Now, $U(x^k) = \sum_{i\geq 1} u_i x^{ik}$, and therefore

$$\begin{aligned}\sum_{k\geq 1} \frac{U(x^k)}{k} &= \sum_{k\geq 1} \frac{1}{k} \sum_{i\geq 1} u_i x^{ik} = \sum_{i\geq 1} u_i \sum_{k\geq 1} \frac{(x^i)^k}{k} \\ &= \sum_{i\geq 1} (-u_i \log(1 - x^i)) \\ &= \sum_{i\geq 1} \log(1 - x^i)^{-u_i} ,\end{aligned}$$

which gives with (18) the formula

$$U(x) = x \prod_{i\geq 1} (1 - x^i)^{-u_i} . \tag{19}$$

Self-complementary Patterns.
There is another natural question that can be effectively dealt with using the Pólya-Redfield approach. Let N be an n-set, $R = \{0, 1\}$, and G a group acting on N. We call $f : N \to \{0, 1\}$ *self-complementary* if f and $h \circ f$ belong to the same pattern, where h is the involution $0 \leftrightarrow 1$ on R. What is the number of self-complementary patterns? Interpreted as black/white colorings, we are asking for the number of patterns that are invariant under color

exchange. Of course, this can happen only when n is even, and $|f^{-1}(0)| = |f^{-1}(1)|$.

Suppose f is self-complementary, that is,

$$h \circ f = f \circ \overline{k}_f \quad \text{for some } k_f \in G\,.$$

Now if $f' \underset{G}{\sim} f$, $f' = f \circ \overline{g}$, then

$$h \circ f' = (h \circ f) \circ \overline{g} = f \circ (\overline{k}_f \circ \overline{g}) = f \circ \overline{k_f g}\,,$$

which means that $h \circ f' \underset{G}{\sim} f$, and thus $h \circ f' \underset{G}{\sim} f'$. Any f' equivalent to a self-complementary map is therefore also self-complementary, and we can unambiguously speak of a *self-complementary pattern*. Suppose $\mathcal{F} \subseteq \mathrm{Map}(N, R)$ is a family closed under G, and $\mathcal{F}^c$ the set of self-complementary mappings in $\mathcal{F}$. Let $\mathcal{M}_{\mathcal{F}^c}$ be the set of self-complementary patterns in $\mathcal{F}$.

Proposition 6.8. *Let N be an n-set, $R = \{0, 1\}$, G a group acting on N, and $\mathcal{F} \subseteq \mathrm{Map}(N, R)$ a G-closed class. Then with $h : 0 \leftrightarrow 1$ in R,*

$$w(\mathcal{F}^c; G) = \sum_{M \in \mathcal{M}_{\mathcal{F}^c}} w(M) = \frac{1}{|G|} \sum_{g \in G} \left(\sum_{f \in \mathcal{F} : h \circ f = f \circ \overline{g}} w(f) \right).$$

Proof. Consider the action of G on $\mathcal{F}^c$ defined by

$$f \to f' = h \circ f \circ \overline{g} \quad (g \in G)\,.$$

Now, $f' = h \circ f \circ \overline{g}$ implies $f' = f \circ \overline{k_f g}$. Conversely, if $f' = f \circ \overline{g}$, then $f' = h \circ (h \circ f) \circ \overline{g} = h \circ f \circ \overline{k_f g}$. Hence the patterns of $\mathcal{F}^c$ under G and the new action are the same, and the proposition follows from (5). □

Let us consider $\mathcal{F} = \mathrm{Map}(N, R)$. As in the Pólya–Redfield theorem we want to relate the inner sum $\sum_{f : h \circ f = f \circ \overline{g}} w(f)$ to the cycle index of G. Consider $a \in N$, $f(a) = b \in R$, and assume $h \circ f = f \circ \overline{g}$. Suppose a lies in a k-cycle of $\overline{g}$. Applying f to the cycle containing a we infer

$$\begin{array}{ccccc} a & \overline{g}(a) & \overline{g}^2(a) & \cdots & \overline{g}^{k-1}(a) \\ {\scriptstyle f}\downarrow & \downarrow & \downarrow & & \downarrow \\ b & h(b) & h^2(b) & \cdots & h^{k-1}(b) \end{array}$$

It follows that $(f \circ \overline{g}^k)(a) = f(a) = b = h^k(b)$, and thus $2 \mid k$. Since we may choose as $b = f(a)$ both 0 and 1, we conclude that

$$\sum_{f:h\circ f = f\circ \overline{g}} w(f) = \lambda_1^{c_1(\overline{g})} \lambda_2^{c_2(\overline{g})} \cdots \lambda_n^{c_n(\overline{g})}$$

with

$$\lambda_k = \begin{cases} 2(x_0x_1)^{k/2} & k \text{ even}, \\ 0 & k \text{ odd}, \end{cases}$$

and thus the following result.

Corollary 6.9. *The weight enumerator of the self-complementary patterns of* $\mathrm{Map}(N,R)$ *is given by*

$$w(\mathrm{Map}(N,R)^c; G) = Z(G; \lambda_1, \ldots, \lambda_n),$$

with

$$\lambda_k = \begin{cases} 2(x_0x_1)^{k/2} & k \textit{ even}, \\ 0 & k \textit{ odd}. \end{cases}$$

In particular,

$$|\mathrm{Map}(N,R)^c| = Z(G; 0, 2, 0, 2, \ldots).$$

Example. For the necklace problem with $r = 2$ we have according to (11),

$$|\mathcal{M}_n^c| = \frac{1}{n} \sum_{d \mid \frac{n}{2}} \varphi(2d) 2^{n/2d}.$$

In particular, when $n = 2^m$, this gives

$$|\mathcal{M}_{2^m}^c| = \frac{1}{2^m} \sum_{d \mid 2^{m-1}} \varphi(2d) 2^{2^{m-1}/d} = \frac{1}{2^m} \sum_{k=0}^{m-1} \varphi(2^{k+1}) 2^{2^{m-1-k}}$$

$$= \frac{1}{2^m} \sum_{k=0}^{m-1} 2^{k+2^{m-1-k}} = \sum_{k=0}^{m-1} 2^{2^k - k - 1}.$$

The figure shows the four self-complementary necklaces for $n = 8$:

Exercises

6.19 Let G be the symmetry group of the cube. Compute the cycle index when N is the set of edges, and when N is the set of vertices.

▷ **6.20** Consider the cycle index of $S(n)$. Without computing it deduce that $\sum_{k=0}^{n} s_{n,k}x^k = x^{\overline{n}}$, where $s_{n,k}$ is the Stirling number.

6.21 How many 2-colored face patterns of the cube exist when white and black appear equally often? The same question with edges and vertices.

6.22 Consider molecules analogous to alcohols with OH- and H-atoms as before, but 3-valent D-atoms. Determine the functional equation for the generating function $D(x) = \sum_{n\geq 0} d_n x^n$, d_n = # molecules with n D-atoms, and d_n for $n \leq 8$.

6.23 Let N be an n-set, $G \leq S(N)$. Consider subset equivalence as in Exercise 6.11, and let m_k be the number of k-subset patterns. Show that $\sum_{k=0}^{n} m_k x^k = Z(G; 1+x, 1+x^2, \ldots, 1+x^n)$.

▷ **6.24** Derive from the previous exercise the following formulas:

a. $\sum_{k=0}^{n} \binom{n}{k} x^k = (1+x)^n$,
b. $(n+1)! = \sum_{\sigma \in S(n)} 2^{c(\sigma)} = \sum_{k=0}^{n} s_{n,k} 2^k$,
c. $\sum_{d|n} \varphi(d) = n$.

6.25 Let $G \leq S(M)$, $H \leq S(N)$, where M and N are disjoint. The *product* $G \cdot H$ is the permutation group on $M \cup N$, $G \cdot H = \{g \cdot h : g \in G, h \in H\}$, with

$$(g \cdot h)(a) = \begin{cases} g(a) & \text{if } a \in M, \\ h(a) & \text{if } a \in N. \end{cases}$$

Prove that $Z(G \cdot H) = Z(G) \cdot Z(H)$.

6.26 What do you get in (17) for $R(x) = 1 + x$?

6.27 Determine the number of different dice that bear the numbers $1, \ldots, 6$ as usual. How many have, as is customary, 1 and 6, 2 and 5, 3 and 4 on opposite faces? Hint: N = faces, $R = \{1, \ldots, 6\}$, $\mathcal{F} = \mathrm{Inj}(N, R)$.

* * *

6.28 Determine the symmetry groups of the tetrahedron and octahedron, and compute the cycle indices for the set of faces.

▷ **6.29** Compute the cycle index of the dihedral group D_n, and determine the number of n-necklaces, colored black and white, that are self-complementary under D_n.

6.30 For some k we have $Z(C_n; 2, \ldots, 2) = Z(D_n; 2, \ldots, 2)$ for all $n \leq k$, and $Z(C_n) \neq Z(D_n)$ for $n > k$. Determine k.

6.31 Find the number of self-equivalent subset patterns of $\{1,\ldots,12\}$ when $G = C_{12}$ and $G = D_{12}$.

6.32 Let $x_1,\ldots,x_r$ be variables. A polynomial $p(x_1,\ldots,x_r)$ is said to be *symmetric* if $p(x_{\sigma(1)},\ldots,x_{\sigma(r)}) = p(x_1,\ldots,x_r)$ for all $\sigma \in S(r)$. Examples are the *elementary* symmetric functions $e_n(x_1,\ldots,x_r) = \sum_{i_1<\cdots<i_n} x_{i_1}x_{i_2}\cdots x_{i_n}$, and the power functions $p_k(x_1,\ldots,x_r) = \sum_{j=1}^{r} x_j^k$. Consider the closed class $\mathrm{Inj}(N,R)$ and prove

$$w(\mathrm{Inj}(N,R);G) = \frac{n!}{|G|} e_n(x_1,\ldots,x_r).$$

What follows for $x_1 = \cdots = x_r = 1$, and $G = \{\mathrm{id}\}$, $G = S(n)$?

▷ **6.33** Clearly, $\sum_{n\ge 0} e_n(x_1,\ldots,x_r)z^n = \prod_{j=1}^{r}(1 + x_j z)$. Now set $x_j = j$ $(j = 1,\ldots,r)$, and deduce $s_{n,m} = \sum_{1\le k_1<\cdots<k_{n-m}\le n-1} k_1 k_1 \cdots k_{n-m}$, where $s_{n,m}$ is the Stirling number.

▷ **6.34** With $e_n(x_1,\ldots,x_r)$ and $p_k(x_1,\ldots,x_r)$ as defined in Exercise 6.32 use (16) to prove the following formula of Waring:

$$e_n(x_1,\ldots,x_r) = Z(S(n); p_1, -p_2, \ldots, (-1)^{n-1}p_n).$$

6.35 Let $U(x) = \sum_{n\ge 1} u_n x^n$ be the generating function of isomorphism classes of rooted trees. Prove that

$$u_n = \sum_{\lambda\in\mathrm{Par}(n-1)} \frac{u_{\lambda_1}^{\overline{i_1}} u_{\lambda_2}^{\overline{i_2}} \cdots u_{\lambda_t}^{\overline{i_t}}}{i_1! i_2! \cdots i_t!},$$

where $\lambda \in \mathrm{Par}(n-1)$ is written as

$$\underbrace{\lambda_1\ldots\lambda_1}_{i_1}\underbrace{\lambda_2\ldots\lambda_2}_{i_2}\ldots\underbrace{\lambda_t\ldots\lambda_t}_{i_t}, \quad \lambda_1 > \lambda_2 > \cdots > \lambda_t.$$

▷ **6.36** Suppose trees are rooted at an *edge*. Call two such trees isomorphic if the root edge is carried onto the root edge by an isomorphism. Show for the corresponding generating function $U^{(e)}(x) = \sum_{n\ge 1} u_n^{(e)} x^n$ of isomorphism classes that $U^{(e)}(x) = \frac{1}{2}[U(x)^2 + U(x^2)]$, with $U(x)$ as in the previous exercise.

6.37 A *planted* tree is a rooted tree whose root has degree 1. Two such trees are isomorphic if they are transformed into each other by a continuous motion in the plane. For instance,

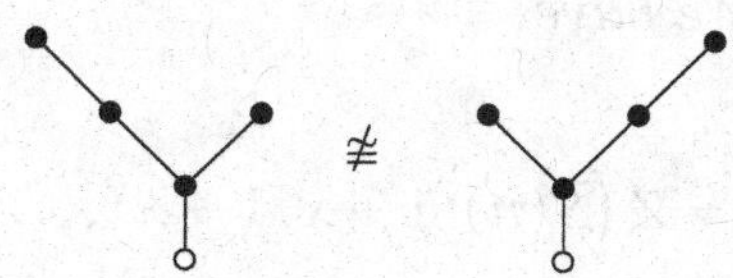

Let k_n be the number of isomorphism classes of planted trees with n edges ($= n+1$ vertices) whose vertices all have degree 1 or 3. Prove for the generating function $K(x) = \sum_{n\geq 1} k_n x^n$:
a. $K(x) = x + K(x)^2$, b. $k_{2n} = 0$, $k_{2n+1} = C_n$ (Catalan number).

6.3 Cycle Index

We have seen in the last section that the cycle index $Z(G)$ contains all the information needed to enumerate patterns of mappings under the action of the group G. It is therefore important to have some general results on the cycle index at hand.

Enumeration of Graphs.
As our first example let us consider the enumeration of graphs, where in this section a graph is always assumed to be simple. We are given a fixed vertex set V with n vertices. Clearly, there are $2^{\binom{n}{2}}$ graphs on V, and $\binom{\binom{n}{2}}{k}$ such graphs with k edges. Now we want to count non-isomorphic graphs. Two graphs $G = (V,E)$ and $G' = (V,E')$ are *isomorphic* if there is a permutation $\sigma \in S(V)$ such that $\{\sigma u, \sigma v\} \in E' \iff \{u,v\} \in E$.

Let $g(n)$ be the number of non-isomorphic graphs on n vertices, and more precisely, $g(n,k)$ the number of non-isomorphic graphs with k edges. We want to determine the polynomial $g(x) = \sum_{k=0}^{\binom{n}{2}} g_{n,k}x^k$.

To apply the Pólya–Redfield theorem we identify each graph $G = (V,E)$ with the mapping $f : \binom{V}{2} \to \{0,1\}$, where $f(\{u,v\}) = 1$ means that $\{u,v\} \in E$, and $f(\{u,v\}) = 0$ otherwise. Hence $N = \binom{V}{2}$, $R = \{0,1\}$, and $S(n)$ acts on $\binom{V}{2}$ through $\overline{\sigma}\{u,v\} = \{\sigma u, \sigma v\}$. Clearly, $(V,E) \cong (V,E')$ if and only if $f' = f \circ \overline{\sigma}$ holds for some $\sigma \in S(n)$. The induced group $\overline{S}(n) = \{\overline{\sigma} : \sigma \in S(n)\}$ is called the *pair group*. Assigning as usual the weights $w(1) = x$, $w(0) = 1$, the Pólya–Redfield theorem implies the following result:

Proposition 6.10. *We have*

$$\sum_{k=0}^{\binom{n}{2}} g_{n,k}x^k = Z\left(\overline{S}(n); 1+x, 1+x^2, \ldots, 1+x^{\binom{n}{2}}\right),$$

and in particular,

$$g(n) = Z(\overline{S}(n); 2, 2, \ldots, 2) .$$

Thus our task consists in computing the cycle index $Z(\overline{S}(n))$. Let $\sigma \in S(n)$ have type $t(\sigma) = 1^{c_1} 2^{c_2} \ldots n^{c_n}$, and consider a cycle $A = (a_1, \ldots, a_k)$ of σ. Let us check what happens to the set of pairs $\binom{A}{2}$ under $\overline{\sigma}$. We have

$$\{a_1, a_i\} \xrightarrow{\overline{\sigma}} \{a_2, a_{i+1}\} \longrightarrow \cdots \longrightarrow \{a_k, a_{i+k-1}\} \longrightarrow \{a_1, a_i\} \pmod{k} .$$

If $\{a_\ell, a_{i+\ell-1}\} = \{a_m, a_{i+m-1}\}$, then $\ell \equiv i + m - 1$, $m \equiv i + \ell - 1 \pmod{k}$, hence $i + m - 1 \equiv -i + m + 1 \pmod{k}$, which means that $2(i-1) \equiv 0 \pmod{k}$. Hence if k is odd, this cannot happen, and if k is even then $i = \frac{k}{2} + 1$. We conclude that for odd k, $\binom{A}{2}$ decomposes into $\frac{k-1}{2}$ cycles of length k, and for even k, $\binom{A}{2}$ splits into $\frac{k}{2} - 1$ cycles of length k and one cycle of length $\frac{k}{2}$.

Now consider two cycles $A = (a_1, \ldots, a_k)$, $B = (b_1, \ldots, b_\ell)$ of σ. The pair $\{a_i, b_j\}$ is carried into $\{a_{i+1}, b_{j+1}\}$, and so forth. The cycle length in $\overline{\sigma}$ is therefore $\operatorname{lcm}(k, \ell)$, and we obtain $\frac{k\ell}{\operatorname{lcm}(k,\ell)} = \gcd(k, \ell)$ cycles, each of length $\operatorname{lcm}(k, \ell)$.

Example. For $n = 4$ we have

$$Z(S(4); z_1, \ldots, z_4) = \frac{1}{24}(z_1^4 + 6z_1^2 z_2 + 8z_1 z_3 + 3z_2^2 + 6z_4) .$$

By our analysis,

$$z_1^4 \to z_1^6, \quad z_1^2 z_2 \to z_1^2 z_2^2, \quad z_1 z_3 \to z_3^2, \quad z_2^2 \to z_1^2 z_2^2, \quad z_4 \to z_2 z_4 ;$$

thus

$$Z(\overline{S}(4); z_1, \ldots, z_6) = \frac{1}{24}(z_1^6 + 9z_1^2 z_2^2 + 8z_3^2 + 6z_2 z_4) .$$

Substituting $z_k \mapsto 1 + x^k$, this gives

$$\begin{aligned} \sum_{k=0}^{6} g_{4,k} x^k &= \frac{1}{24}[(1+x)^6 + 9(1+x)^2(1+x^2)^2 + 8(1+x^3)^2 \\ &\qquad + 6(1+x^2)(1+x^4)] \\ &= 1 + x + 2x^2 + 3x^2 + 2x^4 + x^5 + x^6 . \end{aligned}$$

The figure shows these 11 non-isomorphic graphs:

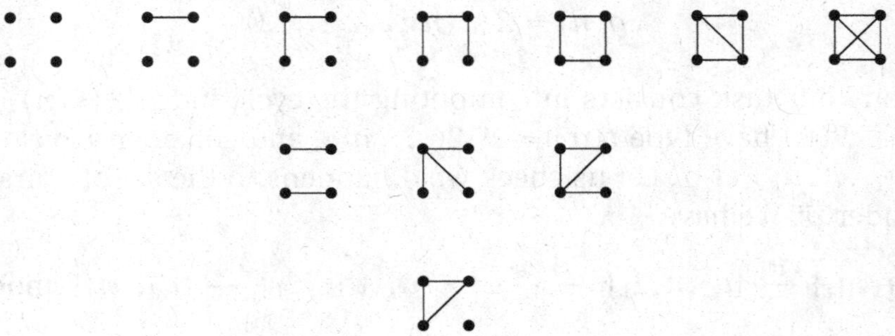

Applying Corollary 6.9 we immediately obtain the number of non-isomorphic self-complementary graphs, that is, $(V,E) \cong (V,\binom{V}{2}\setminus E)$.

Corollary 6.11. *The number of non-isomorphic self-complementary graphs on n vertices is given by*

$$Z(\overline{S}(n);0,2,0,2,\ldots)\,.$$

In the example $n = 4$ we have $Z(\overline{S}(4);0,2,0,2,0,2) = \frac{1}{24}[6\cdot 2\cdot 2] = 1$, with the path •—•—•—• being the only such graph.

The proposition can easily be generalized to the enumeration of subgraphs of a given graph $G = (V,E)$. Let S_G be the group of edge-preserving permutations of V. Then S_G induces a group $\overline{S}_G = \{\overline{\sigma} : \sigma \in S_G\}$ on the edge set E, called the *edge group* of G, just as $S(n)$ induced $\overline{S}(n)$ on the complete graph K_n. Notice that two subgraphs of G may be isomorphic graphs, but non-isomorphic as *subgraphs* of G.

Example. The subgraphs G_1 and G_2 of G are clearly isomorphic but non-isomorphic as subgraphs, since every isomorphism of G must fix a.

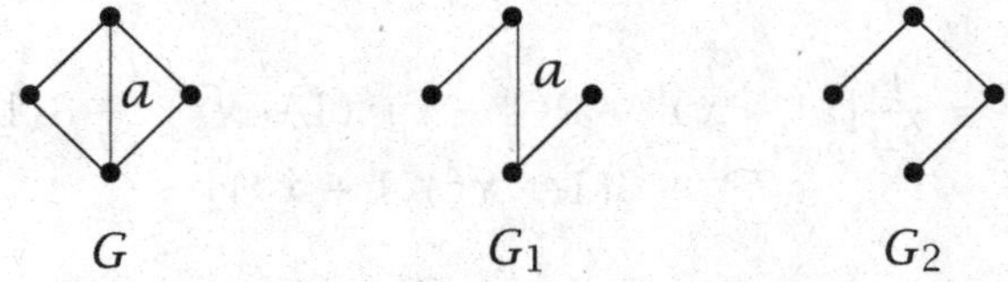

Corollary 6.12. *Let $G = (V, E)$ be a graph, and $g_k(G)$ the number of non-isomorphic subgraphs of G with k edges. Then*

$$\sum_{k\geq 0} g_k(G)x^k = Z(\overline{S}_G; 1+x, 1+x^2, \ldots).$$

The number of self-complementary subgraphs is $Z(\overline{S}_G; 0, 2, 0, 2, \ldots)$.

Example. Consider the graph G from above:

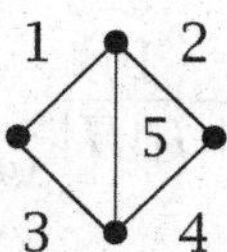

The edge group consists of the four permutations id, $(5)(12)(34)$, $(5)(13)(24)$, $(5)(14)(23)$ (the Klein 4-group), and we compute

$$Z(\overline{S}_G; z_1, \ldots, z_5) = \frac{1}{4}[z_1^5 + 3z_1z_2^2].$$

Thus

$$\sum_{k=0}^{5} g_k(G)x^k = \frac{1}{4}[(1+x)^5 + 3(1+x)(1+x^2)^2]$$
$$= 1 + 2x + 4x^2 + 4x^3 + 2x^4 + x^5.$$

The graphs are shown in the figure:

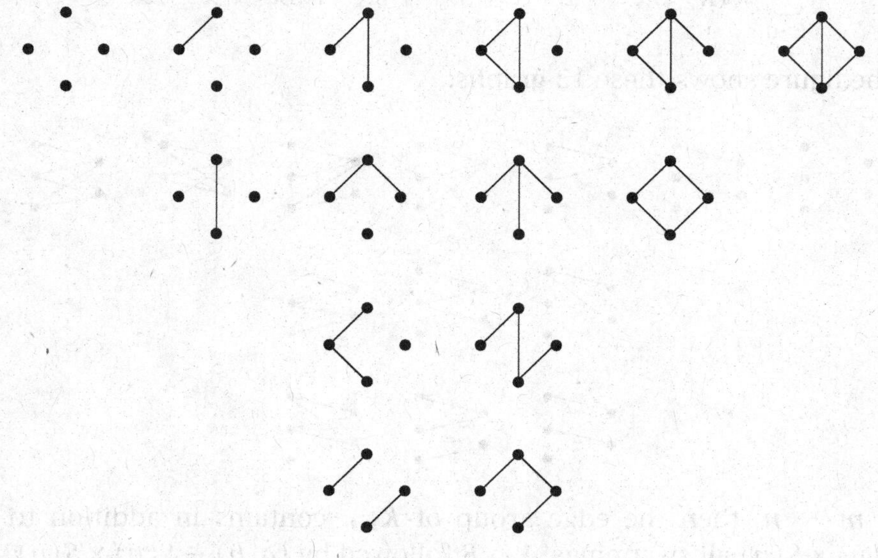

Operations on Groups.
Let A and B be two disjoint sets, $|A| = m$, $|B| = n$, and $G \le S(A)$, $H \le S(B)$. There is a natural product $G \times H$ acting on $A \times B$, obtained by defining

$$(g,h)(a,b) = (ga, hb), \; g \in G, h \in H; \; a \in A, b \in B.$$

You are asked in the exercises to show that

$$Z(G \times H; z_1, \ldots, z_{mn}) = \frac{1}{|G||H|} \sum_{(g,h)} \prod_{k,\ell=1}^{m,n} z_{\mathrm{lcm}(k,\ell)}^{\gcd(k,\ell) c_k(g) c_\ell(h)}. \qquad (1)$$

Consider as an example the complete bipartite graph $G = K_{m,n}$ with $m \ne n$. The group acting on the edge set is clearly $S(m) \times S(n)$, and thus

$$\sum_{k \ge 0} g_k(K_{m,n}) x^k = Z(S(m) \times S(n); 1 + x, \ldots, 1 + x^{mn}) \quad (m \ne n).$$

Example. For $K_{2,3}$ we obtain by (1),

$$Z(S(2) \times S(3); z_1, \ldots, z_6) = \frac{1}{12}(z_1^6 + 3z_1^2 z_2^2 + 4z_2^3 + 2z_3^2 + 2z_6),$$

hence

$$\sum_{k \ge 0} g_k(K_{2,3}) x^k = 1 + x + 3x^2 + 3x^3 + 3x^4 + x^5 + x^6.$$

The figure shows these 13 graphs:

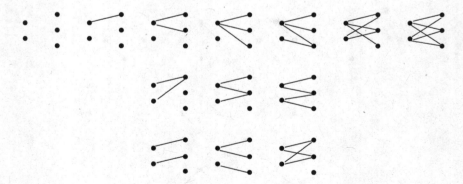

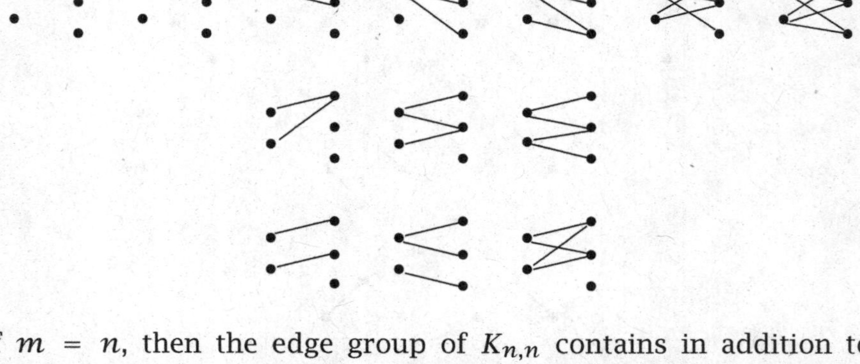

If $m = n$, then the edge group of $K_{n,n}$ contains in addition to $S(n) \times S(n)$ all swappings $A \leftrightarrow B$ followed by $(g,h) \in S(n) \times S(n)$.

This group is called the exponentiation group $[S(n)]^{S(2)}$ (see Exercise 6.48).

Just as we have a product $G \times H$ there is a natural composition of two permutation groups. Let A and B be two sets $A = \{a_1, \dots, a_n\}$, $B = \{b_1, \dots, b_p\}$, and $G \leq S(A)$, $H \leq S(B)$.

Definition. The composition $G[H]$ is the permutation group on $A \times B$ whose elements are all $(n+1)$-tuples

$$[g; h_1, \dots, h_n] \quad \text{with } g \in G, h_i \in H,$$

where

$$[g; h_1, \dots, h_n](a_i, b_j) = (ga_i, h_i b_j).$$

It is easily checked that $G[H]$ is indeed a group of size $|G||H|^n$.

The typical situation for the composition arises when we are given n copies of a structure with p elements. The composition $G[H]$ first permutes the n copies according to $g \in G$, and then applies individual permutations $h_i \in H$ to each copy.

Proposition 6.13. *Let $G \leq S(A)$, $H \leq S(B)$, $|A| = n$, $|B| = p$. Then*

$$Z(G[H]; z_1, \dots, z_{np}) \\ = Z(G; Z(H; z_1, \dots, z_p), Z(H; z_2, z_4, \dots, z_{2p}), \dots), \quad (2)$$

that is, z_k is replaced by $Z(H; z_k, z_{2k}, \dots, z_{pk})$.

Proof. To establish (2), we represent each side as the enumerator of a certain set of patterns (after the substitution $z_k \mapsto \sum x_j^k$), and construct then a weight-preserving bijection between the pattern sets. Since the x_j are variables, the two polynomials must be identical.

Let $A = \{a_1, \dots, a_n\}$, $B = \{b_1, \dots, b_p\}$, and R an r-set disjoint from both A and B. To each $j \in R$ we assign as usual the weight x_j. The enumerator of the $G[H]$-patterns $\mathcal{M}$ of $\mathrm{Map}(A \times B, R)$ is by Pólya-Redfield

$$w(\mathrm{Map}(A \times B, R); G[H]) = Z\left(G[H]; \sum x_j, \sum x_j^2, \dots, \sum x_j^{np}\right). \quad (3)$$

Similarly, the enumerator of all H-patterns $\mathcal{B}$ of $\mathrm{Map}(B, R)$ is given by

$$w(\mathrm{Map}(B,R);H) = Z\left(H;\sum x_j, \sum x_j^2, \ldots, \sum x_j^p\right). \tag{4}$$

Next we compute the enumerator $w(\mathrm{Map}(A,\mathcal{B});G)$ of all G-patterns $\mathcal{N}$ of mappings $f : A \to \mathcal{B}$, where the weight is given by $w(f) = \prod_{i=1}^n w(f(a_i))$:

$$w(\mathrm{Map}(A,\mathcal{B});G) = Z\left(G; \sum_{F\in\mathcal{B}} w(F), \sum_{F\in\mathcal{B}} w^2(F), \ldots, \sum_{F\in\mathcal{B}} w^n(F)\right). \tag{5}$$

By assigning the weight x_j^k to $j \in R$ we obtain

$$\sum_{F\in\mathcal{B}} w^k(F) = Z\left(H;\sum x_j^k, \sum x_j^{2k}, \ldots, \sum x_j^{pk}\right),$$

and thus

$$\begin{aligned} &w(\mathrm{Map}(A,\mathcal{B});G) \\ &\quad = Z\Big(G; Z\left(H;\sum x_j, \sum x_j^2, \ldots\right), Z\left(H;\sum x_j^2, \sum x_j^4, \ldots\right), \ldots\Big), \end{aligned}$$

which is precisely the right-hand side of (2) after the substitution $z_k \mapsto \sum x_j^k$.

It remains to find a weight-preserving bijection $\phi : \mathcal{M} \to \mathcal{N}$. This is accomplished as follows. For $M \in \mathcal{M}$ with representative $f : A\times B \to R$ define $\widehat{f} : A \to \mathrm{Map}(B,R)$ by

$$\widehat{f}(a_i)(b_j) = f(a_i, b_j) \in R \quad (i = 1, \ldots, n; j = 1, \ldots, p).$$

The mapping $\widehat{f}$ induces another mapping $\widehat{F} : A \to \mathcal{B}$, setting

$$\widehat{F}(a_i) = H\text{-pattern of } \widehat{f}(a_i) \in \mathcal{B}.$$

Finally, we set

$$\phi M = G\text{-pattern of } \widehat{F} \in \mathcal{N}.$$

It is readily seen that ϕ is independent of the chosen representative function $f \in M$. The proof that ϕ is a weight-preserving bijection is left to the exercises. □

Example. Consider three necklaces A, B, C with four beads, each of them colored red or blue. Two colorings F and F' are considered equal if F' arises from F by some permutation of the necklaces followed by a rotation. The group involved is therefore $S(3)[C_4]$. Now

$$Z(S(3)) = \frac{1}{6}(z_1^3 + 3z_1z_2 + 2z_3),\ Z(C_4) = \frac{1}{4}(z_1^4 + z_2^2 + 2z_4),$$

and with $Z(C_4; 2, \ldots, 2) = 6$ we obtain for the total number of colorings

$$Z(S(3); 6, 6, 6) = \frac{1}{6}(6^3 + 3 \cdot 6^2 + 2 \cdot 6) = 56\,.$$

Exercises

6.38 Determine the cycle index of $S(n)$ and of the alternating group $A(n)$ of all even permutations in $S(n)$.

6.39 Compute the generating polynomial $\sum_{k=0}^{10} g_{5,k}x^k$ of all subgraphs of K_5. How many graphs are self-complementary?

▷ **6.40** Let C_n be the circuit graph of n vertices. Determine the number of non-isomorphic subgraphs of C_n.

6.41 Prove formula (1) for $Z(G \times H)$.

6.42 Consider two disjoint n-sets $A = \{a_1, \ldots, a_n\}$, $B = \{b_1, \ldots, b_n\}$ and color $A \cup B$ with black or white. Two colorings are equal if one is transformed into the other by the involution $a_i \mapsto b_i$ $(i = 1, \ldots, n)$. Compute $\sum_{k=0}^{2n} m_k x^k$, where m_k is the number of colorings with k black elements. How many are self-complementary?

▷ **6.43** Suppose we are given m n-sets whose elements are colored with two colors. Two colorings are equal if they arise from each other by a permutation of the m sets followed by permutations of the individual sets. Determine the number of different colorings.

* * *

6.44 We are given n cubes whose faces are colored red or blue. Equivalence of colorings is defined by permutations of the cubes and symmetries of the individual cubes. How many color patterns are there? Show that $Z(S(n); 2, 10, 2, 10, \ldots)$ counts the number of patterns invariant under color-exchange. Example: For $n = 4$ we obtain 27 different patterns.

6.45 Find the group of symmetries for digraphs, i.e., there are no loops and at most one arrow $u \to v$ in each direction for each pair $u \neq v$. Compute the number of non-isomorphic digraphs on n vertices for $n \leq 4$.

6.46 Complete the proof of Proposition 6.13.

▷ **6.47** Let e_n and o_n, respectively, count the isomorphism classes of graphs on n vertices having an even resp. odd number of edges. Prove that $e_n \geq o_n$ for all n with equality when $n \equiv 2$ or $3 \pmod 4$.

6.48 Determine the edge group of $K_{3,3}$, and give the polynomial $\sum_{k=0}^{9} g_k(K_{3,3})x^k$.

▷ **6.49** Let s_n be the number of self-complementary graphs on n vertices, and $\vec{s_n}$ the corresponding number for directed graphs as in Exercise 6.45, where the complement of a digraph has an edge $u \to v$ if and only if $u \to v$ does not appear in the original digraph. Prove that $s_{4n} = \vec{s_{2n}}$.

6.4 Symmetries on N and R

At the end of Section 1.5 we classified mappings $f : N \to R$ as to whether the elements of N and R are distinguishable or not. In the language of this chapter, distinguishable (labeled) means that the identity group acts on N (or R), whereas indistinguishable (unlabeled) refers to the group $G = S(N)$ or $H = S(R)$. The Pólya-Redfield theorem adresses the general problem in which $G \leq S(N)$ is an arbitrary permutation group, and $H = \{\mathrm{id}\}$. It is natural to consider the most general setting with two arbitrary groups $G \leq S(N)$ and $H \leq S(R)$.

As usual, N is an n-set, R an r-set, and we are given groups $G \leq S(N)$, $H \leq S(R)$, and a set $\mathcal{F} \subseteq \mathrm{Map}(N,R)$ closed under G and H. That is, $f \in \mathcal{F}$ implies $h \circ f \circ g \in \mathcal{F}$ for all $g \in G$ and $h \in H$. The typical examples of closed sets are, of course, $\mathrm{Map}(N,R)$, $\mathrm{Inj}(N,R)$, and $\mathrm{Surj}\,(N,R)$. We call f, f' *equivalent* if $f' = h \circ f \circ g$ for some $g \in G$, $h \in H$, and our goal is to enumerate the (weighted) patterns of mappings in $\mathcal{F}$.

Let us treat first a single permutation h of R. We say that $f \in \mathcal{F}$ is *h-invariant* if $h \circ f \underset{G}{\sim} f$, that is,

$$h \circ f = f \circ k_f \quad \text{for some } k_f \in G. \tag{1}$$

Let $\mathcal{F}(h)$ be the set of h-invariant mappings in $\mathcal{F}$. Note that this generalizes the notion of self-complementary maps in Section 6.2, with $R = \{0,1\}$ and h being the involution $0 \leftrightarrow 1$. The proof of

Proposition 6.8 goes through without change, yielding the following result:

Proposition 6.14. *Let $G \le S(N)$, $h \in S(R)$. Then*

$$w(\mathcal{F}(h);G) = \sum_{M \in \mathcal{M}_{\mathcal{F}(h)}} w(M) = \frac{1}{|G|} \sum_{g \in G} \left(\sum_{f \in \mathcal{F}: h \circ f = f \circ g} w(f) \right). \quad (2)$$

Now we look at the general case $G \le S(N)$, $H \le S(R)$. It is clear that any G-pattern of $\mathcal{F}$ is contained in a G,H-pattern. If we assign as usual variables x_j to $j \in R$, then G-equivalent mappings have the same weight, but G,H-equivalence will, in general, not preserve the weight. To remedy this fact we proceed as follows.

Let $P_1, \ldots, P_m$ be the H-patterns (orbits) of R. To each P_j we assign a variable x_j, and define then the weight of $f \in \mathcal{F}$ by

$$w(f) = x_1^{|f^{-1}(P_1)|} \cdots x_m^{|f^{-1}(P_m)|}, \quad (3)$$

where $f^{-1}(P_i) = \{a \in N : f(a) \in P_i\}$.

Now if $f' = h \circ f \circ g$, then $f'(a) \in P_i \iff h \circ f(g(a)) \in P_i \iff f(g(a)) \in P_i$, and thus $w(f') = w(f)$. We can therefore unambiguously speak of the weight of a G,H-pattern $w(M) = w(f)$ for $f \in M$.

Notice that for $H = \{\mathrm{id}\}$, all H-orbits of R consist of single elements, and the definition (3) reduces to the previous case.

Theorem of de Bruijn.

We denote again by $\mathcal{M}_{\mathcal{F}}$ the set of G,H-patterns in $\mathcal{F}$, and can now state the theorem of de Bruijn, generalizing Theorem 6.4.

Theorem 6.15 (de Bruijn). *Let $\mathcal{F} \subseteq \mathrm{Map}(N,R)$ be a G,H-closed class. Then*

$$\begin{aligned} w(\mathcal{F};G,H) &= \frac{1}{|H|} \sum_{h \in H} w(\mathcal{F}(h);G) \\ &= \frac{1}{|G||H|} \sum_{g \in G, h \in H} \left(\sum_{f \in \mathcal{F}: h \circ f = f \circ g} w(f) \right). \end{aligned} \quad (4)$$

In particular,

$$|\mathcal{M}_{\mathcal{F}}| = \frac{1}{|G||H|} \sum_{g \in G, h \in H} |\{f \in \mathcal{F} : h \circ f = f \circ g\}|. \quad (5)$$

Proof. Let $M \in \mathcal{M}_{\mathcal{F}}$. The G,H-pattern M splits into disjoint G-patterns, which we denote by fG. Regarding them as new elements, we obtain by the same argument as in the proof of Theorem 6.4,

$$w(M) = \frac{1}{|H|} \sum_{h \in H} \left(\sum_{fG \subseteq M : h \circ fG = fG} w(fG) \right),$$

and by summing over M,

$$w(\mathcal{F}; G, H) = \frac{1}{|H|} \sum_{h \in H} \left(\sum_{fG \subseteq \mathcal{F} : h \circ fG = fG} w(f) \right).$$

Now, the inner sum is precisely $w(\mathcal{F}(h); G)$, and the result follows from the previous proposition. □

Consider $\mathcal{F} = \mathrm{Map}(N, R)$. As in the Pólya–Redfield theorem we want to relate the inner sum $\sum_{f : h \circ f = f \circ g} w(f)$ to the cycle index of G. Let $a \in N$, $f(a) = b \in R$, and assume $h \circ f = f \circ g$. Suppose a lies in a k-cycle of g, and b in a j-cycle of h. Applying f to the cycle we infer

$$\begin{array}{ccccc} a & g(a) & g^2(a) & \cdots & g^{k-1}(a) \\ {\scriptstyle f}\downarrow & \downarrow & \downarrow & & \downarrow \\ b & h(b) & h^2(b) & \cdots & h^{k-1}(b) \end{array} \tag{6}$$

It follows that $(f \circ g^k)(a) = f(a) = b = h^k(b)$, and thus $j \mid k$. Conversely, if we choose for every k-cycle of g a j-cycle of h with $j \mid k$, and determine f according to (6), then $h \circ f = f \circ g$ holds. Since we may pick as image of a any of the j elements of the chosen j-cycle of h, we conclude that

$$\sum_{f : h \circ f = f \circ g} w(f) = \lambda_1^{c_1(g)} \lambda_2^{c_2(g)} \cdots \lambda_n^{c_n(g)}.$$

with

$$\lambda_k = \sum_{j \mid k} j \sum_{j\text{-cycles of } h} (x_i x_{h(i)} \cdots x_{h^{j-1}(i)})^{k/j},$$

and thus obtain the following result.

Corollary 6.16. *Let $G \leq S(N)$, $H \leq S(R)$, $|N| = n$, $|R| = r$, and $P_1, \dots, P_m$ the H-orbits of R. Any $h \in H$ decomposes uniquely into a product $h = h_1 \cdots h_m$ with $h_i \in S(P_i)$. We have*

$$w(\mathrm{Map}(N,R); G, H) = \frac{1}{|H|} \sum_{h \in H} Z(G; \lambda_1(h), \dots, \lambda_n(h)), \tag{7}$$

where

$$\lambda_k(h) = \sum_{i=1}^{m} \left(\sum_{j|k} j c_j(h_i) \right) x_i^k \quad (k = 1, \dots, n).$$

In particular,

$$|\mathcal{M}| = \frac{1}{|H|} \sum_{h \in H} Z(G; \lambda_1(h), \dots, \lambda_n(h)) \tag{8}$$

with

$$\lambda_k(h) = \sum_{j|k} j c_j(h).$$

Examples.

1. Suppose $G = \{\mathrm{id}\}$, and let $m_n(H)$ be the number of $\{\mathrm{id}\}, H$-patterns of $\mathrm{Map}(N, R)$. Then

$$m_n(H) = \frac{1}{|H|} \sum_{h \in H} (c_1(h))^n. \tag{9}$$

For $H = S(r)$, this gives

$$\begin{aligned} \sum_{n \geq 0} m_n(S(r)) \frac{z^n}{n!} &= \frac{1}{r!} \sum_{h \in S(r)} \left(\sum_{n \geq 0} \frac{(c_1(h) z)^n}{n!} \right) \\ &= \frac{1}{r!} \sum_{h \in S(r)} e^{c_1(h) z} = Z(S(r); e^z, 1, \dots, 1). \end{aligned}$$

With the identity (16) in Section 6.2 we get

$$\begin{aligned} \sum_{r \geq 0} Z(S(r); e^z, 1, \dots, 1) y^r &= \exp\left(e^z y + \frac{y^2}{2} + \frac{y^3}{3} + \cdots \right) \\ &= \frac{e^{y(e^z - 1)}}{1 - y}, \end{aligned}$$

and thus

$$\sum_{n,r\geq 0} m_n(S(r))\frac{z^n}{n!}y^r = \frac{e^{y(e^z-1)}}{1-y}.$$

We know from the list at the end of Section 1.5 that $m_n(S(r)) - m_n(S(r-1))$ counts precisely all partitions of an n-set into r blocks, and we have re-proved our old result

$$\sum_{n,r\geq 0} S_{n,r}\frac{z^n y^r}{n!} = e^{y(e^z-1)}.$$

2. Let $H = S(r)$, and G arbitrary. In this case, the patterns may be thought of as *partition patterns* of N under G. For example, for $r = 2$, this gives partition patterns containing at most two blocks. Let $m(G, \leq 2)$ be their number. By (8),

$$m(G;\leq 2) = \frac{1}{2}(Z(G;2,\ldots,2) + Z(G;0,2,0,2,\ldots)).$$

Since $Z(G;2,\ldots,2)$ counts *all* patterns, $Z(G;0,2,0,2,\ldots)$ enumerates those that are self-complementary, reaffirming our earlier result.

Injective Patterns.
Finally, we look at the closed set $\mathrm{Inj}(N,R)$. Arguing as in (6), it follows from the injectivity of f that $b = f(a)$ must also be in a k-cycle. Thus, the $c_k(g)$ k-cycles of g are mapped injectively into the $c_k(h)$ k-cycles of h, and within each k-cycle of g we have k choices for the image of a. In conclusion, for given $g \in G$, $h \in H$ we get

$$|\{f : h\circ f = f\circ g\}| = \prod_{k=1}^{n} k^{c_k(g)}(c_k(g))!\binom{c_k(h)}{c_k(g)}, \qquad (10)$$

and thus the following result.

Corollary 6.17. *For $G \leq S(N)$, $H \leq S(R)$, we have*

$$|\mathcal{M}_{\mathrm{Inj}}| = \frac{1}{|G||H|}\sum_{g\in G,h\in H}\prod_{k=1}^{n} k^{c_k(g)}(c_k(g))!\binom{c_k(h)}{c_k(g)}. \qquad (11)$$

Formula (11) can be conveniently described by the so-called *cap product* of Redfield. Let $z_1^{a_1}z_2^{a_2}\cdots z_n^{a_n}$ be a monomial appearing in the cycle index $Z(G)$, and $z_1^{b_1}z_2^{b_2}\cdots z_r^{b_r}$ one of $Z(H)$. Set

$$z_1^{a_1}\cdots z_n^{a_n} \cap z_1^{b_1}\cdots z_r^{b_r} = \prod_{k=1}^{n} k^{a_k}(a_k!)\binom{b_k}{a_k}, \tag{12}$$

and extend the product linearly to $Z(G)$ and $Z(H)$. Then

$$|\mathcal{M}_{\mathrm{Inj}}| = Z(G) \cap Z(H)\,.$$

Example. Suppose $G = C_3$, $H = C_6$; then $Z(G) = \frac{1}{3}(z_1^3 + 2z_3)$, $Z(H) = \frac{1}{6}(z_1^6 + z_2^3 + 2z_3^2 + 2z_6)$. The only cap products that do not vanish are

$$z_1^3 \cap z_1^6 = 3!\binom{6}{3} = 120, \quad z_3 \cap z_3^2 = 3\binom{2}{1} = 6\,,$$

and we get

$$|\mathcal{M}_{\mathrm{Inj}}| = \frac{1}{18}(120 + 4\cdot 6) = 8\,.$$

A particularly interesting case arises when $|N| = |R|$, that is, for bijective mappings. It follows immediately from (12) that the cap product vanishes unless $a_k = b_k$ for all k. Now suppose $G = H$; then

$$|\mathcal{M}_{\mathrm{Bij}}| = Z(G) \cap Z(G)\,.$$

Example. Consider a graph $G = (V,E)$ with edge-group $\overline{S}_G$. We may interpret the bijective mappings $E \to E$ as *superpositions* of the graph G onto itself. Then $\mathcal{M}_{\mathrm{Bij}}$ is the set of patterns of these superpositions under the edge group. Take as example the 5-circuit. The edge group is clearly the dihedral group D_5. By Exercise 6.29,

$$Z(D_5) = \frac{1}{10}(z_1^5 + 4z_5 + 5z_1z_2^2)\,,$$

and thus

$$|\mathcal{M}_{\mathrm{Bij}}| = \frac{1}{100}(5! + 16\cdot 5 + 25\cdot 4\cdot 2) = 4\,.$$

The figure shows the four different superpositions:

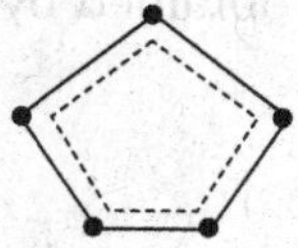 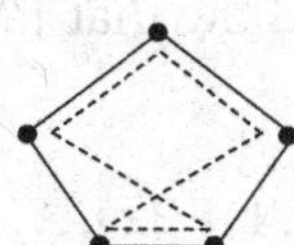 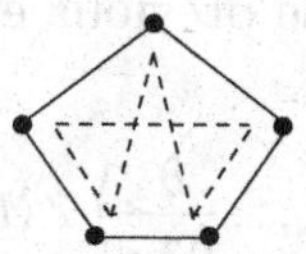 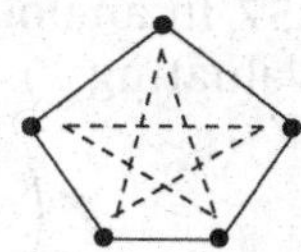

Exercises

6.50 Suppose we color the faces of the cube with 6 colors $F_1, \ldots, F_6$. The colorings are considered equal if they are transformed into each other by the permutation $\binom{F_1\ F_2\ F_3\ F_4\ F_5\ F_6}{F_2\ F_1\ F_4\ F_3\ F_6\ F_5}$. How many colorings are there? How many that use all 6 colors?

▷ **6.51** Let $G = S(n)$, $H \le S(r)$ arbitrary, and denote by $m(S(n), H)$ the number of $S(n), H$-patterns. Prove that

$$\sum_{n \ge 0} m(S(n), H) z^n = Z\left(H; \frac{1}{1-z}, \frac{1}{1-z^2}, \ldots, \frac{1}{1-z^r}\right).$$

Specialize to $H = \{\mathrm{id}\}$, and compare the resulting generating function to previous results.

6.52 Let $G = S(n)$, $H = S(r)$. Compute $\sum_{n,r \ge 0} m(S(n), S(r)) z^n y^r$. Compare the result to generating functions of number-partitions.

6.53 Consider $m_n(S(r))$ as in (9). What do you get for the ordinary generating function $\sum_{n \ge 0} m_n(S(r)) z^n$?

▷ **6.54** Compute the number of patterns of $\mathrm{Inj}(N, R)$ for $G = S(n)$, H arbitrary. You will get as answer the coefficient of x^r in $Z(H; 1+x, \ldots, 1+x^r)$. Can you see this directly?

6.55 Determine the number of G, H-patterns of $\mathrm{Map}(N, R)$ when $|N| = |R| = p$, p a prime, and $G = H = C_p$.

* * *

▷ **6.56** Show that the total number of G, H-patterns of $\mathrm{Map}(N, R)$ is obtained by evaluating

$$Z\left(G; \frac{\partial}{\partial z_1}, \ldots, \frac{\partial}{\partial z_n}\right) Z\left(H; e^{z_1+z_2+\cdots}, e^{2(z_2+z_4+\cdots)}, e^{3(z_3+z_6+\cdots)}, \ldots\right)$$

at $z_1 = z_2 = \cdots = z_r = 0$. The expression $Z\left(G; \frac{\partial}{\partial z_1}, \frac{\partial}{\partial z_2}, \ldots\right)$ is the differential operator obtained from the substitution $z_i \mapsto \frac{\partial}{\partial z_i}$. Example: $G = H = \{\mathrm{id}\}$. Then $Z(G) = z_1^n$, $Z(H) = z_1^r$, and hence

$$\left(\frac{d}{dz_1}\right)^n e^{r(z_1+z_2+\ldots)}\Big|_{z_i=0} = r^n e^{r(z_1+z_2+\cdots)}\Big|_{z_i=0} = r^n.$$

6.57 In analogy to the previous exercise show that $|\mathcal{M}_{\mathrm{Inj}}|$ is obtained by evaluating

$$Z\left(G; \frac{\partial}{\partial z_1}, \ldots, \frac{\partial}{\partial z_n}\right) Z(H; 1+z_1, 1+2z_2, \ldots, 1+rz_r)$$

at $z_1 = \cdots = z_r = 0$. Show that for $r = n$ this can be simplified to

$$|\mathcal{M}_{\mathrm{Bij}}| = Z\left(G; \frac{\partial}{\partial z_1}, \ldots, \frac{\partial}{\partial z_n}\right) Z(H; z_1, 2z_2, \ldots, nz_n)$$

at $z_1 = \cdots = z_n = 0$.

6.58 Generalizing (11) show that

$$w(\mathrm{Inj}(N,R); G, H) = \frac{1}{|G||H|} \sum_{h \in H} \left(\sum_{g \in G} \mu_1(g) \cdots \mu_n(g) \right),$$

where

$$\mu_k(g) = k^{c_k(g)} (c_k(g)!) \sum_{\substack{(r_1,\ldots,r_m) \\ \sum r_i = c_k(g)}} \binom{c_k(h_1)}{r_1} \cdots \binom{c_k(h_m)}{r_m} x_1^{r_1 k} \cdots x_m^{r_m k},$$

and $h = h_1 \cdots h_m$ is the decomposition of h into the H-orbits of R.

▷ **6.59** Compute the cap product $Z(D_p) \cap Z(D_p)$, where D_p is the dihedral group, $p \geq 3$ a prime. Deduce that $4p \mid (p-1)! + (p-1)^2$ for $p > 3$.

6.60 Determine the number of superpositions of the 6-circuit, and draw them.

6.61 Find the number of injective patterns when $|N| = n$, $|R| = n+1$, and $G = C_n$, $H = C_{n+1}$. Can you see the result directly?

Highlight: Patterns of Polyominoes

A polyomino is, as the name suggests, a generalization of a 1×2-domino. It consists of a set of 1×1-cells glued together along their edges. A cell has four possible neighbors, and we call the polyomino P *connected* if one can pass from any cell to any other along neighboring cells. The figure shows the five types of connected polyominoes with four cells:

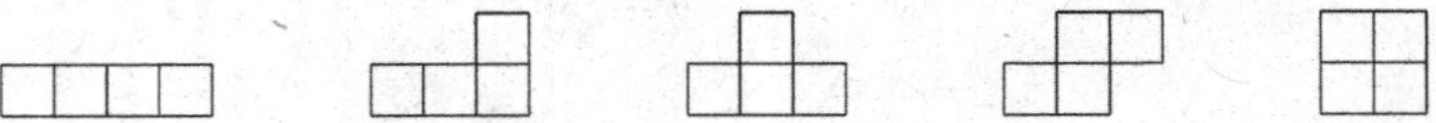

What we are interested in is to enumerate the types of polyominoes P satisfying the following two conditions:

A. P fits into an $n \times n$-square and touches all four sides of the square.
B. P is connected and has the minimal number of cells under condition A.

Clearly, this minimum number is $2n - 1$, since there must be $n - 1$ neighbor relations from left to right, and also from top to bottom. The figure shows the six types for $n = 3$:

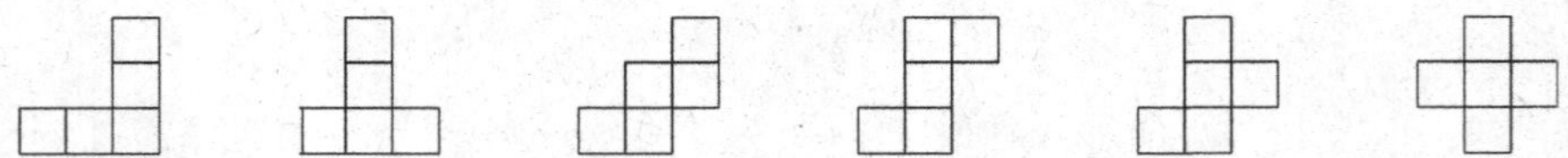

Our task is to find a formula for general n.

Lattice Configurations.
It is convenient to rephrase the problem in terms of lattice configurations, which are easier to handle. Replace each cell by a lattice point, and join two points if the corresponding cells have an edge in common. The polyominoes above correspond then to the following lattice configurations:

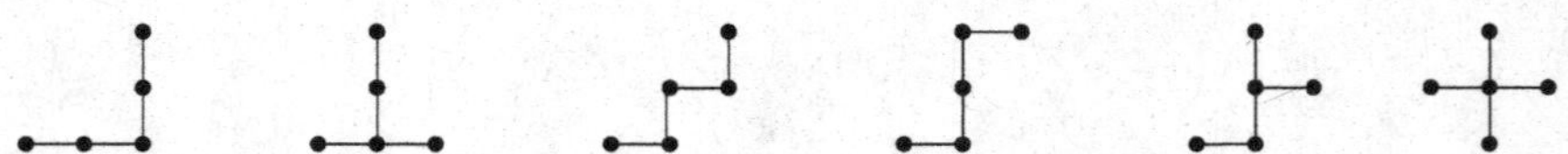

We regard the configurations as embedded in the grid, where $(0,0)$ is the lower left corner and (n,n) the upper right corner. Let $\mathcal{L}(n)$ be the set of all these configurations. Conditions A and B then say that $L \in \mathcal{L}(n)$ if L is connected and contains *exactly* one step $(x,y) \mapsto (x+1,y)$ to the right for $x = 0,1,\ldots,n-1$, and one step $(x,y) \mapsto (x,y+1)$ up for $y = 0,1,\ldots,n-1$. Thus L has n horizontal and n vertical steps.

Our goal is to compute the number p_n of *patterns* in $\mathcal{L}(n)$ under the dihedral group D_4. The original question of patterns of polyominoes fitting in the $n \times n$-square refers then to p_{n-1}.

The following facts are easy to see. They all follow from the uniqueness property of the steps.

1. If L contains two points P,Q on a boundary, then it contains all points between P and Q and at least one corner point of the boundary:

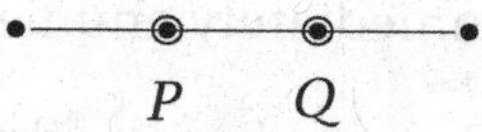

2. If L contains a boundary point P and all its neighbors, then it contains the whole boundary and the line orthogonal to the boundary at P:

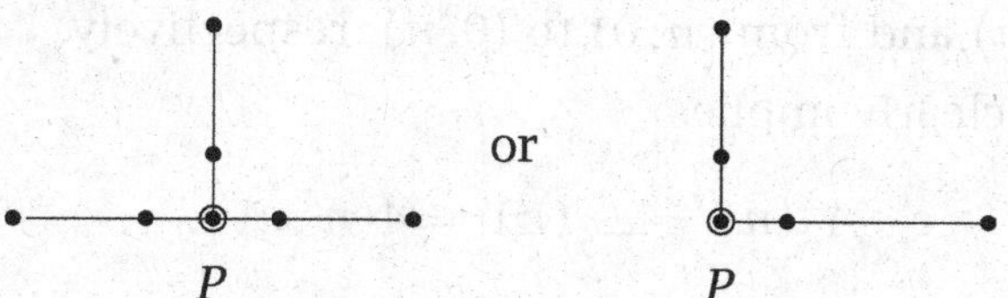

3. If L contains an interior point P and three neighbors of P, then it contains two straight lines emanating from P to the boundaries:

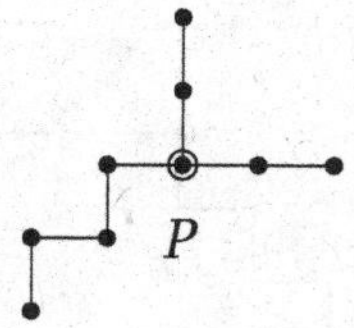

4. If L contains an interior point P together with all its four neighbors, then it contains all four lines to the boundaries, and is thus uniquely determined:

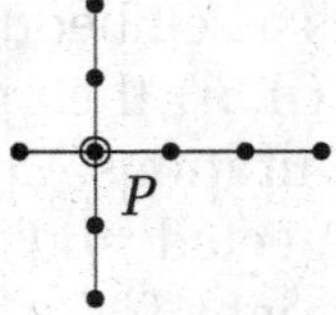

The Total Number of Lattice Configurations.
Let $A(n) = |\mathcal{L}(n)|$. It is convenient to partition $\mathcal{L}(n)$ into the following classes: $\mathcal{L}_1(n), \mathcal{L}_2(n), \mathcal{L}_3(n)$, and $\mathcal{L}_4(n)$, where

$$\mathcal{L}_1(n) = \{L : L \text{ contains two opposite corners}\},$$

$$\mathcal{L}_2(n) = \{L : L \text{ contains two corners, but no opposite corners}\},$$

$$\mathcal{L}_3(n) = \{L : L \text{ contains precisely one corner}\}$$

$$\mathcal{L}_4(n) = \{L : L \text{ contains no corner}\}.$$

The first two classes are immediately enumerated:

$$A_1(n) = |\mathcal{L}_1(n)| = 2\binom{2n}{n}, \tag{1}$$

since the configurations in $\mathcal{L}_1(n)$ correspond to lattice paths from $(0,0)$ to (n,n) and from $(n,0)$ to $(0,n)$, respectively.

Fact 1 above clearly implies

$$A_2(n) = |\mathcal{L}_2(n)| = 4(n-1). \tag{2}$$

Suppose $L \in \mathcal{L}_3(n)$ with $(0,0) \in L$. Then there must be an interior point $P \in L$ with three neighbors in L. Hence by fact 3, L must look as follows:

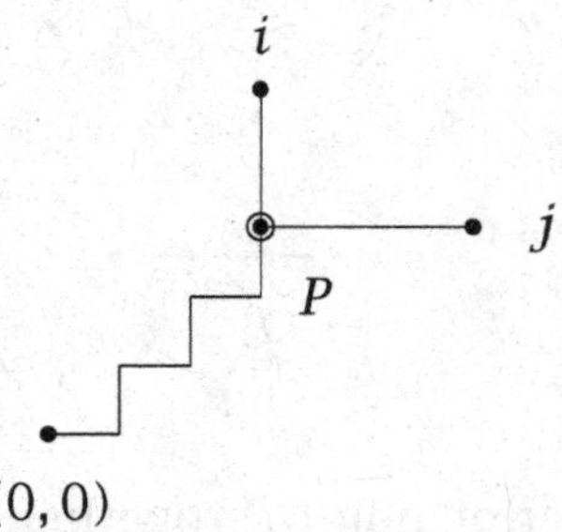

Suppose P has coordinates (i,j). Then L consists of an ordinary lattice path from $(0,0)$ to (i,j) together with the two straight lines emanating from P. It follows that the total number of $L \in \mathcal{L}_3(n)$ with $(0,0) \in L$ is given by $\sum_{i=1}^{n-1} \sum_{j=1}^{n-1} \binom{i+j}{j}$. Now recall the binomial identity

$$\sum_{j=0}^{n} \binom{m+j}{j} = \binom{m+n+1}{n},$$

which will be used several times in the sequel.

This gives

$$\begin{aligned}
\sum_{i=1}^{n-1} \sum_{j=1}^{n-1} \binom{i+j}{j} &= \sum_{i=1}^{n-1} \left[\binom{n+i}{n-1} - 1 \right] = \sum_{i=1}^{n-1} \binom{n+i}{i+1} - (n-1) \\
&= \sum_{i=2}^{n} \binom{n-1+i}{i} - (n-1) \\
&= \binom{2n}{n} - 1 - n - (n-1) \\
&= \binom{2n}{n} - 2n,
\end{aligned}$$

and so

$$A_3(n) = |\mathcal{L}_3(n)| = 4 \left[\binom{2n}{n} - 2n \right]. \qquad (3)$$

Now to the set $\mathcal{L}_4(n)$. A configuration $L \in \mathcal{L}_4(n)$ is by the facts given above of the form

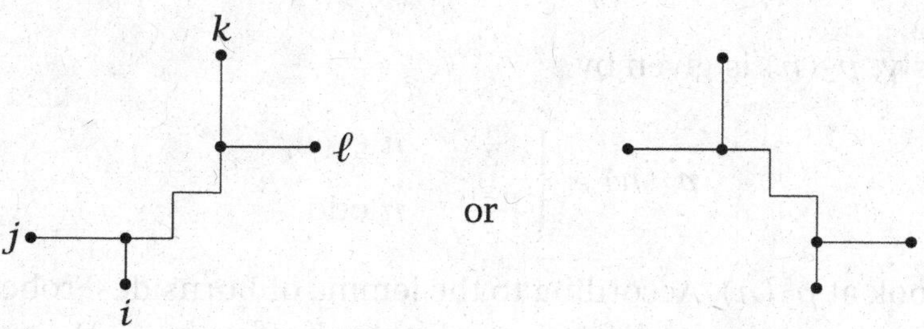

where $i \le k$ and $j \le \ell$. The configurations of the first type are thus enumerated by

$$\sum_{i=1}^{n-1} \sum_{j=1}^{n-1} \sum_{k=i}^{n-1} \sum_{\ell=j}^{n-1} \binom{k+\ell-i-j}{\ell-j},$$

which is easily computed to

$$\binom{2n}{n} - n^2 - 1 .$$

The total number is therefore twice this number minus the number of configurations with $i = k$, $j = \ell$, which were counted twice. These are the $(n-1)^2$ configurations that have an interior point with all its four neighbors. Hence

$$A_4(n) = |\mathcal{L}_4(n)| = 2\binom{2n}{n} - 3n^2 + 2n - 3 . \tag{4}$$

The total number of lattice configurations is therefore

$$A(n) = 8\binom{2n}{n} - 3n^2 - 2n - 7 . \tag{5}$$

The Number of Patterns.
Since any two configurations in different classes $\mathcal{L}_i(n)$ are clearly inequivalent under the group D_4, we may compute the number of patterns $p_i(n)$ separately. The number $p_1(n)$ of patterns in $\mathcal{L}_1(n)$ was already the subject of Exercise 6.5 with the result

$$p_1(n) = \begin{cases} \frac{1}{4}\binom{2n}{n} + 2^{n-2} + \frac{1}{4}\binom{n}{n/2} & n \text{ even,} \\ \frac{1}{4}\binom{2n}{n} + 2^{n-2} & n \text{ odd.} \end{cases} \tag{1'}$$

Obviously, $p_2(n)$ is given by

$$p_2(n) = \begin{cases} \frac{n}{2} & n \text{ even,} \\ \frac{n-1}{2} & n \text{ odd.} \end{cases} \tag{2'}$$

Let us look at $p_3(n)$. According to the lemma of Burnside–Frobenius we must determine the fixed-point sets X_g for $g \in D_4$, where $X = \mathcal{L}_3(n)$. Nontrivial rotations clearly have no fixed configurations, since the corner is moved. Similarly, no reflection leaves the configuration fixed, and so

$$p_3(n) = \frac{1}{8}A_3(n) = \frac{1}{2}\binom{2n}{n} - n . \tag{3'}$$

Finally, we consider the set $\mathcal{L}_4(n)$. A rotation of $\pm 90°$ forces a fixed configuration to satisfy $i = j = k = \ell$ (see the figure above). Hence n must be even, and only the cross configuration

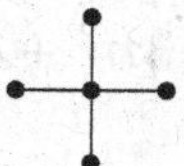

is left fixed. A rotation about 180° implies $k = n - i$, $\ell = n - j$, and a fixed configuration must contain the center $(\frac{n}{2}, \frac{n}{2})$. Hence n must be even, and the number of fixed configurations is

$$\sum_{i=1}^{n/2}\sum_{j=1}^{n/2}\binom{n-i-j}{\frac{n}{2}-j} = \sum_{i=1}^{n/2}\sum_{j=0}^{n/2-1}\binom{\frac{n}{2}-i+j}{j} = \sum_{i=1}^{n/2}\binom{n-i}{\frac{n}{2}-1}$$

$$= \sum_{i=1}^{n/2}\binom{n-i}{\frac{n}{2}+1-i} = \sum_{i=1}^{n/2}\binom{\frac{n}{2}-1+i}{i} = \binom{n}{\frac{n}{2}} - 1 .$$

The total number left fixed under a rotation of 180° is therefore twice this number minus 1, since the cross configuration was counted twice. Hence this gives $2\binom{n}{n/2} - 3$ fixed configurations.

Now to the reflections. The horizontal or vertical reflections leave precisely the configurations

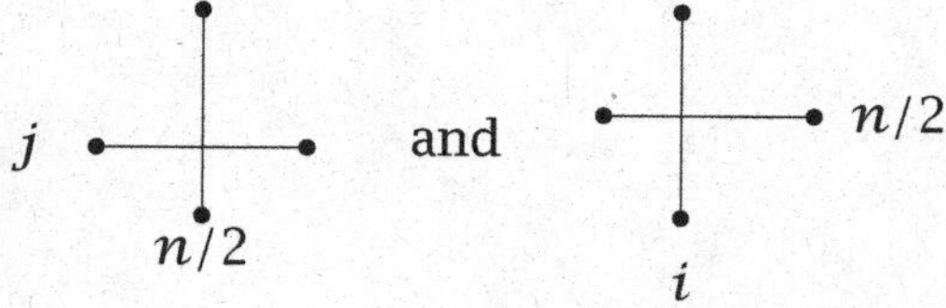

fixed. Thus we obtain $2(n-1)$ fixed configurations for n even, and none for odd n. Finally, it is easily seen that each diagonal reflection leaves $2^n - n - 1$ configurations fixed. The contribution is therefore $2(2^n - n - 1)$, and we obtain by the lemma of Burnside–Frobenius

$$p_4(n) = \frac{1}{8}\left[2\binom{2n}{n} - 3n^2 + 2n - 3 + 2(2^n - n - 1) + \left(2\binom{n}{n/2} + 2n - 3\right)[n \text{ even}]\right]. \quad (4')$$

Summing the expressions for $p_i(n)$ yields the exact result for the total number of patterns:

$$p_n = \binom{2n}{n} + 2^{n-1} + \begin{cases} \frac{1}{2}\binom{n}{n/2} - \frac{1}{8}(3n^2 + 2n + 8) & n \text{ even,} \\ -\frac{1}{8}(3n^2 + 4n + 9) & n \text{ odd.} \end{cases}$$

As small examples, we obtain

$$p(1) = 1, \quad p(2) = 6, \quad p(3) = 18, \quad p(4) = 73, \quad p(5) = 255,$$
$$p(6) = 950, \quad p(7) = 3473, \quad \text{and} \quad p(8) = 13006\,.$$

Notes and References

The main Lemma 6.2 is in the English literature usually called Burnside's lemma. This is one of many examples of a theorem being attributed to the wrong person. It was first proved by Frobenius and earlier in a special case by Cauchy. The pioneering work in this area was the paper by Pólya, although again he was preceded by Redfield. The book of Pólya and Read contains the English translation of Pólya's paper along with an account of the history since then. Another very readable introduction is the article by de Bruijn. For further reading, the book by Kerber is recommended. The standard reference for graphical enumeration is the book by Harary and Palmer. The result of the highlight is due to Knuth.

1. N.G. de Bruijn (1964): Pólya's theory of counting. In: Beckenbach, ed., *Applied Combinatorial Mathematics*, 144–184. Wiley, New York.
2. G. Frobenius (1887): Über die Congruenz nach einem aus zwei endlichen Gruppen gebildeten Doppelmodul. *Crelle's Journal* 101, 273–299.
3. F. Harary and E.M. Palmer (1973): *Graphical Enumeration*. Academic Press, New York.
4. A. Kerber (1991): *Combinatorics via Finite Group Actions*. BI Wissenschaftsverlag, Mannheim.
5. D.E. Knuth (2001): Animals in a cage. *Amer. Math. Monthly* 108, 469, problem 10875; solution in 110(2003), 243–245.
6. G. Pólya (1937): Kombinatorische Anzahlbestimmungen für Gruppen, Graphen und chemische Verbindungen. *Acta Sci. Math.* 68, 145–254.
7. G. Pólya and R.C. Read (1987): *Combinatorial Enumeration of Groups, Graphs, and Chemical Compounds*. Springer, Berlin.
8. J.H. Redfield (1927): The theory of group-reduced distributions. *Amer. J. Math.* 49, 433–455.

Part III: Topics

7 The Catalan Connection

We begin with a six-fold description of the Catalan numbers C_n.

A. Recurrence. We have

$$C_{n+1} = \sum_{k=0}^{n} C_k C_{n-k} \quad (n \geq 0), \; C_0 = 1 .$$

This is the combinatorial definition given in Section 3.1.

B. Integral Representation.

$$C_n = \frac{2^{2n+1}}{\pi} \int_{-1}^{1} x^{2n}(1-x^2)^{1/2} dx ,$$

which is easily evaluated to $C_n = \frac{1}{n+1}\binom{2n}{n}$. We will see how this description via an integral arises quite naturally.

C. Continued Fraction. Let $C(z) = \sum_{n \geq 0} C_n z^n$ be the Catalan series. Then $C(z)$ can be expressed as the continued fraction

$$C(z) = \cfrac{1}{1 - \cfrac{z}{1 - \cfrac{z}{1 - \ddots}}}$$

Looking at this expression we see that $C(z) = \frac{1}{1-zC(z)}$ holds, from which the defining equation $C(z) = 1 + zC^2(z)$ results, derived in Section 3.1. How such a continued fraction is declared formally will be explained later.

D. Lattice Path. C_n is the number of Catalan paths of length $2n$ discussed in Section 5.3. We will take up this idea, considering more general paths.

E. Hankel Matrix. We have seen in Section 5.4 that C_n is the unique sequence such that the determinants of the Hankel matrices H_n and $H_n^{(1)}$ are all equal to 1.

F. Catalan Matrix. This last description is our starting point. Consider the following infinite lower triangular matrix $A = (a_{n,k})$ given by the recurrence

$$a_{0,0} = 1, \quad a_{0,k} = 0 \qquad (k > 0),$$
$$a_{n,k} = a_{n-1,k-1} + a_{n-1,k+1} \qquad (n \geq 1).$$

The first rows and columns look as follows:

n \ k	0	1	2	3	4	5	6	7	8
0	<u>1</u>								
1	0	1							
2	<u>1</u>	0	1						
3	0	2	0	1					
4	<u>2</u>	0	3	0	1				
5	0	5	0	4	0	1			
6	<u>5</u>	0	9	0	5	0	1		
7	0	14	0	14	0	6	0	1	
8	<u>14</u>	0	28	0	20	0	7	0	1

It appears that $C_n = a_{2n,0}$ holds for all n. In fact, this is just another way to represent the Catalan paths in matrix form.

7.1 Catalan Matrices and Orthogonal Polynomials

Our algebraic object are infinite lower triangular matrices indexed by $\mathbb{N}_0$, with the ordinary matrix sum and product as operations. We know from Section 2.4 that any such matrix is invertible if and only if the main diagonal consists of nonzero elements.

Catalan Matrices.
Let $\sigma = (s_0, s_1, s_2, \ldots)$, $\tau = (t_1, t_2, t_3, \ldots)$ be two sequences of complex numbers (or in any field of characteristic 0), with $t_k \neq 0$ for all k. We define the matrix $A = A^{\sigma,\tau}$ by the recurrence

$$\begin{cases} a_{0,0} = 1, \ a_{0,k} = 0 \quad (k > 0), \\ a_{n,k} = a_{n-1,k-1} + s_k a_{n-1,k} + t_{k+1} a_{n-1,k+1} \quad (n \geq 1). \end{cases} \tag{1}$$

Definition. A matrix A is called *Catalan matrix* if $A = A^{\sigma,\tau}$ holds for a pair of sequences σ and τ. The numbers $B_n = B_n^{\sigma,\tau} = a_{n,0}$ are called the *Catalan numbers* associated with σ, τ. A sequence $(B_0, B_1, B_2, \ldots)$ is called a *Catalan sequence* if $B_n = B_n^{\sigma,\tau}$ for some σ and τ.

Example. Our starting example in F is thus the case $\sigma \equiv 0$, $\tau \equiv 1$. To avoid confusion, we will henceforth call C_n the *ordinary Catalan numbers.*

It follows from (1) that any Catalan matrix is lower triangular with main diagonal equal to 1, and thus invertible. With induction it is immediate that

$$a_{n+1,n} = s_0 + s_1 + \cdots + s_n \quad (n \geq 0). \tag{2}$$

For $\tau = (t_1, t_2, t_3, \ldots)$ with $t_k \neq 0$ for all k, we set

$$T_n = t_1 t_2 \cdots t_n, \quad T_0 = 1. \tag{3}$$

The following result is the fundamental lemma for Catalan matrices.

Lemma 7.1. *Let* $A = A^{\sigma,\tau} = (a_{n,k})$, *then we have for all* $m, n \geq 0$,

$$\sum_{k \geq 0} a_{m,k} a_{n,k} T_k = B_{m+n} \quad (= a_{m+n,0}). \tag{4}$$

Proof. We use induction on m. For $m = 0$,

$$\sum_{k \geq 0} a_{0,k} a_{n,k} T_k = a_{0,0} a_{n,0} = B_n.$$

Assume that (4) holds for $m - 1$ and all n. By (1),

$$\begin{aligned}
\sum_k a_{m,k} a_{n,k} T_k &= \sum_{k \geq 0} (a_{m-1,k-1} + s_k a_{m-1,k} + t_{k+1} a_{m-1,k+1}) a_{n,k} T_k \\
&= \sum_{j \geq 0} (a_{n,j+1} T_{j+1} + s_j a_{n,j} T_j + a_{n,j-1} T_j) a_{m-1,j} \\
&= \sum_{j \geq 0} a_{m-1,j} T_j (t_{j+1} a_{n,j+1} + s_j a_{n,j} + a_{n,j-1}) \\
&= \sum_{j \geq 0} a_{m-1,j} a_{n+1,j} T_j = a_{m+n,0} = B_{m+n},
\end{aligned}$$

by induction. □

We may write formula (4) in compact form. Let

$$T = \begin{pmatrix} T_0 & & & \\ & T_1 & & 0 \\ & & T_2 & \\ 0 & & & \ddots \end{pmatrix}, \quad H = \begin{pmatrix} B_0 & B_1 & B_2 & \dots \\ B_1 & B_2 & B_3 & \dots \\ B_2 & B_3 & B_4 & \dots \\ & \dots & & \end{pmatrix}.$$

The matrix H is again called the (infinite) *Hankel matrix* of the Catalan numbers B_n. Thus

$$T = (T_i \delta_{i,j}), \quad H = (B_{i+j}),$$

and (4) takes on the form

$$ATA^T = H, \tag{5}$$

where A^T is the transpose of A. The next result characterizes Catalan matrices.

Proposition 7.2. *A lower triangular matrix A with main diagonal 1 is a Catalan matrix if and only if $ATA^T = H$ for some diagonal matrix $T = (T_i \delta_{i,j})$ with $T_0 = 1$, $T_k \neq 0$ for all k, and some Hankel matrix $H = (B_{i+j})$. If $ATA^T = H$, then we have $A = A^{\sigma,\tau}$ with*

$$s_k = a_{k+1,k} - a_{k,k-1}, \quad t_k = \frac{T_k}{T_{k-1}}.$$

Proof. We have already seen one direction, so assume $ATA^T = H$, that is,

$$\sum_{k \geq 0} a_{m,k} a_{n,k} T_k = B_{m+n} \quad \text{for all } m, n. \tag{6}$$

First we note that $m = 0$ in (6) implies $a_{n,0} = B_n$. Next we claim that $ATA^T = H$ uniquely determines A. Since $A^{\sigma,\tau}$ with $s_k = a_{k+1,k} - a_{k,k-1}$, $t_k = \frac{T_k}{T_{k-1}}$ is such a matrix, this will imply $A = A^{\sigma,\tau}$.

We construct A row by row. Row 0 is given by $a_{0,0} = 1$, and we know that all $a_{n,0} = B_n$. Assume inductively that rows $0, 1, \dots, n$ are uniquely determined. From

$$a_{n+1+i,0} = B_{n+1+i} = \sum_{k=0}^{i} a_{n+1,k} a_{i,k} T_k \quad (i = 1, \dots, n)$$

we compute step by step $a_{n+1,1}, a_{n+1,2}, \ldots, a_{n+1,n}$. Thus row $n+1$ is uniquely determined (with $a_{n+1,0} = B_{n+1}, a_{n+1,n+1} = 1$), and the result follows. □

Suppose $A = A^{\sigma,\tau}$ is a Catalan matrix. Since A is invertible with main diagonal equal to 1, the inverse matrix $U = A^{-1} = (u_{n,k})$ is again lower triangular with diagonal 1. You are asked in the exercises to prove that U satisfies the following recurrence,

$$\begin{cases} u_{0,0} = 1, \; u_{0,k} = 0 \quad (k > 0), \\ u_{n+1,k} = u_{n,k-1} - s_n u_{n,k} - t_n u_{n-1,k} \quad (n \geq 0), \end{cases} \tag{7}$$

and that conversely, (7) implies recurrence (1) for $A = U^{-1} = (a_{n,k})$.

Now we define the polynomials $p_n(x) = \sum_{k=0}^{n} u_{n,k} x^k$ corresponding to the rows of U; $p_n(x)$ has degree n with leading coefficient 1, and (7) means that

$$\begin{cases} p_0(x) = 1, \\ p_{n+1}(x) = (x - s_n)p_n(x) - t_n p_{n-1}(x) \quad (n \geq 0). \end{cases} \tag{8}$$

Proposition 7.2 can thus be summarized in the following diagram:

$$\begin{array}{ccc} A = A^{\sigma,\tau} & \Longleftrightarrow & ATA^T = H \\ \Updownarrow & & \Updownarrow \\ (p_n(x)) \text{ satisfies (8)} & & UHU^T = T \end{array} \tag{9}$$

Orthogonal Polynomials.
The following famous theorem of Favard characterizes all polynomial sequences $(p_n(x))$ that satisfy a recurrence (8) in an unexpected way. As usual, by a polynomial sequence $(p_n(x))$ we mean that $\deg p_n(x) = n$ and the leading coefficient is 1.

Definition. A polynomial sequence $(p_n(x))$ over $\mathbb{C}$ is called an *orthogonal polynomial system* (OPS) if there exists a linear operator $L : \mathbb{C}[x] \longrightarrow \mathbb{C}$ with

$$L(p_m(x)p_n(x)) = \begin{cases} \lambda_n \neq 0 & \text{for } m = n \;\; (\lambda_0 = 1), \\ 0 & \text{for } m \neq n. \end{cases} \tag{10}$$

Lemma 7.3. *The sequence $(p_n(x))$ is an OPS with L and λ_n as in (10) if and only if*

$$L(x^k p_n(x)) = \begin{cases} \lambda_n & k = n \quad (\lambda_0 = 1), \\ 0 & k < n. \end{cases} \tag{11}$$

Proof. Since $(p_n(x))$ constitutes a basis of $\mathbb{C}[x]$, we have $x^k = \sum_{i=0}^{k} c_{k,i} p_i(x)$ with $c_{k,k} = 1$, and hence

$$\begin{aligned} L(x^k p_n(x)) &= \sum_{i=0}^{k} c_{k,i} L(p_i(x) p_n(x)) = \sum_{i=0}^{k} c_{k,i} \delta_{i,n} \lambda_n \\ &= \begin{cases} \lambda_n & k = n, \\ 0 & k < n. \end{cases} \end{aligned}$$

If, conversely, (11) holds, and $m \le n$, $p_m(x) = \sum_{k=0}^{m} u_{m,k} x^k$, we obtain with $u_{m,m} = 1$,

$$L(p_m(x) p_n(x)) = \sum_{k=0}^{m} u_{m,k} L(x^k p_n(x)) = \begin{cases} \lambda_n & m = n, \\ 0 & m \ne n. \end{cases} \qquad \square$$

Theorem 7.4 (Favard). *A polynomial sequence $(p_n(x))$ is an OPS if and only if $(p_n(x))$ satisfies recurrence (8) for some pair of sequences $\sigma = (s_k)$, $\tau = (t_k)$.*

Proof. Suppose $(p_n(x))$ satisfies (8), where $A = A^{\sigma,\tau}$, $U = A^{-1} = (u_{n,k})$, $p_n(x) = \sum_{k=0}^{n} u_{n,k} x^k$, and where $T = (T_i \delta_{i,j})$, $H = (B_{i+j})$ are declared as before. We define the operator $L : \mathbb{C}[x] \longrightarrow \mathbb{C}$ by $Lx^n = B_n$ and linear extension. Because of $UHU^T = T$ we have for all m and n,

$$\sum_{i,k} u_{m,i} u_{n,k} B_{i+k} = \delta_{m,n} T_n ,$$

that is,

$$\begin{aligned} L(p_m(x) p_n(x)) &= L\left(\sum u_{m,i} x^i \cdot \sum u_{n,k} x^k\right) \\ &= \sum_{i,k} u_{m,i} u_{n,k} B_{i+k} = \delta_{m,n} T_n . \end{aligned}$$

It follows that $(p_n(x))$ is an OPS for L with $\lambda_n = T_n$.

Now suppose $(p_n(x))$ is an OPS with $p_n(x) = \sum_{k=0}^{n} u_{n,k}x^k$ and $Lx^n = B_n$. We perform our reasoning backward and obtain from (10)

$$L\left(\sum u_{m,i}x^i \cdot \sum u_{n,k}x^k\right) = \delta_{m,n}\lambda_n\,,$$

and thus

$$\sum_{i,k} u_{m,i}u_{n,k}B_{i+k} = \delta_{m,n}\lambda_n \text{ for all } m \text{ and } n\,.$$

But this means precisely that $UHU^T = T$ with $T_n = \lambda_n$. Hence $(p_n(x))$ satisfies recurrence (8) for some sequences σ and τ by (9). □

Example. Let us look again at our starting example $\sigma \equiv 0$, $\tau \equiv 1$, with $B_{2m} = C_m$, $B_{2m+1} = 0$. The corresponding OPS satisfies the recurrence

$$p_{n+1}(x) = xp_n(x) - p_{n-1}(x), \quad p_0(x) = 1\,.$$

These are the Chebyshev polynomials $c_n(x)$ considered in Section 3.1. One calculates the operator L (see Exercise 7.10) as

$$L(q(x)) = \frac{2}{\pi}\int_{-1}^{1} q(2x)(1-x^2)^{1/2}dx\,.$$

With $q(x) = x^n$, this gives the expression

$$Lx^n = \frac{2^{2m+1}}{\pi}\int_{-1}^{1} x^{2m}(1-x^2)^{1/2}dx = \begin{cases} \frac{1}{m+1}\binom{2m}{m} & n = 2m\,,\\ 0 & n = 2m+1\,,\end{cases}$$

announced in B above.

The Chebyshev polynomials thus satisfy the orthogonality relation

$$\int_{-1}^{1} c_m(2x)c_n(2x)(1-x^2)^{1/2}dx = \begin{cases} \frac{\pi}{2} & m = n\,,\\ 0 & m \neq n\,.\end{cases}$$

Catalan Sequences.
Finally, we give a characterization of Catalan sequences (B_n). Let $A = A^{\sigma,\tau}$, $H = (B_{i+j})$ as before, and consider the principal submatrices with rows 0 to n, $A_n = (a_{i,j})_{i,j=0}^n$, $H_n = (B_{i+j})_{i,j=0}^n$. Since A_n has main diagonal 1, we infer from (5) that

$$\det H_n = T_0 T_1 \cdots T_n \neq 0\,. \tag{12}$$

Theorem 7.5. *A sequence (B_n) is a Catalan sequence if and only if $\det H_n \neq 0$ for all n, where $H_n = (B_{i+j})_{i,j=0}^n$ is the Hankel matrix.*

Proof. We have already seen one half. Suppose then $\det H_n \neq 0$ for all n. We define the sequence $(p_n(x))$ by $p_0(x) = 1$, and for $n \geq 1$,

$$p_n(x) = \frac{\det\begin{pmatrix} B_0 & B_1 & \dots & B_n \\ B_1 & B_2 & \dots & B_{n+1} \\ & & \dots & \\ B_{n-1} & B_n & \dots & B_{2n-1} \\ 1 & x & \dots & x^n \end{pmatrix}}{\det H_{n-1}}\,. \tag{13}$$

Let L be the linear operator defined by $Lx^n = B_n$.

Claim. $(p_n(x))$ is OPS for L.

We note first that $p_n(x)$ has degree n and leading coefficient 1. Consider $k \leq n$; then

$$L(x^k p_n(x)) = \frac{\det\begin{pmatrix} B_0 & B_1 & \dots & B_n \\ & & \dots & \\ B_{n-1} & B_n & \dots & B_{2n-1} \\ B_k & B_{k+1} & \dots & B_{k+n} \end{pmatrix}}{\det H_{n-1}}\,.$$

For $k < n$, the matrix in the numerator has two equal rows, and hence det = 0, while for $k = n$ we obtain

$$L(x^n p_n(x)) = \frac{\det H_n}{\det H_{n-1}} \neq 0$$

by assumption. The claim follows from Lemma 7.3, and shows that (B_n) is a Catalan sequence. □

It is easy to see that every OPS possesses a unique operator L, and our results show that precisely those operators L belong to an OPS for which the images $Lx^n = B_n$ constitute a Catalan sequence. The study of orthogonal systems over $\mathbb{R}$ goes back to Stieltjes and Chebyshev; we will encounter several classical OPS later on. The central question was how these operators L (or equivalently the orthogonality relations) can be described analytically. That is, what can we say about $Lq(x)$, where $q(x)$ is an arbitrary polynomial, given that $Lx^n = B_n$? Stieltjes proved the following: Let $H_n = (B_{i+j})$ and $H_n^{(1)} = (B_{i+j+1})$ be the Hankel matrices as before. Then

$$\det H_n > 0 \ (\forall n) \quad \Longrightarrow \quad Lq(x) = \int_{-\infty}^{\infty} q(x)d\psi(x),$$

$$\det H_n > 0, \det H_n^{(1)} > 0 \ (\forall n) \quad \Longrightarrow \quad Lq(x) = \int_{0}^{\infty} q(x)d\psi(x),$$

where $\psi(x)$ is a certain function of bounded variation.

Exercises

7.1 Why are the descriptions D and F of the ordinary Catalan numbers equivalent?

7.2 Verify the recurrence (7) for $U = A^{-1} = (u_{n,k})$, and show that (7) implies (1) for $A = U^{-1}$.

▷ **7.3** Prove that if $(p_n(x))$ is an OPS, then the operator L (and the numbers λ_n) are uniquely determined. Show, conversely, that if L has an OPS, then it is uniquely determined.

7.4 Consider $\sigma \equiv 0$, $\tau = (t_k)$ arbitrary, $t_k \neq 0$ for all k. Show first that the odd-indexed Catalan numbers B_{2n+1} are 0. Now define the new sequences $\hat{\sigma} = (\hat{s}_k)$, $\hat{\tau} = (\hat{t}_k)$ by $\hat{s}_0 = t_1$, $\hat{s}_k = t_{2k} + t_{2k+1}$ $(k \geq 1)$, and $\hat{t}_k = t_{2k-1}t_{2k}$ $(k \geq 1)$. Prove that for the Catalan number $\hat{B}_n$ associated with $\hat{\sigma}, \hat{\tau}$, $\hat{B}_n = B_{2n}$. What do you get for our starting example $\sigma \equiv 0$, $\tau \equiv 1$?

▷ **7.5** The *central trinomial numbers* are defined as follows: Every number is the sum of the three numbers above it (left, above, right), and the central trinomial numbers Tr_n are the numbers in the middle column: $(Tr_n) = (1, 1, 3, 7, 19, \ldots)$. Show that $Tr_n = B_n^{\sigma,\tau}$ with $\sigma \equiv 1$, $\tau = (2, 1, 1, 1, \ldots)$.

$$
\begin{array}{ccccccccc}
 & & & & \underline{1} & & & & \\
 & & & 1 & \underline{1} & 1 & & & \\
 & & 1 & 2 & \underline{3} & 2 & 1 & & \\
 & 1 & 3 & 6 & \underline{7} & 6 & 3 & 1 & \\
1 & 4 & 10 & 16 & \underline{19} & 16 & 10 & 4 & 1
\end{array}
$$

Show further that Tr_n equals the central coefficient in the expansion of $(1+x+x^2)^n$.

7.6 Given $\sigma \equiv 0$, $\tau = (t_k)$ arbitrary, consider the new sequences $\hat{\sigma} \equiv 0$, $\hat{\tau} = (ct_k)$, $0 \neq c \in \mathbb{C}$. How are the Catalan numbers B_n and $\hat{B}_n$ related?

7.7 What are the Catalan numbers for $\sigma \equiv 2$, $\tau \equiv 1$?

* * *

7.8 Suppose the sequence (B_n) is given by the recurrences $B_0 = 1$, $B_{2n+2} = c^n B_{2n}$, $c \in \mathbb{C}$, $B_{2n+1} = 0$. Show that (B_n) is a Catalan sequence if and only if $c \neq 0$, and c is not a root of unity, that is, $c^n \neq 1$ for all $n \in \mathbb{N}$.

7.9 Consider the sequences a. $\sigma = (1,0,0,0,\ldots)$, $\tau \equiv 1$, b. $\sigma = (3,2,2,2,\ldots)$, $\tau \equiv 1$. Calculate the first rows of the Catalan matrices. Can you formulate conjectures as to what the Catalan numbers might be? Try to prove your conjecture.

7.10 The Chebyshev polynomials $U_n(x)$ were originally defined as $U_n(x) = \frac{\sin(n+1)\vartheta}{\sin\vartheta}$, $x = \cos\vartheta$. Derive the recurrence $U_0(x) = 1$, $U_{n+1}(x) = 2xU_n(x) - U_{n-1}(x)$, and conclude that $U_n(x) = c_n(2x)$, where $c_n(x)$ are the Chebyshev polynomials used in the text. Use the definition of $U_n(x)$ to derive the orthogonality relation $\int_{-1}^{1} U_m(x)U_n(x)(1-x^2)^{1/2}dx = 0$ $(m \neq n)$, thus proving B in the beginning of the section.

▷ **7.11** In analogy to the definition of the Chebyshev polynomials as a determinant, consider the polynomial sequence

$$
U_n(x,\alpha) = \det \begin{pmatrix}
x & 1 & & & & 0 \\
\alpha & x & 2 & & & \\
 & \alpha-1 & x & 3 & & \\
 & & \ddots & \ddots & \ddots & \\
0 & & & & & n-1 \\
 & & & & \alpha-n+2 & x
\end{pmatrix}, \quad U_0(x,\alpha) = 1,
$$

where $\alpha \notin \mathbb{N}_0$. Derive the recurrence $U_{n+1}(x,\alpha) = xU_n(x,\alpha) - n(\alpha - n + 1)U_{n-1}(x,\alpha)$. $U_n(x,\alpha)$ is called a *Cayley continuant*. Consider the exponential generating function $\hat{U}(x,\alpha;z) = \sum_{n\geq 0} U_n(x,\alpha)\frac{z^n}{n!}$ and prove that

$$
\hat{U}(x,\alpha;z) = \frac{(1+z)^{\frac{x+\alpha}{2}}}{(1-z)^{\frac{x-\alpha}{2}}} .
$$

Deduce from this the convolution formula

$$U_n(x_1 + x_2, \alpha_1 + \alpha_2) = \sum_{k=0}^{n} \binom{n}{k} U_k(x_1, \alpha_1) U_{n-k}(x_2, \alpha_2),$$

and the reciprocity law

$$U_n(-x, \alpha) = (-1)^n U_n(x, \alpha).$$

Hint: Convert the recurrence into a differential equation for $\hat{U}(x, \alpha; z)$.

7.12 Let $U_n(x, \alpha)$ be as in the previous exercise. Show that

$$U_n(x, \alpha) = \sum_{k=0}^{n} \binom{n}{k} \left(\frac{x+\alpha}{2}\right)^{\underline{k}} \left(\frac{x-\alpha}{2}\right)^{\overline{n-k}}.$$

7.13 Define the polynomial $s_n(\alpha, \beta) = \sum c_{i,j} \alpha^i \beta^j$, where $c_{i,j}$ is the number of permutations $\sigma \in S(n)$ with i even cycles and j odd cycles. Generalize Exercise 3.53 to find the exponential generating function $\hat{S}(z)$ of $s_n(\alpha, \beta)$, and compare the resulting expression for $s_n(\alpha, \beta)$ with $U_n(x, \alpha)$.

▷ **7.14** Consider the following recurrence: $b_{0,0} = 1$, $b_{0,k} = 0$ $(k > 0)$, $b_{n,k} = b_{n-1,k-1} + b_{n-1,k} + \cdots + b_{n-1,n-1}$ $(n \geq 1)$. The numbers $b_{n,k}$ are called *ballot numbers.* Prove $b_{n,0} = C_n$ (ordinary Catalan number). Hint: Look at the generating function of the k-th column $B_k(z)$, and set $B_k(z) = B_0(z)F(z)^k$, $F(0) = 0$.

7.15 Use the Lagrange inversion formula to derive the formula $b_{n,k} = \frac{k+1}{n+1}\binom{2n-k}{n}$ for the numbers $b_{n,k}$ of the previous exercise. Consult Exercise 5.50 to justify the name ballot numbers.

7.16 Continuing the exercise, show that $b_{n,i} = b_{n-1,i-1} + b_{n,i+1}$ for $i \geq 1$, and deduce $b_{n+k,2k} = a_{n,k}$, where $A = (a_{n,k})$ is the Catalan matrix corresponding to $\sigma = (1, 2, 2, \ldots)$, $\tau \equiv 1$.

▷ **7.17** Let $B_k(z)$ be the ordinary generating function of the k-th column of the ballot table $(b_{n,k})$ as in Exercise 7.14. Prove that $B_k(z) = q_k(z)C(z) - r_k(z)$, where $C(z)$ is the ordinary Catalan series, and $q_k(z)$, $r_k(z)$ are polynomials of degrees at most $\frac{k}{2}$ and $\frac{k-1}{2}$, respectively. Deduce that $q_k(z) = \sum_{i \geq 0} (-1)^i \binom{k-i}{i} z^i$, and from this the formula

$$b_{n,k} = \textstyle\sum_{i \geq 0} (-1)^i \binom{k-i}{i} C_{n-i}\,.$$

7.2 Catalan Numbers and Lattice Paths

Let $A = A^{\sigma,\tau} = (a_{n,k})$ be a Catalan matrix as before, hence

$$a_{n,k} = a_{n-1,k-1} + s_k a_{n-1,k} + t_{k+1} a_{n-1,k+1} \quad (n \geq 1). \tag{1}$$

Now turn the matrix A by 90°, and let $a_{n,k}$ correspond to the lattice point (n, k):

$$\begin{array}{rrrr} & & & a_{3,3} \cdots \\ & & a_{2,2} & a_{3,2} \cdots \\ & a_{1,1} & a_{2,1} & a_{3,1} \cdots \\ a_{0,0} & a_{1,0} & a_{2,0} & a_{3,0} \cdots \end{array}$$

Our recurrence thus reads pictorially

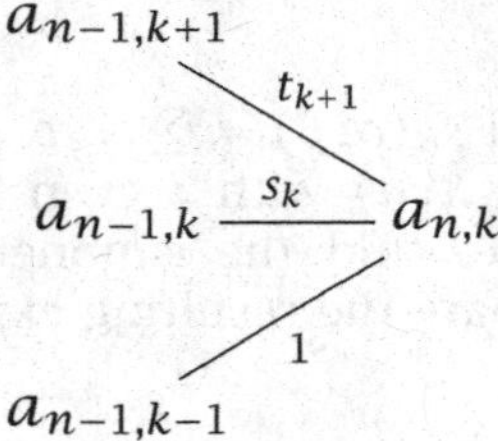

Definition. A *Motzkin path P* of length n is a lattice path from $(0,0)$ to $(n,0)$ with all steps horizontally or diagonally up or down to the right that never falls below the axis $y = 0$. The *Motzkin number* M_n is the number of Motzkin paths of length n. The figure shows the Motzkin paths up to length 3; thus $M_0 = M_1 = 1$, $M_2 = 2$, $M_3 = 4$:

Now we associate weights to the individual steps:

1 $\quad$ s_k $\quad y = k$ $\quad$ t_{k+1} $\quad y = k+1$, $y = k$

and define the weight $w(P)$ of a path as in Section 5.4:

$$w(P) = \prod w(\text{steps}).$$

Given sequences σ and τ, we thus have the Catalan matrix $A^{\sigma,\tau}$ and the Motzkin paths $\mathcal{P}_n$ weighted according to σ and τ.

Recurrence (1) clearly translates into

$$B_n = \sum_{P \in \mathcal{P}_n} w(P), \tag{2}$$

and more generally,

$$a_{n,k} = \sum_{P} w(P),$$

where P runs through all paths that start at $(0,0)$ and terminate in (n,k).

Our starting example leads because of $\sigma \equiv 0$ to all Motzkin paths *without* horizontal steps, that is, to ordinary Catalan paths.

Determinants of Hankel matrices.
To compute the determinants of the Hankel matrices

$$H_n = (B_{i+j})_{i,j=0}^{n}, \quad H_n^{(1)} = (B_{i+j+1})_{i,j=0}^{n},$$

we use the lemma of Gessel–Viennot of Section 5.4. Set as before $T_n = t_1 t_2 \cdots t_n$, $T_0 = 1$, and consider the following graph:

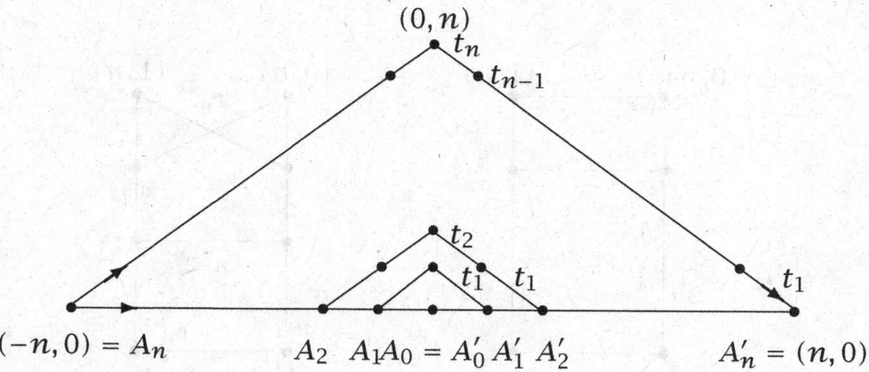

with $A_i = (-i, 0)$, $A_j' = (j, 0)$. Let $\mathcal{A} = \{A_0, \ldots, A_n\}$, $\mathcal{A}' = \{A_0', \ldots, A_n'\}$. The path matrix $M = (m_{ij})$ from $\mathcal{A}$ to $\mathcal{A}'$ is then $m_{ij} = B_{i+j}$ according to (2), that is, $M = H_n$. The only vertex-disjoint path system $\mathcal{P} : A_i \to A_i'$ $(i = 0, \ldots, n)$ is the one depicted in the figure with $w(\mathcal{P}) = T_1 T_2 \cdots T_n$, and the lemma of Gessel–Viennot implies

$$\det H_n = T_1 T_2 \cdots T_n. \tag{3}$$

To compute $\det H_n^{(1)}$ we consider the same lattice graph, this time with $A_i = (-i, 0)$, $A'_j = (j+1, 0)$:

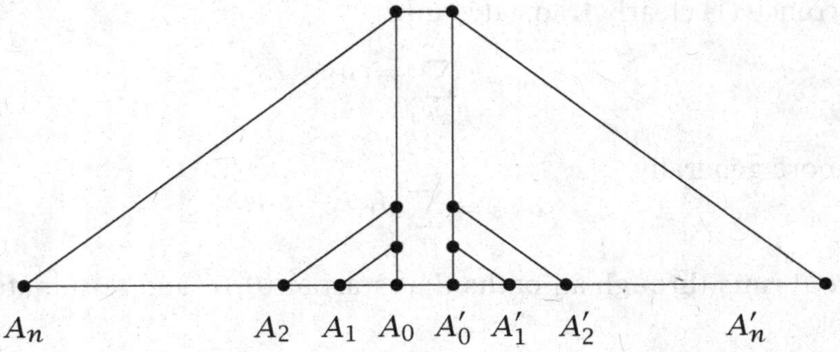

The path matrix is now $M = H_n^{(1)} = (B_{i+j+1})$. The vertex-disjoint path systems are again of the form $\mathcal{P} : A_i \to A'_i$, and they are uniquely determined outside the strip $0 \le x \le 1$. The weight outside the strip is again $T_1 T_2 \cdots T_n$; hence

$$\det H_n^{(1)} = r_n \det H_n .$$

To compute r_n let us classify the path systems $\mathcal{P}$ according to the top step:

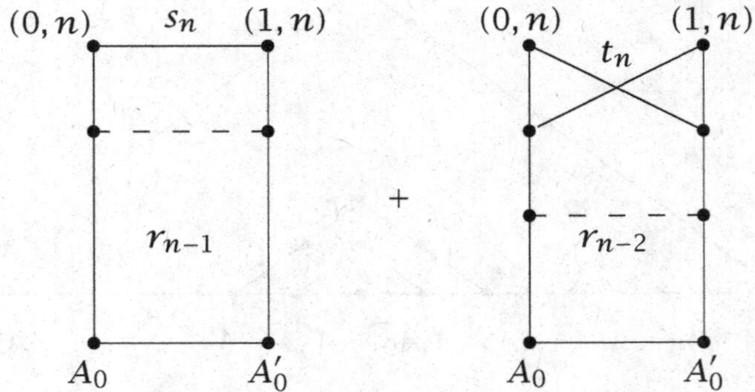

This gives the recurrence

$$r_n = s_n r_{n-1} - t_n r_{n-2}, \quad r_{-1} = 1, \quad r_0 = s_0 .$$

Note the minus sign, because of the transposition $(n-1, n)$ in the second case.

In sum, we have proved the following result.

Theorem 7.6. *Let $A = A^{\sigma,\tau}$, and H_n, $H_n^{(1)}$ the Hankel matrices of the Catalan numbers B_n. Then*

$$\det H_n = T_0 T_1 \cdots T_n, \quad \det H_n^{(1)} = r_n \det H_n ,$$

with $r_{-1} = 1$, $r_0 = s_0$, and

$$r_n = s_n r_{n-1} - t_n r_{n-2} \quad (n \geq 1) . \tag{4}$$

Example. For $\sigma \equiv 0$, $\tau \equiv 1$, $\det H_n = 1$, and $\det H_n^{(1)} = r_n$, where $r_n = -r_{n-2}$ by (4). Hence with the starting values $r_{-1} = 1$, $r_0 = 0$ we get $r_{2n} = 0$, $r_{2n+1} = (-1)^{n+1}$. As an example,

$$\det H_3^{(1)} = \det \begin{pmatrix} 0 & 1 & 0 & 2 \\ 1 & 0 & 2 & 0 \\ 0 & 2 & 0 & 5 \\ 2 & 0 & 5 & 0 \end{pmatrix} = 1 .$$

Given any sequence (B_n), generalizing H_n and $H_n^{(1)}$, we may consider the Hankel matrix $H_n^{(k)} = (B_{i+j+k})_{i,j=0}^{n}$ of k-th order, thus $H_n = H_n^{(0)}$. Let $d_n^{(k)}$ be the determinants. There is an interesting recurrence relating the numbers $d_n^{(k)}$ that follows from a classical matrix theorem due to Jacobi.

Recall the following theorem of linear algebra. Suppose M is an $n \times n$-matrix. By $M_{i,j}$ we denote the submatrix where the i-th row and j-th column have been deleted. Now consider the *cofactor matrix* $\overline{M} = (\overline{m_{ij}})$, where $\overline{m_{ij}} = (-1)^{i+j} \det M_{j,i}$. Then we have

$$M\overline{M} = \begin{pmatrix} \det M & & & \\ & \det M & & 0 \\ & 0 & \ddots & \\ & & & \det M \end{pmatrix} . \tag{5}$$

Lemma 7.7 (Jacobi). *Let M be an $n \times n$-matrix, $n \geq 2$. Then*

$$\det M \cdot \det M_{1,n;1,n} = \det M_{1,1} \cdot \det M_{n,n} - \det M_{1,n} \cdot \det M_{n,1} , \tag{6}$$

where $M_{1,n;1,n}$ is the submatrix with rows $1, n$ and columns $1, n$ deleted, where we set $\det M_{1,n;1,n} = 1$ *if* $n = 2$.

Proof. We regard the entries of M as variables. To shorten the notation, let us set $|M| = \det M$. Consider the matrix

$$M^* = \begin{pmatrix} |M_{1,1}| & 0 \; 0 \ldots 0 & (-1)^{n+1}|M_{n,1}| \\ -|M_{1,2}| & 1 \; 0 \ldots 0 & (-1)^{n}|M_{n,2}| \\ |M_{1,3}| & 0 \; 1 \ldots 0 & (-1)^{n-1}|M_{n,3}| \\ \vdots & \ldots & \ldots \\ (-1)^{n}|M_{1,n-1}| & 0 \; 0 \ldots 1 & -|M_{n,n-1}| \\ (-1)^{n+1}|M_{1,n}| & 0 \; 0 \ldots 0 & |M_{n,n}| \end{pmatrix}.$$

Multiplying M by M^*, we get by (5),

$$MM^* = \begin{pmatrix} |M| & m_{12} & m_{13} & \ldots & 0 \\ 0 & m_{22} & m_{23} & \ldots & 0 \\ 0 & m_{32} & m_{33} & \ldots & 0 \\ \vdots & & \ldots & & \\ 0 & m_{n2} & m_{n3} & \ldots & |M| \end{pmatrix}.$$

Now $|MM^*| = |M|^2|M_{1,n;1,n}|$, $|M^*| = |M_{1,1}||M_{n,n}| - |M_{1,n}||M_{n,1}|$, and $|MM^*| = |M||M^*|$ imply (6). □

Let us apply the lemma to the Hankel matrices $H_n^{(k)}$ of a sequence (B_n). For $M = H_n^{(k)}$ we clearly have $M_{1,1} = H_{n-1}^{(k+2)}$, $M_{n,n} = H_{n-1}^{(k)}$, $M_{1,n} = M_{n,1} = H_{n-1}^{(k+1)}$, and $M_{1,n;1,n} = H_{n-2}^{(k+2)}$. We thus get the following result.

Corollary 7.8. *Let (B_n) be a sequence, and $d_n^{(k)} = \det H_n^{(k)}$. Then*

$$d_n^{(k)} d_{n-2}^{(k+2)} = d_{n-1}^{(k)} d_{n-1}^{(k+2)} - (d_{n-1}^{(k+1)})^2 \quad (n \geq 1, k \geq 0), \tag{7}$$

with $d_{-1}^{(k)} = 1$, or shifting the index,

$$d_n^{(k-2)} d_n^{(k)} = d_{n+1}^{(k-2)} d_{n-1}^{(k)} + (d_n^{(k-1)})^2 \quad (n \geq 0, k \geq 2). \tag{8}$$

Example. We know for the ordinary Catalan numbers C_n that $d_n^{(0)} = d_n^{(1)} = 1$. Hence (8) yields for $k = 2$,

$$d_n^{(2)} = d_{n-1}^{(2)} + 1,$$

and thus $d_n^{(2)} = \det H_n^{(2)} = n + 2$, since $d_0^{(2)} = \det H_0^{(2)} = C_2 = 2$. Similarly, for $k = 3$ we get

$$d_n^{(3)} = d_{n-1}^{(3)} + (n+2)^2 = (n+2)^2 + (n+1)^2 + \cdots + 3^2 + d_0^{(3)},$$

which gives $d_n^{(3)} = \det H_n^{(3)} = \sum_{i=1}^{n+2} i^2$, since $d_0^{(3)} = \det H_0^{(3)} = C_3 = 5 = 2^2 + 1^2$. As an example,

$$\det H_2^{(3)} = \det \begin{pmatrix} 5 & 14 & 42 \\ 14 & 42 & 132 \\ 42 & 132 & 429 \end{pmatrix} = 30 = 1^2 + 2^2 + 3^2 + 4^2 .$$

Exercise 7.25 will show that the recurrence (8) implies, in fact, a nice general formula for $\det H_n^{(k)}$ when $(B_n) = (C_n)$ is the ordinary Catalan sequence.

Exercises

7.18 Compute $\det H_n$, and $\det H_n^{(1)}$ for $\sigma = (1, 2, 2, \ldots)$, $\tau \equiv 1$. What follows for the Catalan numbers B_n? Compare the result to Exercise 7.4.

7.19 Compute $\det H_n$, $\det H_n^{(1)}$ for $\sigma \equiv 2$, $\tau \equiv 1$ (see Exercise 7.7).

▷ **7.20** The Catalan numbers for $\sigma \equiv 1$, $\tau \equiv 1$ are the Motzkin numbers M_n. Calculate $\det H_n$ and $\det H_n^{(1)}$.

7.21 Consider the trinomial numbers of Exercise 7.5, and compute $\det H_n$, $\det H_n^{(1)}$.

7.22 Determine $\det H_n^{(2)}$ for the sequence of Motzkin numbers.

* * *

▷ **7.23** Let $A = A^{\sigma,\tau}$ with Catalan sequence (B_n). Use the lemma of Gessel–Viennot to prove $\det H_n^{(2)} = (T_0 T_1 \cdots T_{n+1}) \sum_{j=-1}^{n} \frac{r_j^2}{T_{j+1}}$, where $T_{j+1} = t_1 \cdots t_{j+1}$ as before, and r_j defined as in (4). Verify the result also with (8).

7.24 Use the previous exercise to compute $\det H_n^{(2)}$ for $\sigma \equiv 0$, $\tau \equiv 1$, and $\sigma = (1, 2, 2, \ldots)$, $\tau \equiv 1$.

▷ **7.25** Let C_n be the ordinary Catalan number. Show that

$$\det H_n^{(k)} = \prod_{1 \le i \le j \le k-1} \frac{i + j + 2n + 2}{i + j}$$

for $k \ge 0$, and derive from this a formula for C_k. Hint: Show that the expression satisfies the recurrence (8).

7.26 Consider the sequence $\left(B_n = \binom{2n+1}{n}\right)$. You were asked in Exercise 7.9 to prove that this is the Catalan sequence for $\sigma = (3, 2, 2, \ldots)$, $\tau \equiv 1$. Now prove an analogous statement to the previous exercise: $\det H_n^{(k)} = \prod_{1 \le i \le j \le k} \frac{i+j+2n+1}{i+j-1}$ $(n, k \ge 0)$.

7.27 Compare the general formula in Exercise 7.23 with the two preceeding exercises for $k = 2$; what do you get?

▷ **7.28** Consider the sequences $\sigma \equiv 0$, $\tau = (\alpha, \beta, \alpha, \beta, \ldots)$, that is, $t_{2k-1} = \alpha$, $t_{2k} = \beta$. Interpret B_{2n} as a sum of weighted Catalan paths of length $2n$, and prove $B_{2n} = \sum_{k=0}^{n-1} N(n,k)\alpha^{n-k}\beta^k$ with $N(n,k) = \frac{1}{n}\binom{n}{k}\binom{n}{k+1}$. Show also that $N(n,k)$ counts the Catalan paths with exactly $k+1$ peaks ⋰⋱. The numbers $N(n,k)$ are called *Narayana numbers*. Hint: Interpret the Catalan paths as ballot lists as in Exercise 3.9. A peak is then a subsequence AB.

7.3 Generating Functions and Operator Calculus

Let us now take a closer look at Catalan numbers, using generating functions. We begin as usual with $A = A^{\sigma,\tau} = (a_{n,k})$, $\sigma = (s_k)$, $\tau = (t_k)$, and the recurrence

$$\begin{cases} a_{0,0} = 1, \ a_{0,k} = 0 \quad (k > 0), \\ a_{n,k} = a_{n-1,k-1} + s_k a_{n-1,k} + t_{k+1} a_{n-1,k+1}. \end{cases} \tag{1}$$

Now set $t_n = q_n u_n$ $(n \ge 1)$, $Q_n = q_1 \cdots q_n$, $Q_0 = 1$, and consider Q-generating functions as in Section 2.2 with our three calculi: $q_n = 1$ (ordinary generating functions), $q_n = n$ (exponential generating functions), $q_n = \frac{1-q^n}{1-q}$ (q-generating functions).

Let $A_k(z)$ be the Q-generating function of the k-th column of A, that is,

$$A_k(z) = \sum_{n \ge 0} a_{n,k} \frac{z^n}{Q_n}, \tag{2}$$

where in particular,

$$B(z) = A_0(z) = \sum_{n \ge 0} B_n \frac{z^n}{Q_n} \tag{3}$$

is the generating function of the Catalan numbers B_n. With the derivative Δ of Section 2.2, recurrence (1) translates into the system

$$\begin{cases} \Delta A_k = A_{k-1} + s_k A_k + t_{k+1} A_{k+1}, \ A_k(0) = 0 \quad (k \geq 1), \\ \Delta A_0 = a A_0 + b A_1, \ A(0) = 1, \end{cases} \tag{4}$$

with $a = s_0$, $b = t_1$ (a and b will play a special role).

Note that (4) implies $\deg A_k(z) = k$ (in the sense of Section 2.3) with $a_{k,k} = 1$. Hence we can say that if $A_0(z), A_1(z), \ldots$ is a sequence of functions that satisfy (4), then $B(z) = A_0(z)$ is the generating function of the Catalan numbers corresponding to $\sigma = (s_k)$, $\tau = (t_k)$.

Continued Fractions.
For ordinary generating functions there is another way to describe the system (4) by means of continued fractions. Suppose $(A_k(z))$ is a sequence of ordinary generating functions with $\deg A_k(z) = k$, $[z^k]A_k(z) = 1$. Then $\frac{A_k(z)}{z^k}$ is invertible. Now set

$$C_k(z) = \frac{A_k(z)}{z^k} \Big/ \frac{A_{k-1}(z)}{z^{k-1}} \quad (k \geq 1),$$
$$C_0(z) = A_0(z).$$

Then $C_k(0) = 1$ and $A_k(z) = zA_{k-1}(z)C_k(z)$; thus

$$A_k(z) = z^k \prod_{i=0}^{k} C_i(z). \tag{5}$$

Lemma 7.9. *Suppose $(A_k(z))$ and $(C_k(z))$ with $C_k(0) = 1$ are connected as in (5). Then $(A_k(z))$ satisfies (4) if and only if*

$$C_k(z) = \frac{1}{1 - s_k z - t_{k+1} z^2 C_{k+1}(z)} \quad (k \geq 0). \tag{6}$$

Proof. Suppose (4) holds for $(A_k(z))$. Then

$$\begin{cases} \dfrac{A_0 - 1}{z} = s_0 A_0 + t_1 A_1, \\ \dfrac{A_k}{z} = A_{k-1} + s_k A_k + t_{k+1} A_{k+1} \quad (k \geq 1). \end{cases}$$

Hence by (5),

$$C_0 = A_0 = 1 + s_0 z A_0 + t_1 z A_1 = 1 + s_0 z C_0 + t_1 z^2 C_0 C_1,$$

that is,

$$C_0(z) = \frac{1}{1 - s_0 z - t_1 z^2 C_1(z)}.$$

For $k \geq 1$ we get

$$\begin{aligned} A_{k-1}C_k = \frac{A_k}{z} &= A_{k-1} + s_k A_k + t_{k+1} A_{k+1} \\ &= A_{k-1} + s_k z A_{k-1} C_k + t_{k+1} z^2 A_{k-1} C_k C_{k+1}, \end{aligned}$$

which gives

$$C_k(z) = \frac{1}{1 - s_k z - t_{k+1} z^2 C_{k+1}(z)}.$$

The converse is just as easily seen, using (5) again. □

In sum, we can say that if $(C_k(z))$ is a sequence of ordinary generating functions with $C_k(0) = 1$ for all k, satisfying (6), then $B(z) = C_0(z)$ is the generating function of the Catalan numbers associated with $\sigma = (s_k)$, $\tau = (t_k)$. Iterating (6), we obtain the continued fraction expansion

$$B(z) = C_0(z) = \cfrac{1}{1 - s_0 z - \cfrac{t_1 z^2}{1 - s_1 z - \cfrac{t_2 z^2}{1 - s_2 z - \ddots}}} \tag{7}$$

Such an expression is then to be understood as a sequence of functions $C_k(z)$ satisfying equation (6).

Example. Our starting example $\sigma \equiv 0$, $\tau \equiv 1$ leads to

$$B(z) = C_0(z) = \cfrac{1}{1 - \cfrac{z^2}{1 - \cfrac{z^2}{1 - \ddots}}}$$

Replacing z^2 by z we obtain from (6),

$$C_0(z) = \frac{1}{1 - zC_0(z)} \quad \text{or} \quad C_0(z) = 1 + zC_0^2(z),$$

and we get precisely the ordinary Catalan series.

Sheffer Matrices.

Let us return to Q-generating functions $A_k(z)$. Solving system (4) presents, in general, great difficulties. Here is the key idea. We restrict ourselves to Catalan matrices $A = A^{\sigma,\tau}$ whose Q-generating functions $A_k(z)$ of the k-th column have the following form: Let $B(z) = A_0(z)$. Then there exists a function $F(z) = \sum_{n\ge 0} F_n \frac{z^n}{Q_n}$ with $F(0) = 0$ such that

$$A_k(z) = B(z)\frac{F(z)^k}{Q_k} \quad \text{for all } k. \tag{8}$$

Matrices $A = A^{\sigma,\tau}$ with this property are called *Sheffer matrices.* Note that (8) implies $F_1 = 1$ because of $q_1 = 1$. Plugging (8) into the system (4), we get

$$\begin{cases} \Delta(BF^k) = q_k BF^{k-1} + s_k BF^k + u_{k+1} BF^{k+1} & (k \ge 1), \\ \Delta B = aB + bBF. \end{cases} \tag{9}$$

Our goal is to compute $B(z)$ from (9), and to determine the associated OPS $(p_n(x))$. In the following we concentrate on $q_n = 1$ (ordinary generating functions) and $q_n = n$ (exponential generating functions), leaving the case $q_n = \frac{1-q^n}{1-q}$ to the exercises.

Ordinary Generating Functions.

Here we have $\Delta = D_0$, $D_0 A(z) = \frac{A(z)-A(0)}{z}$, and we know from Exercise 2.14 that

$$D_0(BF^k) = B(D_0 F^k) = BF^{k-1}(D_0 F).$$

Substituting this into (9) and canceling gives the system

$$\begin{cases} D_0 F = 1 + s_k F + u_{k+1} F^2 & (k \ge 1), \\ D_0 B = aB + bBF. \end{cases} \tag{10}$$

For (10) to be solvable we must therefore have $s_k = s$, $u_{k+1} = u$ for $k \ge 1$. In other words, we obtain as the only possible sequences in the Sheffer case $\sigma = (a, s, s, s, \ldots)$, $\tau = (b, u, u, u, \ldots)$, and hence the system

$$\begin{cases} D_0 F = 1 + sF + uF^2, \\ D_0 B = aB + bBF. \end{cases} \tag{11}$$

Comparison of the coefficients of z^n translates (11) into the convolution recurrences

$$\begin{cases} F_{n+1} = sF_n + u \sum\limits_{k=1}^{n-1} F_k F_{n-k} \quad (n \geq 1),\ F_1 = 1\,, \\ B_{n+1} = aB_n + b \sum\limits_{k=0}^{n-1} B_k F_{n-k} \quad (n \geq 0),\ B_0 = 1\,. \end{cases} \tag{12}$$

In Section 7.4 we will interpret these coefficients combinatorially for various choices of the parameters a, b, s, and u.

Exponential Generating Functions.
Here Δ is the usual derivative $DA(z) = A'(z)$ with $(BF^k)' = B'F^k + kBF^{k-1}F'$. Canceling F^{k-1}, (9) takes on the form

$$\begin{cases} B'F + kBF' = kB + s_k BF + u_{k+1} BF^2\,, \\ B' = aB + bBF\,. \end{cases}$$

Substituting B' into the first equation and canceling, we get

$$F' = 1 + \frac{s_k - a}{k} F + \frac{u_{k+1} - b}{k} F^2 \quad (k \geq 1)\,.$$

Hence $s_k = a + ks$, $u_{k+1} = b + ku$ for fixed s and u, which means that in the Sheffer case, the only possible sequences are $\sigma = (s_k = a + ks)$, $\tau = (t_k = k(b + (k-1)u))$. The resulting system is

$$\begin{cases} F' = 1 + sF + uF^2\,, \\ B' = aB + bBF\,, \end{cases} \tag{13}$$

or in terms of coefficients,

$$\begin{cases} F_{n+1} = sF_n + u \sum\limits_{k=1}^{n-1} \binom{n}{k} F_k F_{n-k} \quad (n \geq 1),\ F_1 = 1\,, \\ B_{n+1} = aB_n + b \sum\limits_{k=0}^{n-1} B_k F_{n-k} \quad (n \geq 0),\ B_0 = 1\,. \end{cases} \tag{14}$$

Again we will interpret (14) in combinatorial terms in Section 7.4.

The Associated OPS.
Now we turn to the orthogonal polynomial system $(p_n(x))$ corresponding to $A = A^{\sigma,\tau}$ in the Sheffer case. Let $A = A^{\sigma,\tau}$, $U = A^{-1} = (u_{n,k})$, and $p_n(x) = \sum_{k=0}^{n} u_{n,k} x^k$ the orthogonal system. As before, $A_k(z) = \sum_{n\geq 0} a_{n,k} \frac{z^n}{Q_n}$, $U_k(z) = \sum_{n \geq 0} u_{n,k} \frac{z^n}{Q_n}$. Let us abbreviate the Sheffer relation $A_k = B \frac{F^k}{Q_k}$ by

$$A \leftrightarrow (B, F)\,.$$

Lemma 7.10. *If $A \leftrightarrow (B, F)$, then we have*

$$U \leftrightarrow \left(\frac{1}{B(F^{\langle -1\rangle})}, F^{\langle -1\rangle}\right),$$

where $F^{\langle -1\rangle}$ is the compositional inverse of F.

Proof. We have to show that

$$U_k(z) = \frac{1}{B(F^{\langle -1\rangle}(z))}\frac{(F^{\langle -1\rangle}(z))^k}{Q_k}.$$

From $AU = I$ follows

$$u_{0,k}A_0(z) + u_{1,k}A_1(z) + \cdots = \frac{z^k}{Q_k} \quad (k \geq 0),$$

and hence

$$\sum_{n\geq 0} u_{n,k}B(z)\frac{F(z)^n}{Q_n} = \frac{z^k}{Q_k}.$$

The substitution $z \mapsto F^{\langle -1\rangle}(z)$ therefore gives

$$\sum_{n\geq 0} u_{n,k}B(F^{\langle -1\rangle}(z))\frac{z^n}{Q_n} = \frac{(F^{\langle -1\rangle}(z))^k}{Q_k},$$

and thus

$$U_k(z) = \sum_{n\geq 0} u_{n,k}\frac{z^n}{Q_n} = \frac{1}{B(F^{\langle -1\rangle}(z))}\frac{(F^{\langle -1\rangle}(z))^k}{Q_k}. \qquad \square$$

With this lemma it is an easy matter to compute the generating function of the sequence $(p_n(x))$.

Proposition 7.11. *Let $A \leftrightarrow (B, F)$, and $(p_n(x))$ the corresponding* OPS. *Then we have*

$$\sum_{n\geq 0} p_n(x)\frac{z^n}{Q_n} = \frac{1}{B(F^{\langle -1\rangle}(z))}\sum_{k\geq 0}\frac{(xF^{\langle -1\rangle}(z))^k}{Q_k}. \tag{15}$$

Proof. Let us define the infinite column-vectors $Z = \left(\frac{z^n}{Q_n}\right)$ and $X = (x^n)$. We obtain

$$Z^T(UX) = \left(1, \frac{z}{Q_1}, \frac{z^2}{Q_2}, \ldots\right) \begin{pmatrix} p_0(x) \\ p_1(x) \\ \vdots \end{pmatrix} = \sum_{n\geq 0} p_n(x) \frac{z^n}{Q_n},$$

and on the other hand,

$$(Z^T U)X = (U_0(z), U_1(z), \ldots) \begin{pmatrix} 1 \\ x \\ x^2 \\ \vdots \end{pmatrix}$$

$$= \left(\frac{1}{B(F^{\langle -1\rangle}(z))}, \frac{1}{B(F^{\langle -1\rangle}(z))} \frac{F^{\langle -1\rangle}(z)}{Q_1}, \ldots\right) \begin{pmatrix} 1 \\ x \\ x^2 \\ \vdots \end{pmatrix}$$

$$= \frac{1}{B(F^{\langle -1\rangle}(z))} \sum_{k\geq 0} \frac{(xF^{\langle -1\rangle}(z))^k}{Q_k},$$

and the result follows. □

Example. Let us once more look at our starting example $\sigma \equiv 0$, $\tau \equiv 1$. Using ordinary generating functions we have to solve the system (11),

$$\begin{cases} D_0 F = \frac{F}{z} = 1 + F^2, \\ D_0 B = \frac{B-1}{z} = BF. \end{cases} \tag{16}$$

Hence $F(z) = z + zF^2(z)$. From $y = z + zy^2$ follows $z = \frac{y}{1+y^2}$, that is, $F^{\langle -1\rangle}(z) = \frac{z}{1+z^2}$. We already know $B(z)$, and it can, of course, be easily computed from (16), $B(z) = \frac{1-\sqrt{1-4z^2}}{2z^2}$. This implies

$$B(F^{\langle -1\rangle}(z)) = \frac{1 - \sqrt{1 - \frac{4z^2}{(1+z^2)^2}}}{2\frac{z^2}{(1+z^2)^2}} = \frac{1 - \frac{1-z^2}{1+z^2}}{2\frac{z^2}{(1+z^2)^2}} = 1 + z^2,$$

and thus by (15) for the Chebyshev polynomials $c_n(x)$,

$$\sum_{n\geq 0} c_n(x) z^n = \frac{1}{1+z^2} \frac{1}{1 - \frac{xz}{1+z^2}} = \frac{1}{1 - xz + z^2},$$

a result that we already obtained from the recurrence $c_{n+1}(x) = xc_n(x) - c_{n-1}(x)$ in Section 3.2 (see Exercise 3.37).

Operator Calculus.
Now we study the q-calculus of generating functions in greater depth, thereby providing easy proofs of some classical results about orthogonal polynomial systems. We start with three easy lemmas. As before, we use Q-generating functions $H(z) = \sum_{n\geq 0} h_n \frac{z^n}{Q_n}$. Let Δ be the derivative. Then $H(\Delta)$ denotes the operator

$$H(\Delta) = \sum_{n\geq 0} h_n \frac{\Delta^n}{Q_n}.$$

Lemma 7.12. *Let* $H(z) = \sum_{n\geq 0} h_n \frac{z^n}{Q_n}$.

(i) *We have* $h_n = H(\Delta)x^n|_{x=0}$ *for all* n.

(ii) *If* $p(x) = \sum_{i=0}^{k} p_i x^i$, *then* $H(\Delta)p(x)|_{x=0} = \sum_{i=0}^{k} p_i h_i$.

Proof. Recall the notation $\left[{n \atop m} \right] = \frac{Q_n}{Q_m Q_{n-m}}$. Now $\Delta x^n = q_n x^{n-1}$, and thus $\Delta^m x^n = q_n q_{n-1} \cdots q_{n-m+1} x^{n-m}$, or

$$\frac{\Delta^m x^n}{Q_m} = \left[{n \atop m} \right] x^{n-m}. \tag{17}$$

It follows that

$$H(\Delta)x^n|_{x=0} = \sum_{m\geq 0} h_m \frac{\Delta^m x^n}{Q_m}|_{x=0} = \sum_{m=0}^{n} h_m \left[{n \atop m} \right] x^{n-m}|_{x=0}$$
$$= h_n.$$

For (ii) we obtain

$$H(\Delta)\sum_{i=0}^{k} p_i x^i|_{x=0} = \sum_{i=0}^{k} p_i H(\Delta)x^i|_{x=0} = \sum_{i=0}^{k} p_i h_i. \qquad \square$$

Lemma 7.13. *Let* $A = (a_{n,k})$ *be invertible, and* $A_k(z) = \sum_{n\geq 0} a_{n,k} \frac{z^n}{Q_n}$. *If for two polynomials* $p(x)$, $q(x)$, *we have* $A_k(\Delta)p(x)|_{x=0} = A_k(\Delta)q(x)|_{x=0}$ *for all* $k \geq 0$, *then* $p(x) = q(x)$.

Proof. The previous lemma shows that

$$A_k(\Delta)p(x)\,|_{x=0} = \sum_i p_i a_{i,k},\ A_k(\Delta)q(x)\,|_{x=0} = \sum_i q_i a_{i,k}.$$

In other words, $p^T A = q^T A$ holds for the coefficient vectors $p^T = (p_0, p_1, p_2, \ldots)$, $q^T = (q_0, q_1, q_2, \ldots)$. Since A is invertible, $p(x) = q(x)$ follows. □

Lemma 7.14. *Let A be invertible, and $(p_n(x))$ the polynomial sequence corresponding to $U = A^{-1}$, $p_n(x) = \sum_{i=0}^n u_{n,i}x^i$. Then we have*

$$A_k(\Delta)p_n(x)\,|_{x=0} = \begin{cases} 1 & n = k, \\ 0 & n \neq k. \end{cases} \tag{18}$$

In particular, the sequence $(p_n(x))$ is uniquely determined by (18).

Proof. By Lemma 7.12 and $U = A^{-1}$,

$$A_k(\Delta)p_n(x)\,|_{x=0} = \sum_i a_{i,k}u_{n,i} = \delta_{n,k}. \qquad \square$$

With these preliminary results it is an easy matter to state the following expansion theorem (see Exercise 7.34).

Proposition 7.15. *Let A be invertible, and $(p_n(x))$ the polynomial sequence corresponding to $U = A^{-1}$. For an arbitrary generating function $H(z) = \sum_{n\geq 0} h_n \frac{z^n}{Q_n}$,*

$$H(z) = \sum_{k\geq 0} (H(\Delta)p_k(x)\,|_{x=0})A_k(z).$$

So far, the results hold for arbitrary invertible matrices. Now we turn to the Sheffer case $A = A^{\sigma,\tau}$, $A \leftrightarrow (B, F)$, that is, $A_k(z) = B(z)\frac{F(z)^k}{Q_k}$, $Q_n = q_1 q_2 \cdots q_n$.

Proposition 7.16. *Let $A \leftrightarrow (B, F)$, and $(p_n(x))$ the corresponding orthogonal polynomial system. Then we have*

$$F(\Delta)p_n(x) = q_n p_{n-1}(x). \tag{19}$$

Proof. By (18),

$$\begin{aligned} A_k(\Delta)(F(\Delta)p_n(x))\,|_{x=0} &= B(\Delta)\frac{F(\Delta)^{k+1}}{Q_k}p_n(x)\,|_{x=0} \\ &= q_{k+1}A_{k+1}(\Delta)p_n(x)\,|_{x=0} \\ &= q_{k+1}\delta_{n,k+1} = q_n\delta_{n,k+1}\,. \end{aligned}$$

On the other hand, again by (18),

$$A_k(\Delta)q_np_{n-1}(x)\,|_{x=0} = q_n\delta_{n-1,k} = q_n\delta_{n,k+1}\,,$$

and the uniqueness result of Lemma 7.13 finishes the proof. □

Suppose $A \leftrightarrow (B,F)$; then we define the matrix $\widetilde{A}$ by $\widetilde{A} \leftrightarrow (1,F)$. In other words, $\widetilde{A}_k(z) = \frac{F(z)^k}{Q_k}$. Because of $F_0 = 0$, $F_1 = 1$, $\widetilde{A}$ is again lower triangular with main diagonal 1. Set $\widetilde{U} = \widetilde{A}^{-1} = (\widetilde{u}_{n,k})$, and define the polynomials $\widetilde{p}_n(X) = \sum_{k=0}^n \widetilde{u}_{n,k}x^k$ as before. Lemma 7.14 holds then for $\widetilde{A}$ and $\widetilde{p}_n(x)$ as well.

Proposition 7.17. *Suppose $A \leftrightarrow (B,F)$. Then*

$$\widetilde{p}_n(x) = B(\Delta)p_n(x) \quad \textit{for all } n\,. \tag{20}$$

Proof. From (18) we get $\frac{F(\Delta)^k}{Q_k}\widetilde{p}_n(x)\,|_{x=0} = \delta_{n,k}$, and on the other hand, again by (18),

$$\frac{F(\Delta)^k}{Q_k}(B(\Delta)p_n(x))\,|_{x=0} = A_k(\Delta)p_n(x)\,|_{x=0} = \delta_{n,k}\,.$$

The uniqueness result of Lemma 7.13 implies $\widetilde{p}_n(x) = B(\Delta)p_n(x)$. □

A particularly beautiful "binomial" property is the so-called *Sheffer identity* for exponential generating functions, which is $q_n = n$, $Q_n = n!$. Let us first note the following property of the exponential function e^z. Let $a \in \mathbb{C}$ and D the usual derivative. Then using (17),

$$e^{aD}x^n = \sum_{k\ge 0} a^k\frac{D^k}{k!}x^n = \sum_{k=0}^n \binom{n}{k}a^kx^{n-k} = (x+a)^n\,,$$

and hence by linearity,

$$e^{aD}p(x) = p(x+a) \quad \text{for any } p(x) \in \mathbb{C}[x]\,. \tag{21}$$

This, in turn, implies

$$e^{aD}p(x)\,|_{x=0} = p(a)\,, \tag{22}$$

the evaluation of $p(x)$ at $x = a$.

Proposition 7.18. *Let $A \leftrightarrow (B,F)$, $\widetilde{A} \leftrightarrow (1,F)$ in the exponential calculus, with the associated polynomial sequences $(p_n(x))$ and $(\widetilde{p}_n(x))$. Then we have*

$$p_n(x+y) = \sum_{k=0}^{n}\binom{n}{k}p_k(x)\widetilde{p}_{n-k}(y)\,. \tag{23}$$

Proof. By the expansion theorem applied to $\widetilde{A}$ and $H(z) = e^{yz}$ and (22) we obtain

$$e^{yz} = \sum_{k\geq 0} e^{yD}\widetilde{p}_k(x)\,|_{x=0}\frac{F(z)^k}{k!} = \sum_{k\geq 0}\widetilde{p}_k(y)\frac{F(z)^k}{k!}\,.$$

Now we apply e^{yD} to $p_n(x)$. According to (21), we have $e^{yD}p_n(x) = p_n(x+y)$, and on the other hand, by (19),

$$e^{yD}p_n(x) = \left(\sum_{k\geq 0}\widetilde{p}_k(y)\frac{F(D)^k}{k!}\right)p_n(x) = \sum_{k=0}^{n}\widetilde{p}_k(y)\binom{n}{k}p_{n-k}(x)\,,$$

and the result follows. □

Some Classical Examples.
Now it is time for some examples; they are all in the exponential calculus.

Example 1. Charlier Polynomials. Consider $\sigma = (s_k = a + k)$, $\tau = (t_k = ak)$. In our previous notation we have $s = 1$, $a = b$, $u = 0$, which means that we have to solve the system

$$\begin{cases} F' = 1 + F\,, \\ B' = aB(1+F)\,. \end{cases}$$

Clearly $F(z) = e^z - 1$, and $\frac{B'}{B} = a(1+F) = ae^z$ implies $B(z) = e^{a(e^z-1)}$, since $B(0) = 1$. We have already seen this function in Section 3.3; it is the exponential generating function of

$$B_n = \sum_{k=0}^{n} S_{n,k} a^k \quad \text{(Stirling polynomial of the second kind).}$$

In particular, for $a = 1$, the Bell numbers $\mathrm{Bell}(n)$ result. The Bell numbers are therefore the Catalan numbers corresponding to $\sigma = (s_k = k+1)$, $\tau = (t_k = k)$. Let us check the first rows of $A = A^{\sigma,\tau}$:

τ		1	2	3	4
σ	1	2	3	4	5
	$\underline{1}$				
	$\underline{1}$	1			
	$\underline{2}$	3	1		
	$\underline{5}$	10	6	1	
	$\underline{15}$	37	31	10	1
	$\underline{52}$				

Let us next compute the determinants of the Hankel matrices for the Bell numbers. We have $T_n = n!$, and hence $\det H_n = n!! := 1!2!\cdots n!$. For $H_n^{(1)}$, the recurrence is $r_n = (n+1)r_{n-1} - nr_{n-2}$ with starting values $r_{-1} = 1, r_0 = 1$, which gives $r_n = 1$ by induction, and thus $\det H_n^{(1)} = n!!$. The Bell numbers are thus the unique sequence with $\det H_n = \det H_n^{(1)} = n!!$ for all n.

The corresponding orthogonal polynomials are the *Charlier polynomials* $C_n^{(a)}(x)$. From $F(z) = e^z - 1$ results $F^{\langle -1\rangle}(z) = \log(1+z)$, and thus $B(F^{\langle -1\rangle}(z)) = e^{az}$. According to (15) the generating function is given by

$$\begin{aligned}\sum_{n\ge 0} C_n^{(a)}(x)\frac{z^n}{n!} &= e^{-az}\sum_{k\ge 0}\frac{(x\log(1+z))^k}{k!} = e^{-az}(1+z)^x \\ &= e^{-az}\sum_{k\ge 0} x^{\underline{k}}\frac{z^k}{k!}.\end{aligned}$$

With convolution we get the final expression

$$C_n^{(a)}(x) = \sum_{k=0}^{n}\binom{n}{k}(-a)^{n-k}x^{\underline{k}}.$$

The matrix $\tilde{A} \leftrightarrow (1, F = e^z - 1)$ corresponds to the case $a = 0$ in $B(z)$; hence $\tilde{p}_n(x) = C_n^{(0)}(x) = x^{\underline{n}}$ and the Sheffer formula yields

$$C_n^{(a)}(x+y) = \sum_{k=0}^{n} \binom{n}{k} C_k^{(a)}(x) y^{\underline{n-k}}.$$

We can also apply (19). With $F(z) = e^z - 1$ we get $F(D) = e^D - I$ and thus by (21),

$$F(D)C_n^{(a)}(x) = C_n^{(a)}(x+1) - C_n^{(a)}(x).$$

Proposition 7.16 therefore yields

$$C_n^{(a)}(x+1) - C_n^{(a)}(x) = nC_{n-1}^{(a)}(x).$$

Example 2. Laguerre Polynomials. Consider $\sigma = (s_k = a + 2k)$, $\tau = (t_k = k(a + k - 1))$, that is, $s = 2$, $u = 1$, $a = b$. The differential equations (13) are

$$\begin{cases} F' = 1 + 2F + F^2 = (1+F)^2, \\ B' = aB(1+F). \end{cases}$$

One easily computes $F(z) = \frac{z}{1-z}$ and $F^{\langle -1 \rangle}(z) = \frac{z}{1+z}$. Furthermore, $(\log B)' = \frac{B'}{B} = a(1+F) = \frac{a}{1-z}$, and thus

$$B(z) = \frac{1}{(1-z)^a} = \sum_{n \geq 0} a^{\overline{n}} \frac{z^n}{n!},$$

according to our list of functions in Section 2.1. The Catalan sequence is therefore

$$B_n = a^{\overline{n}} = \sum_{k=0}^{n} s_{n,k} a^k \quad \text{(Stirling polynomial of the first kind).}$$

In particular, for $a = 1$, $B_n = n!$ results. The factorials $n!$ are thus the Catalan numbers for $\sigma = (s_k = 2k+1)$, $\tau = (t_k = k^2)$.

The corresponding OPS are the *Laguerre polynomials* $L_n^{(a)}(x)$. From

$$B(F^{\langle -1 \rangle}(z)) = \frac{1}{(1 - \frac{z}{1+z})^a} = (1+z)^a$$

follows for the generating function according to (15),

$$\sum_{n \geq 0} L_n^{(a)}(x) \frac{z^n}{n!} = (1+z)^{-a} \sum_{k \geq 0} \frac{(x\frac{z}{1+z})^k}{k!} = (1+z)^{-a} e^{x\frac{z}{1+z}}.$$

A simple computation (see Exercise 7.44) yields the expression

$$L_n^{(a)}(x) = (-1)^n n! \sum_{k=0}^{n} \binom{a+n-1}{n-k} \frac{(-x)^k}{k!}.$$

The matrix $\widetilde{A} \leftrightarrow (1, F(z) = \frac{z}{1-z})$ corresponds again to the case $a = 0$, and we obtain the Sheffer identity

$$L_n^{(a)}(x+y) = \sum_{k=0}^{n} \binom{n}{k} L_k^{(a)}(x) L_{n-k}^{(0)}(y),$$

with

$$L_n^{(0)}(x) = (-1)^n n! \sum_{k=1}^{n} \binom{n-1}{k-1} \frac{(-x)^k}{k!}.$$

Finally, $F(z) = \frac{z}{1-z} = \sum_{n\geq 1} z^n$ and (19) yield the recurrence

$$L_n^{(a)}(x)' + L_n^{(a)}(x)'' + \cdots + L_n^{(a)}(x)^{(n)} = nL_{n-1}^{(a)}(x).$$

Example 3. Hermite Polynomials. Our final example is $\sigma \equiv a$, $\tau = (t_k = bk)$. Here $s = u = 0$, and system (13) reads

$$F' = 1, \quad B' = B(a + bF).$$

We obtain $F(z) = z$, and $(\log B)' = \frac{B'}{B} = a + bz$; thus $B(z) = e^{az + \frac{bz^2}{2}}$. In particular, for $a = b = 1$ we obtain the exponential generating function of the involution numbers i_n considered in Section 3.3. The numbers i_n are thus the Catalan numbers with respect to the sequences $\sigma \equiv 1$, $\tau = (t_k = k)$.

The associated OPS are the *Hermite polynomials* $H_n^{a,b}(x)$. Since $F(z) = z$, we have $F^{\langle -1 \rangle}(z) = z$, and thus the generating function

$$\sum_{n\geq 0} H_n^{a,b}(x) \frac{z^n}{n!} = e^{-az - \frac{bz^2}{2} + xz} = e^{(x-a)z - \frac{bz^2}{2}}.$$

To compute $H_n^{a,b}(x)$ we use Proposition 7.17. From $\widetilde{A} \leftrightarrow (1, z)$ follows $\widetilde{A} = \widetilde{U} = I$, and thus $\widetilde{H}_n^{a,b}(x) = x^n$. Let D be the derivative; then $H_n^{a,b}(x) = B(D)^{-1} x^n = e^{-aD - \frac{bD^2}{2}} x^n$. With (21) this gives

$$\begin{aligned}
H_n^{a,b}(x) &= e^{-b\frac{D^2}{2}} e^{-aD} x^n = e^{-\frac{b}{2}D^2}(x-a)^n \\
&= \sum_{k\geq 0} \frac{n^{\underline{2k}}}{k!}\left(-\frac{b}{2}\right)^k (x-a)^{n-2k} \\
&= \sum_k \frac{n^{\underline{2k}}}{k!}\left(-\frac{b}{2}\right)^k \sum_{j=0}^{n-2k}\binom{n-2k}{j}(-a)^j x^{n-2k-j} \\
&= \sum_k \frac{n^{\underline{2k}}}{k!}\left(-\frac{b}{2}\right)^k \sum_{j=2k}^{n}\binom{n-2k}{j-2k}(-a)^{j-2k} x^{n-j},
\end{aligned}$$

and thus

$$H_n^{a,b}(x) = \sum_{j=0}^{n}\left(\sum_{k\geq 0}\frac{(-1)^{j-k}a^{j-2k}b^k}{(j-2k)!2^k k!}\right) n^{\underline{j}} x^{n-j}.$$

In particular, $a = 0$ gives

$$H_n^{(b)} := H_n^{0,b}(x) = \sum_{k=0}^{\lfloor n/2\rfloor}\left(-\frac{b}{2}\right)^k \frac{n^{\underline{2k}}}{k!} x^{n-2k}.$$

The Sheffer formula reads

$$H_n^{a,b}(x+y) = \sum_{k=0}^{n}\binom{n}{k} H_k^{a,b}(x) y^{n-k},$$

and $F(D) = D$ yields

$$H_n^{a,b}(x)' = nH_{n-1}^{a,b}(x).$$

Example. Suppose $a = 0$, $b = 1$. Then $H_n^{(1)}(x) = \sum_{k=0}^{\lfloor n/2\rfloor}(-\frac{1}{2})^k \frac{n^{\underline{2k}}}{k!} x^{n-2k}$ with the first polynomials $H_0^{(1)}(x) = 1$, $H_1^{(1)}(x) = x$, $H_2^{(1)}(x) = x^2 - 1$, $H_3^{(1)}(x) = x^3 - 3x$, $H_4^{(1)}(x) = x^4 - 6x^2 + 3$.

Exercises

7.29 Suppose $A \leftrightarrow (B, z)$, that is, $F(z) = z$. Show that there is no solution of (11) for ordinary generating functions, and that for exponential generating functions the only solutions for (13) have $s = u = 0$. Thus we obtain $\sigma \equiv a$, $\tau = (t_k = bk)$ considered in example 3 above.

▷ **7.30** Consider the q-calculus, $q_n = \frac{1-q^n}{1-q}$, for $A \leftrightarrow (B, z)$. Show that the sequences σ, τ for which (9) is solvable are $\sigma = (s_k = aq^k)$, $\tau = (t_k = \frac{1-q^k}{1-q} bq^{k-1})$.

7.31 Consider the Motzkin numbers M_n, that is, $\sigma \equiv 1$, $\tau \equiv 1$, and determine the generating function $\sum_{n\geq 0} M_n z^n$ using continued fractions.

7.32 Suppose $A \leftrightarrow (B, F)$ is a Sheffer matrix for ordinary generating functions. Show that for the functions $C_k(z)$ used in the continued fraction expansion, $C_k(z) = \frac{F(z)}{z}$ for $k \geq 1$.

7.33 We know that the ordinary Catalan numbers C_n form the Catalan sequence belonging to $\sigma = (1, 2, 2, \ldots)$, $\tau \equiv 1$ (see Exercise 7.4). Use the previous exercise to show that the ordinary generating function $A_k(z)$ of the k-th column equals $z^k C(z)^{2k+1}$, where $C(z)$ is the ordinary Catalan series. Make use of Exercise 3.59 to find a formula for $a_{n,k}$.

▷ **7.34** Prove the Expansion Theorem 7.15.

7.35 Let D be the differential operator $Dp(x) = p'(x)$. Prove the operator identity $(I-D)^n = (-1)^n e^x D^n e^{-x}$, meaning that both sides give the same result when applied to any polynomial. Here I is the identity operator.

7.36 Show that $x^{-m}(I - D)^n x^{n+m} = (I - D)^{m+n} x^n$ $(m \in \mathbb{N}_0)$.

7.37 Prove for the Hermite polynomial $H_n^{(1)}(x)$ that

$$H_n^{(1)}(0) = (-1)^{n/2}(n-1)(n-3)\cdots 3\cdot 1$$

if n is even, and $= 0$ if n is odd.

▷ **7.38** Let $A \leftrightarrow (B, F)$ be a Sheffer matrix in the ordinary calculus, $A = (a_{n,k})$, and let $H(z) = \sum_{n\geq 0} h_n z^n$ be arbitrary. Prove that

$$[z^n]B(z)H(F(z)) = \sum_{k=0}^{n} a_{n,k} h_k .$$

Define $\tilde{H}(z) = H(F^{\langle -1\rangle}(z)) = \sum_{n\geq 0} \tilde{h}_n z^n$ and show that $[z^n]B(z)H(z) = \sum_{k=0}^{n} a_{n,k} \tilde{h}_k$.

7.39 Continuing the exercise suppose that $\alpha_{n,k}$ and $\beta_{n,k}$ are numbers related by $\sum_k \alpha_{n,k} a_{k,j} = \beta_{n,j}$ for all $n, j \geq 0$. Prove that

$$\sum_{k=0}^{n} \alpha_{n,k}[z^k]B(z)H(F(z)) = \sum_{j\geq 0} h_j \beta_{n,j} ,$$

and $\sum_{k=0}^{n} \alpha_{n,k}[z^k]B(z)H(z) = \sum_{j\geq 0} \tilde{h}_j \beta_{n,j}$.

▷ **7.40** Continuing the exercise let $B(z) = \frac{1}{1-z}$, $F(z) = \frac{z}{1-z}$. Show that $a_{n,k} = \binom{n}{k}$ for all $n, k \geq 0$. Now consider $\alpha_{n,k} = 1$ for all n, k, and derive identities when $H(z) = \frac{1}{1-z}$ or $H(z) = \frac{z^m}{(1-z)(1-2z)\cdots(1-mz)}$. Consider also $\alpha_{n,k} = H_k$ (harmonic number), and derive identities when $H(z)$ is as above.

* * *

7.41 Suppose $\sigma = (a, s, s, \ldots)$, $\tau \equiv 1$. Determine the ordinary generating function of the Catalan numbers B_n using continued fractions.

7.42 Let $A = A^{\sigma,\tau}$, and suppose that $U = A^{-1}$ is also a Catalan matrix $U = U^{\sigma',\tau'}$. Prove that $\sigma' \equiv -\sigma$, that is, $s'_k = -s_k$ for all k.

▷ **7.43** Continuing the exercise, suppose that $A = A^{\sigma,\tau}$ is a Sheffer matrix in the exponential calculus, and $U = A^{-1}$ with $U = U^{\sigma',\tau'}$. Consider the cases $\tau' \equiv -\tau$ and $\tau' \equiv \tau$. Show that a. $\tau' \equiv -\tau$ implies $\sigma \equiv a$, $\tau = (t_k = bk)$, which gives the Hermite polynomials. b. $\tau' \equiv \tau$ implies $\sigma = (s_k = a + 2mk)$, $\tau = (t_k = k(am + m^2(k-1))$ for some m. This yields the generalized Laguerre polynomials $L_n^{(a,m)}(x)$.

7.44 Prove the formula for the Laguerre polynomials $L_n^{(a)}(x)$ in the text.

7.45 Show that $L_n^{(a)}(x) = (I-D)^{a+n-1}x^n$ $(a \in \mathbb{N}_0)$, and further $L_n^{(a+b)}(x) = (I-D)^b L_n^{(a)}(x)$ $(b \in \mathbb{N}_0)$.

▷ **7.46** Let $C_n^{(a)}(x)$ be the Charlier polynomials and $L_n^{(a)}(x)$ the Laguerre polynomials. Prove the identity $C_n^{(x)}(a+n-1) = (-1)^n L_n^{(a)}(x)$, and deduce $C_n^{(a+b)}(x+n) = \sum_{k=0}^{n} \binom{n}{k} C_k^{(a)}(x+k) C_{n-k}^{(b)}(n-k-1)$.

7.47 Prove for the Hermite polynomials the binomial theorem $H_n^{(a+b)}(x+y) = \sum_{k=0}^{n} \binom{n}{k} H_k^{(a)}(x) H_{n-k}^{(b)}(y)$.

7.48 Prove the so-called Rodriguez formula for the Hermite polynomials: $H_n^{(b)}(x) = (-1)^n e^{\frac{x^2}{2b}} b^n D^n e^{-\frac{x^2}{2b}}$. Hint: Use the recurrence for $H_n^{(b)}(x)$ to show first that $H_n^{(b)}(x) = (x - bD)H_{n-1}^{(b)}(x)$.

▷ **7.49** Consider the sequence $\sigma \equiv 0$, $\tau = \left(t_k = \frac{k^2}{(2k-1)(2k+1)}\right)$ in the ordinary calculus, with associated OPS $(p_n(x))$. The polynomials $P_n(x) = \frac{1}{2^n}\binom{2n}{n} p_n(x)$ are called the *Legendre polynomials*. Derive the recurrence $(n+1)P_{n+1}(x) = (2n+1)xP_n(x) - nP_{n-1}(x)$, $P_0(x) = 1$, and compute from this the generating function $\sum_{n \geq 0} P_n(x) z^n$.

7.50 Continuing the exercise, prove the following orthogonality relation for the Legendre polynomials: $\int_{-1}^{1} P_m(x)P_n(x)dx = \frac{2}{2n+1}\delta_{m,n}$.

▷ **7.51** Given the sequences $\sigma = (s_k = aq^k)$, $\tau = (t_k = \frac{1-q^k}{1-q}bq^{k-1})$ as in Exercise 7.30 in the q-calculus, use (9) to show that the Catalan numbers satisfy the recurrence

$$B_{n+1} = aB_n + b\frac{1-q^n}{1-q}B_{n-1}\,.$$

Now set $a = \lambda + 1$, $b = \lambda(q-1)$, and show that $B_n = \sum_{k=0}^{n}\left[{n \atop k}\right]_q \lambda^k$ (Galois polynomial). Hence for $a = 2$, $b = q - 1$, we obtain $B_n = G_n$ (Galois number). Hint: Consider the Gauss polynomials $g_n(x)$, and the operator L defined by $L : g_n(x) \to \lambda^n$.

7.4 Combinatorial Interpretation of Catalan Numbers

Let us ask what combinatorial coefficients are in fact Catalan numbers. We have already seen several examples such as the ordinary Catalan numbers, the Stirling polynomials, number of involutions, and the Galois numbers. There are two methods at hand. First we may compute the generating function $\sum_{n\geq 0} B_n z^n$ and compare it to known series. Or we interpret the recurrence for B_n in a combinatorial setting.

Ordinary Generating Functions.
Let $A = A^{\sigma,\tau}$ be a Sheffer matrix $A \leftrightarrow (B, F)$ as in the previous section. We know that the possible sequences are $\sigma = (a, s, s, \ldots)$, $\tau = (b, u, u, \ldots)$; we abbreviate this to $\sigma = (a, s)$, $\tau = (b, u)$. To compute $B(z)$ we use continued fractions. According to Exercise 7.32 we have $B(z) = C_0(z)$, $C_1(z) = C_2(z) = \cdots = \frac{F(z)}{z} =: G(z)$, and hence by Lemma 7.9 of the previous section,

$$B(z) = \frac{1}{1 - az - bz^2G(z)}, \quad G(z) = \frac{1}{1 - sz - uz^2G(z)}\,. \tag{1}$$

Solving the quadratic equation for $G(z)$, we obtain

$$G(z) = \frac{1 - sz - \sqrt{1 - 2sz + (s^2 - 4u)z^2}}{2uz^2}\,. \tag{2}$$

Plugging this into the first equation, one easily computes $B(z)$ as

$$B(z) = \frac{(2u - b) + (bs - 2au)z - b\sqrt{1 - 2sz + (s^2 - 4u)z^2}}{2(u - b) + 2(bs - 2au + ab)z + 2(a^2u - abs + b^2)z^2}\,. \tag{3}$$

From formula (3) we can now extract a convolution recurrence for the Catalan numbers B_n. Write $B(z)$ in the form

$$B(z) = \frac{\alpha - \sqrt{\beta}}{2\gamma};$$

thus

$$\begin{aligned} \alpha &= (2u - b) + (bs - 2au)z, \\ \beta &= b^2(1 - 2sz + (s^2 - 4u)z^2), \\ \gamma &= (u - b) + (bs - 2au + ab)z + (a^2u - abs + b^2)z^2. \end{aligned} \tag{4}$$

Therefore $B(z)$ satisfies the quadratic equation

$$B^2 - \frac{\alpha}{\gamma}B + \frac{\alpha^2 - \beta}{4\gamma^2} = 0.$$

Looking at (4) we immediately get $\frac{\alpha^2 - \beta}{4\gamma} = u$, and thus the equation

$$\gamma B^2 - \alpha B + u = 0.$$

A short computation (check it!) yields the recurrence

$$\begin{aligned} bB_{n+1} = abB_n + (u-b)\sum_{k=0}^{n-1} B_{k+1}B_{n-k} + (bs - 2au + ab)\sum_{k=0}^{n-1} B_k B_{n-k} \\ + (a^2u - abs + b^2)\sum_{k=0}^{n-1} B_k B_{n-1-k} \quad (n \geq 0). \end{aligned} \tag{5}$$

Examples.

1. Suppose $b = u = 1$, that is, $\tau \equiv 1$. Then

$$B(z) = \frac{1 + (s - 2a)z - \sqrt{1 - 2sz + (s^2 - 4)z^2}}{2(s - a)z + 2(a^2 - as + 1)z^2}$$

with recurrence

$$B_{n+1} = aB_n + (s-a)\sum_{k=0}^{n-1} B_k B_{n-k} + (a^2 - as + 1)\sum_{k=0}^{n-1} B_k B_{n-1-k} \quad (n \geq 0).$$

For $a = s = 0$ we get $B(z) = \frac{1-\sqrt{1-4z^2}}{2z^2} = C(z^2)$, which is, of course, the example we started out with. The case $a = s = 1$ gives the Motzkin numbers M_n with generating function $\sum_{n\geq 0} M_n z^n = \frac{1-z-\sqrt{1-2z-3z^2}}{2z^2}$ and the recurrence

$$M_{n+1} = M_n + \sum_{k=0}^{n-1} M_k M_{n-1-k}\,.$$

The choice $a = s = 2$ yields $B(z) = \frac{1-2z-\sqrt{1-4z}}{2z^2} = \left(\frac{1-\sqrt{1-4z}}{2z} - 1\right)\frac{1}{z}$, and we obtain the shifted ordinary Catalan numbers $(C_1, C_2, C_3, \ldots)$ as the sequence.

2. Another interesting case arises for $a = s + 1$, $b = u = 1$. Here

$$B(z) = \frac{1-(s+2)z-\sqrt{1-2sz+(s^2-4)z^2}}{-2z+2(s+2)z^2}\,.$$

For $s = 0$ we get

$$B(z) = \frac{1-2z-\sqrt{1-4z^2}}{-2z+4z^2}\,,$$

which is the generating function of $B_n = \binom{n}{\lfloor n/2\rfloor}$ according to Exercise 2.8. The case $s = 2$ results in

$$B(z) = \frac{1-4z-\sqrt{1-4z}}{-2z+8z^2}\,,$$

which was again treated in Exercise 2.8 with $B_n = \binom{2n+1}{n}$.

3. Suppose $b = 2u$, and $a = s$. Then

$$B(z) = \frac{-b\sqrt{1-2sz+(s^2-2b)z^2}}{-b+2bsz+(2b^2-bs^2)z^2} = \frac{1}{\sqrt{1-2sz+(s^2-2b)z^2}}\,.$$

For $b = s = 2$, we have $B(z) = \frac{1}{\sqrt{1-4z}}$ with Catalan number $B_n = \binom{2n}{n}$, and for $b = 4$, $s = 3$, $B(z) = \frac{1}{\sqrt{1-6z+z^2}}$ results, with the central Delannoy numbers $D_{n,n}$ appearing as Catalan numbers.

4. Instead of comparing $\sum_{n\geq 0} B_n z^n$ to known generating functions we may also look directly at the recurrence (5). We know that the ordinary Catalan number C_n counts the lattice paths from $(0,0)$ to (n,n) with steps $(1,0)$ and $(0,1)$ that never go above the diagonal $y = x$. The *Schröder number* Sch_n counts the paths that, in addition, also use diagonal steps $(1,1)$ and do not cross the diagonal. You are asked in the exercises to prove the recurrence

$$Sch_{n+1} = Sch_n + \sum_{k=0}^{n} Sch_k Sch_{n-k} \quad (n \geq 0)$$

with $Sch_0 = 1$. As next values we get $Sch_1 = 2$, $Sch_2 = 6$, $Sch_3 = 22$.

Looking at (5) it is easily seen that the Schröder recurrence corresponds to the case $a = 2$, $s = 3$, $b = u = 2$. Hence (Sch_n) is the Catalan sequence for $\sigma = (2,3)$, $\tau = (2,2)$. The following table verifies the first values:

	2	2	2	2
2	3	3	3	3
1				
2	1			
6	5	1		
22	23	8	1	
90	107	49	11	1

From the Catalan matrix it follows that the Schröder numbers Sch_n are all congruent to 2 (mod 4) for $n \geq 1$ (why?).

To summarize our findings so far let us make a table of some important Catalan sequences:

$\tau \equiv 1 \; (b = u = 1)$:

a	s		
0	0	$(C_0, 0, C_1, 0, C_2, 0, C_3, \ldots)$	Catalan
1	1	$M_n = (1, 1, 2, 4, 9, 21, 51, \ldots)$	Motzkin
2	2	$C_{n+1} = (1, 2, 5, 14, 42, 132, \ldots)$	shifted Catalan
1	0	$\binom{n}{\lfloor n/2 \rfloor} = (1, 1, 2, 3, 6, 10, 20, \ldots)$	middle binomials
3	2	$\binom{2n+1}{n} = (1, 3, 10, 35, 126, \ldots)$	central binomials
1	2	$C_n = (1, 1, 2, 5, 14, 42, 132, \ldots)$	Catalan

$b = 2, u = 1:$

$a = s = 1:$	$Tr_n = (1, 1, 3, 7, 19, 51, \ldots)$	central trinomials
$a = s = 2:$	$\binom{2n}{n} = (1, 2, 6, 20, 70, \ldots)$	central binomials
$\left.\begin{array}{l} b = 2, u = 2 \\ a = 2, s = 3 \end{array}\right\}$	$Sch_n = (1, 2, 6, 22, 90, 394, \ldots)$	Schröder
$\left.\begin{array}{l} b = 4, u = 2 \\ a = s = 3 \end{array}\right\}$	$D_{n,n} = (1, 3, 13, 63, 321, \ldots)$	central Delannoy

Two further cases give very interesting numbers:

$\tau \equiv 1:$

$a = 0,\ s = 1$	$R_n = (1, 0, 1, 1, 3, 6, 15, \ldots)$	Riordan
$a = 0,\ s = 2$	$Fi_n = (1, 0, 1, 2, 6, 18, 57, \ldots)$	Fine

Binomial Formulas.
Catalan numbers are connected by binomial relations, as we now show. Let $A = A^{\sigma,\tau}$, and $P = \left(\binom{n}{k}\right)$ the Pascal matrix. The (easy) proof of the following lemma is left to the excercises.

Lemma 7.19. *Suppose $A = A^{\sigma,\tau} = (a_{n,k})$.*

(i) *$\widetilde{A} = PA$ is the Catalan matrix corresponding to the sequences $\widetilde{\sigma} \equiv \sigma + 1 = (s_0 + 1, s_1 + 1, s_2 + 1, \ldots)$, $\widetilde{\tau} = \tau$.*

(ii) *$\overline{A} = (\overline{a}_{n,k})$ with $\overline{a}_{n,k} = (-1)^{n-k} a_{n,k}$ is the Catalan matrix with respect to $\overline{\sigma} = -\sigma = (-s_0, -s_1, -s_2, \ldots)$, $\overline{\tau} = \tau$.*

We know that the ℓ-th power of P has as (n,k)-entry $\binom{n}{k}\ell^{n-k}$, $\ell \in \mathbb{Z}$ (see Exercise 2.38). Thus if $\widetilde{A} = \widetilde{A}^{\sigma+\ell,\tau}$, then $\widetilde{A} = P^{\ell} A$, that is, $\widetilde{a}_{n,k} = \sum_i \binom{n}{i} \ell^{n-i} a_{i,k}$. Setting $k = 0$, we thus obtain the following result.

Proposition 7.20. *Let $A = A^{\sigma,\tau}$ with Catalan numbers B_n.*

(i) *If $\widetilde{A} = \widetilde{A}^{\sigma+\ell,\tau}$ with Catalan numbers $\widetilde{B}_n$, then*

$$\widetilde{B}_n = \sum_{k=0}^{n} \binom{n}{k} \ell^{n-k} B_k .$$

(ii) *If* $\overline{A} = A^{-\sigma,\tau}$ *with Catalan numbers* $\overline{B}_n$*, then*

$$\overline{B}_n = (-1)^n B_n .$$

Examples. Looking at our table above, we can deduce a plethora of binomial formulas.

1. $\sigma \equiv 0$, $\tau \equiv 1$, $\ell = 1$ and 2 give

$$M_n = \sum_{k=0}^{n} \binom{n}{2k} C_k, \quad C_{n+1} = \sum_{k\geq 0} \binom{n}{2k} 2^{n-2k}\, C_k = \sum_{k=0}^{n} \binom{n}{k} M_k .$$

2. The choice $\sigma \equiv -1$, $\tau \equiv 1$, and $\ell = 2$ and 3 yields

$$M_n = \sum_{k=0}^{n} (-1)^k \binom{n}{k} 2^{n-k} M_k, \quad C_{n+1} = \sum_{k=0}^{n} (-1)^k \binom{n}{k} 3^{n-k} M_k ,$$

and $\sigma \equiv -2$, $\tau \equiv 1$, $\ell = 4$ gives

$$C_{n+1} = \sum_{k=0}^{n} (-1)^k \binom{n}{k} 4^{n-k} C_{k+1} .$$

3. $\sigma = (-1, 0)$, $\tau \equiv 1$, $\ell = 2$ yields

$$C_n = \sum_{k=0}^{n} (-1)^k \binom{n}{k} 2^{n-k} \binom{k}{\lfloor k/2 \rfloor} ,$$

and $\sigma = (-1, -2, -2, \ldots)$, $\tau \equiv 1$, $\ell = 4$,

$$\binom{2n+1}{n} = \sum_{k=0}^{n} (-1)^k \binom{n}{k} 4^{n-k} C_k .$$

Exponential Generating Functions.
We turn to the second method, interpreting the system (14) of the last section in combinatorial terms. We are given the sequences $\sigma = (s_k = a + ks)$, $\tau = (t_k = k(b + (k-1)u))$ in the exponential calculus, and want to solve the recurrences

$$\begin{cases} F_{n+1} = sF_n + u \sum_{k=1}^{n-1} \binom{n}{k} F_k F_{n-k} \ (n \geq 1),\ F_0 = 0, F_1 = 1 , \\ B_{n+1} = aB_n + b \sum_{k=0}^{n-1} \binom{n}{k} B_k F_{n-k} \ (n \geq 0),\ B_0 = 1 . \end{cases} \tag{6}$$

Here is the idea: We choose a combinatorial structure whose elements depend on a parameter n, choose a proper weighting for the elements, and show that (6) is satisfied. Specialization of the parameters involved will then lead to various counting coefficients as Catalan numbers.

Case A. $u = 0$. Here we have $\sigma = (s_k = a + ks)$, $\tau = (t_k = bk)$, and the first equation clearly yields $F_n = s^{n-1}$ $(n \geq 1)$. Our combinatorial structure is the set $\Pi(n)$ of set-partitions of $\{1, 2, \ldots, n\}$. For $P \in \Pi(n)$ we define the weight

$$W(P) = a^{n_1} b^{n_2} s^{n-n_1-2n_2}, \tag{7}$$

where n_1 is the number of singleton blocks and n_2 the number of blocks of P with size greater than or equal to 2. Thus $b(P) = n_1+n_2$, $b(P) = \#$ blocks of P.

Proposition 7.21. *We have*

$$B_n = \sum_{P \in \Pi(n)} W(P). \tag{8}$$

Proof. We have to check the recurrence

$$\sum_{P \in \Pi(n+1)} W(P) = a \sum_{P \in \Pi(n)} W(P) + b \sum_{k=0}^{n-1} \binom{n}{k} s^{n-k-1} \sum_{P \in \Pi(k)} W(P). \tag{9}$$

Let us classify $\Pi(n+1)$ according to the element $n+1$. If $\{n+1\}$ is a singleton block, then it may be combined with any $P \in \Pi(n)$, resulting in the first summand $a \cdot \sum_{P \in \Pi(n)} W(P)$. Suppose $n+1$ is contained in a block of size $n+1-k \geq 2$; hence $0 \leq k \leq n-1$. The remaining k elements (outside the block of $n+1$) may be chosen in $\binom{n}{k}$ ways. If $P' \in \Pi(k)$ is the corresponding partition with $W(P') = a^{k_1} b^{k_2} s^{k-k_1-2k_2}$, then

$$W(P) = a^{k_1} b^{k_2+1} s^{n+1-k_1-2(k_2+1)},$$

and thus

$$W(P) = b s^{n-k-1} W(P').$$

Summing this expression over all $\binom{n}{k}$ partitions $P' \in \Pi(k)$, the second summand results, and the proposition follows. □

Examples.
1. $a = b$, $s = 1$. In this case $W(P) = a^{b(P)}$, implying $B_n = \sum_{k=0}^{n} S_{n,k} a^k$ (Stirling polynomial), which we have already seen in Section 7.3.

2. $a = b = 1$, $s = 0$. The definition of $W(P)$ in (7) shows that $W(P) = 1$ for $n = n_1 + 2n_2$, and $= 0$ otherwise. Hence we count all partitions whose blocks have size 1 or 2, that is, all involutions, giving $B_n = i_n$. Similarly, for $a = 0$, only fixed-point-free involutions are counted, resulting in $B_{2n} = (2n-1)(2n-3)\cdots 3 \cdot 1$, $B_{2n+1} = 0$.

Case B. $u \neq 0$. Our combinatorial object is the set of permutations of $\{1, 2, \ldots, n\}$. Consider a permutation $\pi = \pi_1 \pi_2 \ldots \pi_n$ in word form, and set $\pi_0 = \pi_{n+1} = n + 1$. The weight $w(\pi)$ is defined as

$$w(\pi) = s^{\ell} u^{m}, \tag{10}$$

where $\ell = \#\{i : \pi_{i-1} < \pi_i < \pi_{i+1}\}$ is the number of *double rises*, and $m = \#\{i : \pi_{i-1} < \pi_i > \pi_{i+1}\}$ is the number of *local maxima*. Let $\overline{S}(n)$ be the subset of $S(n)$ that have no *double falls* $\pi_{i-1} > \pi_i > \pi_{i+1}$ (including $\pi_0 > \pi_1 > \pi_2$).

Claim. $F_n = \sum_{\pi \in \overline{S}(n)} w(\pi)$. (11)

Consider $\pi \in \overline{S}(n+1)$. Then if $\pi_{n+1} = n + 1$, the contribution is sF_n (recall that $\pi_{n+2} = n + 2$). On the other hand, if $\pi_{n+1} \neq n + 1$, then we get $u \sum_{k=1}^{n-1} \binom{n}{k} F_k F_{n-k}$ as seen from the picture (you should fill in the details!):

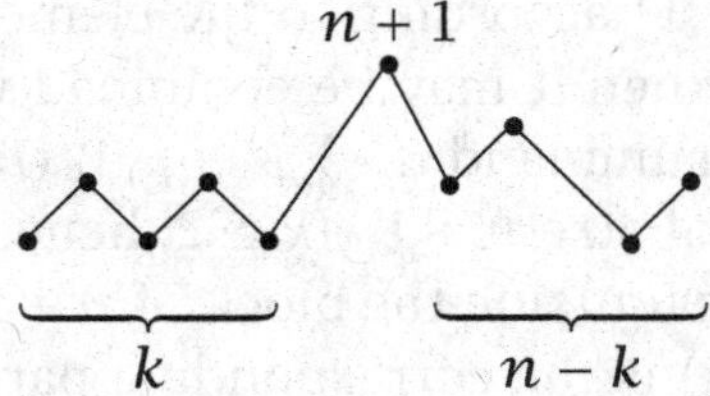

Hence we obtain precisely the first recurrence in (6), and $F_n = \sum_{\pi \in \overline{S}(n)} w(\pi)$ follows.

To compute B_n we use the following refined weight for $\pi \in \overline{S}(n)$,

$$W(\pi) = a^i b^j s^h u^k, \tag{12}$$

with $\pi_0 = 0$, $\pi_{n+1} = n + 1$. The exponents i, j, h, k count the following instances:

i π_m — $\pi_\ell < \pi_m (\forall \ell < m)$

j π_m — $\pi_\ell < \pi_m (\forall \ell < m)$

h π_m — $\exists \ell < m : \pi_\ell > \pi_m$

k π_m — $\exists \ell < m : \pi_\ell > \pi_m$

Proposition 7.22. *We have*

$$B_n = \sum_{\pi \in \overline{S}(n)} W(\pi). \tag{13}$$

The proof of recurrence (6) for $\sum_{\pi \in \overline{S}(n)} W(\pi)$ is analogous, by considering separately the cases $\pi_{n+1} = n+1$ and $\pi_{n+1} \neq n+1$, and is left to the exercises.

Specialization leads again to interesting coefficients appearing as Catalan numbers.

Examples.

1. $a = s = 0$, $b = u = 1$. Looking at the definition (12) of $W(\pi)$, this gives all permutations in $\overline{S}(n)$ without double rises, that is, all *alternating* permutations, starting with a descent (because of $\pi_0 = 0$) and ending with a descent ($\pi_{n+1} = n+1$)

hence $B_{2n+1} = 0$. The associated sequences are $\sigma \equiv 0$, $\tau = (t_k = k^2)$, and the Catalan numbers $B_n = (1,0,1,0,5,0,61,0,\ldots)$ are called the *secant numbers* $B_n = sec_n$ for the following reason.

Look at the differential equation in (13) of the last section:

$$F' = 1 + F^2, \quad B' = BF.$$

One easily computes $F(z) = \tan z$, $\frac{B(z)'}{B(z)} = \tan z$, and hence $B(z) = \sec z = \frac{1}{\cos z}$. We thus get the famous result that in the expansion

$$\frac{1}{\cos z} = \sum_{n \geq 0} sec_n \frac{z^n}{n!},$$

the coefficients sec_n count precisely the alternating permutations of even length starting with a descent. For $n = 4$ we have $sec_4 = 5$ with the permutations

$$2143,\ 3142,\ 3241,\ 4132,\ 4231\,.$$

2. For $s = 2$, $u = 1$, the differential equation for $F(z)$ is $F' = 1 + 2F + F^2$. We have solved this equation already in the last section, $F(z) = \frac{z}{1-z} = \sum_{n\ge 0} z^{n+1} = \sum_{n\ge 1} n!\frac{z^n}{n!}$; hence $F_n = n!$ for $n \ge 1$. Recurrence (6) therefore reads

$$B_{n+1} = aB_n + b\sum_{k=0}^{n-1}\binom{n}{k}B_k(n-k)!\,.$$

Now define the new weight $\overline{W}(\pi)$ for $\pi \in S(n)$ by

$$\overline{W}(\pi) = a^{\ell}b^{m}\,,$$

where ℓ = # fixed points, m = # cycles of length greater than or equal to 2. Then it is an easy matter to show that (see Exercise 7.58)

$$B_n = \sum_{\pi\in S(n)} \overline{W}(\pi)\,. \tag{14}$$

From this we deduce the following examples:

$s = 2$, $u = 1$:

$a = b = 1$	$\sigma = (2k+1)$ $\tau = (k^2)$	$B_n = n!$	all permutations,
$a = 0, b = 1$	$\sigma = (2k)$ $\tau = (k^2)$	$B_n = D_n$	derangements,
$a = b$	$\sigma = (a+2k)$ $\tau = (k(a+k-1))$	$B_n = \sum s_{n,k}a^k$	Stirling polynomial.

3. Suppose $a = s$, $b = 2u$. The differential equations are

$$F' = 1 + sF + uF^2\,,$$
$$B' = sB + 2uBF\,.$$

Differentiating the first equation, we get $F'' = sF' + 2uFF'$, from which $B = F'$ results because of $F_1 = B_0 = 1$, and this means that $B_n = F_{n+1}$. In particular, for $a = s = 0$, $u = 1$, we get $F' = 1 + F^2$, $F(z) = \tan z$, and thus $B(z) = (\tan z)' = \frac{1}{\cos^2 z}$. Looking at (12), we find that $B_n = F_{n+1}$ counts all alternating permutations in $S(n+1)$ starting with a rise ($\pi_0 = n+1$) and ending with a fall ($\pi_{n+2} = n+2$):

The numbers F_n appearing in the expansion of $\tan z$ are called the *tangent numbers* tan_n, with small values (0, 1, 0, 2, 0, 16, 0, 272, 0, ...), and we have

$$B(z) = \frac{1}{\cos^2 z} = \sum_{n\ge 0} tan_{n+1} \frac{z^n}{n!}.$$

For $n = 2$ we obtain the two permutations 132, 231, and for $n = 4$ the 16 alternating permutations

13254	14253	14352	15243	15341
23154	24153	24351	25143	25341
34152	34251	35142	35241	
45132	45231.			

Exercises

7.52 Verify formula (3) for $B(z)$.

7.53 Compute the ordinary generating function $B(z)$ corresponding to $a = s - 1$, $b = u$. Specialize to the Schröder and Riordan numbers.

▷ **7.54** Prove the recurrence for the Schröder numbers in the text, and the identity $Sch_n = \sum_{k\ge 0} \binom{2n-k}{k} C_{n-k}$, C_k = Catalan. Show $Sch_n \equiv 2 \pmod 4$ for $n \ge 1$.

7.55 Prove Lemma 7.19.

7.56 Establish for the central trinomial numbers the formula $Tr_n = \sum_{k=0}^{n} \binom{2k}{k}\binom{n}{2k}$.

7.57 Show the equality $B_n = \sum_{\pi \in \overline{S}(n)} W(\pi)$ in (13).

▷ **7.58** Check the equality $B_n = \sum_{\pi \in S(n)} \overline{W}(\pi)$ in (14).

7.59 Suppose $B(z) = \sum_{n\ge 0} B_n z^n$ for the sequences $\sigma = (0, s)$, $\tau \equiv 1$, and $\widetilde{B}(z) = \sum_{n\ge 0} \widetilde{B}_n z^n$ for $\sigma = (s, s)$, $\tau \equiv 1$. Prove that $\widetilde{B}_n = B_n + sB_{n+1}$, and deduce $M_n = R_n + R_{n+1}$, $C_n = Fi_{n-1} + 2Fi_n$ $(n \ge 1)$.

▷ **7.60** Consider the ballot numbers $b_{n,k}$ of Exercises 7.14 ff. We have proved $B_k(z) = z^k C(z)^{k+1}$. Use this and the previous exercise to show that $Fi_{n+1} = \sum_{k \text{ odd}} b_{n,k}$.

* * *

7.61 Let $\tau \equiv 1$, and consider the sequences $\sigma = (s,s)$, $\sigma = (s+1,s)$, $\sigma = (s-1,s)$. Prove the recurrences

$$B_{n+1}^{(s,s)} = sB_n^{(s,s)} + \sum_{k=0}^{n-1} B_k^{(s,s)} B_{n-1-k}^{(s,s)},$$

$$B_{n+1}^{(s+1,s)} = (s+2)^{n+1} - \sum_{k=0}^{n} B_k^{(s+1,s)} B_{n-k}^{(s+1,s)},$$

$$B_{n+1}^{(s-1,s)} = (s-2)^{n+1} + \sum_{k=0}^{n} B_k^{(s-1,s)} B_{n-k}^{(s-1,s)}.$$

Apply these recurrences to known sequences.

▷ **7.62** Let $B^{(s,s)}(z)$, $B^{(s+1,s)}(z)$, and $B^{(s-1,s)}(z)$ be the series as in the previous exercise, and show that

$$B^{(s+1,s)}(z) = \frac{1}{1-(s+2)z}\left(1 - zB^{(s,s)}(z)\right),$$

$$B^{(s-1,s)}(z) = \frac{1}{1-(s-2)z}\left(1 + zB^{(s,s)}(z)\right).$$

Verify the examples: $\binom{n}{\lfloor n/2 \rfloor} = 2^n - \sum_{k=0}^{\lfloor (n-1)/2 \rfloor} 2^{n-1-2k} C_k$, $R_n = M_{n-1} - M_{n-2} \pm \cdots + (-1)^n M_1$, R_n = Riordan, M_n = Motzkin.

7.63 Use the previous exercise to deduce an identity relating the Catalan and Motzkin numbers: $\sum_{k=1}^{n} \left((-1)^k \binom{n}{k} C_k + M_{k-1}\right) 3^{n-k} = 0 \ (n \ge 1)$.

7.64 Consider the case $\sigma = (a,s)$, $\tau = (b,u)$ with $a = s$, $b = 2u$, treated in the text, with the trinomials, $\binom{2n}{n}$, and the Delannoy numbers as special instances. Prove the general recurrence $(n+1)B_{n+1} = s(2n+1)B_n + n(2b-s^2)B_{n-1}$ for the corresponding Catalan numbers, and deduce the formula $B_n = \frac{1}{2^n} \sum_{k \ge 0} \binom{2k}{k} \binom{k}{n-k} s^{2k-n} (2b-s^2)^{n-k}$. Hint: Proceed as for the Delannoy numbers in Section 3.2.

▷ **7.65** Consider $\sigma \equiv s$, $\tau \equiv u$, and prove the following recurrence for the Catalan numbers: $(n+2)B_n = s(2n+1)B_{n-1} + (4u-s^2)(n-1)B_{n-2}$ $(n \ge 1)$. Deduce that the sequence (M_n) of Motzkin numbers is logarithmically convex, that is, $M_n^2 \le M_{n-1}M_{n+1}$ $(n \ge 1)$. What is $\lim_{n\to\infty} \frac{M_n}{M_{n-1}}$? Hint: Consider $(zB(z))'$.

7.66 Let $\sigma = (\alpha, \alpha+\beta)$, $\tau = (\alpha\beta, \alpha\beta)$ for $\alpha, \beta \ne 0$. Using Exercises 7.28 and 7.4 show that $B_{n+1} = \alpha B_n + \beta \sum_{k=0}^{n-1} B_k B_{n-k}$, and derive $B_n = \sum_{k=0}^{n} N(n,k)\alpha^{n-k}\beta^k$, $N(n,k)$ = Narayana number.

7.67 Let $\sigma = (s_k)$, $\tau \equiv 1$, and $A = A^{(\sigma)}$ the Catalan matrix. The *sum matrix* $S^{(\sigma)} = (t_{n,k})$ is defined by $t_{n,k} = \sum_{i \ge k} a_{n,i}$, thus $S^{(\sigma)} = A^{(\sigma)} J$

where J is the lower triangular matrix of all 1's. Prove that $S^{(\sigma+1)} = PS^{(\sigma)}$, where P is the Pascal matrix.

7.68 Continuing the exercise, consider the case $\sigma = (a,s)$. Prove for $t_{n,k}$ the recurrence

$$t_{n,k} = t_{n-1,k-1} + st_{n-1,k} + t_{n-1,k+1} \quad (n \geq 1),$$
$$t_{n,0} = (a+1)t_{n-1,0} + (s-a+1)t_{n-1,1}.$$

Derive from this the following formulas in the ordinary calculus:

$$S^{(s,s)}(z) = B^{(s+1,s)}(z), \quad S^{(s+1,s)}(z) = \frac{1}{1-(s+2)z},$$
$$S^{(s-1,s)}(z) = \frac{1}{\sqrt{1-2sz+(s^2-4)z^2}}.$$

Use these formulas to prove $S_n^{(-1,0)} = \binom{n}{n/2}[n \text{ even}]$, $S_n^{(0,1)} = \sum_k \binom{n}{k}\binom{n-k}{k} = Tr_n$, $S_n^{(1,2)} = \binom{2n}{n}$, $S_n^{(2,3)} = \sum_k \binom{n}{k}\binom{2k}{k}$.

7.69 Continuing the exercise prove $S^{(-s-1,-s)}(z) = S^{(s-1,s)}(-z)$, and deduce $S_n^{(s-1,s)} = \sum_k (-1)^k \binom{n}{k}(2s)^{n-k} S_k^{(s-1,s)}$. Use this to establish the identities $\binom{2n}{n} = \sum_k (-1)^k \binom{n}{k} 4^{n-k}\binom{2k}{k}$, $\sum_{k=0}^n \binom{n}{k}\binom{2k}{k} = \sum_k (-1)^k \binom{n}{k} 6^{n-k} \sum_i \binom{k}{i}\binom{2i}{i}$, and also $D_{n,n} = \sum_k (-1)^k \binom{n}{k} 6^{n-k} D_{k,k}$, $D_{n,n}$ = central Delannoy.

▷ **7.70** A *lattice walk* in $(d+1)$-dimensional space $\mathbb{Z}^{d+1}$ is a path that starts at the origin $(0,\ldots,0)$, uses steps $(0,\ldots,\pm \overset{\downarrow i}{1},0,\ldots,0)$, $i = 1,\ldots,d+1$, and satisfies $x_{d+1} \geq 0$. Prove that the number of these walks of length n is $S_n = \sum_{k=0}^n \binom{n}{k}(2d-2)^{n-k}\binom{2k+1}{k} = \sum_{k=0}^n \binom{n}{k}(2d)^{n-k}\binom{k}{\lfloor k/2 \rfloor}$. Hint: Let $a_{n,k}$ be the number of walks of length n that end in a point with $x_{d+1} = k$. Prove a recurrence for $a_{n,k}$ and note that $S_n = \sum_k a_{n,k}$. What do you get for $d = 2$?

7.71 Give the following interpretation of the Catalan numbers B_n corresponding to $\sigma = (a,s)$, $\tau = (b,u)$. A *Motzkin tree* is a rooted plane tree, in which every vertex has out-degree less than or equal to 2. Thus for $n = 4$ we have the following four trees:

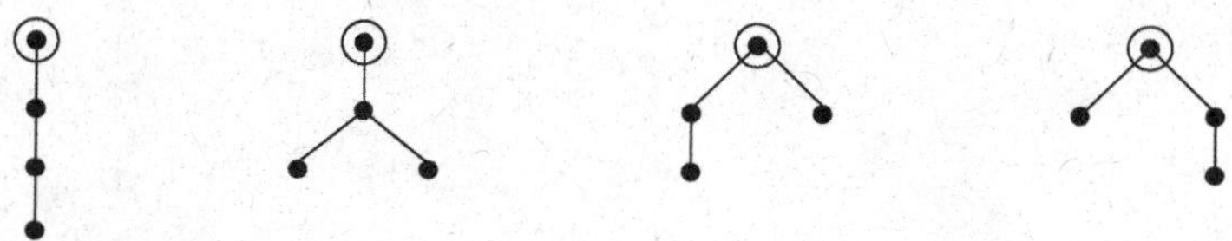

Let $\mathcal{T}_n$ be the set of Motzkin trees with n vertices. For $T \in \mathcal{T}_n$ let $W(T) = a^i b^j s^h u^k$, where i = #{vertices of out-degree 1 on the leftmost branch},

$j = \#\{$vertices of out-degree 2 on the leftmost branch$\}$, $h = \#\{$vertices of out-degree 1 not on the leftmost branch$\}$, $k = \#\{$vertices of out-degree 2 not on the leftmost branch$\}$. Prove $B_n = \sum_{T \in \mathcal{T}_n} W(T)$. Consider $\sigma \equiv 0$, $\tau \equiv 1$.

7.72 Consider the exponential case with equations (13), (14) of the last section. Suppose $s = 2m$, $u = m^2$. Prove that $F_n = n!m^{n-1}$, $B_{n+1} = (mn+a)B_n + (b-am)nB_{n-1}$. Consider the case $m = 1$, and in particular $a = b = 2$, reestablishing results from the text.

▷ **7.73** Consider again exponential series with $a = b = u$, $s = u + 1$. Show that $B_n = uF_n$ $(n \geq 1)$, and deduce $B_n = \sum_{k=0}^{n} A_{n,k} u^k$, where $A_{n,k}$ are the Eulerian numbers (see Exercise 1.48). Hint: Look at the differential equations and use a proper weighting.

Highlight: Chord Diagrams

An old problem asks for the number of ways in which $2n$ points on a circle can be joined in pairs such that the corresponding chords do not intersect. The easy answer is the (ordinary) Catalan number C_n; we will see that in a moment. But consider now the following more general and much more difficult question: How many of these chord diagrams on $2n$ points have exactly k pairs of crossing chords? Let us denote this number by $t_{n,k}$, and by $T_n(q) = \sum_{k\geq 0} t_{n,k}q^k$ the generating polynomial.

This problem has been studied under various names, e.g., "the folding stamp problem." The solution we present goes essentially back to Jacques Touchard and John Riordan and combines in a clever way several ideas from this and earlier chapters.

The result cited above says that $T_n(0) = t_{n,0} = C_n$, and it is clear that there is a unique diagram with the maximum number $\binom{n}{2}$ of crossing pairs; thus $t_{n,\binom{n}{2}} = 1$.

Furthermore, the total number $T_n(1)$ of chord diagrams on $2n$ points is clearly $(2n-1)(2n-3)\cdots 3\cdot 1$. As small examples we have

$$T_0(q) = T_1(q) = 1, \quad T_2(q) = 2 + q, \quad T_3(q) = 5 + 6q + 3q^2 + q^3 .$$

The figure shows the 15 chord diagrams for $n = 3$:

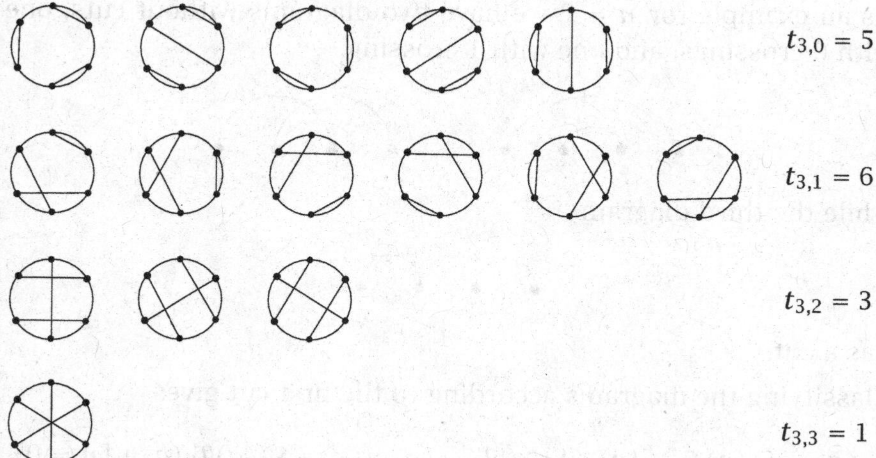

$t_{3,0} = 5$

$t_{3,1} = 6$

$t_{3,2} = 3$

$t_{3,3} = 1$

Linearizing the Problem.
For our purposes it is convenient to consider the equivalent problem, when the points are arranged on a line, and the chords correspond to *arcs* above the line. The figure shows this linearization:

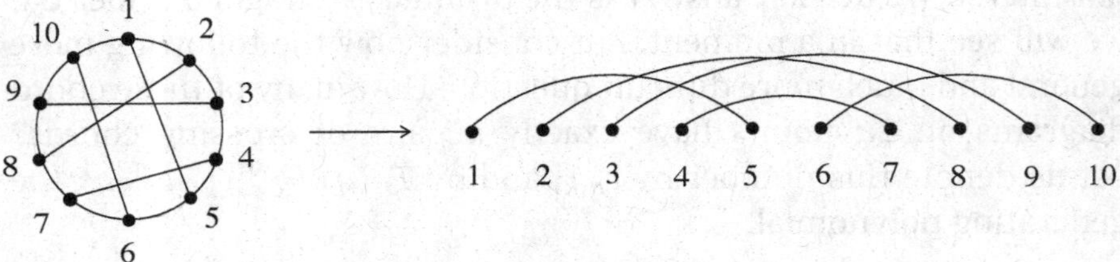

From now on we will consider this latter version. A *cut* in the diagram is a vertical line between points $2k$ and $2k+1$ that meets none of the arcs. In other words, the first k arcs occupy the points 1 to $2k$, and the remaining $n-k$ arcs the right-hand part, as in the figure for $n = 10$, $k = 3$:

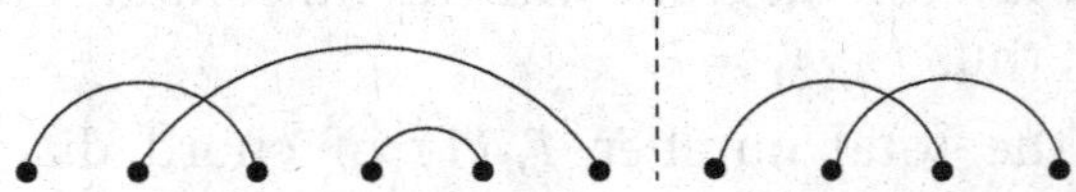

Let $S_n(q)$ be the generating polynomial of diagrams *without* cuts. Thus

$$S_0(q) = S_1(q) = 1, \quad S_2(q) = 1 + q, \quad S_3(q) = 2 + 4q + 3q^2 + q^3 .$$

As an example for $n = 2$ we have two diagrams without cuts, one with 0 crossings, and one with 1 crossing:

while the third diagram

has a cut.

Classifying the diagrams according to the first cut gives

$$T_n(q) = S_1(q)T_{n-1}(q) + S_2(q)T_{n-2}(q) + \cdots + S_n(q)T_0(q) + [n = 0],$$

and thus

$$2T_n(q) = \sum_{i=0}^{n} S_i(q)T_{n-i}(q) + [n = 0].$$

For the generating functions $T(q,z) = \sum_{n\ge0} T_n(q)z^n$, $S(q,z) = \sum_{n\ge0} S_n(q)z^n$ this means

$$2T(q,z) = T(q,z)S(q,z) + 1,$$

or

$$T(q,z) = \frac{1}{2 - S(q,z)}. \qquad (1)$$

Equation (1) reduces the problem to diagrams without cuts, and these are much easier to handle.

A Coding for Diagrams.
Let D be an arbitrary diagram, and suppose that the k-th arc starts at position p_k; thus $k \le p_k \le 2k-1$. Set $w_k = 2k - p_k$; then $w_1 = 1$, $1 \le w_k \le k$, and $p_k = 2k - w_k < p_{k+1} = 2k + 2 - w_{k+1}$ implies $w_{k+1} \le w_k + 1$ $(1 \le k \le n-1)$.

Consider $k \ge 2$. If D has a cut between $2k-2$ and $2k-1$, then $p_k = 2k-1$ and thus $w_k = 1$. Conversely, if $w_k = 1$ for $k \ge 2$, then $p_k = 2k-1$, which implies that there is a cut between $2k-2$ and $2k-1$. In sum, diagrams *without* cuts correspond to code words $w_1w_2\dots w_n$ with

$$w_1 = 1,\quad 2 \le w_{k+1} \le w_k + 1 \quad (k \ge 1). \qquad (2)$$

Call words observing (2) *admissible* codewords, and denote by $\mathcal{W}_n$ the set of admissible code words. It is easily verified that $|\mathcal{W}_n| = C_{n-1}$, Catalan number.

Example. The set $\mathcal{W}_4$ consists of the words

$$1222,\ 1223,\ 1232,\ 1233,\ 1234.$$

The coding $w_1w_2\dots w_n$ is the crucial step toward a description of the polynomial $S_n(q)$. Consider the q-integer $[n]_q = 1 + q + \cdots + q^{n-1}$. For ease of notation we set $a_i = 1 + q + \cdots + q^{i-1}$ $(i \ge 1)$.

Claim 1. $S_n(q) = \sum_{w \in \mathcal{W}_n} a_{w_1} a_{w_2} \cdots a_{w_n}$.

For the proof consider all diagrams without cuts belonging to the word $w_1 w_2 \dots w_n$. We want to show that their contribution to $S_n(q)$ is precisely $a_{w_1} a_{w_2} \cdots a_{w_n}$. Look at the n-th arc starting at $p_n = 2n - w_n$:

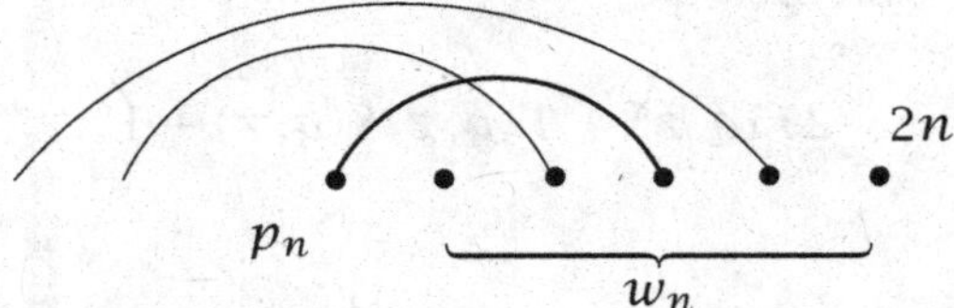

The n-th arc crosses $0, 1, 2, \dots,$ or $w_n - 1$ other arcs, yielding the contribution $1 + q + q^2 + \cdots + q^{w_n - 1} = a_{w_n}$. The result follows then by induction.

The coding gives us even more. Let

$$S_n^{(2)}(q) = \sum_{w \in \mathcal{W}_n} a_{w_1+1} a_{w_2+1} \cdots a_{w_n+1},$$

that is, all indices are increased by 1. As examples, we have $S_1(q) = a_1$, $S_2(q) = a_1 a_2$, and thus $S_1^{(2)}(q) = a_2$, $S_2^{(2)}(q) = a_2 a_3$.

Claim 2. $S_n(q) = S_{n-1}(q) S_1^{(2)}(q) + S_{n-2}(q) S_2^{(2)}(q) + \cdots + S_1(q) S_{n-1}^{(2)}(q)$ *for* $n \geq 2$.

We classify the words $w \in \mathcal{W}_n$ according to the *last* occurrence of 2. Suppose 2 appears last in position $k+1$ $(1 \leq k \leq n-1)$. Then we have the following situation:

$$\overbrace{1 \dots\dots}^{k} \quad \overbrace{2 \dots\dots}^{n-k}$$

$$S_k(q) \qquad S_{n-k}^{(2)}(q)$$

The first part accounts for $S_k(q)$, and the second part for $S_{n-k}^{(2)}(q)$, since condition (2) on the admissible words remains the same.

By the same argument we have the following general result. For $k \geq 1$, let

$$S_n^{(k)}(q) = \sum_{w \in \mathcal{W}_n} a_{w_1+k-1} a_{w_2+k-1} \cdots a_{w_n+k-1},$$

where $S_n^{(1)}(q) = S_n(q)$. Then

$$\begin{cases} S_n^{(k)}(q) = \sum_{i=1}^{n-1} S_i^{(k)}(q) S_{n-i}^{(k+1)}(q) (n \geq 2), \\ S_1^{(k)}(q) = a_k . \end{cases} \tag{3}$$

Continued Fractions.

With (3) in hand we are all set for the continued fractions approach of Section 7.3. For $k \geq 1$ set

$$F_k(q,z) = \sum_{n\geq 0} S_{n+1}^{(k)}(q) z^n ,$$

so in particular,

$$F_1(q,z) = \sum_{n\geq 0} S_{n+1}(q) z^n = \frac{S(q,z)-1}{z} .$$

This implies $1 - zF_1(q,z) = 2 - S(q,z)$, and we obtain by (1),

$$T(q,z) = \frac{1}{1 - zF_1(q,z)} . \tag{4}$$

Furthermore, by (3),

$$F_k(q,z) = zF_k(q,z)F_{k+1}(q,z) + a_k . \tag{5}$$

Finally, with $C_0(q,z) = T(q,z)$, $C_k(q,z) = \frac{F_k(q,z)}{a_k}$ $(k \geq 1)$, (4) and (5) translate into

$$C_k(q,z) = \frac{1}{1 - a_{k+1} z C_{k+1}(q,z)} \quad (k \geq 0) . \tag{6}$$

Now if we make the substitution $z \mapsto z^2$, then Lemma 7.9 yields the beautiful result that $T_n(q)$ is the Catalan number B_{2n} corresponding to the sequences $\sigma \equiv 0$, $\tau = (t_n = 1 + q + \cdots + q^{n-1})$.

Example. If we set $q = 0$, then $T_n(0) = t_{n,0}$ is the Catalan number B_{2n} belonging to $\sigma \equiv 0$, $\tau \equiv 1$. But this was our starting example $B_{2n} = C_n$, which proves $t_{n,0} = C_n$, as promised. For $q = 1$, the sequences $\sigma \equiv 0$, $\tau = (t_n = n)$ result, with $B_{2n} = T_n(1) = (2n-1)(2n-3)\cdots 3 \cdot 1$ (see the examples in Section 7.4).

The Final Step.

Now that we know that the polynomials $T_n(q)$ are Catalan numbers, we may use recurrence (4) in Section 7.3 to compute them. This is no easy task, but a simple trick will help.

Instead of $\sigma \equiv 0$, $\tau = (t_n = a_n)$, consider the sequences $\hat{\sigma} \equiv 0$, $\hat{\tau} = (\hat{t}_n = (1-q)a_n = 1-q^n)$. Exercise 7.6 shows that the associated Catalan numbers are then $\hat{B}_{2n} = T_n(q)(1-q)^n$, $\hat{B}_{2n+1} = 0$. The recurrence (4) in Section 7.3 now reads

$$\begin{cases} A_k = zA_{k-1} + z(1-q^{k+1})A_{k+1} \quad (k \ge 1)\,, \\ \hat{B} = A_0 = 1 + z(1-q)A_1\,, \end{cases} \tag{7}$$

where $A_k = A_k(q,z)$ is as usual the generating function of the k-th column. Setting $U_k(q,z) = \frac{A_k(q,z)}{z^k}$, (7) transforms into

$$\begin{cases} U_k = U_{k-1} + (1-q^{k+1})z^2U_{k+1} \quad (k \ge 1)\,, \\ \hat{B} = U_0 = 1 + (1-q)z^2U_1\,. \end{cases} \tag{8}$$

Every function U_k is even, since $\hat{\sigma} \equiv 0$. Setting $V_k(q,z^2) = U_k(q,z)$, the system (8) becomes

$$\begin{cases} V_k = V_{k-1} + (1-q^{k+1})zV_{k+1}\,, \\ V_0 = 1 + (1-q)zV_1\,, \end{cases} \tag{9}$$

and this is the system we want to solve, where $[z^n]V_0(q,z) = T_n(q)(1-q)^n$.

Set $V_k(q,z) = \sum_{n\ge 0} g_{k,n}z^n$, then (9) reads

$$\begin{cases} g_{k,n} = g_{k-1,n} + (1-q^{k+1})g_{k+1,n-1}\,, \\ g_{0,n} = [n=0] + (1-q)g_{1,n-1}\,. \end{cases} \tag{10}$$

We are going to express $g_{k,n}$ in terms of the ballot numbers $b_{n,i}$ considered in Exercises 7.14 ff, and the numbers $\phi_{n,i} = q^{\binom{i+1}{2}}\left[{n \atop i}\right]_q$. First we recall

$$b_{n,i} = b_{n-1,i-1} + b_{n,i+1}\,, \tag{11}$$

and Exercise 1.75 with the recurrence

$$\phi_{k+i,i} = \phi_{k+i-1,i-1} + \phi_{k+i-1,i} + (q^{k+1}-1)\phi_{k+i,i-1}\,. \tag{12}$$

Using (11) and (12) it is an easy matter to verify the formula

$$g_{k,n} = \sum_{i=0}^{n} (-1)^i b_{k+n+i,k+2i}\phi_{k+i,i}\,.$$

For $k = 0$ this gives

$$T_n(q)(1-q)^n = \sum_{i=0}^{n} (-1)^i b_{n+i,2i} \phi_{i,i} .$$

Now, we know from Exercise 7.15 that

$$b_{n+i,2i} = \frac{2i+1}{n+i+1}\binom{2n}{n-i} = \binom{2n}{n-i} - \binom{2n}{n-i-1} . \tag{13}$$

Since $\phi_{i,i} = q^{\binom{i+1}{2}}$, we have thus found a compact expression for $T_n(q)$:

$$T_n(q)(1-q)^n = \sum_{i=0}^{n} (-1)^i \left[\binom{2n}{n-i} - \binom{2n}{n-i-1}\right] q^{\binom{i+1}{2}} . \tag{14}$$

Example. Formula (14) shows the surprising fact that

$$\sum_{k=0}^{n} (-1)^k \binom{n}{k} t_{n,m-k} = 0 \text{ for } m \neq \binom{i+1}{2},$$

while for $m = \binom{i+1}{2}$,

$$\sum_{k=0}^{n} (-1)^k \binom{n}{k} t_{n,\binom{i+1}{2}-k} = (-1)^i \left[\binom{2n}{n-i} - \binom{2n}{n-i-1}\right] \tag{15}$$

results. For $m = 1$ this gives

$$t_{n,1} - n t_{n,0} = \binom{2n}{n-2} + \binom{2n}{n-1},$$

and hence with $t_{n,0} = C_n = \frac{1}{n+1}\binom{2n}{n}$,

$$t_{n,1} = \binom{2n}{n-2} .$$

Similarly, (15) yields for $m = 2$,

$$t_{n,2} = \frac{n+3}{2}\binom{2n}{n-3} .$$

Finally, moving $(1-q)^n$ in (14) to the right-hand side, a summation formula for $t_{n,k}$ results,

$$t_{n,k} = \sum_{i=0}^{n} (-1)^i \frac{2i+1}{n+i+1}\binom{2n}{n-i}\binom{n+k-\binom{i+1}{2}-1}{n-1},$$

and this is the most explicit expression one could hope for.

Notes and References

The subject of orthogonal polynomials is a classical topic in analysis that attracted some of the greatest names from Legendre to Chebyshev and Jacobi. The standard references are the books by Szegő and Chihara. Our presentation of the combinatorial part of the theory follows the seminal papers by Viennot, emphasizing lattice paths, and by Flajolet, who takes a continued fractions point of view; see also the survey by Aigner. The important Theorem 7.4 is contained in the paper of Favard. A rigorous treatment from an operator standpoint was advanced in the 1970s by Rota under the name "umbral calculus." The book of Roman gives a nice overview for this approach. Yet another equivalent setup, called the Riordan group, was suggested by Shapiro et al.

1. M. Aigner (2001): Catalan and other numbers—a recurrent theme. In: *Algebraic Combinatorics and Computer Science*, Crapo and Senato, eds., 347–390. Springer, Berlin.
2. T.S. Chihara (1978): *An Introduction to Orthogonal Polynomials.* Gordon & Breach, New York.
3. J. Favard (1935): Sur les polynômes de Tchebicheff. *C.R. Acad. Sci. Paris* 200, 2052–2053.
4. P. Flajolet (1980): Combinatorial aspects of continued fractions. *Discrete Math.* 32, 125–161.
5. J. Riordan (1975): The distribution of crossings of chords joining pairs of $2n$ points on a circle. *Math. Computation* 29, 215–222.
6. S. Roman (1984): *The Umbral Calculus.* Academic Press, Orlando.
7. G.-C. Rota (1975): *Finite Operator Calculus.* Academic Press, New York.
8. L.W. Shapiro, S. Getu, W.-J. Woan, and L.C. Woodson (1991): The Riordan group. *Discrete Appl. Math.* 34, 229–239.
9. G. Szegő (1967): Orthogonal Polynomials, 2nd edition. *Amer. Math. Soc. Coll. Publ.*, vol. 23. Amer. Math. Soc., New York.
10. J. Touchard (1952): Sur une problème de configurations et sur les fractions continues. *Canad. J. Math.* 4, 2-25.
11. G. Viennot (1984): *Une théorie combinatoire des polynômes orthogonaux.* Lecture notes, Univ. Quebec.

8 Symmetric Functions

The theory of symmetric functions provides an elegant algebraic framework for many enumeration problems, in particular, as we shall see, for plane partitions. As with generating functions they encode a great deal of information, and algebraic manipulations often provide stupendously simple proofs of seemingly difficult problems.

8.1 Symmetric Polynomials and Functions

Let us start with symmetric polynomials f in n variables.

Definition. A polynomial $f(x_1,\dots,x_n)$ over $\mathbb{C}$ (or any field of characteristic 0) is called *symmetric* if

$$f(x_{\sigma(1)},\dots,x_{\sigma(n)}) = f(x_1,\dots,x_n)$$

holds for all permutations $\sigma \in S(n)$. A polynomial $f(x_1,\dots,x_n)$ is said to be *alternating* if $f(x_{\sigma(1)},\dots,x_{\sigma(n)}) = -f(x_1,\dots,x_n)$ for all *transpositions* $\sigma \in S(n)$. The polynomial f has *degree* d if d is the highest degree of the monomials $cx_1^{i_1}\cdots x_n^{i_n}$, $d = \sum_{j=1}^{n} i_j$, appearing in f. We call f *homogeneous* of degree d if all monomials have degree d.

Since any $\sigma \in S(n)$ is a product of transpositions, an equivalent formulation for alternating is that $f(x_{\sigma(1)},\dots,x_{\sigma(n)}) = -f(x_1,\dots,x_n)$ for all odd permutations.

Example. The classical example is the *elementary symmetric* polynomials, which we have already encountered in Chapter 6. Let $f(x) = \sum_{k=0}^{n} a_k x^{n-k}$ be a polynomial with roots $x_1,\dots,x_n$, and leading coefficient 1. We have $f(x) = (x-x_1)(x-x_2)\cdots(x-x_n)$, and thus for the coefficients

$$\begin{aligned}
a_1 &= -(x_1 + \cdots + x_n)\,, \\
a_2 &= x_1x_2 + x_1x_3 + \cdots + x_{n-1}x_n\,, \\
&\cdots \\
a_k &= (-1)^k \sum_{i_1<\cdots<i_k} x_{i_1}\cdots x_{i_k}\,, \\
&\cdots \\
a_n &= (-1)^n x_1x_2\cdots x_n\,.
\end{aligned}$$

Definition. Let $X = \{x_1, x_2, \ldots, x_n\}$. The *k-th elementary symmetric polynomial* over X is

$$\begin{aligned}
e_k(x_1, x_2, \ldots, x_n) &= \sum_{i_1<\cdots<i_k} x_{i_1}x_{i_2}\cdots x_{i_k} \quad (k \geq 1)\,, \\
e_0(x_1, x_2, \ldots, x_n) &= 1\,.
\end{aligned} \tag{1}$$

The polynomial e_k is clearly a homogeneous symmetric polynomial of degree k. Note that $e_k(x_1, \ldots, x_n) = 0$ for $k > n$.

The best-known alternating function is the determinant of a matrix, for example, over the columns. An example is the product $\prod_{1\leq i<j\leq n}(x_i - x_j)$, which appeared as the determinant of the Vandermonde matrix $(x_j^{n-i})_{i,j=1}^n$.

Proposition 8.1. *Let $f(x_1, \ldots, x_n)$ be an alternating polynomial of degree d. Then*

$$\frac{f(x_1, \ldots, x_n)}{\prod\limits_{1\leq i<j\leq n} (x_i - x_j)} \tag{2}$$

is a symmetric polynomial of degree $d - \binom{n}{2}$.

Proof. Since both numerator and denominator are alternating, the quotient is symmetric, and the degree is clearly $d - \binom{n}{2}$. So all we have to show is that the quotient in (2) is a *polynomial.*

Consider $f(x_1, \ldots, x_n)$ as a polynomial in x_1 with polynomial coefficients in $x_2, \ldots, x_n$, $f(x_1, \ldots, x_n) = \sum_{k\geq 0} f_k(x_2, \ldots, x_n)x_1^k$. Since exchange of the first two variables switches the sign, we obtain

$$f(x_2, x_2, x_3, \ldots, x_n) = -f(x_2, x_2, x_3, \ldots, x_n)\,,$$

that is, $f(x_2, x_2, \dots, x_n) = 0$. This means that x_2 is a root of the polynomial f in the variable x_1. Similarly, $x_3, \dots, x_n$ are roots, and we obtain

$$f(x_1, \dots, x_n) = g(x_1, \dots, x_n) \prod_{j=2}^{n} (x_1 - x_j), \tag{3}$$

where $g(x_1, \dots, x_n)$ is a polynomial. For $n = 2$, this gives

$$g(x_1, x_2) = \frac{f(x_1, x_2)}{x_1 - x_2},$$

and the quotient is thus a polynomial. Now we use induction on n. Regard the polynomial $g(x_1, \dots, x_n)$ in (3) as a polynomial in $x_2, \dots, x_n$ with polynomial coefficients in x_1. The equality in (3) implies that g is alternating in the variables $x_2, \dots, x_n$, and so by induction

$$\frac{g(x_1, \dots, x_n)}{\prod\limits_{2 \le i < j \le n} (x_i - x_j)} = \frac{f(x_1, \dots, x_n)}{\prod\limits_{1 \le i < j \le n} (x_i - x_j)}$$

is a symmetric polynomial. $\square$

Example. We can now quickly compute the determinant of the Vandermonde matrix

$$D = \det \begin{pmatrix} x_1^{n-1} & \dots & x_n^{n-1} \\ x_1^{n-2} & \dots & x_n^{n-2} \\ & \dots & \\ 1 & \dots & 1 \end{pmatrix}.$$

Since D is an alternating polynomial in $x_1, \dots, x_n$ of degree $0 + 1 + \dots + (n-1) = \binom{n}{2}$, we have that

$$\frac{D}{\prod\limits_{1 \le i < j \le n} (x_i - x_j)}$$

is a symmetric polynomial of degree 0, that is, a constant c. Comparison of the coefficients for $x_1^{n-1} x_2^{n-2} \cdots x_n^0$ yields $c = 1$, and we conclude that $D = \prod_{1 \le i < j \le n} (x_i - x_j)$.

When we pass from a finite set of variables to a possibly countable set we arrive at the general definition of a symmetric function. Let

$X = \{x_1, x_2, \ldots\}$ be a finite or countable set. A *monomial* over X is any expression $c_\alpha x^\alpha = c_\alpha x_1^{\alpha_1} x_2^{\alpha_2} \cdots$, where c_α is a constant, $\alpha_i \ge 0$ for all i, and all but finitely many α_i are 0. The *degree* of the monomial $c_\alpha x^\alpha$ is $\sum_{i\ge1} \alpha_i$.

Definition. Let $X = \{x_1, x_2, \ldots\}$ be a finite or countable set of variables. The function $f(x_1, x_2, \ldots) = \sum_\alpha c_\alpha x^\alpha$ is a *symmetric function* over x if

a. $f(x_{\sigma(1)}, x_{\sigma(2)}, \ldots) = f(x_1, x_2, \ldots)$ over all permutations of the index set,

b. the monomials appearing in f (that is, $c_\alpha \neq 0$) have bounded degree.

The largest degree that appears is called the *degree* of f.

As with generating functions, we regard symmetric functions in a formal way, meaning that $f = g$ if and only if corresponding coefficients for α agree. The following is clear from the definition: Consider $\alpha = (\alpha_1, \alpha_2, \ldots)$. Then $c_\alpha = c_\beta$ where β runs through all *distinct* permutations of α. For example, $\alpha = (1,3,1,0,0,0,\ldots)$ gives rise to $(3,1,0,1,0,\ldots)$, $(1,1,0,0,3,\ldots)$, etc., and the corresponding coefficients must be the same.

As before, we call f *homogeneous* of degree d, if all nonzero monomials of f have degree d. Let us denote by $\Lambda(X)$ the set of symmetric functions over X, and by $\Lambda^m(X)$ the set of homogeneous symmetric functions of degree m. Thus $\Lambda^0(X)$ are just the constants. Clearly, $\Lambda(X)$ and $\Lambda^m(X)$ are vector spaces, and the definition implies that any $f \in \Lambda(X)$ of degree m can be uniquely written as

$$f = f_0 + f_1 + \cdots + f_m, \quad f_i \in \Lambda^i(X).$$

In other words, $\Lambda(X)$ is the direct sum of the vector spaces $\Lambda^m(X)$. Furthermore, it is clear that for $f \in \Lambda^m(X)$, $g \in \Lambda^n(X)$ the product fg is in $\Lambda^{m+n}(X)$. Hence we are mostly interested in homogeneous symmetric functions, and to those we turn in the next section.

Exercises

8.1 Let $x_1, x_2, \ldots, x_n$ be variables. Show that

$$\det\left(x_i^{n-j}\right)_{i,j=1}^{n} = \prod_{1\le i<j\le n}(x_i - x_j),$$

that is, equal to the Vandermonde determinant. What about $\det\left(x_i^{\overline{n-j}}\right)$?

8.2 Compute the determinant of the following matrices:
a. $(x_j^{n-i} - x_j^{2n-i})_{i,j=1}^{n}$, b. $(x_j^{n-i} - x_j^{2n+1-i})_{i,j=1}^{n}$, c. $(x_j^{n-i} + x_j^{2n-1-i})_{i,j=1}^{n}$.

▷ **8.3** Find the determinant of the matrix $\left(\binom{x_i+j}{j}\right)_{i,j=1}^{n}$, $\binom{x_i+j}{j}$ = binomial coefficient. What do you get for $x_i = i$ $(i = 1, \ldots, n)$?

8.4 Show that $\sum_{k\ge 0} e_k(x_1, \ldots, x_n) z^k = \prod_{i=1}^{n}(1 + x_i z)$, where the e_k are the elementary symmetric polynomials.

* * *

▷ **8.5** Prove that

$$\det\left(\frac{1}{1-x_i y_j}\right)_{i,j=1}^{n} = \prod_{1\le i<j\le n}(x_i - x_j)(y_i - y_j)\prod_{i,j=1}^{n}\frac{1}{1-x_i y_j}.$$

8.6 Prove the following extension of the Vandermonde formula due to Krattenthaler. Let $x_1, \ldots, x_n, a_2, \ldots, a_n, b_2, \ldots, b_n$ be variables. Then

$$\det\left((x_i + a_n)\cdots(x_i + a_{j+1})(x_i + b_j)\cdots(x_i + b_2)\right)_{i,j=1}^{n} = \prod_{1\le i<j\le n}(x_i - x_j)\prod_{2\le i\le j\le n}(b_i - a_j).$$

Hint: The determinant is alternating in the x_i's.

8.7 Represent the symmetric polynomial $\prod_{i=1}^{n}(1 + x_i + x_i^2)$ as a polynomial in the elementary symmetric polynomials e_k. Hint: Write $1+x_i+x_i^2 = (1 - \omega x_i)(1 - \omega^2 x_i)$, ω a third root of unity.

▷ **8.8** Prove

$$\det\left(\frac{1-q^{i+j-1}}{1-t^{i+j-1}}\right)_{i,j=1}^{n} = t^{n^3/3-n^2/2+n/6}\prod_{1\le i<j\le n}(1-t^{j-i})^2\cdot\prod_{i,j=1}^{n}\frac{1-qt^{j-i}}{1-t^{i+j-1}}.$$

Hint: Interpret the determinant as polynomial $P(q)$ in q of degree $\le n^2$, and prove the that t^k, t^{-k} are roots of $P(q)$ of multiplicity $n-k$, $k = 0, 1, \ldots, n-1$. Hence $P(q) = (1-q)^n \prod_{k=1}^{n}(t^k - q)^{n-k}(t^{-k} - q)^{n-k} P(0)$. For $P(0)$ use Exercise 8.5.

8.2 Homogeneous Symmetric Functions

Let $X = \{x_1, x_2, \ldots\}$ be a countable set of variables. We are going to construct three natural bases for $\Lambda^m(X)$. A fourth basis, which from a combinatorial point of view is the most interesting, will be discussed in the next section. We write $f(x) = f(x_1, x_2, \ldots)$, and often f and Λ^m when there is no danger of confusion.

Monomial Symmetric Functions.
The most obvious basis for $\Lambda^m(X)$ is obtained by looking at the monomials. Suppose $x^\alpha = x_1^{\alpha_1} x_2^{\alpha_2} \cdots$ appears in $f(x) \in \Lambda^m$. Then we know that every $x^\beta = x_1^{\beta_1} x_2^{\beta_2} \cdots$ must also appear, where β is a permutation of α. Normalizing the vector $\alpha = (\alpha_1, \alpha_2, \ldots)$ in non-increasing fashion, we obtain a number partition $\lambda = \lambda_1 \lambda_2 \ldots \lambda_r$ of m, and arrive at the following definition:

Let $\lambda = \lambda_1 \lambda_2 \ldots \lambda_r \in \mathrm{Par}(m)$. Then the *monomial function* $m_\lambda(x)$ is defined by

$$m_\lambda(x) = \sum_\alpha x^\alpha ,$$

where α runs through all *distinct* permutations of $\lambda = (\lambda_1, \lambda_2, \ldots, \lambda_r, 0, 0, \ldots)$; $m_\lambda(x)$ is a homogeneous symmetric function of degree m.

Example. We have

$$\begin{aligned} m_\emptyset &= 1, \\ m_1 &= \textstyle\sum_i x_i, \\ m_{11} &= \textstyle\sum_{i<j} x_i x_j, \\ m_{21} &= \textstyle\sum_{i\neq j} x_i^2 x_j, \\ &\cdots \\ m_{\underbrace{11\ldots1}_{n}} &= \textstyle\sum_{i_1<\cdots<i_n} x_{i_1} \cdots x_{i_n} . \end{aligned}$$

Every $f \in \Lambda^m(X)$ is a linear combination of the m_λ's, where λ extends over all partitions of m. Since the functions m_λ are clearly linearly independent, the set $\{m_\lambda(x) : \lambda \in \mathrm{Par}(m)\}$ is a basis for $\Lambda^m(X)$, and we conclude that $\dim \Lambda^m(X) = p(m)$.

Elementary Symmetric Functions.
These we already know, extending X to a countable set:

$$e_k(x) = \sum_{i_1<\cdots<i_k} x_{i_1}\cdots x_{i_k} \quad (k \geq 1),$$
$$e_0(x) = 1.$$

Note that $e_n = m_{\underbrace{11\ldots1}_{n}}$. By expanding the right-hand side we obtain for the generating function

$$\sum_{k\geq 0} e_k(x_1,x_2,\ldots)z^k = \prod_{i\geq 1}(1+x_iz). \tag{1}$$

For $\lambda = \lambda_1\lambda_2\ldots\lambda_r \in \mathrm{Par}(m)$ set

$$e_\lambda = e_{\lambda_1}e_{\lambda_2}\cdots e_{\lambda_r}.$$

Then $e_\lambda(x) \in \Lambda^m(x)$, since the degrees add to m.

Example. We have

$$e_{211}(x_1,x_2,\ldots) = (x_1x_2 + x_1x_3 + \cdots)(x_1 + x_2 + \cdots)^2.$$

Since $\{m_\mu : \mu \in \mathrm{Par}(m)\}$ forms a basis of Λ^m, the elementary symmetric functions e_λ ($\lambda \in \mathrm{Par}(m)$) are linear combinations of monomial functions m_μ. The following proposition demonstrates a first combinatorial significance. We write a partition $\mu = \mu_1\ldots\mu_r$ as the vector $(\mu_1,\mu_2,\ldots,\mu_r,0,0,\ldots)$. For a matrix $A = (a_{ij})_{i,j\geq 1}$ with only finitely nonzero elements, we denote by $\mathrm{row}(A) = (r_1,r_2,\ldots)$ the *row-sum vector*, that is, $r_i = \sum_{j\geq 1} a_{ij}$, and similarly by $\mathrm{col}(A)$ the *column-sum vector*.

Proposition 8.2. *Let* $\lambda = \lambda_1\lambda_2\ldots\lambda_r \in \mathrm{Par}(m)$. *Then*

$$e_\lambda = \sum_{\mu\in\mathrm{Par}(m)} M_{\lambda\mu}m_\mu, \tag{2}$$

where $M_{\lambda\mu}$ *is the number of* 0, 1*-matrices* $A = (a_{ij})_{i,j\geq 1}$ *with* $\mathrm{row}(A) = \lambda$, $\mathrm{col}(A) = \mu$.

Proof. We have

$$e_\lambda = e_{\lambda_1}e_{\lambda_2}\cdots = \sum_{i_1<\cdots<i_{\lambda_1}} x_{i_1}\cdots x_{i_{\lambda_1}} \sum_{j_1<\cdots<j_{\lambda_2}} x_{j_1}\cdots x_{j_{\lambda_2}}\cdots.$$

We may represent any product $x_{i_1} \cdots x_{i_{\lambda_1}} x_{j_1} \cdots x_{j_{\lambda_2}} \cdots$ that appears in the sum as a 0, 1-matrix by writing 1's in columns $i_1, \ldots, i_{\lambda_1}$ of row 1, in columns $j_1, \ldots, j_{\lambda_2}$ of row 2, and so on (filling A up with 0's). Hence $\text{row}(A) = \lambda$. This product is a monomial $x_1^{\mu_1} x_2^{\mu_2} \cdots$ if x_i appears μ_i times, that is, if $\text{col}(A) = \mu$. The result follows. □

Example. Consider $\lambda = 311 \in \text{Par}(5)$. The 0, 1-matrices A with $\text{row}(A) = \lambda$, $\text{col}(A) = \mu \in \text{Par}(5)$ are the following, where we write down only the part that contains nonzero entries.

$$\underset{\mu = 311}{\begin{matrix} 1 & 1 & 1 \\ 1 & 0 & 0 \\ 1 & 0 & 0 \end{matrix}} \qquad \underbrace{\begin{matrix} 1 & 1 & 1 \\ 1 & 0 & 0 \\ 0 & 1 & 0 \end{matrix} \quad \begin{matrix} 1 & 1 & 1 \\ 0 & 1 & 0 \\ 1 & 0 & 0 \end{matrix}}_{\mu = 221} \qquad \underbrace{\begin{matrix} 1 & 1 & 1 & 0 \\ 1 & 0 & 0 & 0 \\ 0 & 0 & 0 & 1 \end{matrix} \; \begin{matrix} \text{by exchange} \\ 2 \cdot 3 = 6 \\ \text{possibilities} \end{matrix} \; \begin{matrix} 0 & 1 & 1 & 1 \\ 1 & 0 & 0 & 0 \\ 1 & 0 & 0 & 0 \end{matrix}}_{\mu = 2111}$$

$$\underset{\mu = 11111}{\begin{matrix} 1 & 1 & 1 & 0 & 0 \\ 0 & 0 & 0 & 1 & 0 \\ 0 & 0 & 0 & 0 & 1 \end{matrix}} \quad \text{by exchange } 2 \cdot \binom{5}{3} = 20 \text{ possibilities.}$$

Hence $e_{311} = \mu_{311} + 2\mu_{221} + 7\mu_{2111} + 20\mu_{11111}$. Looking at the transpose, we immediately infer the following corollary.

Corollary 8.3. *We have $M_{\lambda\mu} = M_{\mu\lambda}$ for any $\lambda, \mu \in \text{Par}(m)$.*

The following basic result is sometimes called the fundamental theorem of symmetric functions.

Theorem 8.4. *The set $\{e_\lambda(x) : \lambda \in \text{Par}(m)\}$ is a basis for $\Lambda^m(X)$.*

Proof. We have to show that any monomial function m_μ is in the span $\langle e_\lambda : \lambda \in \text{Par}(m)\rangle$. To accomplish this we first define a total ordering on the set $\text{Par}(m)$. Set

$$\lambda < \mu :\Longleftrightarrow \lambda_i < \mu_i, \text{ where } i \text{ is the first index with } \lambda_i \neq \mu_i .$$

With this order, called the *lexicographic order*, $\text{Par}(m)$ becomes a totally ordered set with the partition $11\ldots1$ as minimal element.

Example. For $m = 5$ we get

$$11111 < 2111 < 221 < 311 < 32 < 41 < 5 .$$

Claim. a. *If $M_{\lambda\mu} > 0$, then $\mu \leq \lambda^*$, where λ^* is the conjugate partition,* b. $M_{\lambda\lambda^*} = 1$.

If $M_{\lambda\mu} > 0$, then there exists a 0,1-matrix A with $\text{row}(A) = \lambda$, $\text{col}(A) = \mu$. According to the definition of lexicographic order, we obtain the *maximal* possible μ with $M_{\lambda\mu} > 0$ by pushing all 1's in each row to the left. But this gives precisely the Ferrers diagram of λ (with the 1's as rows). Hence $\mu = \lambda^*$ is the maximal possible μ, and in this case we clearly have $M_{\lambda\lambda^*} = 1$.

Now we use induction on the lexicographic order. For the minimal partition $\underbrace{11\ldots1}_{m}$ we know that $m_{11\ldots1} = \sum x_{i_1}\cdots x_{i_m} = e_m$; hence $m_{11\ldots1} \in \langle e_\lambda : \lambda \in \text{Par}(m)\rangle$. Assume inductively $m_\nu \in \langle e_\lambda : \lambda \in \text{Par}(m)\rangle$ for $\nu < \mu$. From $e_{\mu^*} = m_\mu + \sum_{\nu<\mu} M_{\mu^*\nu} m_\nu$ follows $m_\mu = e_{\mu^*} - \sum_{\nu<\mu} M_{\mu^*\nu} m_\nu \in \langle e_\lambda : \lambda \in \text{Par}(m)\rangle$, and we are through. □

Remark. Recall the example at the beginning of the chapter: The coefficients of a polynomial are (apart from the sign) the elementary symmetric functions in the roots. The fundamental theorem therefore shows that *any* symmetric function in the roots is a polynomial in the coefficients.

Complete Symmetric Functions.

The natural analogue to the elementary symmetric functions $e_k(x)$ arises when we allow repetitions of the indices. We define the *complete symmetric functions* h_k as

$$h_k(x_1, x_2, \ldots) = \sum_{i_1 \leq \cdots \leq i_k} x_{i_1} x_{i_2} \cdots x_{i_k} \quad (k \geq 1),$$

$$h_0(x_1, x_2, \ldots) = 1.$$

Note that $h_k(x) = \sum_{\mu \in \text{Par}(k)} m_\mu(x)$, and $h_k(x) \in \Lambda^k(X)$.

By expanding the right-hand side we get the following generating function analogous to (1):

$$\sum_{k \geq 0} h_k(x_1, x_2, \ldots) z^k = \prod_{i \geq 1} \frac{1}{1 - x_i z}. \tag{3}$$

Let $\lambda = \lambda_1 \lambda_2 \ldots \lambda_r \in \text{Par}(m)$. As for the functions e_λ, we set

$$h_\lambda = h_{\lambda_1} \cdots h_{\lambda_r} \in \Lambda^m(X).$$

The analogue to Proposition 8.2 is the following result, whose proof is left to the exercises.

Proposition 8.5. *Let* $\lambda = \lambda_1\lambda_2\ldots\lambda_r \in \mathrm{Par}(m)$. *Then*

$$h_\lambda = \sum_{\mu\in\mathrm{Par}(m)} N_{\lambda\mu} m_\mu\,, \tag{4}$$

where $N_{\lambda\mu}$ *is the number of matrices* $A = (a_{ij})_{i,j\ge 1}$ *over* $\mathbb{N}_0$ *with* $\mathrm{row}\,(A) = \lambda$, $\mathrm{col}\,(A) = \mu$. *Furthermore,* $N_{\lambda\mu} = N_{\mu\lambda}$.

Now let us show that the complete symmetric functions h_λ ($\lambda \in \mathrm{Par}(m)$) also form a basis for $\Lambda^m(X)$.

Theorem 8.6. *The set* $\{h_\lambda(x) : \lambda \in \mathrm{Par}(m)\}$ *is a basis for* $\Lambda^m(X)$.

Proof. Since $\{e_\lambda : \lambda \in \mathrm{Par}(m)\}$ is a basis, it suffices to show that any e_λ is in the span $\langle h_\mu : \mu \in \mathrm{Par}(m)\rangle$. First we note from (1) and (3) that

$$\left(\sum_{k\ge 0} e_k z^k\right)\cdot\left(\sum_{k\ge 0} h_k(-z)^k\right) = 1\,,$$

or by convolution,

$$\sum_{k=0}^{n} e_k(-1)^{n-k}h_{n-k} = 0 \quad \text{for } n \ge 1\,. \tag{5}$$

Now, $e_0 = h_0 = 1$, $e_1 = h_1$, so assume $n \ge 2$. Then by (5) and induction,

$$e_n = -\sum_{k=0}^{n-1} e_k(-1)^{n-k}h_{n-k} \in \left\langle h_{j_1}\cdots h_{j_t} : \sum j_i = n\right\rangle\,.$$

We thus conclude for $\lambda = \lambda_1\ldots\lambda_r \in \mathrm{Par}(m)$ that

$$e_\lambda = e_{\lambda_1}\cdots e_{\lambda_r} \in \left\langle h_{j_1}\cdots h_{j_\ell} : \sum j_i = m\right\rangle = \langle h_\mu : \mu \in \mathrm{Par}(m)\rangle\,.$$

□

Example. Consider $\lambda = 311$. We have $e_1 = h_1$, $e_2 = e_1h_1 - e_0h_2 = h_1^2 - h_2$, $e_3 = e_2h_1 - e_1h_2 + e_0h_3 = (h_1^2 - h_2)h_1 - h_1h_2 + h_3 = h_1^3 - 2h_1h_2 + h_3$, and thus $e_{311} = e_3e_1^2 = (h_1^3 - 2h_1h_2 + h_3)h_1^2 = h_1^5 - 2h_1^3h_2 + h_1^2h_3$. Hence $e_{311} = h_{11111} - 2h_{2111} + h_{311}$.

The close relation between the functions e_n and h_n can be captured by the following mapping. Define $\omega : \Lambda \to \Lambda$ by $\omega : e_n \mapsto h_n$ ($n \ge 0$), and extend it to an algebra homomorphism. Thus $\omega(e_\lambda) = h_\lambda$ for any partition λ. It follows from (5) that for $n \geq 1$,

$$0 = \sum_{k=0}^{n} h_k(-1)^{n-k}\omega(h_{n-k}) = \sum_{k=0}^{n} \omega(h_k) \cdot (-1)^{n-k} h_{n-k}.$$

But this implies again by (5) that $\omega(h_k) = e_k$ for all k. Hence ω is an involution, $\omega^2 = \text{id}$. We will make use of this fact later on.

A final remark: When $X = \{x_1, \ldots, x_n\}$ is finite, we may regard this as the special case in which we set $x_i = 0$ for $i \geq n+1$. So for example, $e_k(x_1, \ldots, x_n) = \sum_{i_1 < \cdots < i_k} x_{i_1} \cdots x_{i_k}$ becomes 0 for $k > n$, as we have noted before. The structure of the various bases and the dimension of $\Lambda^m(X)$ are treated in the exercises. Whenever X is finite, then we will specify the variables explicitly; otherwise, X is assumed to be countable.

Exercises

▷ **8.9** Find an expression for the five monomial functions m_λ ($\lambda \in \text{Par}(4)$) in terms of the e_μ ($\mu \in \text{Par}(4)$).

8.10 Prove Proposition 8.5.

8.11 Determine $N_{\lambda\mu}$ for $\lambda = \lambda_1 \ldots \lambda_r \in \text{Par}(m)$, $\mu = \underbrace{11\ldots1}_{m}$.

8.12 Show that $e_{\underbrace{11\ldots1}_{m}} = h_{\underbrace{11\ldots1}_{m}} = \sum_{\mu \in \text{Par}(m)} \binom{m}{\mu_1 \ldots \mu_m} m_\mu$.

▷ **8.13** Let $X = \{x_1, x_2, \ldots\}$, $Y = \{y_1, y_2, \ldots\}$. Show that

$$\prod_{i,j}(1 + x_i y_j) = \sum_{\lambda,\mu \in \text{Par}} M_{\lambda\mu} m_\lambda(x) m_\mu(y) = \sum_{\lambda \in \text{Par}} m_\lambda(x) e_\lambda(y).$$

8.14 Prove analogously

$$\prod_{i,j} \frac{1}{1 - x_i y_j} = \sum_{\lambda,\mu \in \text{Par}} N_{\lambda\mu} m_\lambda(x) m_\mu(y) = \sum_{\lambda \in \text{Par}} m_\lambda(x) h_\lambda(y).$$

8.15 Suppose $X = \{x_1, \ldots, x_n\}$. Prove $\dim \Lambda^m(X) = p(m; \leq n)$ (partitions of m with at most n parts), and find the bases in terms of m_λ, e_λ, and h_λ, by placing various restrictions on the partitions λ.

* * *

8.16 We had as an example in the text that $e_{311} = h_{11111} - 2h_{2111} + h_{311}$. Verify this equality, using the numbers $M_{\lambda\mu}$ and $N_{\lambda\mu}$.

▷ **8.17** When you express e_λ ($\lambda \in \mathrm{Par}(m)$) in terms of h_μ ($\mu \in \mathrm{Par}(m)$), what can you say about the partitions μ with nonzero coefficients in relation to λ?

8.18 Compute $\omega(m_\lambda)$ for $\lambda \in \mathrm{Par}(4)$ in terms of m_μ, where ω is the involution $\omega : e_\lambda \mapsto h_\lambda$ in $\Lambda^4(X)$.

▷ **8.19** Prove that $h_m(1,q,q^2,\ldots,q^n) = \left[{m+n \atop m} \right]_q$ (Gaussian coefficient), and $e_m(1,q,q^2,\ldots,q^{n-1}) = q^{\binom{m}{2}} \left[{n \atop m} \right]_q$.

8.20 The power sums $p_k(x_1,x_2,\ldots) = \sum_{i\ge1} x_i^k$ are clearly homogeneous symmetric functions of degree k. As before set $p_\lambda = p_{\lambda_1} \cdots p_{\lambda_r}$ for $\lambda = \lambda_1 \ldots \lambda_r \in \mathrm{Par}(m)$; thus $p_\lambda(x) \in \Lambda^m(X)$. Set $p_\lambda = \sum_{\mu\in\mathrm{Par}(m)} R_{\lambda\mu} m_\mu$; what does $R_{\lambda\mu}$ count? Hint: $R_{\lambda\mu}$ enumerates some set of ordered set-partitions.

▷ **8.21** Prove that $\{p_\lambda : \lambda \in \mathrm{Par}(m)\}$ is a basis for $\Lambda^m(X)$. Hint: Prove that $R_{\lambda\mu} = 0$ unless $\lambda \le \mu$, and proceed as usual by considering this ordering on $\mathrm{Par}(m)$. Alternatively, you may use Waring's formula of Exercise 6.34.

8.22 When $X = \{x_1,\ldots,x_n\}$ is finite, what partitions $\lambda \in \mathrm{Par}(m)$ give a basis $\{p_\lambda\}$ for $\Lambda^m(X)$?

8.3 Schur Functions

In this section we consider a finite set $X = \{x_1,\ldots,x_n\}$ of variables, and partitions λ with at most n summands. The following functions will provide the most interesting basis.

Definition. Let $\lambda = \lambda_1 \ldots \lambda_n \in \mathrm{Par}$ with possible 0's at the end. The function

$$s_\lambda(x_1,\ldots,x_n) = \frac{\det(x_j^{n-i+\lambda_i})_{i,j=1}^n}{\prod\limits_{1\le i<j\le n} (x_i - x_j)}$$

is called a *Schur function*.

Since the numerator is an alternating polynomial, Proposition 8.1 says that $s_\lambda(x)$ is a homogeneous symmetric function of degree

$$\sum_{i=1}^n (n-i+\lambda_i) - \binom{n}{2} = \sum_{i=1}^n \lambda_i = |\lambda|.$$

The following basic result shows the connection of the Schur functions with the complete symmetric functions. It is called the *Jacobi-Trudi identity*.

Theorem 8.7. *Let* $\lambda = \lambda_1 \dots \lambda_n \in \mathrm{Par}$ *with possible* 0*'s at the end. Then*

$$s_\lambda(x_1, \dots, x_n) = \det(h_{\lambda_i - i + j})_{i,j=1}^n, \tag{1}$$

where we set $h_k = 0$ *for* $k < 0$.

Proof. Let $e_j^{(\ell)}$ be the j-th elementary symmetric function on $X = \{x_1, \dots, x_n\} \setminus x_\ell$. By (1) and (3) of the previous section,

$$\sum_{k \ge 0} h_k z^k \cdot \sum_{j=0}^{n-1} e_j^{(\ell)} (-z)^j = \prod_{i=1}^n \frac{1}{1 - x_i z} \prod_{\substack{m=1 \\ m \ne \ell}}^n (1 - x_m z) = \frac{1}{1 - x_\ell z}$$

$$= 1 + x_\ell z + x_\ell^2 z^2 + \cdots .$$

Comparing coefficients for z^{α_i} we get

$$\sum_{j=0}^{n-1} h_{\alpha_i - j} (-1)^j e_j^{(\ell)} = \sum_{k=1}^n h_{\alpha_i - n + k} (-1)^{n-k} e_{n-k}^{(\ell)} = x_\ell^{\alpha_i} . \tag{2}$$

For $\alpha = (\alpha_1, \dots, \alpha_n)$ define matrices

$$A_\alpha = (x_j^{\alpha_i}), \quad H_\alpha = (h_{\alpha_i - n + j}), \quad E = ((-1)^{n-i} e_{n-i}^{(j)}) .$$

Equation (2) then translates into the matrix equation

$$H_\alpha E = A_\alpha . \tag{3}$$

Now look at the special vector $\overline{\alpha} = (n-1, n-2, \dots, 1, 0)$. In this case $H_{\overline{\alpha}}$ is an upper triangular matrix with diagonal 1, and so

$$\det E = \det A_{\overline{\alpha}} = \det(x_j^{n-i}) = \prod_{1 \le i < j \le n} (x_i - x_j) .$$

For arbitrary vectors α we therefore obtain

$$\det H_\alpha = \frac{\det A_\alpha}{\prod_{1 \le i < j \le n} (x_i - x_j)} . \tag{4}$$

The theorem follows by taking the vector $\alpha = (\lambda_1 + n - 1, \lambda_2 + n - 2, \dots, \lambda_n)$. In this case $\alpha_i = \lambda_i + n - i$; hence $\det A_\alpha = \det(x_j^{n-i+\lambda_i})$, which is equal to the numerator of the Schur function, and $\det H_\alpha = \det(h_{\lambda_i - i + j})$. □

Semistandard Tableaux.
The Jacobi–Trudi identity leads to an unexpected and beautiful combinatorial definition of the Schur functions, with the help of the Lemma of Gessel–Viennot.

Definition. Let $\lambda = \lambda_1 \dots \lambda_r \in \text{Par}$ with $r \leq n$. A *semistandard tableau* (SST) T over $\{1, 2, \dots, n\}$ of *shape* λ is a scheme

$$
\begin{array}{ll}
 & T_{11}\, T_{12} \dots T_{1\lambda_1} \\
T = & T_{21}\, T_{22} \dots T_{2\lambda_2} \qquad\qquad T_{ij} \in \{1, 2, \dots, n\} \\
 & \dots \\
 & T_{r1} \dots T_{r\lambda_r}
\end{array}
$$

with $T_{i1} \leq T_{i2} \leq \cdots \leq T_{i\lambda_i}$ for the rows, and $T_{1j} < T_{2j} < \cdots$ for the columns. Set $\mu_k = \#\{T_{ij} = k\}$; then $\mu = (\mu_1, \mu_2, \dots, \mu_n)$ is called the *type* of T.

Example. Let $n = 6$; then

$$
T = \begin{array}{l}
11334 \\
234 \\
355 \\
46 \\
5 \\
6
\end{array}
$$

is an SST of shape 533211 and type $(2, 1, 4, 3, 3, 2)$.

To an SST T of type μ we associate the monomial $x^T = x_1^{\mu_1} \cdots x_n^{\mu_n}$. We shall let $\mathcal{T}(n, \lambda)$ denote the set of all SSTs over $\{1, \dots, n\}$ of shape λ.

Example. Consider $n = 3$, $\lambda = 21$. The tableaux are as follows:

(shape 21)	11 2	11 3	12 2	12 3	13 2	13 3	22 3	23 3
	$x_1^2x_2$	$x_1^2x_3$	$x_1x_2^2$	$x_1x_2x_3$	$x_1x_2x_3$	$x_1x_3^2$	$x_2^2x_3$	$x_2x_3^2$

Now by an easy computation we obtain for the Schur function

$$s_{21}(x_1,x_2,x_3) = \frac{\det\begin{pmatrix} x_1^4 & x_2^4 & x_3^4 \\ x_1^2 & x_2^2 & x_3^2 \\ 1 & 1 & 1 \end{pmatrix}}{(x_1-x_2)(x_1-x_3)(x_2-x_3)}$$

$$= x_1^2x_2 + x_1^2x_3 + x_2^2x_3 + x_1x_2^2 + x_1x_3^2 + x_2x_3^2 + 2x_1x_2x_3\,,$$

and this is precisely $\sum_{T\in\mathcal{T}(3,\lambda)} x^T$.

That this is no coincidence is the content of the following fundamental result.

Theorem 8.8. *Let* $\lambda = \lambda_1 \dots \lambda_r \in \mathrm{Par}$, $r \le n$. *Then*

$$s_\lambda(x_1,\dots,x_n) = \sum_{T\in\mathcal{T}(n,\lambda)} x^T\,. \tag{5}$$

Proof. Consider the usual lattice graph with steps up and to the right. All horizontal steps have weight 1, and a vertical step along $x = k$ has weight x_k, $k = 1,\dots,n$. Now we apply the lemma of Gessel–Viennot to the vertex sets $\mathcal{A} = \{A_1,\dots,A_n\}$, $\mathcal{B} = \{B_1,\dots,B_n\}$, with $A_i = (1, n-i)$, $B_j = (n, n+\lambda_j - j)$.

Example. Consider $n = 4$, $\lambda = 211$:

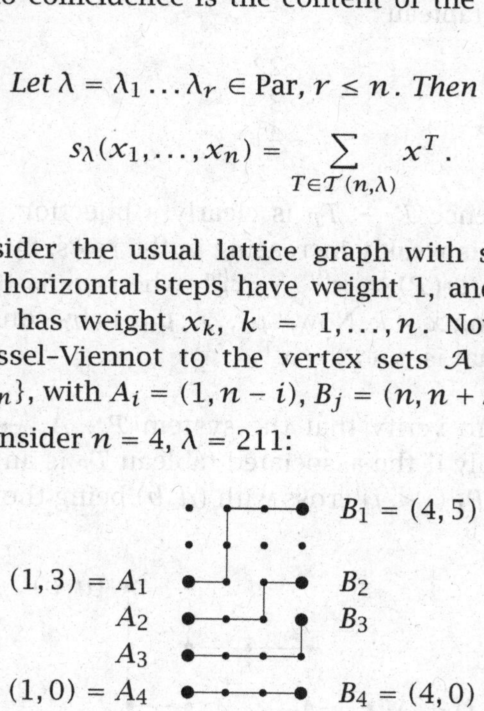

$B_1 = (4,5)$

$(1,3) = A_1$ B_2

A_2 B_3

A_3

$(1,0) = A_4$ $B_4 = (4,0)$

We find for the path matrix $M = (m_{ij})$,

$$m_{ij} = \sum_{\substack{(i_1,\dots,i_n)\\ \sum i_k = \lambda_j - j + i}} x_1^{i_1}\cdots x_n^{i_n} = h_{\lambda_j-j+i}(x_1,\dots,x_n)\,,$$

and so by Jacobi–Trudi,

$$\det M = \det(h_{\lambda_j-j+i}) = \det(h_{\lambda_i-i+j}) = s_\lambda(x_1,\dots,x_n)\,.$$

It is plain that only the identity permutation allows vertex-disjoint path systems from $\mathcal{A}$ to $\mathcal{B}$. Thus it remains to show that

$$\sum_{\mathcal{P}\in VD} w(\mathcal{P}) = \sum_{T\in\mathcal{T}(n,\lambda)} x^T . \tag{6}$$

Consider any system $\mathcal{P}: A_i \to B_i$. We associate to $P_i : A_i \to B_i$ the i-th row of a tableau $T_{\mathcal{P}}$ with the x-coordinates of the vertical steps as entries, as they appear in P_i. For the path system in the example above we get the tableau

$$\begin{matrix} 22 \\ 3 \\ 4 \end{matrix}$$

This correspondence $\mathcal{P} \to T_{\mathcal{P}}$ is clearly a bijection. Note that $T_{\mathcal{P}}$ has shape λ, and is weakly increasing in the rows. By the definition of the weighting, $w(\mathcal{P}) = x_1^{\mu_1} \cdots x_n^{\mu_n}$, where μ_k is the number of vertical steps along $x = k$. Now, $(\mu_1, \ldots, \mu_n)$ is by construction also the type of $T_{\mathcal{P}}$, that is, $w(\mathcal{P}) = x^T$.

It thus remains to verify that the system $\mathcal{P}: A_i \to B_i$ is vertex-disjoint if and only if the associated tableau $T_{\mathcal{P}}$ is an SST. Suppose the paths P_i and P_j $(i < j)$ cross with (a, b) being the first common point:

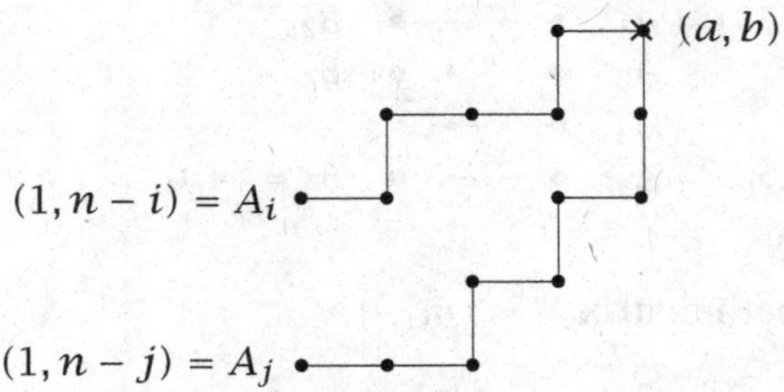

For the tableau $T_{\mathcal{P}}$, we have $T_{j,b-n+j} = a \le T_{i,b-n+i+1}$. Set $\ell = b - n + j$, $k = b - n + i + 1$; then $\ell \ge k$ (because of $j > i$), thus $T_{j,\ell} \le T_{i,k} \le T_{i,\ell}$, contradicting the strict monotonicity on the columns. The converse is similarly shown. $\square$

Example. For $\lambda = 11\ldots 1$ there is only one SST of shape λ, namely $\begin{matrix}1\\2\\\vdots\\n\end{matrix}$, and we get $s_{11\ldots 1}(x_1,\ldots,x_n) = x_1x_2\cdots x_n$.

Note that the theorem implies the astonishing result that given $\mu = \mu_1\ldots\mu_n$ there are equally many SSTs of the same shape and type equal to any permutation of μ.

Corollary 8.9. *Let $\lambda \in \mathrm{Par}(n)$, $\lambda = \lambda_1\ldots\lambda_n$, with possible 0's at the end. Then*

$$s_\lambda(x_1,\ldots,x_n) = \sum_{\mu\in\mathrm{Par}(n)} K_{\lambda\mu}m_\mu\,, \tag{7}$$

where $K_{\lambda\mu}$ is the number of SSTs $T \in \mathcal{T}(n,\lambda)$ of type $\mu = (\mu_1 \geq \cdots \geq \mu_n)$. The $K_{\lambda\mu}$'s are called Kostka numbers.

Proof. A monomial $x^T = x_1^{\mu_1}\cdots x_n^{\mu_n}$ appears as often as there are SSTs of type μ. $\square$

Example. For $n = 4$, $\lambda = 31$ we get

shape							
$\begin{smallmatrix}\square\square\square\\\square\end{smallmatrix}$	$\begin{matrix}111\\2\end{matrix}$	$\begin{matrix}112\\2\end{matrix}$	$\begin{matrix}112\\3\end{matrix}$	$\begin{matrix}113\\2\end{matrix}$	$\begin{matrix}123\\4\end{matrix}$	$\begin{matrix}124\\3\end{matrix}$	$\begin{matrix}134\\2\end{matrix}$
	m_{31}	m_{22}	m_{211}		m_{1111}		

Hence $s_{31}(x_1,x_2,x_3,x_4) = m_{31} + m_{22} + 2m_{211} + 3m_{1111}$.

With (7) we can now easily show that the Schur functions s_λ ($\lambda \in \mathrm{Par}(n)$) give another basis for $\Lambda^n(X)$, $X = \{x_1,\ldots,x_n\}$.

Theorem 8.10. *The set $\{s_\lambda : \lambda \in \mathrm{Par}(n)\}$ is a basis for $\Lambda^n(X)$, $X = \{x_1,\ldots,x_n\}$.*

Proof. We proceed as in the proof of Theorem 8.4. This time we choose the so-called *dominance order* on $\mathrm{Par}(n)$. Let $\lambda = \lambda_1\ldots\lambda_n$, $\mu = \mu_1\ldots\mu_n$ in $\mathrm{Par}(n)$ with possible 0's at the end. Then we define

$$\lambda \preceq \mu :\Longleftrightarrow \sum_{i=1}^{k}\lambda_i \leq \sum_{i=1}^{k}\mu_i \ \text{ for } k = 1,\ldots,n\,.$$

With the dominance ordering, $\mathrm{Par}(n)$ becomes a poset, which, in general, is not a linear order.

Example. For $n = 6$ we get

$$111111 \prec 21111 \prec 2211 \prec \begin{matrix} 222 \\ 3111 \end{matrix} \prec 321 \prec \begin{matrix} 33 \\ 411 \end{matrix} \prec 42 \prec 51 \prec 6 .$$

The partitions 222 and 3111 are unrelated, as are 33 and 411.

Claim. *If $K_{\lambda\mu} > 0$ then $\mu \preceq \lambda$, and furthermore, $K_{\lambda\lambda} = 1$.*

Suppose $T \in \mathcal{T}(n, \lambda)$ has type $\mu = (\mu_1 \geq \mu_2 \geq \cdots \geq \mu_n)$. If $T_{ij} = k$ with $i > k$, then by strict monotonicity of the columns,

$$1 \leq T_{1j} < T_{2j} < \cdots < T_{kj} < \cdots < T_{ij} = k,$$

which contradicts $i > k$. This means that all numbers $1, 2, \ldots, k$ must appear in the first k rows. But this implies $\mu_1 + \cdots + \mu_k \leq \lambda_1 + \cdots + \lambda_k$ for all k, and so $\mu \preceq \lambda$. If $\lambda = \mu$, then T contains λ_1 1's in the first row, λ_2 2's in the second row, and so on. Hence T is uniquely determined, and we get $K_{\lambda\lambda} = 1$.

Now we use induction on the dominance order. The minimal element is $11\ldots1$ with $s_{11\ldots1} = x_1 x_2 \cdots x_n = m_{11\ldots1}$; hence $m_{11\ldots1} \in \langle s_\lambda : \lambda \in \mathrm{Par}(n) \rangle$. Assume inductively $m_\nu \in \langle s_\lambda : \lambda \in \mathrm{Par}(n) \rangle$ for $\nu \prec \mu$. Then by the claim, $s_\mu = m_\mu + \sum_{\nu \prec \mu} K_{\mu\nu} m_\nu$, and thus $m_\mu = s_\mu - \sum_{\nu \prec \mu} K_{\mu\nu} m_\nu \in \langle s_\lambda : \lambda \in \mathrm{Par}(n) \rangle$. □

Remark. We may use the combinatorial interpretation of the Schur functions to extend $X = \{x_1, x_2, \ldots, x_n\}$ to the countable case. Set $s_\lambda(x_1, x_2, \ldots) = \sum_T x^T$ over all SSTs T of shape λ. Then s_λ is a homogeneous symmetric function of degree $|\lambda|$. Note that $s_\lambda(x_1, \ldots, x_n) = 0$ when λ has more than n parts, since in this case no SST on $\{1, \ldots, n\}$ can exist.

Plane Partitions.
One of the striking successes of the theory of Schur functions concerns the enumeration of plane partitions. Recall from Section 5.4 that a plane partition of n is an array of integers $\lambda_{ij} \geq 1$ with $\sum \lambda_{ij} = n$ such that any row and column is non-increasing. We denoted by $pp(n; r, s, t)$ the number of plane partitions of n with at most r rows, at most s columns, and $\max = \lambda_{11} \leq t$.

Our goal is to prove the following remarkable formula for their generating function.

Theorem 8.11. *We have*

$$\sum_{n\geq 0} pp(n;r,s,t)q^n = \prod_{i=1}^{r}\prod_{j=1}^{s}\prod_{k=1}^{t}\frac{1-q^{i+j+k-1}}{1-q^{i+j+k-2}}\,. \tag{8}$$

Proof. We work with the partition $\lambda = \underbrace{ss\ldots s}_{r}\underbrace{00\ldots 0}_{t}$. Theorem 8.8 then gives

$$s_\lambda(x_1,x_2,\ldots,x_{t+r}) = \sum_{T\in\mathcal{T}(t+r,\lambda)} x^T \quad \text{on} \quad \{1,2,\ldots,t+r\}\,. \tag{9}$$

In a semistandard tableau T the rows and columns are increasing, while in a plane partition they are decreasing. Suppose $T = (T_{ij})$ is an SST of shape λ. The map

$$T_{ij} \mapsto \lambda_{ij} = t + r + 1 - T_{ij} \tag{10}$$

turns T into a plane partition φT with *exactly* r rows, *exactly* s columns, and max $\leq t+r$, where in addition the columns are strictly decreasing. Let us call these plane partitions *column-strict*. Conversely, any such strict plane partition corresponds via (10) to an SST of shape λ on $\{1,2,\ldots,t+r\}$. So these numbers are the same.

Example. $r=4$, $s=5$, $t=3$,

$$T = \begin{matrix} 1\,1\,1\,2\,2 \\ 2\,3\,3\,4\,4 \\ 3\,4\,4\,5\,6 \\ 6\,6\,7\,7\,7 \end{matrix} \quad\longrightarrow\quad \varphi T = \begin{matrix} 7\,7\,7\,6\,6 \\ 6\,5\,5\,4\,4 \\ 5\,4\,4\,3\,2 \\ 2\,2\,1\,1\,1 \end{matrix}.$$

Here is the crucial observation. In (9), set $x_1 = q^{t+r}$, $x_2 = q^{t+r-1}$, $\ldots, x_i = q^{t+r+1-i}, \ldots, x_{t+r} = q$, and suppose $T \in \mathcal{T}(t+r,\lambda)$ has type $(\mu_1,\ldots,\mu_{t+r})$. Then $x^T = q^n$, where $n = \sum_{k=1}^{t+r}\mu_k(t+r+1-k)$. Looking at (10) we conclude that n equals $\sum\lambda_{ij}$ for the associated strict plane partition φT. Hence we have proved

$$s_\lambda(q^{t+r},\ldots,q) = \sum_{n\geq 0} spp(n;=r,=s,t+r)q^n\,, \tag{11}$$

where $spp(n;=r;=s;t+r)$ denotes the number of strict plane partitions of n with r rows, s columns, and max $\leq t+r$.

To pass from strict to ordinary plane partitions we proceed as follows. Subtract 1 from all elements in row r, 2 from all elements in row $r-1$, and so on, and finally r from all elements in row 1. This gives an ordinary plane partition ψT with at most r rows, at most s columns, and max $\le t$, and the correspondence is clearly bijective.

Example. Our partition above gives

$$\varphi T = \begin{matrix} 7\,7\,7\,6\,6 \\ 6\,5\,5\,4\,4 \\ 5\,4\,4\,3\,2 \\ 2\,2\,1\,1\,1 \end{matrix} \quad \longrightarrow \quad \psi T = \begin{matrix} 3\,3\,3\,2\,2 \\ 3\,2\,2\,1\,1 \\ 3\,2\,2\,1 \\ 1\,1 \end{matrix}\,.$$

The total number subtracted is $s\binom{r+1}{2}$, and we obtain

$$\sum_{n\ge 0} pp(n;r,s,t)q^n = q^{-s\binom{r+1}{2}} s_\lambda(q^{t+r},\dots,q)\,. \tag{12}$$

It remains to compute the right-hand side, and for this we use the original definition of the Schur function s_λ,

$$s_\lambda(q^{t+r},\dots,q) = \frac{\det\left((q^{t+r-j+1})^{t+r-i+\lambda_i}\right)_{i,j=1}^{t+r}}{\det\left((q^{t+r-j+1})^{t+r-i}\right)_{i,j=1}^{t+r}}\,. \tag{13}$$

Let X be the numerator and Y the denominator on the right-hand side of (13). Factoring $q^{t+r-i+\lambda_i}$ out of the i-th row in X, we obtain with $\lambda_1 = \cdots = \lambda_r = s$, $\lambda_{r+1} = \cdots = \lambda_{t+r} = 0$,

$$X = q^{rs+\binom{t+r}{2}} \det\left(q^{(t+r-j)(t+r-i+\lambda_i)}\right)_{i,j=1}^{t+r}\,.$$

With $a_i = t+r-i+\lambda_i$ we see that the inner matrix $((q^{a_i})^{t+r-j})$ is a Vandermonde matrix, and so

$$X = q^{rs+\binom{t+r}{2}} \prod_{1\le i<j\le t+r} (q^{t+r-i+\lambda_i} - q^{t+r-j+\lambda_j})\,. \tag{14}$$

Similarly, in Y we factor out q^{t+r-i} from the i-th row, and get by Vandermonde

$$Y = q^{\binom{t+r}{2}} \prod_{1\le i<j\le t+r} (q^{t+r-i} - q^{t+r-j})\,. \tag{15}$$

Altogether we obtain for the right-hand side of (12),

$$q^{-s\binom{r}{2}} \prod_{1 \le i < j \le t+r} \frac{(q^{t+r-i+\lambda_i} - q^{t+r-j+\lambda_j})}{(q^{t+r-i} - q^{t+r-j})}. \tag{16}$$

To finish the proof we distinguish three cases:

Case 1: $i < j \le r$: Here $\lambda_i = \lambda_j = s$. The quotient in (16) is then q^s. With the $\binom{r}{2}$ pairs $i < j \le r$ we get $q^{s\binom{r}{2}}$, so that the factor in front is canceled out.

Case 2: $r < i < j$: Then $\lambda_i = \lambda_j = 0$, and we get 1 as contribution.

Case 3: $i \le r < j$: Here $\lambda_i = s$, $\lambda_j = 0$, resulting in

$$\prod_{i=1}^{r} \prod_{j=r+1}^{r+t} \frac{q^{t+r-i+s} - q^{t+r-j}}{q^{t+r-i} - q^{t+r-j}}.$$

Multiplying the inner quotient by $\frac{q^{j-t-r}}{q^{j-t-r}}$ yields

$$\prod_{i=1}^{r} \prod_{j=r+1}^{t+r} \frac{1 - q^{j-i+s}}{1 - q^{j-i}},$$

and with index transformation $i \mapsto r + 1 - i$, $j \mapsto k + r$,

$$\prod_{i=1}^{r} \prod_{k=1}^{t} \frac{1 - q^{i+k+s-1}}{1 - q^{i+k-1}}. \tag{17}$$

Writing

$$\frac{1 - q^{i+k-1+s}}{1 - q^{i+k-1}} = \frac{1 - q^{i+k-1+s}}{1 - q^{i+k-1+(s-1)}} \cdot \frac{1 - q^{i+k-1+(s-1)}}{1 - q^{i+k-1+(s-2)}} \cdots \frac{1 - q^{i+k-1+1}}{1 - q^{i+k-1}},$$

we see that (17) becomes the product

$$\prod_{i=1}^{r} \prod_{j=1}^{s} \prod_{k=1}^{t} \frac{1 - q^{i+j+k-1}}{1 - q^{i+j+k-2}},$$

and we are done. □

With this generating function in hand it is only a small step to prove the following magnificent formula of MacMahon.

Corollary 8.12 (MacMahon). *We have*

$$\sum_{n \ge 0} pp(n) q^n = \prod_{i \ge 1} \frac{1}{(1 - q^i)^i}.$$

Proof. All we have to do is to let r, s, and t go to infinity in (8). Let $n = i+j+k-1$ or $n+1 = i+j+k$, that is, $n+1$ is an *ordered partition* into three parts. We know from Section 1.5 that there are precisely $\binom{n}{2}$ such ordered partitions. Letting r, s, t go to ∞, the expression (8) thus becomes

$$\prod_{n\geq 2}\left(\frac{1-q^n}{1-q^{n-1}}\right)^{\binom{n}{2}} = \frac{1-q^2}{1-q}\cdot\left(\frac{1-q^3}{1-q^2}\right)^3\cdots\left(\frac{1-q^n}{1-q^{n-1}}\right)^{\binom{n}{2}}\left(\frac{1-q^{n+1}}{1-q^n}\right)^{\binom{n+1}{2}}\cdots$$
$$= \frac{1}{1-q}\cdot\frac{1}{(1-q^2)^2}\cdots\frac{1}{(1-q^n)^n}\cdots,$$

and this is MacMahon's formula. □

Exercises

8.23 Derive $s_{11\dots1}(x_1,\dots,x_n) = x_1\cdots x_n = e_n(x_1,\dots,x_n)$ directly from the definition.

8.24 Show that $s_n(x_1,\dots,x_n) = \sum_{\mu\in\mathrm{Par}(n)} m_\mu = h_n(x_1,\dots,x_n)$ by looking at the coefficients $K_{\lambda\mu}$ when $\lambda = n \in \mathrm{Par}(n)$.

▷ **8.25** Let $\lambda,\mu \in \mathrm{Par}(n)$. Show that $\lambda \prec \mu \iff \mu^* \prec \lambda^*$ (conjugate partitions). Show further that $\lambda \prec \mu$ implies $\lambda < \mu$ in the lexicographic ordering.

8.26 Let $\lambda = a\underbrace{1\dots1}_{b} \in \mathrm{Par}(n)$, $s_\lambda = \sum_\mu K_{\lambda\mu}m_\mu$. Compute $K_{\lambda\mu}$ for $\mu = c\,1\dots1 \in \mathrm{Par}(n)$, $c \leq a$.

8.27 Find all semistandard tableaux over $\{1,\dots,12\}$ of shape $\lambda = 4431$ and type $(4,2,2,2,2)$.

8.28 What is the coefficient of $x_1^4x_2^3x_3^3x_4^2x_5$ in the Schur function $s_{5431}(x_1,\dots,x_5)$?

▷ **8.29** We know that $K_{\lambda\mu} > 0$ implies $\mu \preceq \lambda$, $\lambda,\mu \in \mathrm{Par}(n)$. Consider $\lambda = \lambda_1\lambda_2 \in \mathrm{Par}(n)$ with two summands, and show conversely that $K_{\lambda\mu} > 0$ whenever $\mu \preceq \lambda$. Note: The converse holds in general.

* * *

8.30 Prove $e_n = \det(h_{1-i+j})_{i,j=1}^n$, and $h_n = \det(e_{1-i+j})_{i,j=1}^n$.

▷ **8.31** Construct a suitable lattice graph to show that $s_{\lambda^*}(x_1,\dots,x_n) = \det(e_{\lambda_i-i+j})$, $\lambda = \lambda_1\dots\lambda_n$, λ^* = conjugate partition.

8.32 Let $\omega : \Lambda^n \to \Lambda^n$ be the involution $\omega : e_\lambda \mapsto h_\lambda$. Use the previous exercise to show that $\omega(s_\lambda) = s_{\lambda^*}$, $\lambda \in \text{Par}(n)$.

8.33 We have expressed $s_\lambda(q^{t+r}, \dots, q)$ in (12) as the generating function of plane partitions (apart from the factor in front), where $\lambda = ss\dots s$. Consider now $\lambda = \lambda_1 \dots \lambda_r$ arbitrary. For which set of plane partitions is $s_\lambda(q^m, q^{m-1}, \dots, q)$ the generating function?

▷ **8.34** Let λ, μ be partitions with at most n parts. Show that

$$s_\lambda(q^{\mu_1+n-1}, q^{\mu_2+n-2}, \dots, q^{\mu_n}) = s_\mu(q^{\lambda_1+n-1}, \dots, q^{\lambda_n}) \cdot \frac{s_\lambda(1, q, \dots, q^{n-1})}{s_\mu(1, q, \dots, q^{n-1})}.$$

8.35 Find the Schur expansion of $\sum_{\mu \in \text{Par}(n)} q^{b(\mu)-1} m_\mu$, where $b(\mu)$ is the number of parts of μ. Hint: The answer is $\sum_{k=0}^{n-1} (q-1)^k s_{n-k\underbrace{1\dots1}_{k}}$.

8.36 Let $\lambda = \lambda_1 \dots \lambda_r \in \text{Par}$, $r \le n$, $\lambda_1 \le t$, and define the partition $\tilde{\lambda} = t - \lambda_n, t - \lambda_{n-1}, \dots, t - \lambda_1$. Use the definition of Schur functions and also the combinatorial description to show that $(x_1 x_2 \cdots x_n)^t s_\lambda(x_1^{-1}, \dots, x_n^{-1}) = s_{\tilde{\lambda}}(x_1, \dots, x_n)$.

▷ **8.37** A plane partition λ is *symmetric* if $\lambda_{ij} = \lambda_{ji}$ for all i, j. Show that the number of symmetric plane partitions of n with at most r rows (and columns) and max $\le t$ equals the number of column-strict plane partitions of n with at most r rows, at most t columns, max $\le 2r-1$, and for which all λ_{ij} are odd. Hint: Find a clever decomposition of a symmetric plane partition reminiscent of the decomposition of a self-conjugate partition $\lambda = \lambda^*$ into odd parts (see Exercise 1.53).

8.4 The RSK Algorithm

We come to a remarkable combinatorial correspondence between integer matrices and pairs of semistandard tableaux, named after Robinson, Schensted, and Knuth. This correspondence, which is achieved through a simple algorithmic procedure, certainly ranks among the finest discoveries in all of combinatorics.

The elementary operation is the insertion of an element k into a given semistandard tableau T. It goes as follows:

(1) Replace in row 1 of T the smallest element a that is larger than k by the new element k. If all elements in row 1 are less than or equal to k, set k at the end of row 1, and stop.

(2) If a has been "bumped" by k, insert a into row 2 according to rule (1), and continue.

Example. Suppose we are given the SST T, and the element 3 is to be inserted. The insertion of 3 results in the tableau $T \leftarrow 3$ on the right, where the inserted elements in each row are drawn in boldface.

$$T = \begin{array}{l} 1\,2\,2\,4\,5\,6\,6 \\ 2\,3\,3\,6\,7 \\ 4\,4\,7\,7\,8 \\ 6\,7 \\ 7\,8 \end{array} \qquad T \leftarrow 3 = \begin{array}{l} 1\,1\,2\,\mathbf{3}\,5\,6\,6 \\ 2\,3\,3\,\mathbf{4}\,7 \\ 4\,4\,\mathbf{6}\,7\,8 \\ 6\,7\,\mathbf{7} \\ 7\,8 \end{array}$$

Let us call the path of the bold elements the *insertion path.* Observe that the strict monotonicity on the columns of T implies that the insertion path goes down and to the left, but never to the right.

Lemma 8.13. *If T is an* SST, *then so is $T \leftarrow k$.*

Proof. The rows in $T \leftarrow k$ are obviously monotone, and rule (1) implies that the columns are strictly increasing. □

Note that the insertion path always stops at the end of a row and column. The following result, whose easy proof is left to the exercises, shows the effect of two insertions.

Lemma 8.14. *Suppose we perform $T \leftarrow k$, and then $(T \leftarrow k) \leftarrow \ell$ with $\ell \geq k$. Then the insertion path of ℓ runs strictly to the right of the insertion of k.*

Example. For $(T \leftarrow 3) \leftarrow 3$ in the example above we get

$$(T \leftarrow 3) \leftarrow 3: \quad \begin{array}{l} 1\,2\,2\,3\,\mathbf{3}\,6\,6 \\ 2\,3\,3\,4\,\mathbf{5} \\ 4\,4\,6\,7\,\mathbf{7} \\ 6\,7\,7\,\mathbf{8} \\ 7\,8 \end{array}$$

By means of this insertion procedure we now construct from a given integer matrix a pair of SSTs of the same shape, and this is the RSK algorithm.

Let $A = (a_{ij})_{i,j\ge1}$ be a matrix over $\mathbb{N}_0$ with only finitely many nonzero elements, $\sum_{i,j} a_{ij} = n$. We associate to A a $2 \times n$-scheme as follows:

$$\begin{pmatrix} 1\dots1\,2\dots2\,3\dots3\dots \\ 1\dots\quad 1\dots\quad 1\dots \end{pmatrix},$$

where column $\binom{i}{j}$ appears a_{ij} times. The first row is increasing $i_1 \le i_2 \le \dots \le i_m$, and below each i the indices j are arranged in increasing fashion.

Example.

$$A = \begin{pmatrix} 1&0&2&1 \\ 0&2&0&0 \\ 1&0&1&0 \end{pmatrix} \longrightarrow \begin{pmatrix} 1&1&1&1&2&2&3&3 \\ 1&3&3&4&2&2&1&3 \end{pmatrix}.$$

It is clear that we may uniquely recover A from the $2 \times n$-scheme. Note that in the $2 \times n$-scheme

$$\begin{aligned} &i \text{ appears } \textstyle\sum_{j\ge1} a_{ij} \text{ times in the 1st row,} \\ &j \text{ appears } \textstyle\sum_{i\ge1} a_{ij} \text{ times in the 2nd row.} \end{aligned} \tag{1}$$

Let us identify from now on the matrix A with its associated $2 \times n$-scheme

$$A = \begin{pmatrix} i_1\ i_2 \dots i_n \\ j_1\ j_2 \dots j_n \end{pmatrix}.$$

The Algorithm.
The RSK algorithm associates to a given matrix A a pair (P, Q) of SSTs by inserting elements step by step. At the beginning $P(0) = Q(0) = \emptyset$. Suppose $(P(t), Q(t))$ has been constructed. Then

(A) $P(t+1) = P(t) \leftarrow j_{t+1}$,
(B) $Q(t+1)$ arises from $Q(t)$ by putting i_{t+1} in that position such that $Q(t+1)$ has the same shape as $P(t+1)$. The other elements of $Q(t)$ remain unchanged.

Example. With $A = \begin{pmatrix} 1&1&1&1&2&2&3&3 \\ 1&3&3&4&2&2&1&3 \end{pmatrix}$ we get

$P(t)$:	1	13	133	1334	1234 3	1224 33	1124 23 3	1123 234 3
$Q(t)$:	1	11	111	1111	1111 2	1111 22	1111 22 3	1111 223 3

Theorem 8.15. *The* RSK *algorithm gives a bijection between the matrices A over* $\mathbb{N}_0$ *(with finitely many nonzero elements) and the ordered pairs* (P, Q) *of semistandard tableaux of the same shape. Furthermore, type* $(P) = \operatorname{col}(A)$*, type* $(Q) = \operatorname{row}(A)$.

Proof. Suppose $A \overset{\text{RSK}}{\longrightarrow} (P, Q)$. By Lemma 8.13, P is an SST, and Q has by construction the same shape as P. So we have to show that Q is also an SST. Since in the $2 \times n$-scheme the first row $i_1 \leq \cdots \leq i_n$ is increasing, Q is certainly monotone in the rows and columns, since the new element is always placed at the end of a row and column. To prove strict monotonicity of the columns we have to verify that $i_k = i_\ell$ $(k < \ell)$ never appear in the same column of Q. Suppose $i_k = i_{k+1}$; then $j_k \leq j_{k+1}$ by the setup of the $2 \times n$-scheme. According to Lemma 8.14, the insertion path of j_{k+1} is strictly to the right of the insertion path of j_k. But this means that i_{k+1} must land in a position to the right of i_k (not necessarily in the same row), and so Q is an SST.

This last property is the basis for the inverse construction $(P, Q) \to A$. Equal elements $i_k = i_{k+1} = \cdots = i_\ell$ appear in Q strictly from left to right. Suppose we are given a pair (P, Q) of the same shape, $P = P(n)$, $Q = Q(n)$. Let Q_{rs} be the *rightmost* entry of the largest element in Q. Then we know that this is the position where the last insertion path in $P = P(n)$ ended. We can now uniquely recover this insertion path. Indeed, P_{rs} was bumped by the rightmost entry in row $r - 1$, which is smaller than P_{rs} (or $r = 1$, in which case we are finished). From row $r - 1$ we go back to row $r - 2$ until we find the element a that was inserted in $P(n-1)$. Now set $Q(n-1) = Q(n) \smallsetminus Q_{rs}$, and continue. The last assertion is clear from (1). □

Example. Consider $P = P(7) = \begin{matrix} 1\,1\,2\,2 \\ 2\,2 \\ 3 \end{matrix}$, $Q = Q(7) = \begin{matrix} 1\,1\,1\,2 \\ 2\,2 \\ 3 \end{matrix}$

In Q, 3 was placed last. Hence in P, 3 was bumped by $P_{22} = 2$, 2 was bumped by $P_{12} = 1$ in row 1, and thus

$$P(6) = \begin{matrix} 1\,2\,2\,2 \\ 2\,3 \end{matrix}, \quad Q(6) = \begin{matrix} 1\,1\,1\,2 \\ 2\,2 \end{matrix}, \quad i_7 = 3,\ j_7 = 1.$$

The other backward steps are

$$\begin{aligned}
&P(5) = \begin{matrix} 1\,2\,2 \\ 2\,3 \end{matrix}, && Q(5) = \begin{matrix} 1\,1\,1 \\ 2\,2 \end{matrix}, && i_6 = 2,\ j_6 = 2,\\
&P(4) = \begin{matrix} 1\,2\,3 \\ 2 \end{matrix}, && Q(4) = \begin{matrix} 1\,1\,1 \\ 2 \end{matrix}, && i_5 = 2,\ j_5 = 2,\\
&P(3) = 2\,2\,3, && Q(3) = 1\,1\,1, && i_4 = 2,\ j_4 = 1,\\
&P(2) = 2\,2, && Q(2) = 1\,1, && i_3 = 1,\ j_3 = 3,\\
&P(1) = 2, && Q(1) = 1, && i_2 = 1,\ j_2 = 2,\\
&P(0) = \emptyset, && Q(0) = \emptyset, && i_1 = 1,\ j_1 = 2.
\end{aligned}$$

This gives the scheme $\begin{pmatrix} 1&1&1&2&2&2&3 \\ 2&2&3&1&2&2&1 \end{pmatrix}$, or the matrix $A = \begin{pmatrix} 0&2&1 \\ 1&2&0 \\ 1&0&0 \end{pmatrix}$.

Applications to Symmetric Functions.

The RSK algorithm can be used to give quick proofs of some classical theorems about symmetric functions.

Corollary 8.16 (Cauchy). *Let* $X = \{x_1, x_2, \ldots\}$, $Y = \{y_1, y_2, \ldots\}$. *Then*

$$\prod_{i,j} \frac{1}{1 - x_i y_j} = \sum_{\lambda \in \mathrm{Par}} s_\lambda(x) s_\lambda(y). \tag{2}$$

Proof. We write

$$\prod_{i,j} \frac{1}{1 - x_i y_j} = \prod_{i,j} \left(\sum_{a_{ji} \geq 0} (x_i y_j)^{a_{ji}} \right). \tag{3}$$

Expanding the right-hand side of (3), we see that a monomial $x^\alpha y^\beta = (x_1^{\alpha_1} x_2^{\alpha_2} \cdots)(y_1^{\beta_1} y_2^{\beta_2} \cdots)$ corresponds to a matrix $A = (a_{ij})$ with $\mathrm{col}\,(A) = \alpha$ and $\mathrm{row}\,(A) = \beta$. In other words, the coefficient of $x^\alpha y^\beta$ in (3) is precisely the number of these matrices. On the other hand, Theorem 8.8 says that the coefficient of $x^\alpha y^\beta$ in $\sum_\lambda s_\lambda(x) s_\lambda(y)$ is the number of ordered pairs (P, Q) of SSTs of the same shape, and with $\mathrm{type}(P) = \alpha$ and $\mathrm{type}(Q) = \beta$. The result follows now from the RSK correspondence. □

Corollary 8.17. *Let $\mu, \nu \in \operatorname{Par}(n)$. Then*

$$\sum_{\lambda \in \operatorname{Par}(n)} K_{\lambda\mu} K_{\lambda\nu} = N_{\mu\nu}\,, \tag{4}$$

where $K_{\lambda,\mu}, N_{\mu\nu}$ are defined as in the previous sections.

Proof. According to Exercise 8.14,

$$\prod_{i,j} \frac{1}{1 - x_i y_j} = \sum_{\mu,\nu \in \operatorname{Par}} N_{\mu\nu} m_\mu(x) m_\nu(y)\,.$$

So for $\mu, \nu \in \operatorname{Par}(n)$ the coefficient of $x^\mu y^\nu$ is $N_{\mu\nu}$. On the other hand, by the result just proved this is also the coefficient of $x^\mu y^\nu$ in $\sum_\lambda s_\lambda(x) s_\lambda(y) = \sum_\lambda \left(\sum_\mu K_{\lambda\mu} m_\mu \cdot \sum_\nu K_{\lambda\nu} m_\nu \right)$, that is, $\sum_\lambda K_{\lambda\mu} K_{\lambda\nu}$. □

Symmetry of the RSK Algorithm.
The RSK algorithm enjoys a remarkable symmetry property, which, in turn, leads to elegant proofs of some theorems about symmetric functions. Our goal is to show that

$$A \overset{\text{RSK}}{\longrightarrow} (P, Q) \;\text{ implies }\; A^T \overset{\text{RSK}}{\longrightarrow} (Q, P)\,, \tag{5}$$

where A^T is the transpose of A.

In order to prove this we proceed in several steps. First we show that it suffices to verify (5) when A is an $n \times n$ permutation matrix. That is, A contains precisely one entry 1 in each row and column, and 0's elsewhere.

Let us note the following invariance property of the insertion algorithm. Suppose we are given a sequence $\alpha = a_1 a_2 \dots a_n$ of positive integers. We denote by P_α the SST obtained by successively inserting $a_1, a_2, \dots, a_n$. We linearly order the a_i's according to magnitude from left to right. That is, $a_i \prec a_j$ if $a_i < a_j$ or if $a_i = a_j$ and $i < j$, and we denote by $r(a_i)$ the rank of a_i in this ordering. Thus, the smallest element gets rank 1, the second smallest rank 2, and so on. The resulting permutation $\pi_\alpha = r(a_1) \dots r(a_n) \in S(n)$ is called the *order permutation* of α.

Example. For $\alpha = 351126533$ we obtain $\pi_\alpha = 471239856$.

Lemma 8.18. *Suppose* $\alpha = a_1 \dots a_n$ *and* $\beta = b_1 \dots b_n$ *have associated tableaux* P_α *and* P_β. *If the order permutations are identical,* $\pi_\alpha = \pi_\beta$, *then* P_β *arises from* P_α *by replacing each* a_i *in* P_α *by* b_i, *and conversely,* P_α *arises from* P_β *by replacing each* b_i *in* P_β *by* a_i.

Proof. All we have to notice is that at every stage of the algorithm the insertion path of a_{t+1} in $P_\alpha(t)$ to produce $P_\alpha(t+1)$ is identical to the insertion path of b_{t+1} in $P_\beta(t)$. But this is clear by rule (1) of the algorithm. □

Suppose now we are given $A = (a_{ij})$ over $\mathbb{N}_0$ with $2 \times n$ scheme $S = \begin{pmatrix} i_1 \, i_2 \, \dots \, i_n \\ j_1 \, j_2 \, \dots \, j_n \end{pmatrix}$. We associate to S a new scheme $\widetilde{S} = \begin{pmatrix} 1 & 2 & \dots & n \\ \widetilde{j}_1 & \widetilde{j}_2 & \dots & \widetilde{j}_n \end{pmatrix}$, where $\widetilde{j}_1\widetilde{j}_2 \dots \widetilde{j}_n$ is the order permutation of $j_1 j_2 \dots j_n$. The matrix $\widetilde{A}$ belonging to $\widetilde{S}$ is thus an $n \times n$ permutation matrix.

Example. $S = \begin{pmatrix} 112223566 \\ 231156144 \end{pmatrix} \longrightarrow \widetilde{S} = \begin{pmatrix} 123456789 \\ 451289367 \end{pmatrix}$.

Since $j_1 j_2 \dots j_n$ and $\widetilde{j}_1\widetilde{j}_2 \dots \widetilde{j}_n$ have by construction the same order permutation (namely $\widetilde{j}_1\widetilde{j}_2 \dots \widetilde{j}_n$), the preceding lemma immediately yields the following result.

Lemma 8.19. *Suppose* $A \overset{\text{RSK}}{\longrightarrow} (P,Q)$ *and* $\widetilde{A} \overset{\text{RSK}}{\longrightarrow} (\widetilde{P},\widetilde{Q})$. *Then we get* P *from* $\widetilde{P}$ *by replacing each* $\widetilde{j}_k$ *in* $\widetilde{P}$ *by* j_k, *and similarly* Q *arises from* $\widetilde{Q}$ *by replacing each* $\widetilde{i}_k = k$ *in* $\widetilde{Q}$ *by* i_k.

We can express the content of the lemma compactly as a "commuting" diagram:

$$\begin{array}{ccccc} A & \xrightarrow{\sim} & \widetilde{A} & \xrightarrow{\text{RSK}} & (\widetilde{P},\widetilde{Q}) \\ & \searrow^{\text{RSK}} & & \swarrow_{\sim^{-1}} & \\ & & (P,Q) & & \end{array} \tag{6}$$

It is straightforward (see Exercise 8.40) that

$$A \xrightarrow{\sim} \widetilde{A} \quad \text{implies} \quad A^T \xrightarrow{\sim} \widetilde{A}^T, \tag{7}$$

and this leads to the first step of the proof of (5).

Lemma 8.20. *If the* RSK *algorithm observes the symmetry property (5) for all permutation matrices, then it holds in general.*

Proof. Suppose $A \stackrel{\text{RSK}}{\longrightarrow} (P, Q)$, $\widetilde{A} \longrightarrow (\widetilde{P}, \widetilde{Q})$. Then by (6) and (7),

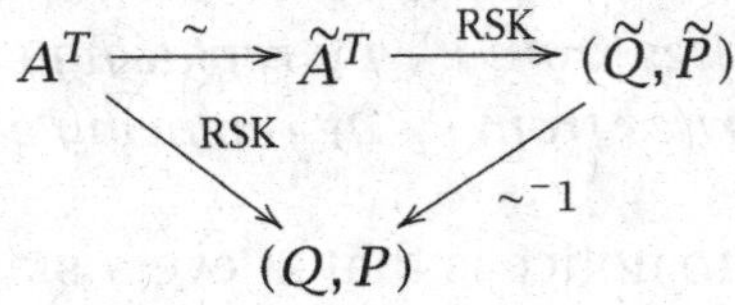

and we are done. □

Let us therefore look at permutation matrices A with $2 \times n$ scheme $\binom{1 \ldots n}{\pi_1 \ldots \pi_n}$, $\pi = \pi_1 \ldots \pi_n \in S(n)$. We then write $A = A(\pi)$ with $(P(\pi), Q(\pi))$ as associated pair. Now, $A(\pi)^T = A(\pi^{-1})$, since $\pi(i) = j$ implies $\pi^{-1}(j) = i$. Hence we have to show that

$$A(\pi) \stackrel{\text{RSK}}{\longrightarrow} (P, Q) \text{ implies } A(\pi^{-1}) \stackrel{\text{RSK}}{\longrightarrow} (Q, P). \tag{8}$$

From now on we identify $A(\pi)$ with π, and write $\pi \stackrel{\text{RSK}}{\longrightarrow} (P(\pi), Q(\pi))$. The semistandard tableaux $P(\pi)$ and $Q(\pi)$ contain the numbers $1, 2, \ldots, n$ exactly once each, that is, they have type $(1, 1, \ldots, 1)$. This suggests the following definition.

Definition. A semistandard tableau T over $\{1, 2, \ldots, n\}$ is called a *standard tableau* ST if every $k \in \{1, \ldots, n\}$ appears exactly once in T.

The following idea is the key to the proof of (8). Consider $\pi = \pi_1 \ldots \pi_n \in S(n)$. A *subword* of π is a sequence $\pi_{i_1}, \ldots, \pi_{i_k}$ with $i_1 < \cdots < i_k$. The subword is *increasing* if $\pi_{i_1} < \cdots < \pi_{i_k}$, and *decreasing* if $\pi_{i_1} > \cdots > \pi_{i_k}$. Denote by $is(\pi_j)$ the length of a longest increasing subword *ending* in π_j.

Consider now the pair of STs (P, Q) associated to π by the RSK algorithm. We wish to determine the first rows of P and Q. Call the sequence $\pi_{i_1}, \ldots, \pi_{i_k}$ successively inserted into the j-th box of the first row of P the j-th *fundamental sequence.* The rules of the algorithm clearly imply the following properties:

(i) Every number is in exactly one fundamental sequence.

(ii) Every fundamental sequence is a *decreasing* subword $\pi_{i_1} > \cdots > \pi_{i_k}$.

(iii) When a is inserted into the j-th box of row 1 of π, then the number in the adjacent box $j - 1$ of row 1 is *smaller* than a and comes *before* a in π.

It helps to write the j-th fundamental sequence as $2 \times k$ scheme. According to property (ii),

$$\begin{pmatrix} i_1 < i_2 < \cdots < i_k \\ \pi_{i_1} > \pi_{i_2} > \cdots > \pi_{i_k} \end{pmatrix}, \tag{9}$$

and we infer that $P_{1j} = \pi_{i_k}$ and $Q_{1j} = i_1$.

The main idea toward the proof that $\pi^{-1} \overset{\text{RSK}}{\longrightarrow} (Q, P)$ is the following description of the fundamental sequences. Denote by $fs(a)$ the index of the fundamental sequence which contains a.

Lemma 8.21. *Given $\pi \in S(n)$, then $fs(a) = is(a)$ for all $a \in \{1, \ldots, n\}$.*

Proof. Property (iii) immediately implies $fs(a) \leq is(a)$. Now let $b_1 < b_2 < \cdots < b_{is(a)} = a$ be a longest increasing subword ending in a. We clearly have $is(b_k) = k$, $1 \leq k \leq is(a)$, and therefore $fs(b_k) \leq k$. On the other hand, we infer from property (ii) that the b_k's must appear in different fundamental sequences. But this means $fs(b_1) = 1, fs(b_2) = 2, \ldots, fs(a) = is(a)$. □

Example. Consider $\pi = 72483615$. The tableaux are

$$P = \begin{array}{l} 135 \\ 26 \\ 48 \\ 7 \end{array}, \quad Q = \begin{array}{l} 134 \\ 26 \\ 58 \\ 7 \end{array},$$

with fundamental sequences

$$FS_1 = \begin{pmatrix} 127 \\ 721 \end{pmatrix}, \quad FS_2 = \begin{pmatrix} 35 \\ 43 \end{pmatrix}, \quad FS_3 = \begin{pmatrix} 468 \\ 865 \end{pmatrix}.$$

The next observation relates π to π^{-1}. Look at the diagram of π as in Section 1.4:

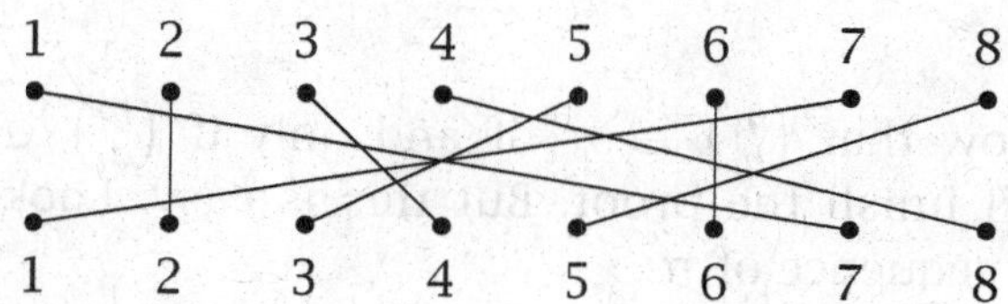

It is immediate that $is(\pi_h) = j$, where j is the maximum number of *nonintersecting* edges to the left and including the edge (h, π_h).

But the same diagram read from bottom up gives π^{-1}. Accordingly, $is(h,\pi_h) = j$ in π if and only if $is(\pi_h,h) = j$ in π^{-1}. By the lemma this means that

$$\text{If } \begin{pmatrix} \boldsymbol{i_1} < \cdots < i_k \\ \pi_{i_1} > \cdots > \boldsymbol{\pi_{i_k}} \end{pmatrix} \text{ is the } j\text{-th fundamental sequence of } \pi$$

$$\text{then } \begin{pmatrix} \boldsymbol{\pi_{i_k}} < \cdots < \pi_{i_1} \\ i_k > \cdots > \boldsymbol{i_1} \end{pmatrix} \text{ is the } j\text{-th fundamental sequence of } \pi^{-1}. \tag{10}$$

With these preparations we can now prove the desired result.

Proposition 8.22. *Let* $\pi \in S(n)$. *If* $\pi \overset{\text{RSK}}{\longrightarrow} (P,Q)$, *then* $\pi^{-1} \overset{\text{RSK}}{\longrightarrow} (Q,P)$.

Proof. Let $\pi = \pi_1 \dots \pi_n$, and $\pi^{-1} \overset{\text{RKS}}{\longrightarrow} (\widehat{P},\widehat{Q})$. We use induction on the number of rows in P. We know the first rows of P and Q, $P_{1j} = \pi_{i_k}$, $Q_{1j} = i_1$ from (9), and similarly $\widehat{P}_{1j} = i_1$, $\widehat{Q}_{1j} = \pi_{i_k}$. Hence $\widehat{P}, Q$ agree in the first row, as do $\widehat{Q}$ and P. Now look at the tableaux P', Q' of P and Q below row 1. The elements of P' are those that get bumped down from row 1. So this is the set $\{1,\dots,n\}$ minus the *last* elements π_{i_k} of each fundamental sequence. Let $b_1,\dots,b_s$ be the order in which they are bumped. Similarly, the available elements for Q' comprise the set $\{1,\dots,n\}$ minus the *first* elements i_1 in each fundamental sequence, with ordering $a_1 < \cdots < a_s$. Thus for $\pi' = \binom{a_1a_2 \dots a_s}{b_1b_2 \dots b_s}$ we have $\pi' \overset{\text{RSK}}{\longrightarrow} (P',Q')$.

Example. In the example above we obtain

$$\pi' = \begin{pmatrix} 25678 \\ 74826 \end{pmatrix} \quad \text{with} \quad \pi' \overset{\text{RSK}}{\longrightarrow} \begin{pmatrix} 26 & 26 \\ 48 \; , & 58 \\ 7 & 7 \end{pmatrix}.$$

If we can show that $\binom{a_i}{b_i} \in \pi'$ if and only if $\binom{b_i}{a_i} \in \pi'^{-1}$, then induction will finish the proof. But this is easy. Look at the j-th fundamental sequence of π:

$$\begin{pmatrix} \boldsymbol{i_1} < i_2 < \cdots < i_k \\ \pi_{i_1} > \pi_{i_2} > \cdots > \boldsymbol{\pi_{i_k}} \end{pmatrix}.$$

Since π_{i_2} bumps π_{i_1}, π_{i_3} bumps π_{i_2}, and so on, we see that the bumped elements of the j-th fundamental sequence appear in the order $\pi_{i_1} > \pi_{i_2} > \cdots > \pi_{i_{k-1}}$ in π', and similarly the available elements in Q' appear in the order $i_2 < \cdots < i_k$. Thus π' contains the pairs

$$\begin{pmatrix} i_2 & < \cdots < & i_k \\ \pi_{i_1} & > \cdots > & \pi_{i_{k-1}} \end{pmatrix}.$$

By the same argument, π'^{-1} contains the pair

$$\begin{pmatrix} \pi_{i_{k-1}} & < \cdots < & \pi_{i_1} \\ i_k & > \cdots > & i_2 \end{pmatrix}.$$

Since by property (i) of fundamental sequences all elements in π' and π'^{-1} are captured in this way, the proof is complete. □

With this result we have proved the main theorem.

Theorem 8.23. *If* $A \stackrel{\text{RSK}}{\longrightarrow} (P,Q)$, *then* $A^T \stackrel{\text{RSK}}{\longrightarrow} (Q,P)$.

Corollary 8.24. *A matrix A is symmetric if and only if* $A \stackrel{\text{RSK}}{\longrightarrow} (P,P)$ *for some semistandard tableau P. Furthermore,* $\text{row}(A) = \text{col}(A) = \text{type}(P)$. *The insertion algorithm thus furnishes a bijection between all symmetric matrices A with* $\text{row}(A) = \alpha$ *and all semistandard tableaux P of type α.*

Example. Consider the symmetric matrices A with $\text{row}(A) = (2,2)$. There are three such matrices with the correspondence $A \to P$:

$$\begin{pmatrix} 2 & 0 \\ 0 & 2 \end{pmatrix} \longrightarrow 1122, \quad \begin{pmatrix} 0 & 2 \\ 2 & 0 \end{pmatrix} \longrightarrow \begin{matrix} 11 \\ 22 \end{matrix}, \quad \begin{pmatrix} 1 & 1 \\ 1 & 1 \end{pmatrix} \longrightarrow \begin{matrix} 112 \\ 2 \end{matrix}.$$

To finish we derive another famous identity involving Schur functions, which will prove useful in the enumeration of plane partitions (see Exercises 8.50 ff).

Corollary 8.25. *We have*

$$\frac{1}{\prod_{i \ge 1} (1 - x_i) \prod_{1 \le i < j} (1 - x_i x_j)} = \sum_{\lambda \in \text{Par}} s_\lambda(x). \tag{11}$$

Proof. The product on the left is

$$\left(\sum_{k\ge 0} x_1^k\right)\left(\sum_{k\ge 0} x_2^k\right)\cdots\left(\sum_{k\ge 0} (x_1x_2)^k\right)\cdots.$$

It follows that the coefficient of $x^\alpha = x_1^{\alpha_1}x_2^{\alpha_2}\cdots$ is equal to the number of *symmetric* matrices A with $\text{row}(A) = \alpha$. On the other hand, by Theorem 8.8,

$$\sum_\lambda s_\lambda(x) = \sum_\lambda \sum_T x^T,$$

where the inner sum extends over all SSTs T of shape λ. The coefficient of a monomial x^α is therefore the number of all SSTs T with type $(T) = \alpha$. The correspondence $A = A^T \to T \in \text{SST}$ (RSK algorithm) finishes the proof. □

Exercises

8.38 Prove Lemma 8.14.

8.39 Find the associated tableaux (P,Q) for $A = \begin{pmatrix} 2 & 1 & 0 & 3\\ 1 & 0 & 0 & 1\\ 0 & 2 & 1 & 2\end{pmatrix}$.

8.40 Prove (7).

▷ **8.41** Show that the number of SSTs with type (r,r) equals $r+1$, and of those with type (r,r,r), that is, r 1's, 2's, and 3's, is given by $\frac{1}{16}[4r^3 + 18r^2 + 28r + 15 + (-1)^r]$.

8.42 Give the bijection $A \to P \in \text{SST}$ for all symmetric matrices A with row $(A) = (3,1,2)$.

▷ **8.43** Show that $h_\mu = \sum_\lambda K_{\lambda\mu} s_\lambda$, $e_\mu = \sum_\lambda K_{\lambda^*\mu} s_\lambda$, λ^* = conjugate partition, h_μ = complete symmetric function, e_μ = elementary symmetric function. Use either the RSK algorithm or Corollary 8.16.

* * *

8.44 Let A be a matrix with entries 0, 1, and $S = \binom{i_1 \dots i_n}{j_1 \dots j_n}$ the $2\times n$-scheme. The *dual* RSK algorithm, denoted by RSK^d, proceeds as before, except that a new element k bumps the leftmost element greater than or equal to k (rather than the leftmost element greater than k). Show that we get a pair (P,Q), where each row of P is strictly increasing. Show further that RSK and RSK^d agree when A is a permutation matrix.

8.45 Prove that the dual RSK algorithm gives a bijection between 0, 1-matrices and pairs (P,Q) such that P^* (conjugate of P) and Q are SSTs of the same shape. Show further that $\mathrm{col}(A) = \mathrm{type}(P)$, $\mathrm{row}(A) = \mathrm{type}(Q)$.

8.46 Use the RSK^d algorithm to derive the analogue to Corollary 8.16: $\prod_{i,j}(1+x_iy_j) = \sum_\lambda s_\lambda(x)s_{\lambda^*}(y)$, λ^* = conjugate partition.

▷ **8.47** Use the previous exercise to re-prove $\omega s_\lambda = s_{\lambda^*}$, where ω is the involution defined in Section 8.2.

8.48 Let A be a symmetric matrix, and suppose $A \overset{\text{RSK}}{\longrightarrow} (P,P)$. Show that the trace of A $(= \sum a_{ii})$ equals the number of columns of P of odd length. Hint: $\mathrm{tr}(A) = \mathrm{tr}(\tilde{A})$.

▷ **8.49** Use the previous exercise to prove

$$\prod_{i\ge1}\frac{1}{1-qx^i}\prod_{1\le i<j}\frac{1}{1-x_ix_j} = \sum_{\lambda\in\mathrm{Par}} q^{o(\lambda^*)}s_\lambda(x),$$

where $o(\lambda^*)$ denotes the number of odd parts of λ^*.

8.50 Recall the definition of column-strict plane partitions, and let $spp(n)$ be the number of these partitions of n. Prove that the generating function is given by

$$\sum_{n\ge0} spp(n)q^n = \prod_{i\ge1}\frac{1}{1-q^i}\prod_{1\le i<j}\frac{1}{1-q^{i+j}} = \prod_{k\ge1}\frac{1}{(1-q^k)^{\lceil k/2\rceil}}.$$

Hint: Use Corollary 8.25.

8.51 Use Corollary 8.25 to prove that the generating function for column-strict plane partitions for which every entry is *odd* is given by

$$\prod_{i\ge1}\frac{1}{1-q^{2i-1}}\prod_{1\le i<j}\frac{1}{1-q^{2i+2j-2}} = \prod_{k\ge1}\frac{1}{(1-q^k)^{\rho(k)}},$$

where $\rho(k) = 1$ if k is odd, and $\rho(k) = \lfloor \frac{k}{4} \rfloor$ if k is even.

▷ **8.52** Combine Exercises 8.37 and 8.51 to prove that the generating function of *symmetric* plane partitions with at most r rows and max $\le t$ is given by

$$\sum_\lambda s_\lambda(q^{2r-1}, q^{2r-3}, \ldots, q^3, q),$$

where the summation extends over all $\lambda \in \mathrm{Par}(;\le r;\le t)$.

8.5 Standard Tableaux

We know from the previous section that any permutation $\pi = \pi_1 \dots \pi_n \in S(n)$ corresponds to a pair (P, Q) of standard tableaux of the same shape, and conversely that to every such pair there is a unique permutation. Let $f(\lambda)$ be the number of standard tableaux of shape λ.

Example. The standard tableaux for $n = 4$ are

$$\begin{array}{llllllllll} 1234 & 123 & 124 & 134 & 12 & 13 & 12 & 13 & 14 & 1 \\ & 4 & 3 & 2 & 34 & 24 & 3 & 2 & 2 & 2 \\ & & & & & & 4 & 4 & 3 & 3 \\ & & & & & & & & & 4 \end{array}$$

It is clear that the *conjugate* tableau T^* of a standard tableau is also a standard tableau; hence $f(\lambda) = f(\lambda^*)$.

Now there are $n!$ permutation matrices $A(\pi)$, and $A(\pi)^T = A(\pi^{-1})$ implies that $A(\pi)$ is symmetric if and only if π is an involution. The main theorems of the last section imply therefore the following result.

Theorem 8.26. *We have*

a. $\sum_{\lambda \in \mathrm{Par}(n)} f(\lambda)^2 = n!$,
b. $\sum_{\lambda \in \mathrm{Par}(n)} f(\lambda) = i_n$ *(number of involutions).*

Looking at the example above we obtain $f(4) = 1$, $f(31) = 3$, $f(22) = 2$, $f(211) = 3$, $f(1111) = 1$, which agrees with $1^2 + 3^2 + 2^2 + 3^2 + 1^2 = 24$ and $1 + 3 + 2 + 3 + 1 = 10 = i_4$. Note also that $f(\lambda)$ is the Kostka number $K_{\lambda\overline{\mu}}$, where $\overline{\mu} = 11 \dots 1$.

Since permutations π correspond via the insertion algorithm to standard tableaux $P(\pi)$, we will expect that some combinatorial properties of π are reflected in $P(\pi)$. Let us write $\pi \stackrel{ins}{\longrightarrow} P(\pi)$. We already know one such example proved by means of fundamental sequences. For $\pi = \pi_1 \dots \pi_n \in S(n)$ let $is(\pi)$ be the length of a longest increasing subword of π; thus $is(\pi) = \max_{1 \le a \le n} is(a)$. Appealing to Lemma 8.21 we have the following result.

Proposition 8.27. *Let* $\pi \stackrel{ins}{\longrightarrow} P$, *and* $\lambda = \lambda_1 \dots \lambda_r$ *the shape of* P. *Then*
a. $is(\pi) = \lambda_1$,

b. $\#\{\pi \in S(n) : is(\pi) = t\} = \sum_{\lambda \in \mathrm{Par}(n;;t)} f(\lambda)^2$.

It is natural to look at decreasing subwords as well. Let $ds(\pi)$ be the length of a longest decreasing subword of π. We suspect that $ds(\pi) = \lambda_1^* =$ number of rows of λ, and this is indeed the case.

If $\pi = \pi_1 \dots \pi_n$, then $\pi^* = \pi_n \dots \pi_1$ is called the *reversed permutation.* We are going to show that if $\pi \overset{ins}{\longrightarrow} P$, then $\pi^* \overset{ins}{\longrightarrow} P^*$, which will clearly imply the assertion. For the proof we need the following basic lemma, whose (elementary) proof is only sketched. Recall that $T \leftarrow k$ denotes row insertion. Dually, we may insert $\ell \rightarrow T$ columnwise, observing the usual rules of the insertion algorithm. Thus $\ell \rightarrow T = (T^* \leftarrow \ell)^*$.

Lemma 8.28. *Let T be an* SST *with distinct entries, and $k \neq \ell$ two numbers not contained in T. Then*

$$\ell \rightarrow (T \leftarrow k) = (\ell \rightarrow T) \leftarrow k. \tag{1}$$

In other words, the operations $\leftarrow$ and $\rightarrow$ commute.

Proof. Let $W_1 : k < x_1 < \cdots < x_r$ be the insertion path when k is inserted in T. Thus x_1 is bumped by k, x_2 is bumped by x_1, and so on. The element x_i is therefore in row i of T, and in row $i+1$ of $T \leftarrow k$. Similarly, $W_2 : \ell < y_1 < \cdots < y_s$ is the insertion path in $\ell \rightarrow (T \leftarrow k)$, where y_j is in column j of $T \leftarrow k$, and in column $j+1$ of $\ell \rightarrow (T \leftarrow k)$. If W_1 and W_2 never meet, then (1) plainly holds.

Suppose $x = x_i$ is the first position where W_2 hits $\{k, x_1, \dots, x_r\}$ in $T \leftarrow k$. Set $w = x_{i-1}$, $z = x_{i+1}$, and let $y = y_j$ be the element in W_2 that bumps x.

Case 1. z is to the left of x in T. Then we have the following situation:

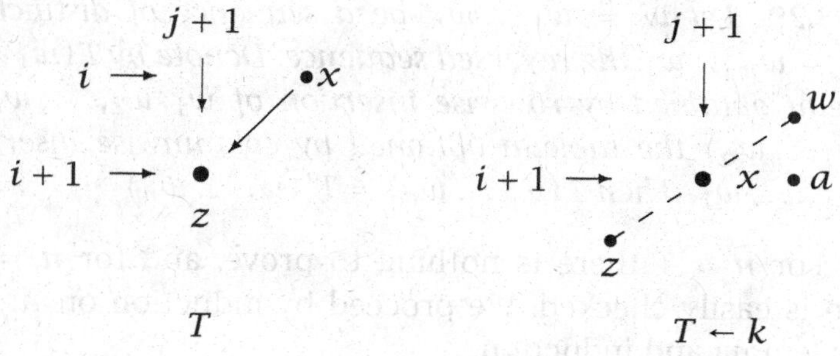

Since $w < x$, x is bumped by y in $\ell \to (T \leftarrow k)$ to the right, and so on, and then a starts an insertion path that may also go up. But note that W_2 never touches W_1 again, since $y = y_j < x = y_{j+1} < y_{j+2} < \cdots$ is an increasing sequence, whereas $x = x_i > w = x_{i-1} > \cdots$ is decreasing. The picture in $\ell \to (T \leftarrow k)$ looks therefore as follows:

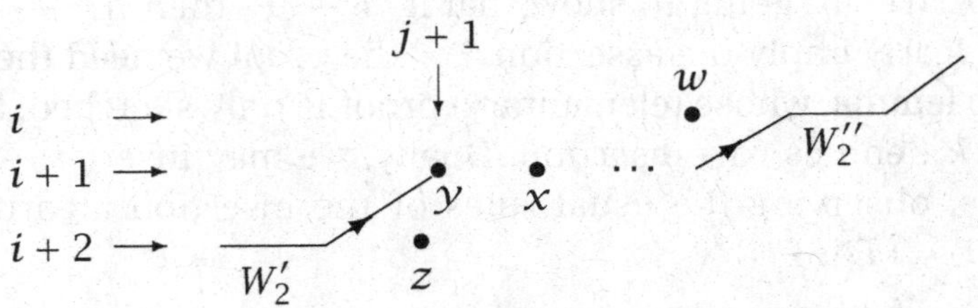

Now we interchange the insertions. Up to y_{j-1} the path W_2 is the same. Since $z > x > y$, $y = y_j$ bumps z in $\ell \to T$ (in column $j+1$), z moves to the right, and so on, and a is bumped along W_2'' as before. The situation is therefore

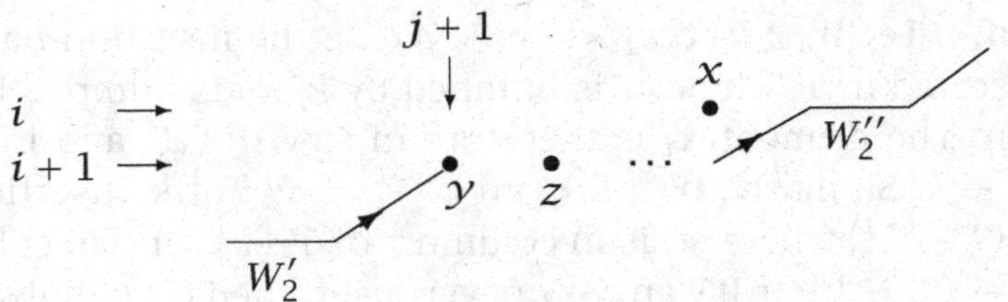

Now we insert k in $\ell \to T$. Up to x the path W_1 proceeds as before (since it does not touch W_2''), x bumps z because of $y < x < z$, and z continues the insertion path W_1 below row $i + 1$ as before.

Case 2. x is above z. This is settled in an analogous fashion. □

Lemma 8.29. *Let $w = w_1 \dots w_n$ be a sequence of distinct numbers, $w^* = w_n \dots w_1$ the reversed sequence. Denote by $T(w_1 \dots w_n)$ the tableau obtained by rowwise insertion of $w_1, w_2, \dots, w_n$, and by $T'(w_1 \dots w_n)$ the tableau obtained by columnwise insertion of $w_n, w_{n-1}, \dots, w_1$. Then $T(w_1 \dots w_n) = T'(w_1 \dots w_n)$.*

Proof. For $n = 1$ there is nothing to prove, and for $n = 2$ the assertion is easily checked. We proceed by induction on n. By the previous lemma and induction,

$$\begin{aligned}T(w_1\dots w_n) &= T(w_1\dots w_{n-1}) \leftarrow w_n = T'(w_1\dots w_{n-1}) \leftarrow w_n\\ &= (w_1 \rightarrow T'(w_2\dots w_{n-1})) \leftarrow w_n\\ &= w_1 \rightarrow (T'(w_2\dots w_{n-1}) \leftarrow w_n)\\ &= w_1 \rightarrow (T(w_2\dots w_{n-1}) \leftarrow w_n)\\ &= w_1 \rightarrow T(w_2\dots w_{n-1}w_n) = w_1 \rightarrow T'(w_2\dots w_n)\\ &= T'(w_1\dots w_n),\end{aligned}$$

and we are done. □

Proposition 8.30. *Let* $\pi = \pi_1\dots\pi_n \in S(n)$ *with* $\pi \xrightarrow{ins} P$. *Then* $\pi^* \xrightarrow{ins} P^*$ *for the reversed permutation* $\pi^* = \pi_n\dots\pi_1$.

Proof. By the lemma, $P = T(\pi_1\dots\pi_n) = T'(\pi_1\dots\pi_n)$, where $\pi^* = \pi_n\dots\pi_1$ is inserted columnwise. Hence $\pi^* \xrightarrow{ins} P^*$ when π^* is inserted rowwise. □

Since increasing subwords in π^* corresponds to decreasing subwords in π, we obtain the following companion result to Proposition 8.27:

Corollary 8.31. *Let* $\pi \xrightarrow{ins} P$ *with shape* $\lambda = \lambda_1\dots\lambda_r$. *Then*

a. $ds(\pi) = r = \lambda_1^*$,
b. $\#\{\pi \in S(n) : ds(\pi) = s\} = \sum_{\lambda\in\mathrm{Par}(n;s;)} f(\lambda)^2$,
c. $\#\{\pi \in S(n) : ds(\pi) = s, is(\pi) = t\} = \sum_{\lambda\in\mathrm{Par}(n;s;t)} f(\lambda)^2$.

Example. For $n = 4$ we have noted that $f(\lambda) = 3$ for $\lambda = 31$, and $\lambda = 31$ is the only partition with $\lambda_1 = 3$, $\lambda_1^* = 2$. Hence there are 9 permutations $\pi \in S(4)$ with $is(\pi) = 3$, $ds(\pi)$=2:

1243, 1324, 1342, 1423, 2134, 2314, 2341, 3124, 4123.

Exercises

8.53 Let $\pi = \pi_1\dots\pi_n \in S(n)$. Show that $is(\pi) = k$ if and only if π can be decomposed into k decreasing subwords.

8.54 Let $\pi \in S(n)$ with $\pi \xrightarrow{ins} P$. For $a \in \{1,\dots,n\}$ let $r(a)$ be the index of the row in P that contains a, and similarly $c(a)$ the column index. Show that $c(a) \le is(a)$, $r(a) \le ds(a)$, where $is(a)$ is the length of a longest increasing subword ending in a, and $ds(a)$ the length of a longest decreasing subword starting in a.

8.55 Compute $f(\lambda)$ for $\lambda = \underbrace{22\ldots2}_{n}$ by a direct combinatorial argument.

▷ **8.56** Let π be any permutation of $\{1, 2, \ldots, mn+1\}$. Show that π contains an increasing subword of length $m+1$ or a decreasing subword of length $n+1$ (or both). Does the conclusion still hold for mn?

8.57 How many STs of shape $\lambda = \underbrace{nn\ldots n}_{n}$ have main diagonal $(1, 4, 9, \ldots, n^2)$?

* * *

8.58 Show that the number of $\pi \in S(n)$ with $is(\pi) \le 2$ is C_n (Catalan). Hint: Show more generally that the ballot number $b_{n,k}$ equals the number of $\pi \in S(n+1)$ with $is(\pi) \le 2$, and where $n+1 = \pi_{k+1}$ $(0 \le k \le n)$.

▷ **8.59** For $\lambda = \lambda_1 \ldots \lambda_r \in \mathrm{Par}(n)$ and for $1 \le i \le r$ denote by $f(\lambda, -i)$ the number of STs with shape $\lambda_1 \ldots \lambda_{i-1} \lambda_i - 1 \lambda_{i+1} \ldots \lambda_r$. If the monotonicity condition is violated, that is, $\lambda_i - 1 < \lambda_{i+1}$, set $f(\lambda, -i) = 0$. Similarly, $f(\lambda, +i)$ is the number of STs with shape $\lambda_1 \ldots \lambda_{i-1} \lambda_i + 1 \lambda_{i+1} \ldots \lambda_r$ or $\lambda_1 \ldots \lambda_r 1$ for $i = r+1$. Again, $f(\lambda, +i) = 0$ if $\lambda_{i-1} < \lambda_i + 1$. Re-prove Theorem 8.26 (a) using the following steps: a. $f(\lambda) = \sum_{i=1}^{r} f(\lambda, -i)$, b. $(n+1) f(\lambda) = \sum_{i=1}^{r+1} f(\lambda, +i)$, c. $\sum_{\lambda \in \mathrm{Par}(n)} f(\lambda)^2 = n!$. Hint: Use $f(\lambda, -i) = 0 \iff f(\lambda, +(i+1)) = 0$.

8.60 Let $O(n)$ be the number of $\lambda \in \mathrm{Par}(n)$ such that $f(\lambda)$ is odd. Write $n = a_0 + a_1 2 + a_2 2^2 + \cdots$ in binary expansion, and denote by $P(z) = \prod_{i \ge 1} \frac{1}{1-z^i}$ the partition function. Prove $O(n) = \prod_{j \ge 0} [z^{a_j}] P(z)^{2^j}$. Example: $n = 3 = 1+2$, $a_0 = a_1 = 1$, $O(3) = [z]P(z) \cdot [z]P(z)^2 = 1 \cdot 2 = 2$, with $f(3) = 1$, $f(21) = 2$, $f(111) = 1$.

▷ **8.61** Show that the number of STs over $\{1, \ldots, n\}$ with at most two rows is $\binom{n}{\lfloor n/2 \rfloor}$, and with at most three rows is the Motzkin number M_n (see Section 7.4).

8.62 Complete the details of Lemma 8.28.

8.63 Give a direct proof for $\sum_{\lambda \in \mathrm{Par}(n)} f(\lambda) = i_n$ (involution number) by showing that the recurrence $i_{n+1} = i_n + n i_{n-1}$ also holds for $\sum_{\lambda \in \mathrm{Par}(n)} f(\lambda)$.

▷ **8.64** Take $\lambda \in \mathrm{Par}$ and consider the Ferrers diagram with the hook lengths inscribed in the cells. Show that the number of odd hook lengths minus the number of even hook lengths is always equal to $\binom{m+1}{2}$ for some m.

Highlight: Hook-Length Formulas

Perhaps the most beautiful formula expressing the number $f(\lambda)$ of standard tableaux of shape λ is the hook-length formula of Frame, Thrall, and Robinson.

The Number of Standard Tableaux.
Let $\lambda = \lambda_1 \ldots \lambda_r \in \text{Par}(n)$, and let (i, j) be a cell in the Ferrers diagram of λ, i = row number, j = column number. The *hook length* h_{ij} in (i, j) is the number of cells below and to the right of (i, j), including (i, j) itself. Hence if $\lambda^* = \lambda_1^* \ldots \lambda_t^*$ is the conjugate partition, then

$$h_{ij} = (\lambda_i - j) + (\lambda_j^* - i) + 1 . \tag{1}$$

Example. Consider $\lambda = 644211 \in \text{Par}(18)$. The hook lengths are written in the cells:

11	8	6	5	2	1
8	5	3	2		
7	4	2	1		
4	1				
2					
1					

We set $\mu_i = h_{i1} = \lambda_i + r - i \quad (i = 1, \ldots, r)$. The following result is the key to the proof.

Lemma. *Let $\lambda = \lambda_1 \ldots \lambda_r \in \text{Par}(n)$. Then for all i, the sequence of the $\mu_i = \lambda_i + r - i$ numbers*

$$\mu_i = h_{i1}, h_{i2}, \ldots, h_{i\lambda_i}, \mu_i - \mu_{i+1}, \mu_i - \mu_{i+2}, \ldots, \mu_i - \mu_r \tag{2}$$

is a permutation of $\{1, 2, \ldots, \mu_i\}$.

Proof. First it is clear that the hook lengths decrease strictly in every row and column. Hence

$$\mu_i = h_{i1} > h_{i2} > \cdots > h_{i\lambda_i} \geq 1 .$$

Next we note that

$$1 \leq \mu_i - \mu_{i+1} < \mu_i - \mu_{i+2} < \cdots < \mu_i - \mu_r < \mu_i .$$

All numbers in (2) are therefore between 1 and μ_i, and it remains to show that no two numbers h_{ij} and $\mu_i - \mu_k$ $(k > i)$ are equal.

Case 1. $j \le \lambda_k$. Then $\lambda_j^* \ge k$, and by (1),

$$h_{ij} \ge (\lambda_i - \lambda_k) + (k - i) + 1 > (\lambda_i - \lambda_k) + (k - i) = \mu_i - \mu_k .$$

Case 2. $j > \lambda_k$. Then $\lambda_j^* \le k - 1$, and hence

$$h_{ij} \le (\lambda_i - \lambda_k - 1) + (k - 1 - i) + 1 < (\lambda_i - \lambda_k) + (k - i) = \mu_i - \mu_k .$$

The result follows. □

We conclude that for fixed i,

$$\prod_{j=1}^{\lambda_i} h_{ij} = \frac{\mu_i!}{\prod\limits_{i<j} (\mu_i - \mu_j)}$$

holds, and therefore

$$\prod_{i,j} h_{ij} = \frac{\prod\limits_{i=1}^{r} (\mu_i!)}{\prod\limits_{1\le i<j\le r} (\mu_i - \mu_j)} . \tag{3}$$

There are many proofs for the following "hook-length formula." We look at one that uses the results on symmetric functions discussed in Section 8.3.

Theorem. *Let* $\lambda = \lambda_1 \dots \lambda_r \in \mathrm{Par}(n)$. *Then the number of standard tableaux of shape* λ *is given by*

$$f(\lambda) = \frac{n!}{\prod\limits_{i,j} h_{ij}} ,$$

where h_{ij} *are the hook lengths.*

Proof. Set $X = \{x_1, \dots, x_n\}$. We know from Corollary 8.9 in Section 8.3 that

$$s_\lambda(x_1, \dots, x_n) = \sum_{\mu \in \mathrm{Par}(n)} K_{\lambda\mu} m_\mu , \tag{4}$$

where $K_{\lambda\mu}$ is the number of SSTs of shape λ and type μ. In particular, as noted before, $f(\lambda) = K_{\lambda\overline{\mu}}$ is the coefficient of $m_{\overline{\mu}}$,

$\overline{\mu} = 11\ldots1$. Define the ring homomorphism $\phi : \Lambda^n(X) \longrightarrow \mathbb{Q}$ by $\phi m_{11\ldots1} = 1$, and $\phi m_\mu = 0$ for $\mu \neq \overline{\mu}$. Then by (4),

$$\phi s_\lambda = f(\lambda)\,.$$

Now by the Jacobi–Trudi identity,

$$s_\lambda = \det(h_{\lambda_i - i + j})_{i,j=1}^{n}\,, \tag{5}$$

where as usual $\lambda_{r+1} = \cdots = \lambda_n = 0$, and $h_j = 0$ for $j < 0$. This means for $i > r$ that $h_{\lambda_i} = 1$ and $h_{\lambda_i - k} = 0$ for $k > 0$. The matrix in (5) has thus the form

$$\begin{pmatrix} \begin{array}{cccc|c} h_{\lambda_1} & h_{\lambda_1+1} & \ldots & h_{\lambda_1+r-1} & \\ h_{\lambda_2-1} & h_{\lambda_2} & \ldots & h_{\lambda_2+r-2} & * \\ & & \cdots & & \\ h_{\lambda_r-(r-1)} & & \ldots & h_{\lambda_r} & \\ \hline & & & & 1 \\ & & 0 & & \ddots \\ & & & & & 1 \end{array} \end{pmatrix},$$

and we obtain

$$s_\lambda = \det \begin{pmatrix} h_{\lambda_1} & \ldots & h_{\lambda_1+r-1} \\ & \cdots & \\ h_{\lambda_r-(r-1)} & \ldots & h_{\lambda_r} \end{pmatrix}. \tag{6}$$

Every term of the determinant in (6) is therefore of the form $\pm h_{\rho_1} h_{\rho_2} \cdots h_{\rho_r} = \pm h_\rho$, $\rho = \rho_1 \ldots \rho_r \in \mathrm{Par}(n)$. Now we apply the map ϕ to equation (6). According to Exercise 8.11, the coefficient of $m_{11\ldots1}$ in h_ρ is $\binom{n}{\rho_1 \ldots \rho_r} = \frac{n!}{\rho_1! \cdots \rho_r!}$, and we conclude that

$$f(\lambda) = \phi s_\lambda = n! \det \begin{pmatrix} \frac{1}{\lambda_1!} & \frac{1}{(\lambda_1+1)!} & \cdots & \frac{1}{(\lambda_1+r-1)!} \\ \frac{1}{(\lambda_2-1)!} & \frac{1}{\lambda_2!} & \cdots & \frac{1}{(\lambda_2+r-2)!} \\ & & \cdots & \\ \frac{1}{(\lambda_r-(r-1))!} & & \cdots & \frac{1}{\lambda_r!} \end{pmatrix},$$

where we set $\frac{1}{k!} = 0$ for $k < 0$. But this last determinant is easily computed. First we note that the last column contains precisely $\frac{1}{\mu_1!}, \frac{1}{\mu_2!}, \ldots, \frac{1}{\mu_r!}$, with μ_i defined as in the lemma. Factoring these numbers out, we get

$$f(\lambda) = \frac{n!}{\prod\limits_{i=1}^{r}(\mu_i)!} \det \begin{pmatrix} \mu_1^{\underline{r-1}} & \mu_1^{\underline{r-2}} & \dots & \mu_1^{\underline{0}} \\ \mu_2^{\underline{r-1}} & \mu_2^{\underline{r-2}} & \dots & \mu_2^{\underline{0}} \\ & & \dots & \\ \mu_r^{\underline{r-1}} & \mu_r^{\underline{r-2}} & \dots & \mu_r^{\underline{0}} \end{pmatrix},$$

and this latter determinant is equal to $\prod_{1\le i<j\le r}(\mu_i - \mu_j)$, as was proven in Exercise 8.1. So we finally arrive at

$$f(\lambda) = n! \frac{\prod\limits_{1\le i<j\le r}(\mu_i - \mu_j)}{\prod\limits_{i=1}^{r}(\mu_i)!},$$

and the result follows from (3). □

Example. Consider $\lambda = 3221$, with the hook lengths inscribed in the cells:

6	4	1
4	2	
3	1	
1		

Thus $f(\lambda) = \dfrac{8!}{6 \cdot 4 \cdot 4 \cdot 3 \cdot 2} = 70$.

Remark. Writing $\frac{f(\lambda)}{n!} = \prod_{i,j} \frac{1}{h_{ij}}$ we see that $\frac{1}{\prod\limits_{i,j} h_{ij}}$ is the probability for obtaining a standard tableau when we fill the n cells of the Ferrers diagram of λ randomly with 1 to n. Using ideas from probability theory such a proof for the hook-length formula has indeed been given by Greene, Nijenhuis, and Wilf.

The Number of Semistandard Tableaux.
In the same spirit we can also give a hook-length formula for the number of semistandard tableaux of shape λ. Again let $\lambda = \lambda_1 \dots \lambda_r$ be a partition of n. We know from Theorem 8.8 that

$$\#\mathrm{SST}_\lambda = s_\lambda(1, 1, \dots, 1),$$

where s_λ is the Schur function. The computation proceeds in two steps: First we determine $s_\lambda(q^{n-1}, \dots, q, 1)$ and then we set $q = 1$.

By the definition of the Schur function,

$$s_\lambda(q^{n-1},\dots,q,1) = \frac{\det(q^{(n-j)(n-i+\lambda_i)})_{i,j=1}^n}{\prod\limits_{1\le i<j\le n}(q^{n-i}-q^{n-j})}.$$

Taking the transpose of the matrix in the numerator, we see that it is again a Vandermonde matrix, which gives

$$s_\lambda(q^{n-1},\dots,q,1) = \prod_{1\le i<j\le n} \frac{(q^{n-i+\lambda_i}-q^{n-j+\lambda_j})}{q^{n-i}-q^{n-j}}.$$

For brevity let us set $v_i = n-i+\lambda_i$, where clearly $v_1 \ge v_2 \ge \cdots \ge v_n$. For each pair $i<j$ we factor out q^{v_j} in the numerator, and q^{n-j} in the denominator. This gives for some exponent K,

$$s_\lambda(q^{n-1},\dots,q,1) = q^K \frac{\prod\limits_{i<j}(1-q^{v_i-v_j})}{\prod\limits_{i<j}(1-q^{j-i})}.$$

Now divide numerator and denominator by $(1-q)^{\binom{n}{2}}$, and set $q=1$ to obtain

$$\#\mathrm{SST}_\lambda = s_\lambda(1,1,\dots,1) = \frac{\prod\limits_{i<j}(v_i-v_j)}{\prod\limits_{i<j}(j-i)}. \tag{7}$$

The denominator in (7) clearly equals $\prod_{i=1}^n (n-i)!$, so let us look at the numerator. For $j>r$ we have $\lambda_j = 0$, that is, $v_j = n-j$. On the other hand, for $1\le i,j\le r$, $v_i - v_j = \mu_i - \mu_j$, where the μ_i's are defined as in the lemma. In summary,

$$\prod_{i<j}(v_i-v_j) = \prod_{1\le i<j\le r}(\mu_i-\mu_j)\cdot\prod_{i=1}^{r}\prod_{j=r+1}^{n}(\lambda_i-i+j)\cdot\prod_{i=r+1}^{n}(n-i)!. \tag{8}$$

Multiply numerator and denominator in (7) by $\prod_{i=1}^r((\lambda_i-i+r)!) = \sum_{i=1}^r(\mu_i!)$ and use (3) to obtain

$$\#\mathrm{SST}_\lambda = \frac{\prod\limits_{1\le i<j\le r} (\mu_i - \mu_j) \prod\limits_{i=1}^{r} ((\lambda_i - i + n)!)}{\prod\limits_{i=1}^{r} (\mu_i!) \prod\limits_{i=1}^{r} (n-i)!}$$

$$= \frac{\prod\limits_{i=1}^{r} ((\lambda_i - i + n)!)}{\prod\limits_{i=1}^{r} (n-i)!} \frac{1}{\prod\limits_{i,j} h_{ij}} .$$

Finally,

$$\prod_{i=1}^{r} \frac{(\lambda_i + n - i)!}{(n-i)!} = \prod_{i=1}^{r} (\lambda_i + n - i) \cdots (n + 1 - i) = \prod_{i=1}^{r} \prod_{j=1}^{\lambda_i} (n + j - i) ,$$

and we arrive at the hook-length formula:

Theorem. *Let* $\lambda \in \mathrm{Par}(n)$*; then*

$$\#\mathrm{SST}_\lambda = \prod_{i,j} \frac{n + j - i}{h_{ij}} .$$

For the example $\lambda = 3221$ above this gives $\#\mathrm{SST}_\lambda = 14700$.

Notes and References

The starting point for the study of symmetric functions was the problem of finding a formula for the roots of a polynomial, leading to what is called the fundamental Theorem 8.4. The further development of symmetric functions is closely related to the study of the symmetric group, more precisely to the characters of representations of $S(n)$. Schur and Frobenius used the functions that today are called Schur functions (but which were known already to Cauchy) for research on these characters. A nice introduction to symmetric functions is presented in the book by Bressoud. More advanced and comprehensive are the books by Macdonald and Stanley. Plane partitions were invented and intensively studied by MacMahon. In an important paper Bender and Knuth pointed out the connection between the theory of symmetric functions and the enumeration of plane partitions. The references list the contributions of Robinson, Knuth, and Schensted to the RSK algorithm. The hook-length formula for standard tableaux appears in a paper by Frame, Thrall, and Robinson. An elegant probabilistic proof was given by Greene, Nijenhuis, and Wilf.

1. E.A. Bender and D.E. Knuth (1972): Enumeration of plane partitions. *J. Combinatorial Theory* 13, 225–245.

2. D.M. Bressoud (1999): Proofs and Confirmations. *The Story of the Alternating Sign Matrix Conjecture.* Cambridge Univ. Press, Cambridge.

3. J.S. Frame, G. de B. Robinson, and R.M. Thrall (1954): The hook graphs of S_n. *Canad. J. Math.* 6, 316–324.

4. C. Greene, A. Nijenhuis, and H.S. Wilf (1979): A probabilistic proof of a formula for the number of Young tableaux of a given shape. *Advances Math.* 31, 104–109.

5. D.E. Knuth (1970): Permutations, matrices, and generalized Young tableaux. *Pacific J. Math.* 34, 709–727.

6. I.G. Macdonald (1995): *Symmetric Functions and Hall Polynomials,* 2nd edition. Oxford Univ. Press, Oxford.

7. P.A. MacMahon (1915): *Combinatory Analysis*, 2 vols. Cambridge Univ. Press, Cambridge; reprinted in one volume by Chelsea, New York, 1960.
8. G. de B. Robinson (1938): On representations of S_n. *Amer. J. Math.* 60, 745–760.
9. C.E. Schensted (1961): Longest increasing and decreasing subsequences. *Canad. J. Math.* 13, 179–191.
10. R.P. Stanley (1999): *Enumerative Combinatorics*, vol. 2. Cambridge Univ. Press, Cambridge.

9 Counting Polynomials

So far we have studied polynomials and generating functions whose coefficients have a combinatorial significance. In this chapter we take a different view: Given sets $S_0, S_1, S_2, \ldots$, we want to determine the *counting polynomial* $f(x)$ that at $x = i$ gives $f(i) = |S_i|$.

The classical example is the *chromatic function* $\chi(G;\lambda)$ in the variable λ. Let $G = (V,E)$ be a graph, which may have loops and multiple edges. Then

$$\chi(G;i) := \#i\text{-colorings of } G,$$

where as usual an i-coloring is a mapping $c : V \longrightarrow \{1,\ldots,i\}$ with $c(u) \neq c(v)$ if $\{u,v\} \in E$. As an example, for the complete graph K_n we have

$$\chi(K_n;i) = i(i-1)\cdots(i-n+1),$$

since any i-coloring corresponds to an n-permutation of the colorset. The chromatic function is therefore $\chi(K_n;\lambda) = \lambda(\lambda-1)\cdots(\lambda-n+1)$.

It is not immediately clear that $\chi(G;\lambda)$ is a polynomial. But we shall see shortly that this is indeed always the case; accordingly, we will call $\chi(G;\lambda)$ the *chromatic polynomial*. Note that the chromatic number of G, that is, the minimal number of colors needed to color G, is the smallest positive integer k with $\chi(G;k) > 0$.

9.1 The Tutte Polynomial of Graphs

Let $G = (V,E)$ be a graph. A *loop* is as usual an edge with equal end-vertices. A *bridge* of G is an edge whose removal disconnects the component that contains it (thus the number $k(G)$ of components is increased by 1). The graph of the figure has two bridges e, e':

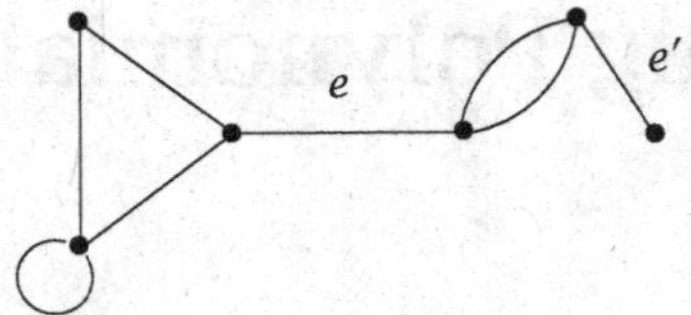

Restriction and Contraction.
There are two operations on G that are basic to all that is to come.

Definition. Let e be an edge. The *restriction* $G \smallsetminus e$ is the graph obtained by removing e. The *contraction* G/e is the graph that results after contracting e to a single vertex. That is, the end-vertices of e are identified, keeping all other adjacencies.

Example.

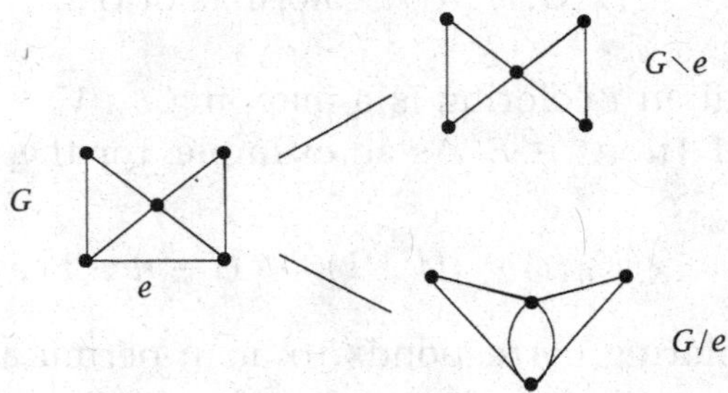

It is clear that $\chi(G;\lambda) = 0$ if G has a loop, and furthermore $\chi(G_1 \,\dot{\cup}\, G_2;\lambda) = \chi(G_1;\lambda)\chi(G_2;\lambda)$, where $G_1 \,\dot{\cup}\, G_2$ is the disjoint union.

The following fundamental recurrence permits in principle the computation of $\chi(G;\lambda)$.

Proposition 9.1. *Suppose $e = \{u, v\}$ is not a loop of G. Then*

$$\chi(G;\lambda) = \chi(G \smallsetminus e;\lambda) - \chi(G/e;\lambda)\,. \tag{1}$$

Proof. Look at the λ-colorings of $G \smallsetminus e$. The colorings c with $c(u) \neq c(v)$ correspond bijectively to the colorings of G, and those with $c(u) = c(v)$ to the colorings of G/e. The result follows. □

Example. We determine $\chi(G;\lambda)$ in the following "graphic" way, singling out e at each stage:

$= \ldots (\lambda - 1)$

$= \left(\ldots - \ldots \right) (\lambda - 1)$

$= \left(\ldots - \ldots - \ldots + \ldots \right) (\lambda - 1)$

$= \ldots (\lambda - 2)(\lambda - 1) + \ldots (\lambda - 1)$ (since multiple edges are irrelevant)

$= \left(\ldots - \ldots \right) (\lambda - 2)(\lambda - 1) + \ldots (\lambda - 1)$

$= \ldots [(\lambda - 1)^2(\lambda - 2) + (\lambda - 1)]$

$= \lambda(\lambda - 1)(\lambda - 2)[(\lambda - 1)^2(\lambda - 2) + (\lambda - 1)]$

$= \lambda(\lambda - 1)^2(\lambda - 2)(\lambda^2 - 3\lambda + 3)\,.$

Thus G has $\chi(G;3) = 36$ 3-colorings and $\chi(G;4) = 504$ 4-colorings.

Apart from the algorithm we get a bonus out of (1). The recurrence immediately implies that $\chi(G;\lambda)$ is a polynomial with integral coefficients. Some further properties are discussed in the exercises.

Proposition 9.2. *Let $G = (V,E)$ be a loopless graph. Then $\chi(G;\lambda) = a_0\lambda^n + a_1\lambda^{n-1} + \cdots + a_n$ is a polynomial in λ with coefficients in $\mathbb{Z}$. Furthermore,*

(i) *$\chi(G;\lambda)$ has degree $n = |V|$,*

(ii) *the coefficients $a_0, a_1, \ldots, a_{n-k(G)}$ are nonzero with alternating sign, $a_0 = 1$, and $a_i = 0$ for $i > n - k(G)$.*

Proof. We use induction on $|E|$. If G has no edges, then $\chi(G;\lambda) = \lambda^n$. Otherwise, pick an edge $e \in E$. Then

$$\chi(G;\lambda) = \chi(G\smallsetminus e;\lambda) - \chi(G/e;\lambda),$$

and all assertions follow immediately by induction. □

Tutte Polynomial.
The deletion/contraction recurrence spelled out in (1) leads to the Tutte polynomial $T(G;x,y)$ in two variables as a natural extension of the chromatic polynomial. However, the Tutte polynomial has a much richer structure and is arguably the most important counting polynomial for graphs.

Definition. The *Tutte polynomial* $T(G;x,y)$ of a graph $G = (V,E)$ is defined recursively as follows:

(i) If $E = \emptyset$, then $T(G;x,y) = 1$.

(ii) If e is a bridge, then $T(G;x,y) = xT(G\smallsetminus e;x,y)$; if e is a loop, then $T(G;x,y) = yT(G\smallsetminus e;x,y)$. (2)

(iii) If e is not a bridge or loop, then

$$T(G;x,y) = T(G\smallsetminus e;x,y) + T(G/e;x,y).$$

Thus $T(G;x,y)$ may be calculated by choosing an ordering of the edges to be removed and repeatedly using the recurrences. At first sight, the polynomial depends on the order in which we delete the edges. It was Tutte's great discovery that in fact, it does not; we always get the same polynomial! This phenomenon will be a recurrent theme for the other polynomials we are going to study.

Example. Let us compute the Tutte polynomial of the graph K_4^- (K_4 minus an edge), deleting the edges in the order $1,\dots,5$:

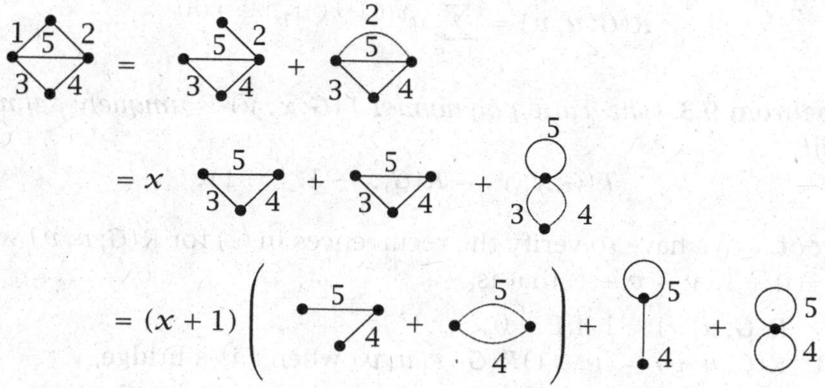

$$= (x+1)(x^2+x+y)+xy+y^2$$

$$= x^3+2x^2+x+2xy+y+y^2\,.$$

The uniqueness of $T(G;x,y)$ can be shown by induction on the order, but here is another and more elegant proof that does it in one stroke. We exhibit a polynomial that observes the defining recurrences, and is *independent* of the order. What's more, this description of $T(G;x,y)$ is the easiest way to deduce some basic evaluations for x and y.

We need some preparations. Let A be a subset of the edges of $G=(V,E)$, and identify A with the subgraph $G_A=(V,A)$. Thus all graphs G_A are taken to be *spanning* subgraphs in the sense that $V(G_A)=V$. We define the rank of A by

$$r(A)=|V|-k(G_A)\,. \tag{3}$$

It is easily seen that $0 \le r(A) \le |A|$ with

$$\begin{aligned} r(A)=0 &\iff A=\emptyset\,,\\ r(A)=|A| &\iff G_A \text{ is a forest.} \end{aligned} \tag{4}$$

Furthermore, we infer from (3) that $A \subseteq B$ implies $r(A)\le r(B)$ and

$$r(A)=r(E) \iff k(G_A)=k(G)\,. \tag{5}$$

Finally, let us set $r(G)=r(E)=|V|-k(G)$.

Definition. The *rank-generating function* of $G=(V,E)$ is the 2-variable polynomial

$$R(G;u,v) = \sum_{A \subseteq E} u^{r(G)-r(A)} v^{|A|-r(A)}. \tag{6}$$

Theorem 9.3. *The Tutte polynomial $T(G;x,y)$ is uniquely defined, with*

$$T(G;x,y) = R(G;x-1,y-1). \tag{7}$$

Proof. We have to verify the recurrences in (2) for $R(G;u,v)$ with $x = u+1$, $y = v+1$, that is,

(i) $R(G;u,v) = 1$ if $E = \emptyset$,

(ii) $R(G;u,v) = (u+1)R(G \setminus e;u,v)$ when e is a bridge,
$R(G;u,v) = (v+1)R(G \setminus e;u,v)$ when e is a loop,

(iii) $R(G;u,v) = R(G \setminus e;u,v) + R(G/e;u,v)$, when e is not a bridge or loop.

It will follow by induction that $T(G;x,y)$ is the evaluation of $R(G;u,v)$ at $u = x-1$, $v = y-1$, and thus uniquely defined.

Assertion (i) is obvious. For (ii) and (iii) we check how the rank function changes after deletion and contraction. Suppose $e \in A$, and let r' and r'' be the rank functions of $A \setminus e$ in $G \setminus e$ and G/e, respectively.

Clearly,

$$r'(A \setminus e) = r(A) - 1 \quad \text{if } e \text{ is a bridge.} \tag{8}$$

For the contraction check that

$$r''(A \setminus e) = r(A) - 1 \quad \text{if } e \text{ is not a loop.} \tag{9}$$

Suppose e is a bridge. Then by (8),

$$\begin{aligned} R(G;u,v) &= \sum_{A \subseteq E \setminus e} u^{r(G)-r(A)} v^{|A|-r(A)} + \sum_{A: e \in A} u^{r(G)-r(A)} v^{|A|-r(A)} \\ &= u \sum_{A \subseteq E \setminus e} u^{r'(G \setminus e)-r'(A)} v^{|A|-r'(A)} \\ &\qquad + \sum_{B = A \setminus e} u^{r'(G \setminus e)+1-(r'(B)+1)} v^{|B|+1-(r'(B)+1)} \\ &= (u+1)R(G \setminus e;u,v). \end{aligned}$$

The case of a loop is settled similarly, and for (iii) we get by (8) and (9),

$$\begin{aligned} R(G;u,v) &= \sum_{A\subseteq E\setminus e} u^{r(G)-r(A)}v^{|A|-r(A)} + \sum_{A:e\in A} u^{r(G)-r(A)}v^{|A|-r(A)} \\ &= \sum_{A\subseteq E\setminus e} u^{r'(G\setminus e)-r'(A)}v^{|A|-r'(A)} \\ &\quad + \sum_{B=A\setminus e} u^{r''(G/e)+1-(r''(B)+1)}v^{|B|+1-(r''(B)+1)} \\ &= R(G\setminus e;u,v)+R(G/e;u,v). \qquad \square \end{aligned}$$

Looking at the recursive definition of $T(G;x,y)$ we have the following corollary, which is rather surprising when we compare it to the expansion of $R(G;x-1,y-1)$ in terms of x and y.

Corollary 9.4. *The Tutte polynomial* $T(G;x,y)=\sum_{i,j}t_{ij}x^iy^j$ *has nonnegative integral coefficients.*

Example. The definition of the rank function implies the following evaluations (see (4),(5)): Let G be connected. Then

$$\begin{aligned} T(G;1,1) &= R(G;0,0) = \#\text{ spanning trees}, \\ T(G;2,1) &= R(G;1,0) = \#\text{ spanning forests}, \\ T(G;1,2) &= R(G;0,1) = \#\text{ connected spanning subgraphs}, \\ T(G;2,2) &= R(G;1,1) = 2^{|E|}. \end{aligned}$$

As illustration, the graph K_4^- considered above contains 8 spanning trees and 14 spanning connected subgraphs altogether.

The Recipe Theorem.

One of the most pleasing features of the Tutte polynomial is that it can be regarded as a universal polynomial for all functions observing the deletion/contraction recurrence.

Definition. A function f that maps any graph into a field of characteristic 0 is called a *chromatic invariant* (also termed a Tutte–Grothendieck invariant) if the following hold:

(i) If G has no edges, then $f(G)=1$.

(ii) Let $f(\text{bridge})=A$, $f(\text{loop})=B$, then

$$\begin{aligned} f(G) &= Af(G\setminus e), \ e \text{ bridge}, \\ f(G) &= Bf(G\setminus e), \ e \text{ loop}. \end{aligned}$$

(iii) There exist constants $\alpha \neq 0$, $\beta \neq 0$ such that whenever e is not a bridge or loop,

$$f(G) = \alpha f(G \setminus e) + \beta f(G/e)\,.$$

The following result is sometimes called the *recipe theorem.*

Theorem 9.5. *Let f be a chromatic invariant with A, B, α, β as above. Then for all graphs $G = (V, E)$,*

$$f(G) = \alpha^{|E|-|V|+k(G)}\beta^{|V|-k(G)}T\left(G; \frac{A}{\beta}, \frac{B}{\alpha}\right). \tag{10}$$

Proof. Formula (10) is certainly true for edgeless graphs. If G consists only of bridges and loops, say ℓ bridges and m loops, then $f(G) = A^\ell B^m$ by (ii) above. Now, $|V|-k(G) = \ell$, $|E|-|V|+k(G) = m$, and $T(G; \frac{A}{\beta}, \frac{B}{\alpha}) = (\frac{A}{\beta})^\ell (\frac{B}{\alpha})^m$, so (10) is again satisfied. Let e be an edge that is neither a bridge nor a loop. By induction and $k(G) = k(G \setminus e) = k(G/e)$, we get

$$\begin{aligned} f(G) &= \alpha f(G \setminus e) + \beta f(G/e) \\ &= \alpha \cdot \alpha^{|E|-1-|V|+k(G)}\beta^{|V|-k(G)}T\left(G \setminus e; \frac{A}{\beta}, \frac{B}{\alpha}\right) \\ &\quad + \alpha^{|E|-|V|+k(G)}\beta \cdot \beta^{|V|-1-k(G)}T\left(G/e; \frac{A}{\beta}, \frac{B}{\alpha}\right) \\ &= \alpha^{|E|-|V|+k(G)}\beta^{|V|-k(G)}T\left(G; \frac{A}{\beta}, \frac{B}{\alpha}\right). \qquad \square \end{aligned}$$

Example. Let us look once more at the chromatic polynomial $\chi(G;\lambda)$. Setting $f(G) = \frac{\chi(G;\lambda)}{\lambda^{k(G)}}$, it is immediately checked that $f(G)$ is a chromatic invariant with $A = \lambda - 1$, $B = 0$, $\alpha = 1$, $\beta = -1$. The recipe theorem yields therefore the following result.

Proposition 9.6. *We have*

$$\chi(G;\lambda) = (-1)^{|V|-k(G)}\lambda^{k(G)}T(G; 1-\lambda, 0)\,. \tag{11}$$

For K_4^- we obtain from $T(K_4^-; x, y) = x^3 + 2x^2 + x + 2xy + y + y^2$,

$$\chi(K_4^-;\lambda) = -\lambda[(1-\lambda)^3 + 2(1-\lambda)^2 + (1-\lambda)] = \lambda(\lambda-1)(\lambda-2)^2\,.$$

The Flow Polynomial.

In a sense dual to colorings we have what are called flows. By this we mean the following. Take a graph $G = (V, E)$ and orient its edges arbitrarily. If $e : u \to v$ we set $u = e^-$, $v = e^+$. Choose any finite commutative group A (written additively) with $|A| \geq 2$. A mapping $\phi : E \to A \setminus \{0\}$ is called an *A-flow* if Kirchhoff's law is satisfied at every vertex v, that is, in-flow equals out-flow:

$$\sum_{e^+=v} \phi(e) = \sum_{e^-=v} \phi(e) . \tag{12}$$

Another way to state (12) is that the *net-flow* $\partial v = \sum_{e^+=v} \phi(e) - \sum_{e^-=v} \phi(e)$ equals 0 for all v. We note that G has no flow if it contains a bridge e. To see this, look at the figure

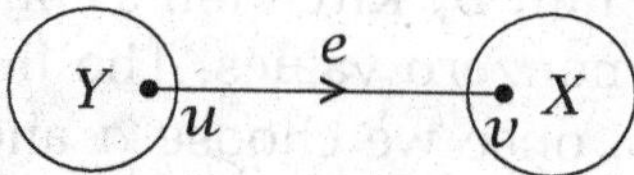

For a flow ϕ we would have $\sum_{x \in X} \partial x = 0$, but on the other hand, since every edge in X is counted once positively and once negatively we get $\sum_{x \in X} \partial x = 0 + \phi(e)$, implying $\phi(e) = 0$, which cannot be. Let $F(G; A)$ be the number of A-flows. When e is a loop, then we may assign any nonzero value to e; hence

$$F(G; A) = (|A| - 1) F(G \setminus e; A) .$$

You are asked in the exercises to provide a proof for the recurrence

$$F(G; A) = F(G/e; A) - F(G \setminus e; A) , \tag{13}$$

when e is not a loop or bridge. Thus $F(G; A)$ is a chromatic invariant with $\alpha = -1$, $\beta = 1$, and the recipe theorem gives the following result:

Proposition 9.7. *Let $F(G; A)$ be the number of A-flows; then*

$$F(G; A) = (-1)^{|E|-|V|+k(G)} T(G; 0, 1 - |A|) . \tag{14}$$

Now this implies the somewhat unexpected result that the number of flows depends only on the size $|A|$, but not on the structure of the group A, nor on the orientation. We may thus give the following definition:

Definition. The *flow polynomial* $F(G;\lambda)$ is the polynomial such that $F(G;k)$ is the number of A-flows for any abelian group A with k elements.

Example. Orient K_4^- as given and assign a flow with λ elements, where the upper two and lower two edges

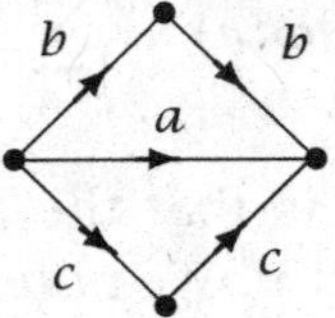

must receive the same flow by Kirchhoff's condition. For a we may take any of the $\lambda - 1$ nonzero values. The flow condition implies $a + b + c = 0$, that is, once we choose a and b, then c is determined. Given a, the values $0, -a$ are forbidden for b, and we obtain $F(K_4^-;\lambda) = (\lambda-1)(\lambda-2)$, in agreement with $(-1)^{|E|-|V|+k(K_4^-)}T(K_4^-; 0, 1-\lambda) = (1-\lambda)^2 + (1-\lambda) = (\lambda-1)(\lambda-2)$.

The simplest case arises for $A = \{0,1\}$ with $1+1=0$. Since $-1 = 1$ the orientation does not matter. Hence a graph $G = (V,E)$ has a 2-flow if and only if all degrees are even. These graphs are called *Eulerian*, since any connected component admits an Eulerian walk, that is, a closed walk using every edge exactly once. In this case there is trivially only one 2-flow, and we obtain the "dual" results

$$\begin{aligned} G \text{ bipartite} &\iff \chi(G;2) = 2^{k(G)} &&\iff T(G;-1,0) = (-1)^{|V|-k(G)}, \\ G \text{ Eulerian} &\iff F(G;2) = 1 &&\iff T(G;0,-1) = (-1)^{|E|-|V|+k(G)}. \end{aligned} \tag{15}$$

Colorings and flows are thus connected by a certain duality relation (interchanging x and y in the Tutte polynomial), but there is an important difference. For the chromatic polynomial the smallest k with $\chi(G;k) > 0$ (the chromatic number) may be arbitrarily high, as witnessed by the complete graphs K_n. But for flows the situation is entirely different. Every graph without bridges is known to have a 6-flow, and a famous unsolved conjecture of Tutte claims that every such graph even has a 5-flow, which in view of Exercise 9.16 would be best possible.

Two Further Evaluations.
The evaluations we have discussed so far were directly inspired by the rank-generating function or colorings and flows. We close this section with two examples that are less obvious.

Example. Suppose we orient the edges of a graph $G = (V, E)$ in such a way that no directed circuits result. We then call the orientation O *acyclic*. A loop has no acyclic orientations, but any loopless graph does. Just label the vertices $1, \ldots, n$ and direct any edge from the smaller to the higher number. We want to determine the number $ac(G)$ of acyclic orientations, where $ac(G) = 1$ by definition if E is empty. Clearly,

$$ac\text{ (bridge)} = 2, \quad ac\text{ (loop)} = 0\,,$$

and furthermore

$$ac(G) = 2ac(G \diagdown e)$$

if e is a bridge, since both orientations on e are allowed.

Proposition 9.8. *We have for $G = (V, E)$,*

$$ac(G) = T(G; 2, 0) = (-1)^{|V|}\chi(G; -1) = |\chi(G; -1)|\,. \tag{16}$$

Proof. With (11) the Recipe Theorem 9.5 will imply (16), once we show that for e not a loop or bridge,

$$ac(G) = ac(G \diagdown e) + ac(G/e)\,. \tag{17}$$

But this is easy. Let $e = \{u, v\}$, and consider an arbitrary acyclic orientation O of $G \diagdown e$. There is always one direction $u \to v$ or $v \to u$ possible such that O can be extended to an acyclic orientation of G. Indeed, if both directions were forbidden, then we would have a directed path in $G \diagdown e$ from u to v and one from v to u, which would yield a directed circuit in $G \diagdown e$. Hence those orientations of $G \diagdown e$ that permit exactly one direction of e are bijectively mapped onto this subset of acyclic orientations of G by assigning the proper direction to e.

Consider now the acyclic orientations of $G \diagdown e$ that allow both directions on e. These are precisely the orientations that induce acyclic orientations on the contracted graph G/e, and (17) follows. □

Example. Our final illustration concerns the evaluation of the Tutte polynomial $T(K_n;x,y)$ of the complete graphs K_n at $x = 1$, $y = -1$. We begin by calculating the exponential generating function

$$\sum_{n\ge1} T_n \frac{z^n}{n!}, \quad T_n = T(K_n;x,y)\,.$$

Since $r(K_n) = n-1$, $r(A) = n - k(A)$, we have by (7),

$$\begin{aligned}\sum_{n\ge1} T_n \frac{z^n}{n!} &= \sum_{n\ge1}\left[\sum_{A\subseteq E}(x-1)^{k(A)-1}(y-1)^{|A|+k(A)-n}\right]\frac{z^n}{n!}\\ &= \frac{1}{x-1}\sum_{n\ge1}\left[\sum_{A\subseteq E}(y-1)^{|A|}\big((x-1)(y-1)\big)^{k(A)}(y-1)^{-n}\right]\frac{z^n}{n!}\,.\end{aligned}$$

Since K_n is the complete graph, the expression in the inner sum refers to *all* graphs on n vertices specified according to the number of edges and components, which we considered in Exercise 3.55. Setting there $\alpha = y-1$, $\beta = (x-1)(y-1)$, $z = \frac{z}{y-1}$, we obtain

$$\sum_{n\ge1} T_n \frac{z^n}{n!} = \frac{1}{x-1}\left[\left(\sum_{n\ge0} y^{\binom{n}{2}}(y-1)^{-n}\frac{z^n}{n!}\right)^{(x-1)(y-1)} - 1\right].$$

Let us denote by $T(z)$ the derivative; hence

$$T(z) = \sum_{n\ge0} T_{n+1}\frac{z^n}{n!} = (y-1)S(y,z)^{(x-1)(y-1)-1}S'(y,z)\,,$$

where

$$S(y,z) = \sum_{n\ge0} y^{\binom{n}{2}}(y-1)^{-n}\frac{z^n}{n!}\,.$$

Now we consider the case $x = 1$, $y = -1$, and set $t_n = T(K_n;1,-1)$. Hence

$$t(z) = \sum_{n\ge0} t_{n+1}\frac{z^n}{n!} = (-2)\frac{S'(z)}{S(z)} \tag{18}$$

with

$$S(z) = \sum_{n\ge0}(-1)^{\binom{n}{2}}(-2)^{-n}\frac{z^n}{n!}\,. \tag{19}$$

From (19) it is easily seen that

$$S(-z) = (-2)S'(z), \tag{20}$$

and so

$$t(z) = \frac{S(-z)}{S(z)}. \tag{21}$$

Differentiation gives with (21) and (20)

$$\begin{aligned} t'(z) &= \frac{-S'(-z)S(z) - S(-z)S'(z)}{S(z)^2} \\ &= \frac{1}{2} + \frac{1}{2}\left(\frac{S(-z)}{S(z)}\right)^2 \\ &= \frac{1}{2} + \frac{1}{2}t(z)^2. \end{aligned}$$

But we know already this differential equation from Section 7.4. Its solution is

$$t(z) = \frac{1}{\cos z} + \tan z,$$

and out comes a beautiful result connecting $T(K_n)$ with the secant and tangent numbers:

$$T(K_{2n+1}; 1, -1) = sec_{2n}, \quad T(K_{2n}; 1, -1) = tan_{2n-1} \quad (n \geq 1).$$

Exercises

9.1 Compute the chromatic polynomial of a tree and the circuit of length n both directly, and by evaluating the Tutte polynomial.

▷ **9.2** Determine the chromatic polynomial and the number of acyclic orientations of the complete bipartite graph $K_{m,n}$.

9.3 Calculate $T(K_n; x, y)$ for $n \leq 6$.

9.4 What is the highest degree of x and y in $T(G; x, y)$?

9.5 Show for a connected graph $G = (V, E)$ that a. $xT(G; 1+x, 1) = \sum_{A \subseteq E \text{ forest}} x^{k(A)}$, b. $y^{n-1}T(G; 1, 1+y) = \sum_{A \subseteq E, k(A)=1} y^{|A|}$.

▷ **9.6** Let $\chi(G; \lambda) = \sum_{i \geq 0} a_i \lambda^{n-i}$ be the chromatic polynomial of a simple graph G (no loops or multiple edges). Show that $a_1 = -|E|$. Can you find an interpretation for a_2?

9.7 For $G = (V, E)$ and $e \in E$ let $T_e(G)$ be the result of the operation spelled out in the recursive definition of the Tutte polynomial. That is, $T_e(G) = xT(G \setminus e)$ if e is a bridge, etc. Prove the uniqueness of $T(G; x, y)$ by showing that $T_f T_e(G) = T_e T_f(G)$ for any two edges e, f.

* * *

9.8 Let $G = (V, E)$ be a connected simple graph on n vertices, and write the chromatic polynomial as $\chi(G; \lambda) = \sum_{i=0}^{n-1} (-1)^i \alpha_i \lambda^{n-i}$, $\alpha_i > 0$ for all i. Consider $\psi(G; \lambda) = \sum_{i=0}^{n-1} \alpha_i \lambda^{n-i}$, and prove

a. $\psi(G; \lambda) = \psi(G \setminus e; \lambda) + \psi(G/e; \lambda)$ for e not a loop.
b. $\psi(G; \lambda) = \lambda \sum_{j=0}^{n-1} t_j (\lambda + 1)^{n-1-j}$ with $t_j \geq 0$.

9.9 Continuing the exercise, show the following for the sequence $\alpha_0, \alpha_1, \ldots, \alpha_{n-1}$ $(n \geq 3)$:

a. $\alpha_k < \alpha_\ell$ for $0 \leq k < \frac{n-1}{2}$ and $k < \ell < n - 1 - k$,
b. $\alpha_k \leq \alpha_{n-1-k}$ for $0 \leq k < \frac{n-1}{2}$
c. $\alpha_0 < \alpha_1 < \cdots < \alpha_{\lfloor \frac{n}{2} \rfloor - 1} \leq \alpha_{\lfloor \frac{n}{2} \rfloor}$.

Hint: Prove first $\alpha_k = \sum_{j=0}^{k} \binom{n-1-j}{k-j} t_j$.

▷ **9.10** Let $G = (V, E)$ be a graph, and denote by $b_i(\lambda)$ the number of λ-colorings with exactly i bad edges, where bad means that the end-vertices are colored alike. Thus $b_0(\lambda)$ is the chromatic polynomial. Set $B(G; \lambda, s) = \sum_{i=0}^{|E|} b_i(\lambda) s^i$, and prove that $\frac{B(G;\lambda,s)}{\lambda^{k(G)}}$ is a chromatic invariant with $B(G; \lambda, s) = \lambda^{k(G)} (s-1)^{|V|-k(G)} T(G; \frac{s+\lambda-1}{s-1}, s)$. The function $B(G; \lambda, s)$ is called the *monochromial* of G.

9.11 Let G be a simple graph with n vertices and m edges, and denote by k and ℓ the unique integers with $m = \binom{k}{2} + \ell, 0 \leq \ell < k$. Prove that $ac(G) \geq (\ell + 1)(k!)$, and show that the bound may be attained for any n and $m \leq \binom{n}{2}$. Hint: Use induction on n.

9.12 Let r be a positive integer, $G = (V, E)$. Show that $r^{k(G)} T(G; r+1, 0) = |\chi(G; -r)|$ equals the number of pairs (g, O), where $g : V \longrightarrow \{1, 2, \ldots, r\}$ and O is an acyclic orientation such that $u \underset{O}{\longrightarrow} v$ implies $g(u) \leq g(v)$.

▷ **9.13** An orientation of a graph G is called *strong* if there is a directed path from any vertex to any other. Show that the number of strong orientations of G is given by $T(G; 0, 2)$.

9.14 Prove recurrence (13) for the flow polynomial.

▷ **9.15** Suppose G is a 4-regular graph. Show that the number of 2-in 2-out orientations of G (at every vertex v two edges point toward v and two away from v) equals the number of 3-flows, that is, $|T(G; 0, -2)|$.

9.16 Show that the Petersen graph of the figure has no 4-flow but a 5-flow:

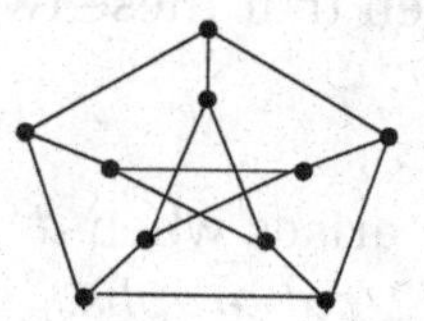

9.17 Set $t_n(y) = T(K_n; 1, y)$ and $t(z) = \sum_{n\geq 0} t_{n+1}(y)\frac{z^n}{n!}$. Establish the equation $t'(z) = \frac{y}{y-1}t(yz)t(z) - \frac{1}{y-1}t(z)^2$, and deduce $t_{n+1}(y) = \sum_{k=0}^{n-1}\binom{n-1}{k}t_{k+1}(y)t_{n-k}(y)(1+y+\cdots+y^{n-k-1})$. Set $y = 1$ and re-prove Cayley's formula n^{n-2} for the number of trees (Theorem 3.7). Hint: Mimic the proof given for $y = -1$, and for the last assertion use Exercise 3.42.

▷ **9.18** Take any spanning tree T on $\{1, \ldots, n\}$ and consider T as rooted at 1. We say that (i, j) is an *inversion* if $i < j$ and j lies on the unique path from 1 to i (j comes before i). Let $\text{inv}(T)$ be the number of inversions; thus $0 \leq \text{inv}(T) \leq \binom{n-1}{2}$.

Example: For $n = 3$, the trees (root 1 with children 2 and 3) and (path 1–2–3) have no inversion; (path 1–3–2) has one. Prove that for the Tutte polynomial $T(K_n; 1, y) = \sum_{T \text{ tree}} y^{\text{inv}(T)}$. How many trees have no inversions? Hint: Apply the exponential formula to show that $\sum y^{\text{inv}(T)}$ satisfies the differential equation of the previous exercise.

9.2 Eulerian Cycles and the Interlace Polynomial

Just as the number of colorings led to the deletion/contraction recurrence and to the Tutte polynomial, we now consider Eulerian cycles and show how again a natural recurrence gives rise to a polynomial.

Let $G = (V, E)$ be a directed graph that may have loops and multiple edges. An *Eulerian cycle* of G is a walk that starts at some vertex u, passes along the (directed) edges and returns to u, using every edge exactly once. A graph G is said to be *Eulerian* if it possesses an Eulerian cycle. Clearly, when G is Eulerian then it must be connected, and every vertex v has equal in- and out-degree, $d^+(v) = d^-(v)$. Conversely, it is easily seen that these two conditions are also sufficient.

The first interesting case arises when $d^+(v) = d^-(v) = 2$ for all v. We call such graphs 2-*in* 2-*out graphs*.

Example. The 2-in 2-out graph of the figure will serve as illustration throughout.

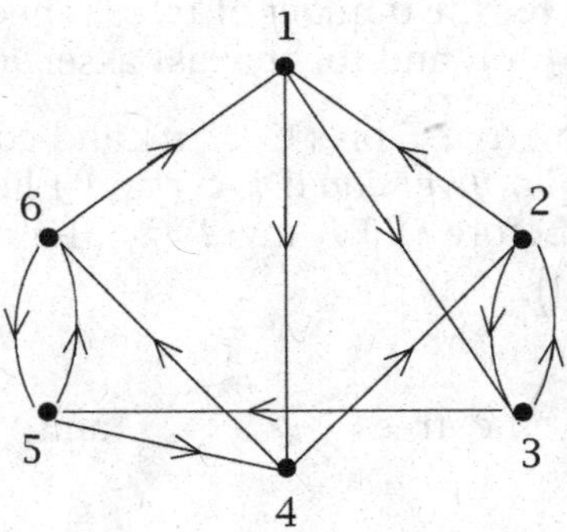

Given a 2-in 2-out graph $G = (V, E)$ we ask the following questions:

1. What is the number $e(G)$ of Eulerian cycles?
2. More generally, what is the number $e_k(G)$ of decompositions of G into k cycles, where a cycle is an Eulerian subgraph? Thus $e_1(G) = e(G)$.
3. How large can m be with $e_m(G) > 0$?
4. Among all 2-in 2-out graphs on n vertices, what is $\max e(G)$ or $\min e(G)$?

We will answer these questions as we go along.

Eulerian Cycles and Interlace Graphs.
Let us start with a useful representation of the graph G. Pick some Eulerian cycle C and draw the sequence of edges clockwise around a circle. In the example above we may choose the following Eulerian cycle C:

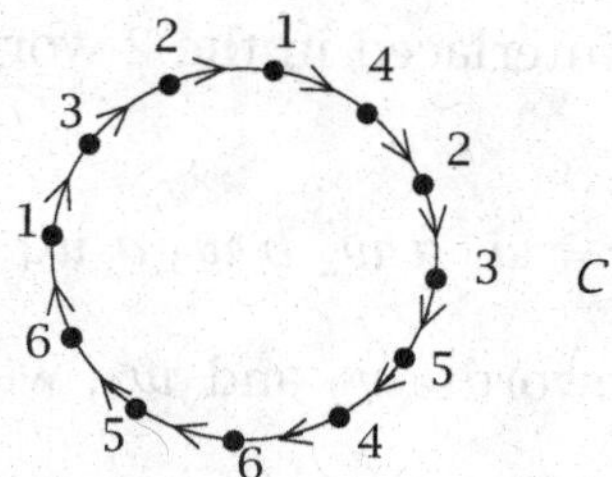

We may also represent G (or C) as a cyclic 2-word, meaning that every vertex appears exactly twice:

$$w = 1\,4\,2\,3\,5\,4\,6\,5\,6\,1\,3\,2.$$

Next we associate to C the following (undirected) graph $H(C)$. In the diagram of C draw a chord between like symbols. The vertices of $H(C)$ are the chords $1, \ldots, n$ with two chords i, j being adjacent if they intersect. For the 2-word w this means that i and j are "interlaced" in the sense that w is of the form $w = w_1 i w_2 j w_3 i w_4 j$. The resulting (simple) graph is called the *interlace graph* $H(C)$ associated with C.

In the example we get

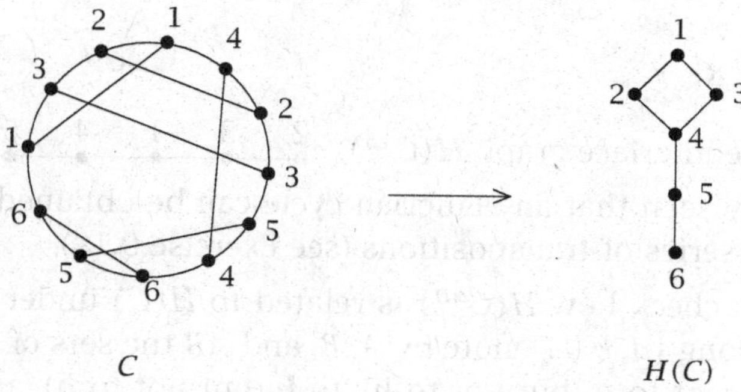

Remember that we enumerated the chord diagrams with respect to the number of crossings (= number of edges in $H(C)$) as highlight to Chapter 7.

Any Eulerian cycle C_i of G gives rise in this way to an interlace graph $H(C_i)$, and these graphs may be quite different, as we shall see in a moment. A natural question is then what these graphs $H(C_i)$ have in common. Before we answer this let us look at the following operation, which allows us to pass from any Eulerian cycle C to any other.

Suppose a and b are interlaced in the 2-word w corresponding to the Eulerian cycle C:

$$w = w_1 \; a \; w_2 \; b \; w_3 \; a \; w_4 \; b \, .$$

Interchanging the subwords w_2 and w_4, we get another Eulerian cycle C^{ab} with word

$$w^{ab} = w_1 \; a \; w_4 \; b \; w_3 \; a \; w_2 \; b \, .$$

We call this operation a *transposition of C along $\{a, b\}$* .

Example. Let us perform the transposition in C along $\{1, 3\}$:

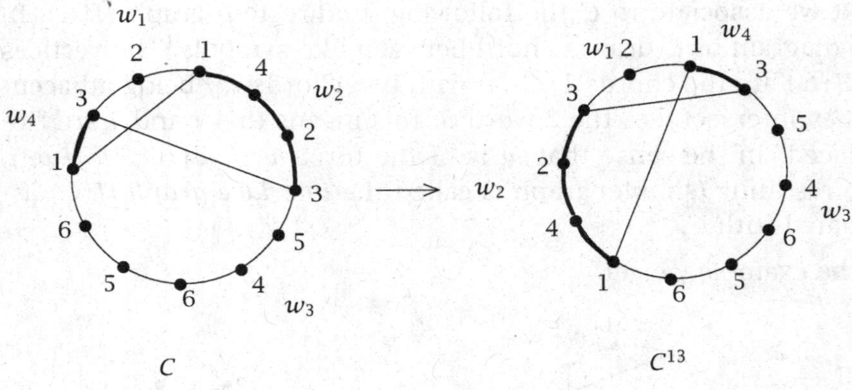

C^{13} has the interlace graph $H(C^{13})$ 2 3 1 4 5 6

It is readily seen that any Eulerian cycle can be obtained from any other by a series of transpositions (see Exercise 9.19).

Now let us check how $H(C^{ab})$ is related to $H(C)$ under the transposition along $\{a, b\}$. Denote by A, B, and AB the sets of vertices in $H(C)$ adjacent to a (but not to b), to b (but not to a), and to both a and b, respectively. Let $N = V \setminus (A \cup B \cup AB \cup \{a, b\})$ denote the remaining set. The easy proof of the following lemma is left to the exercises.

Lemma 9.9. *$H(C^{ab})$ arises from $H(C)$ by the following two operations:*

1. *Switch along $\{a, b\}$: Exchange edges and nonedges between any two different sets A, B, and AB, keeping the rest unchanged (including the edges within A, B, AB, and N).*

2. *Swap of labels a, b: This means that a is adjacent to v in $H(C^{ab})$ if and only if b is adjacent to v in $H(C)$; similarly for b.*

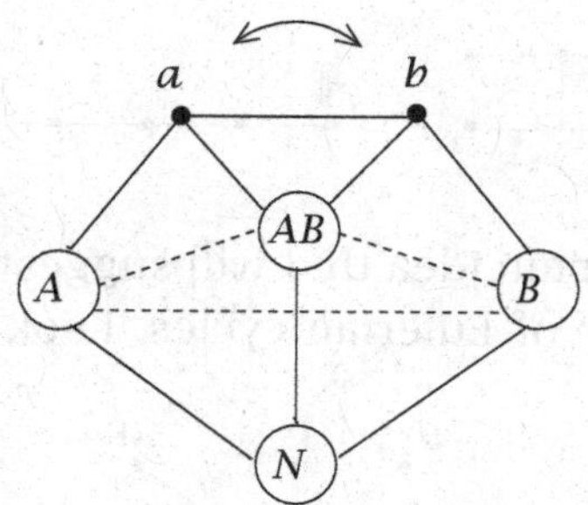

In our example we obtain

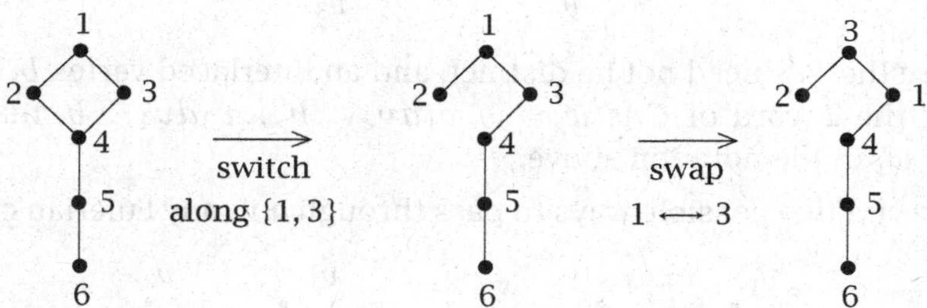

in agreement with what we found above.

The Interlace Polynomial.
Interlace graphs, that is, graphs that correspond to a 2-in 2-out graph, are a small class of graphs; see Exercise 9.21 for a graph that is not an interlace graph. But the switching operation can, of course, be performed on any graph.

Definition. Let $H = (V, E)$ be any simple graph. We say that H' is *switching equivalent* to H if H' is obtained from H by a series of edge switchings (no swaps). We then write $H \approx H'$.

The relation $\approx$ is an equivalence relation, with the equivalence classes called *switching classes*. Note that $(H^{ab})^{ab} = H$. Lemma 9.9 implies that for interlace graphs (apart from swapping labels),

$$H(C)^{ab} = H(C^{ab}) .$$

Any two interlace graphs belonging to the same 2-in 2-out graph are therefore switching equivalent.

Example. Check that the switching class of the interlace graph $H(C)$ in our example consists of the following graphs:

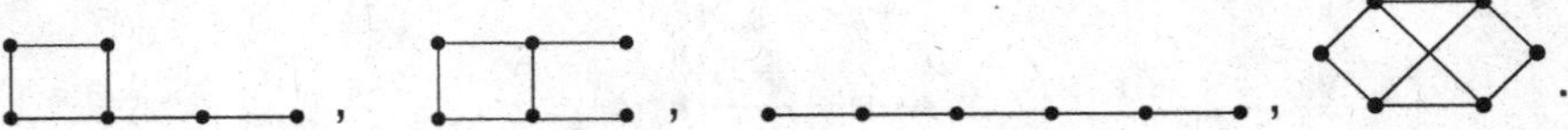

Now we come to the main idea that will suggest a "Tutte-like" recurrence for the number of Eulerian cycles. Look at a vertex a in the 2-in 2-out graph G,

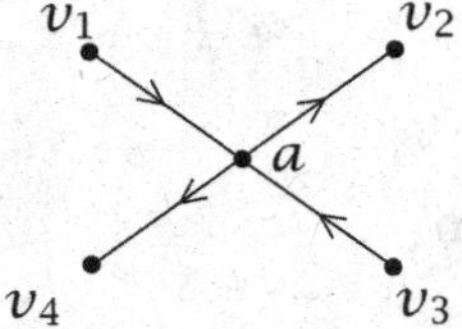

where the v_i's need not be distinct, and an interlaced vertex b. Suppose the 2-word of C is $w = \dots v_1 a v_2 \dots b \dots v_3 a v_4 \dots b$, that is, $v_i \in w_i$ in the notation above.

There are two possible ways to pass through a in any Eulerian cycle:

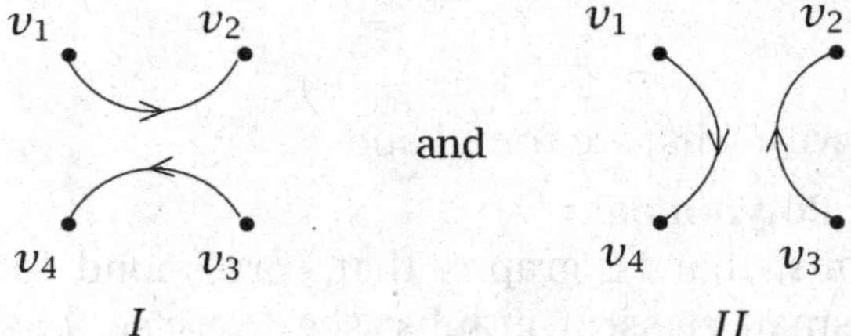

and it is clear that

$$e(G) = e(G \setminus a)_I + e(G \setminus a)_{II}\,, \tag{1}$$

where $(G \setminus a)_I$, $(G \setminus a)_{II}$ are again 2-in 2-out graphs. The interlace graphs corresponding to the merged graphs $(G \setminus a)_I$, $(G \setminus a)_{II}$ are $H(C) \setminus a$ and $H(C^{ab}) \setminus a$, since we remove the chord a and keep the rest as before. Hence if we define the function $f(H(C)) = e(G)$, then it follows from (1) that

$$f(H(C)) = f(H(C) \setminus a) + f(H(C^{ab}) \setminus a)\,,$$

or since $H(C^{ab}) = H(C)^{ab}$ (a and b swapped),

$$f(H(C)) = f(H(C)\setminus a) + f(H(C)^{ab}\setminus b). \tag{2}$$

Recurrence (2) is the basis for the following polynomial, which is defined for any simple graph.

Definition. The *interlace polynomial* $Q(H;x)$ of a simple graph $H = (V,E)$ is defined recursively as follows:

$$\begin{aligned} &\text{(i)} \quad Q(H;x) = x^{|V|}, \text{ if } H \text{ has no edges,} \\ &\text{(ii)} \quad Q(H;x) = Q(H\setminus a;x) + Q(H^{ab}\setminus b;x) \text{ for } \{a,b\} \in E. \end{aligned} \tag{3}$$

As for the Tutte polynomial, it is not at all clear that $Q(H;x)$ is independent of the order of edge-switchings. Uniqueness can be proved (somewhat tediously) by considering directly the effect of two successive switchings. We use a linear-algebraic approach to produce a polynomial that is independent of order, reminiscent of Theorem 9.3.

Fix the vertex set $\{1,\dots.n\}$ of H, and let $A = (a_{ij})$ be the adjacency matrix, that is,

$$a_{ij} = \begin{cases} 1 & \text{if } \{i,j\} \in E, \\ 0 & \text{if } \{i,j\} \notin E. \end{cases}$$

Observe that A is a symmetric matrix and denote by I_n the $n \times n$-identity matrix. Henceforth all matrices will be considered as matrices over the field $GF(2) = \{0,1\}$.

Next, let L be the $n \times 2n$-matrix

$$L = (\underset{1\dots n}{A} \mid \underset{\overline{1}\dots\overline{n}}{I_n})$$

where we label the rows $1,\dots,n$ and the columns $1,\dots,n, \overline{1},\dots,\overline{n}$ as indicated. A column set S is *admissible* if $|S \cap \{i,\overline{i}\}| = 1$ for all i; thus $|S| = n$. Let L_S be the $n \times n$-submatrix of L with columns from S. Let $\mathcal{A}$ denote the set of all admissible column sets; thus $|\mathcal{A}| = 2^n$, and $\text{rk}(M)$ denote the *rank* of any matrix M.

Theorem 9.10. *The interlace polynomial of a simple graph $H = (V,E)$ on n vertices is given by*

$$Q(H;x) = \sum_{S\in\mathcal{A}} (x-1)^{n-\text{rk}(L_S)}. \tag{4}$$

Proof. When H has no edges, we have $L = (O_n|I_n)$, where O_n is the $n\times n$-zero matrix; hence $\text{rk}(L_S) = |S\cap\{\overline{1},\dots,\overline{n}\}|$. It follows that

$$\sum_S (x-1)^{n-\text{rk}(L_S)} = \sum_{k=0}^{n}\binom{n}{k}(x-1)^k = x^n .$$

Suppose without loss of generality that $\{n-1,n\}\in E(H)$. We have to verify recurrence (3) for the right-hand side of (4).

Case 1. $\overline{n}\in S$. Let $S' = S\smallsetminus\overline{n}$. Then the matrix L_S looks as follows:

$$\left(\begin{array}{cc|cc} \overset{S'}{\boxed{}} & \begin{matrix}\vdots\\ 0\end{matrix} & I & \overset{S'}{\boxed{}}\ \begin{matrix}0\\ \vdots\\ 0\\ 1\end{matrix} \end{array}\right) .$$
$$\qquad\qquad n\notin S \qquad\qquad \overline{n}\in S$$

Clearly, $\text{rk}(L_S) = \text{rk}(L_{S'}) + 1$, where $L_{S'}$, is the submatrix with the n-th row removed. Hence $n - \text{rk}(L_S) = (n-1) - \text{rk}(L_{S'})$, and we obtain by induction

$$\sum_{S:\overline{n}\in S}(x-1)^{n-\text{rk}(L_S)} = \sum_{S'}(x-1)^{n-1-\text{rk}(L_{S'})} = Q(H\smallsetminus n; x) . .$$

Case 2. $n\in S$. Write L as

$$L = \left(\begin{array}{ccc|ccc} B & c_1 & c_2 & I_{n-2} & 0 & 0 \\ c_1^T & 0 & 1 & 0^T & 1 & 0 \\ c_2^T & 1 & 0 & 0^T & 0 & 1 \end{array}\right),$$
$$\qquad\qquad n-1 \qquad n \qquad\qquad \overline{n-1} \qquad \overline{n}$$

where c_1, c_2 are column vectors of length $n-2$, and c_1^T, c_2^T the transposes.

Now we multiply L from the left by the matrix C:

$$C = \begin{pmatrix} I_{n-2} & c_2 & c_1 \\ 0 & 1 & 0 \\ 0 & 0 & 1 \end{pmatrix} .$$

Since C is nonsingular, all ranks are preserved, and we get (note that $c_i + c_i = 0$)

$$CL = \left(\begin{array}{ccc|ccc} B + c_1c_2^T + c_2c_1^T & 0 & 0 & I_{n-2} & c_2 & c_1 \\ c_1^T & 0 & 1 & 0^T & 1 & 0 \\ c_2^T & 1 & 0 & 0^T & 0 & 1 \end{array} \right).$$
$$\qquad\qquad\qquad\qquad n-1 \quad n \qquad\qquad \overline{n-1} \quad \overline{n}$$

It is easily seen that $B+c_1c_2^T+c_2c_1^T$ is precisely the adjacency matrix of $H^{(n-1,n)}$ on $V \setminus \{n-1, n\}$. (Check it!) Interchanging columns $n-1 \leftrightarrow \overline{n-1}$, $n \leftrightarrow \overline{n}$ and rows $n-1 \leftrightarrow n$ yields

$$\left(\begin{array}{ccc|ccc} B + c_1c_2^T + c_2c_1^T & c_2 & c_1 & I_{n-2} & 0 & 0 \\ c_2^T & 0 & 1 & 0^T & 1 & 0 \\ c_1^T & 1 & 0 & 0^T & 0 & 1 \end{array} \right).$$
$$\qquad\qquad\qquad\qquad \overline{n-1} \quad \overline{n} \qquad\qquad n-1 \quad n$$

Hence by the same argument as in case 1 we obtain

$$\sum_{S:n \in S} (x-1)^{n-\mathrm{rk}(L_S)} = Q(H^{(n-1,n)} \setminus (n-1); x),$$

and the proof is complete. $\square$

Formula (4) can be rewritten in a more convenient way. Suppose S is admissible with $T = S \cap \{1, \ldots, n\}$, and A_T is the submatrix of A with rows and columns in T. In other words, A_T is the adjacency matrix of the *induced* subgraph of H on T. Since $\mathrm{rk}(L_S) = \mathrm{rk}(A_T) + n - |T|$, we get

$$|T| - \mathrm{rk}(A_T) = n - \mathrm{rk}(L_S),$$

and thus the following result.

Corollary 9.11. *We have*

$$Q(H; x) = \sum_{T \subseteq \{1,\ldots,n\}} (x-1)^{|T| - \mathrm{rk}(A_T)}, \tag{5}$$

with $\mathrm{rk}(A_\emptyset) = 0$ *by definition.*

It follows from recurrence (3) that $Q(H;x)$ is a polynomial with nonnegative integer coefficients, which in view of the expression (4) is rather surprising. Furthermore, (3) implies

$$Q(H^{ab};x) = Q(H^{ab}\setminus b;x) + Q(H\setminus a;x) = Q(H;x),$$

since $(H^{ab})^{ab} = H$, and thus the following result.

Corollary 9.12. *Switching-equivalent graphs have the same interlace polynomial. In particular, any two interlace graphs belonging to the same 2-in 2-out graph possess the same interlace polynomial.*

Example. For the complete graph K_n, switching does not change the rest, that is, $K_n\setminus a = K_n^{ab}\setminus b = K_{n-1}$; thus $Q(K_n;x) = 2Q(K_{n-1};x)$, and so $Q(K_n;x) = 2^{n-1}x$, since $Q(K_1;x) = x$.

Before we apply the results to Eulerian cycles let us look at some evaluations of $Q(H;x)$. For $x = 2$ we get $Q(H;2) = 2^n$. Now let us look at $x = 1$. By (5),

$$Q(H;1) = \#\{T \subseteq \{1,\dots,n\} : \mathrm{rk}(A_T) = |T|\},$$

or equivalently (remember we work over $GF(2)$),

$$Q(H;1) = \#\{T : \det A_T = 1\}. \tag{6}$$

Let G be any graph, and B its adjacency matrix. If we orient the edges, then B becomes a skew-symmetric matrix (over $\mathbb{Z}$). Now recall Section 5.3, where we studied these matrices. If $n = |V(G)|$ is odd, then $\det B = 0$ over $\mathbb{Z}$, and hence also over $GF(2)$, since here $1 = -1$. If n is even, then

$$\det B = (\mathrm{Pf}\, B)^2.$$

Now, $|\mathrm{Pf}\, B|$ counts all perfect matchings of G, and passing to $GF(2)$ we have

$$\mathrm{Pf}\, B = 1 \iff G \text{ has an } \textit{odd} \text{ number of perfect matchings.}$$

Consequently, by (6) we arrive at the following result.

Corollary 9.13. *We have*

$$Q(H;1) = \#\ \textit{induced subgraphs of } H \textit{ with an odd number of perfect matchings (including the empty set).}$$

Since a forest has at most one perfect matching; this yields in particular the following:

Corollary 9.14. *For a forest H, $Q(H;1)$ counts the number of matchings (disjoint edge sets) in H, including the empty set.*

Example. For the complete graph, $Q(K_n;1) = 2^{n-1}$, and it is precisely the complete subgraphs K_{2h} on an *even* number of vertices that have an odd number $(2h-1)(2h-3)\cdots 3\cdot 1$ of perfect matchings.

Counting Eulerian Cycles and Decompositions.
For the interlace polynomial there is a simple recipe theorem. Suppose f is a function on graphs that satisfies

$$\begin{aligned} f(H_1 \dot\cup H_2) &= f(H_1)f(H_2) \text{ for disjoint unions,} \\ f(H) &= f(H\smallsetminus a) + f(H^{ab}\smallsetminus b),\ \{a,b\}\in E\,. \end{aligned} \tag{7}$$

Then (3) implies $f(H) = Q(H;s)$, where $s = f(K_1)$.

Proposition 9.15. *Let G be a 2-in 2-out graph, and H any of its interlace graphs. Then for the number $e(G)$ of Eulerian cycles, and for $e_k(G)$, we have*

$$\begin{aligned} &\text{a. } e(G) = Q(H;1), \\ &\text{b. } e(G;x) := \textstyle\sum_{k\geq 0} e_{k+1}(G)x^k = Q(H;1+x). \end{aligned} \tag{8}$$

Proof. For an interlace graph H of G define $f(H) = e(G)$. We clearly have $f(H_1 \dot\cup H_2) = f(H_1)f(H_2)$, and the recurrence holds because of (2). It remains to consider $H = K_1$, with

$G =$ 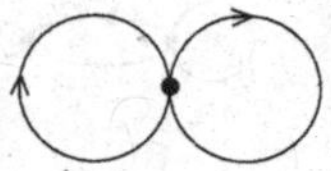 .

Since G has precisely one Eulerian cycle, $f(K_1) = 1$, and (a) follows. Similarly, using $f(H) = \sum_{k\geq 0} e_{k+1}(G)x^k$, we obtain $f(H) = 1 + x$ for $H = K_1$, thus proving (b). □

Example. In our running example take the path P_6 as interlace graph. The interlace polynomial is quickly computed,

$$Q(P_6;x) = 2x + 7x^2 + 4x^3\,,$$

and we get

$$Q(P_6; 1 + x) = 13 + 28x + 19x^2 + 4x^3 .$$

Therefore G has 13 Eulerian cycles, and further, $e_2(G) = 28$, $e_3(G) = 19$, $e_4(G) = 4$. Note also that by Corollary 9.14, P_6 has precisely 13 matchings. A moments's reflection shows that the path P_n on n vertices possesses F_{n+1} matchings (Fibonacci number); hence $Q(P_n; 1) = F_{n+1}$.

The third question raised at the beginning, as to the maximal m with $e_m(G) > 0$, refers to the degree of $Q(H; x)$, with $m = \deg Q + 1$. It is treated in the exercises.

Let us finally look at all 2-in 2-out graphs on n vertices and estimate $\max e(G)$ and $\min e(G)$, or what is the same, $\max Q(H; 1)$ and $\min Q(H; 1)$ for a corresponding interlace graph H. An easy lower bound is provided by Corollary 9.13. Since $\emptyset$ and single edges are induced subgraphs with an odd number of perfect matchings (namely 1), we obtain

$$Q(H; 1) \geq |E(H)| + 1 \text{ for all graphs } H .$$

Hence $e(G) \geq |E| + 1$ and $e(G) = 1$ if and only if the chord diagram has no crossings. We know from the highlight in Chapter 7 that the number of these diagrams is the Catalan number C_n. Two examples of 2-in 2-out graphs G with $e(G) = 1$ are

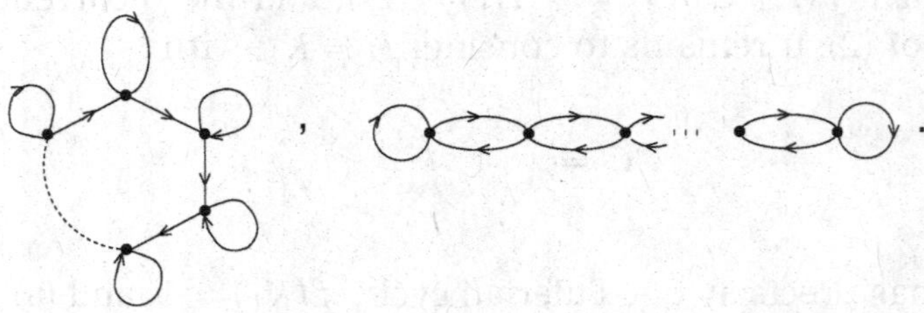

Another lower bound that depends only on $n = |V|$ is contained in Exercise 9.33.

As for an upper bound, recurrence (3) implies inductively

$$Q(H; 1) \leq 2^{n-1} ,$$

and furthermore, $Q(H;1) = 2^{n-1}$ only for $H = K_n$. Thus $e(G) \le 2^{n-1}$, and $e(G) = 2^{n-1}$ only for the 2-in 2-out graph G with 2-word $1\,2\ldots n\,1\,2\ldots n$. This graph is unique (up to labeling) and given by

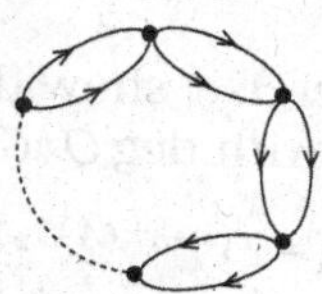

Exercises

9.19 Prove that any Eulerian cycle of a 2-in 2-out graph can be obtained from any other by a series of transpositions.

9.20 Prove Lemma 9.9.

9.21 Show that the 5-wheel is not an interlace graph. In fact, it is the smallest such graph.

▷ **9.22** Compute the interlace polynomial for the graphs $K_{m,n}$ and P_n (path with n vertices). What is $Q(H;1)$ for these graphs?

9.23 Show that bipartiteness is preserved by edge-switching. More precisely, if H is a connected bipartite graph with m and n vertices of each color, then any equivalent graph has this property.

9.24 Show that the trees and have the same interlace polynomial but are not switching equivalent. Hint: Previous exercise.

9.25 Let H have the components $H_1, \ldots, H_{k(H)}$. Show that $Q(H;x) = \prod_{i=1}^{k(H)} Q(H_i;x)$, and deduce that $k(H)$ is the smallest index i for which $q_i > 0$ in $Q(H;x) = \sum q_i x^i$.

9.26 Suppose H is an induced subgraph of a connected graph G. Show that $Q(H;x) \le Q(G;x)$, meaning that for all i, the i-th coefficient of $Q(H;x)$ is less than or equal to that of $Q(G;x)$.

▷ **9.27** Show that the linear coefficient of $Q(C_n;x)$ is $n+1$ if n is odd, and $n-2$ if n is even, and deduce that H is connected and bipartite if $Q(H;x)$ has linear coefficient $q_1 = 2$.

* * *

9.28 Distribute n colored balls around a circle where $n_i \ge 1$ balls are colored with color i, $i = 1, \ldots, m$. A distribution is *admissible* if no two colors interlace. Prove the surprising result that the number of admissible distributions is always $n^{\underline{m-1}}$ regardless of the frequencies n_i.

▷ **9.29** The independence number $\alpha(G)$ of a simple (undirected) graph is the maximal size of a set of pairwise nonadjacent vertices. Show that $\deg Q(G;x) = \max \alpha(H)$, where the maximum is taken over all graphs $H \approx G$.

9.30 In continuation of the exercise, show that $\deg Q(G;x) = \alpha(G)$ when G is a forest. Give an example with $\deg Q(G;x) > \alpha(G)$.

9.31 Prove $Q(H;-1) = (-1)^n(-2)^{n-\mathrm{rk}(A+I_n)}$ with the notation as in Theorem 9.10. Hint: Use a similar argument as in the proof of Theorem 9.10, considering $B + I_{n-2}$.

9.32 Let G be a 2-in 2-out graph and $e_k = e_k(G)$ as before, with $m = \max k$, $e_k > 0$. Deduce from Proposition 9.15:

a. $e_m < e_{m-1} < \cdots < e_{\lceil \frac{m}{2} \rceil}$,
b. $e_i < e_{m-i+1}$ for $i > \frac{m+1}{2}$.

Hint: This proceeds along the lines of Exercise 9.9.

▷ **9.33** Let H be a graph without isolated vertices. Show that $Q(H;1) \geq n = |V|$, with $Q(H;1) = n$ if and only if $H = K_{1,n-1}$, with one exception, which? What do the associated 2-word to $K_{1,n-1}$ and the 2-in 2-out graph look like?

9.34 Let $H = (V, E)$ be a simple graph. We know that $Q(H;1) \geq |E| + 1$. Show that for a graph without isolated vertices, equality holds if and only if $H = K_{h,i,j}$, where one of the indices may be 0.

▷ **9.35** Consider all $(2n-1)(2n-3)\cdots 3\cdot 1$ diagrams with n chords. Any diagram gives rise to a 2-in 2-out graph G. Let $f_k(n)$ be the number of diagrams whose associated graph has exactly k Eulerian cycles, thus $\sum_{k=0}^{\binom{n}{2}} f_k(n) = (2n-1)\cdots 3\cdot 1$. Compute $f_k(n)$ for $k \leq 3$. Hint: Use the bounds for $Q(H;1)$, where H is an interlace graph of G.

9.3 Plane Graphs and Transition Polynomials

Plane Graphs.

You are probably familiar with plane graphs and the 4-color theorem. Let us recall the basic definitions. A graph $G = (V, E)$ is *plane* if the vertices are points of $\mathbb{R}^2$ and the edges Jordan (non-self-intersecting) curves between the points that intersect only in the endpoints. The complement $\mathbb{R}^2 \setminus (V \cup E)$ splits into disjoint regions, called the *faces* F of G. Incidences between vertices, edges, and faces are declared in the natural way. We then write $G = (V, E, F)$. A graph is *planar* if it admits a plane embedding.

Example. The plane graph

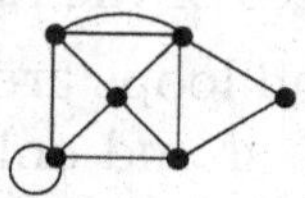

has 6 vertices, 12 edges, and 8 faces (including the outer face).

Here are three fundamental facts which can be found in any graph theory book.

A. The Jordan curve theorem. Any closed simple curve in $\mathbb{R}^2$ divides the plane into precisely two regions, one of which is unbounded.

For a plane graph this means that any circuit partitions the graph into the circuit, the interior, and the exterior.

B. Euler formula. For a connected plane graph $G = (V, E, F)$,

$$|V| - |E| + |F| = 2\,. \tag{1}$$

C. To every plane graph $G = (V, E, F)$ there is a *dual* plane graph $G^* = (V^*, E^*, F^*)$ constructed as follows. Place a point in the interior of every face and join two such points if the corresponding faces in G share an edge e on the boundary, by drawing an edge across e. If the faces have several common boundary edges, then draw a new edge across every such boundary edge. Clearly, this can be done in such a way that the new graph G^* is again plane.

Note that G^* is always connected, and that $G \leftrightarrow G^*$ is an involution among connected plane graphs. We may thus identify $V^* = F$, $E^* = E$, and $F^* = V$.

Example.

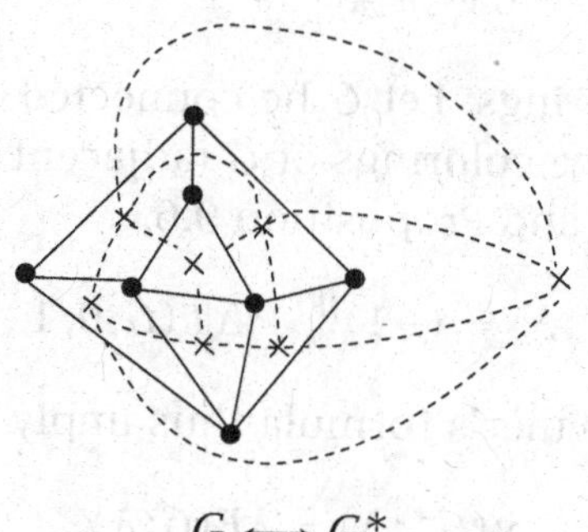

$G \longleftrightarrow G^*$

The following facts should be clear: The degree $d(v^*)$ in the dual graph G^* equals the number of boundary edges of the face f in G with $f = v^*$, and dually the degree $d(v)$ of a vertex v in G equals the number of boundary edges of the face f^* in G^* where $f^* = v$.

Next we note that bridges and loops are dual concepts in the sense that e is a bridge (loop) in G if and only if e^* is a loop (bridge) in G^*. Just note that by the Jordan curve theorem, e is a bridge of G if and only if the two incident faces are the same. Furthermore, we have (see Exercise 9.37) that if e is not a loop or bridge of G, then

$$(G \setminus e)^* = G^*/e, \; (G/e)^* = G^* \setminus e. \tag{2}$$

Hence we obtain the following important result.

Proposition 9.16. *Let G be a plane graph. Then for the Tutte polynomial,*

$$T(G^*; x, y) = T(G; y, x). \tag{3}$$

Example. For graphs that admit a self-dual embedding, that is, $G \cong G^*$, like the two examples of the figure

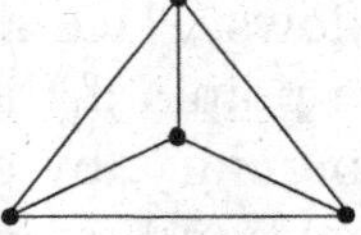
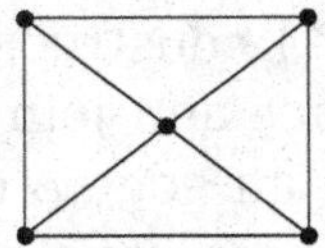

the Tutte polynomial must be invariant under $x \leftrightarrow y$. Indeed, we have for K_4 and the 4-wheel W_4,

$$T(K_4; x, y) = x^3 + 3x^2 + 2x + 4xy + 2y + 3y^2 + y^3,$$
$$T(W_4; x, y) = x^4 + 4x^3 + 6x^2 + 3x + 4x^2y + 9xy + 4xy^2 + 3y + 6y^2 + 4y^3 + y^4.$$

Next we look at colorings. Let G be connected. Vertex-colorings of G^* correspond to face-colorings of G (adjacent faces receive different colors), so by (3) and Proposition 9.6,

$$\chi(G^*; \lambda) = (-1)^{|F|-1} \lambda T(G; 0, 1 - \lambda). \tag{4}$$

Proposition 9.7 and Euler's formula thus imply

$$\chi(G^*; \lambda) = \lambda F(G; \lambda), \tag{5}$$

where $F(G;\lambda)$ is the *flow polynomial* of G.

In particular, (5) gives another pair of dual concepts: The plane graph G is Eulerian if and only if G^* is bipartite. Indeed,

$$G^* \text{ bipartite } \iff \chi(G^*;2) \neq 0 \iff F(G;2) \neq 0 \iff G \text{ Eulerian.}$$

In other words, a plane graph G has a 2-face-coloring if and only if G is Eulerian. Continuing, we see that a plane graph G has a 4-face-coloring if and only if G has a 4-flow. The 4-color theorem (4-CT) thus takes on two equivalent forms:

$$\begin{aligned} \text{4-CT} &\iff \text{every planar graph has a 4-vertex-coloring} \\ &\iff \text{every planar graph has a 4-flow.} \end{aligned}$$

The Medial Graph.

We come to the main definition that will allow us to define counting polynomials on plane graphs G. Let $G = (V, E, F)$ be a plane graph. Every edge e appears on the boundary of two faces (which may be identical if e is a bridge), and we call the four incident edges along these faces the *neighbors* of e as in the figure:

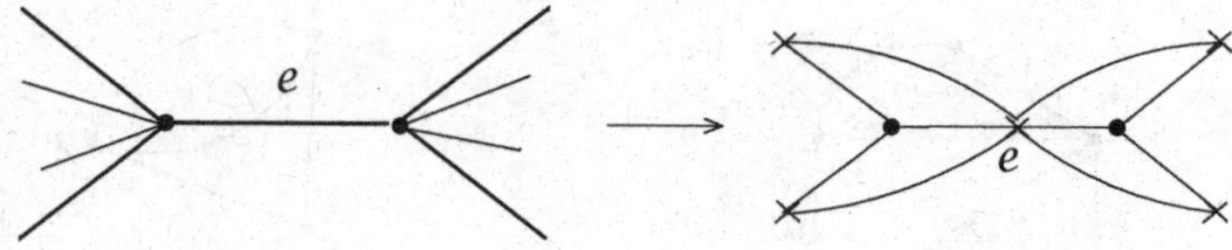

Put a vertex on every edge and join it by a small curve inside the face to each of its four neighbors. The resulting graph $\widetilde{G} = (\widetilde{V}, \widetilde{E}, \widetilde{F})$ is again plane, and 4-regular; $\widetilde{G}$ is called the *medial graph* of G. The graph $\widetilde{G}$ is connected if and only if G is. Since $\widetilde{G}$ is Eulerian, its faces can be 2-colored black and white. We color the outer face white; the rest is then determined by connectivity.

Example. The 4-wheel G and its medial graph $\widetilde{G}$:

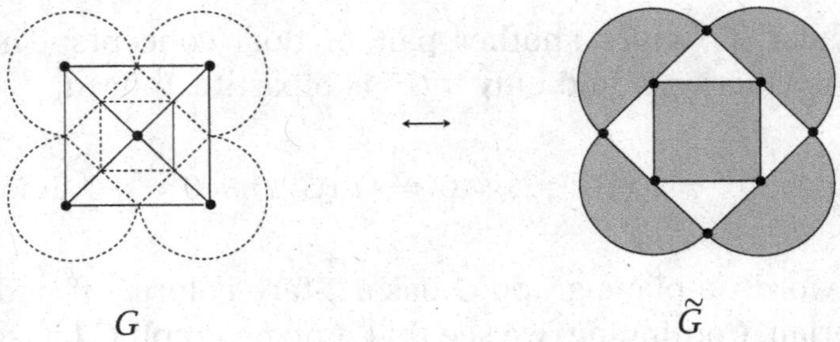

The following facts are immediate from the definition:

1. The *black* faces of $\widetilde{G}$ correspond to the vertices of G, and the number of boundary edges equals the degree of the corresponding vertex.
2. The *white* faces of $\widetilde{G}$ correspond to the faces of G, with the same boundary length.
3. Every edge of $\widetilde{G}$ is incident to a black and a white face.
4. Two vertices of G are adjacent if and only if the corresponding black faces of $\widetilde{G}$ share a common vertex, and dually for the faces of G and the white faces of $\widetilde{G}$.

In sum, a plane graph G determines a 2-colored medial graph that is 4-regular. Conversely, if $\widetilde{G}$ is a 2-colored 4-regular plane graph, then we may uniquely reconstruct the underlying plane graph G. We then call G the *Tait graph* of $\widetilde{G}$.

Example. Suppose we are given $\widetilde{G}$ as in the figure, with the black faces numbered 1 to 4.

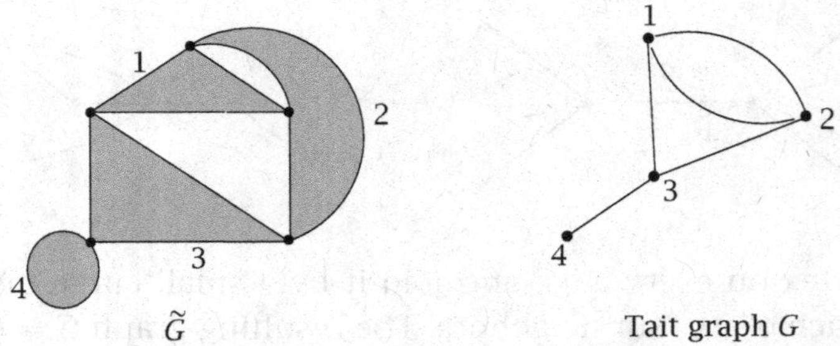

Transition Polynomials.

Consider a connected plane graph $G = (V, E, F)$ and its medial graph $\widetilde{G} = (\widetilde{V}, \widetilde{E}, \widetilde{F})$. We may identify $E = \widetilde{V}$. Take a vertex $e \in \widetilde{V}$.

There are three possibilities to decompose the incident edges at e into two vertices of degree 2:

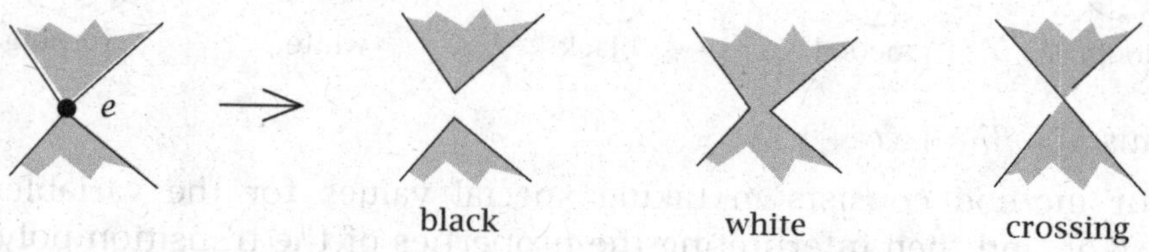

We speak of a *transition* $p(e)$ at e of black, white, or crossing type. If a transition $p(e)$ is chosen at every vertex e of $\widetilde{G}$, then $p = \{p(e) : e \in \widetilde{V}\}$ is called a *transition system*. Hence there are $3^{|E|}$ different transition systems. Every transition system decomposes the medial graph $\widetilde{G}$ into a number $c(p)$ of edge-disjoint cycles (Eulerian subgraphs).

Now assign to every transition type $p(e)$ a weight (variable),

$$W(p(e)) = \begin{cases} \alpha & \text{black}, \\ \beta \quad \text{if } p(e) \text{ is} & \text{white}, \\ \gamma & \text{crossing}, \end{cases} \tag{6}$$

and set

$$W(p) = \prod_{e \in E} W(p(e)) .$$

Definition. The *transition polynomial* $S(\widetilde{G}, W; \lambda)$ with respect to W is

$$S(\widetilde{G}, W; \lambda) = \sum_{p} W(p) \lambda^{c(p)} . \tag{7}$$

Example. Let us look at the two smallest examples with $|E| = 1$.

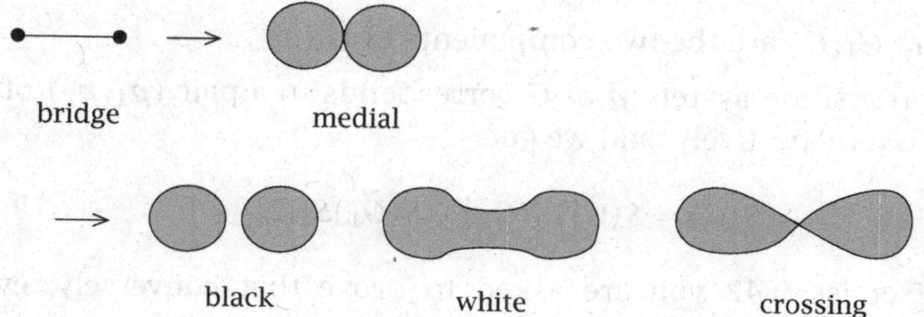

Hence $S = \alpha\lambda^2 + (\beta + \gamma)\lambda$. Similarly, for the loop

loop medial black white crossing

thus $S = \beta\lambda^2 + (\alpha + \gamma)\lambda$.

Our method consists in taking special values for the variables α, β, γ, and then interpreting the properties of the transition polynomial in terms of the underlying Tait graph G. For example, when $\alpha = \beta = \gamma = 1$, then $S(\widetilde{G}, W_{1,1,1}; \lambda) = \sum_p \lambda^{c(p)} = \sum_{k\geq 1} e_k \lambda^k$, where e_k is the number of decompositions of $\widetilde{G}$ into k Eulerian cycles.

Penrose Polynomial.
A very interesting transition polynomial suggested by Penrose comes from the evaluation $\alpha = 0$ (no black transitions), $\beta = 1$, $\gamma = -1$.

Definition. The *Penrose polynomial* of a connected plane graph $G = (V, E, F)$ is

$$P(G; \lambda) = S(\widetilde{G}, W_{0,1,-1}; \lambda) = \sum_p (-1)^{x(p)} \lambda^{c(p)},$$

where $x(p)$ is the number of crossing vertices in p.

First we note that $P(G; \lambda) = 0$ if G contains a bridge. Indeed, a bridge e turns into a cut vertex of $\widetilde{G}$, and we obtain pictorially

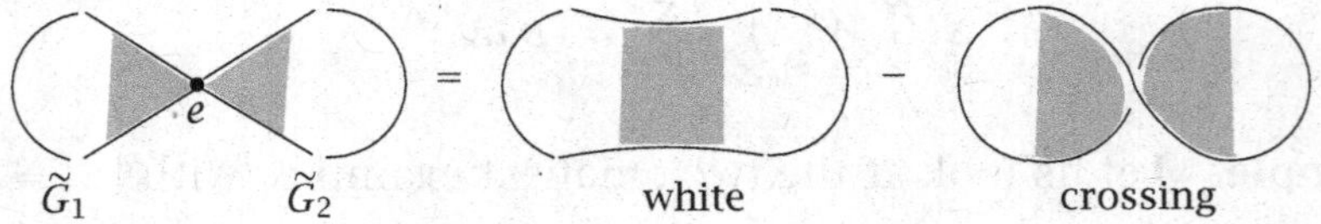

where $\widetilde{G}_1, \widetilde{G}_2$ are the two components of $\widetilde{G} \setminus e$.

Any transition system p of $\widetilde{G}$ corresponds to a pair (p_1, p_2) of $\widetilde{G}_1$ and $\widetilde{G}_2$, respectively, and we get

$$S(\widetilde{G}) = S(\widetilde{G}_1)S(\widetilde{G}_2) - S(\widetilde{G}_1)S(\widetilde{G}_2) = 0.$$

In Exercise 9.42 you are asked to prove that conversely, every bridgeless connected plane graph G has nonzero Penrose polynomial of degree $|F|$, the number of faces in G.

The first question we want to tackle is whether $P(G;\lambda)$ counts something for positive integers k. Call a map $g : \widetilde{E} \longrightarrow \{1, 2, \ldots, k\}$ a k-*valuation* if at every vertex e every integer appears an even number (possibly zero) of times. Hence either the same integer appears at all four edges, or two integers $i \neq j$ appear twice. We therefore have four possibilities for the *type* (g, e):

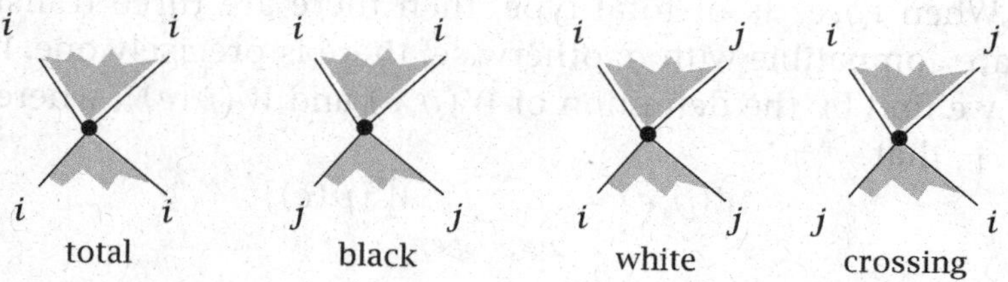

A k-valuation is *admissible* if only the white and crossing types occur.

Proposition 9.17. *Let k be a positive integer, then*

$$P(G;k) = \#\ \textit{admissible}\ k\textit{-valuations}.$$

Proof. The following reasoning is essentially an inclusion-exclusion argument. For a valuation g and a vertex $e \in \widetilde{V}$ set

$$W(g,e) = \begin{cases} 0 & \text{if } (g,e) \text{ is total or black,} \\ 1 & \text{if } (g,e) \text{ is white,} \\ -1 & \text{if } (g,e) \text{ is crossing,} \end{cases}$$

and

$$W(g) = \prod_e W(g,e)\,.$$

If g is not admissible, then $W(g) = 0$ by definition. Suppose g is admissible, and let $x(g)$ be the number of crossing types (g, e); thus $W(g) = (-1)^{x(g)}$. Denote by $C_i = \{y \in \widetilde{E} : g(y) = i\}$ the preimage of i, $i = 1, \ldots, k$. The C_i's form a set of disjoint circuits in $\widetilde{G}$, and the Jordan curve theorem implies that any two C_i, C_j cross an even number of times. This implies that $x(g)$ is even, and so $W(g) = 1$. We conclude therefore that

$$\sum_g W(g) = \#\ \text{admissible}\ k\text{-valuations}.$$

It remains to show that

$$\sum_{g} W(g) = \sum_{p} (-1)^{x(p)} k^{c(p)}. \tag{8}$$

Let g be an arbitrary k-valuation, and p a transition system. We write $p(e) \prec (g,e)$ if the transition $p(e)$ has equal numbers in g at e. When (g,e) is of total type, then there are three transitions that are compatible with g; otherwise, there is precisely one. In any case, we find by the definition of $W(g,e)$ and $\overline{W}(p(e))$, where $\overline{W} = W_{0,1,-1}$, that

$$W(g,e) = \sum_{p(e) \prec (g,e)} \overline{W}(p(e)).$$

Set $p \prec g$ if $p(e) \prec (g,e)$ holds for all e. Then we get

$$\begin{aligned}
\sum_{g} W(g) &= \sum_{g} \prod_{e} W(g,e) = \sum_{g} \prod_{e} \sum_{p(e) \prec (g,e)} \overline{W}(p(e)) \\
&= \sum_{g} \sum_{p \prec g} \prod_{e} \overline{W}(p(e)) = \sum_{g} \sum_{p \prec g} (-1)^{x(p)} \\
&= \sum_{p} (-1)^{x(p)} |\{g : p \prec g\}| .
\end{aligned}$$

For fixed p, we have $p \prec g$ if and only if g is constant on the $c(p)$ cycles induced by p, and we conclude that $|\{g : p \prec g\}| = k^{c(p)}$. This proves (8), and thus the proposition. □

Clearly, $P(G;0) = 0$, and since there are no admissible 1-valuations we also have $P(G;1) = 0$.

Corollary 9.18. *We have*

$$P(G;2) = \begin{cases} 2^{|V|} & \text{if } G \text{ is Eulerian,} \\ 0 & \text{otherwise.} \end{cases}$$

Proof. An admissible 2-valuation must assume alternate numbers around a black face of $\widetilde{G}$. Hence every black face must have an even number of boundary edges, or equivalently G must be Eulerian. Since there are two possibilities for each black face, the result follows. □

We come to one of the main discoveries of Penrose. An *edge-coloring* of any graph is a coloring that assigns different colors to incident edges.

Theorem 9.19. *Suppose G is a connected plane 3-regular graph; then*

$$P(G;3) = \#\, 3\text{-edge-colorings.}$$

Proof. The following picture shows the whole proof. Relate a 3-edge-coloring g of G to a 3-valuation $\overline{g}$ of $\tilde{G}$ as in the figure:

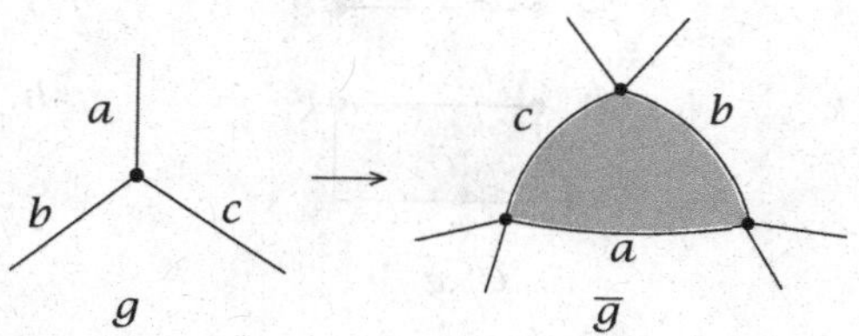

It is easily seen that $\overline{g}$ is an admissible 3-valuation, and that $g \mapsto \overline{g}$ is a bijection. □

One of the first results on the 4-color problem was Tait's theorem stating that the 4-color theorem holds if and only if every 3-regular connected plane graph without loops and bridges has a 3-edge-coloring. Hence another equivalent formulation of 4-CT is that $P(G;3) > 0$ for all 3-regular connected plane graphs without loops or bridges.

The Case of No Crossings.
Another interesting situation is given by the evaluation $y = 0$. Let us denote the weight function by $W_{\alpha,\beta}$. For example, for a bridge or loop we get

$$S(\widetilde{\text{bridge}}, W_{\alpha,\beta};\lambda) = \alpha\lambda^2 + \beta\lambda\,,$$
$$S(\widetilde{\text{loop}}, W_{\alpha,\beta};\lambda) = \beta\lambda^2 + \alpha\lambda\,.$$

The following figure shows what happens when we delete or contract e in G where e is neither a bridge nor a loop.

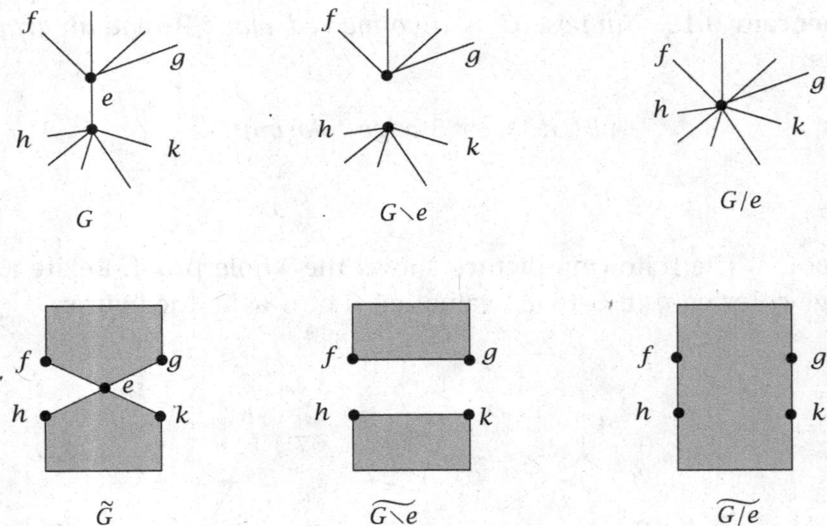

We see that the restriction $G \setminus e$ corresponds to the *black* transition at e, and the contraction to the *white* transition. Hence

$$S(\widetilde{G}, W_{\alpha,\beta}; \lambda) = \alpha S(\widetilde{G \setminus e}, W_{\alpha,\beta}; \lambda) + \beta S(\widetilde{G/e}, W_{\alpha,\beta}; \lambda).$$

From this one readily obtains that $\frac{S(\widetilde{G}, W_{\alpha,\beta}; \lambda)}{\lambda}$ is a chromatic invariant of the Tait graph G, with $A = \alpha\lambda + \beta$, $B = \beta\lambda + \alpha$ in the notation of Theorem 9.5. Together with Euler's formula $|E| - |V| + 1 = |F| - 1$ we have proved the following result.

Theorem 9.20. *Let $G = (V, E, F)$ be a connected plane graph, $\alpha \neq 0$, $\beta \neq 0$. Then*

$$S(\widetilde{G}, W_{\alpha,\beta}; \lambda) = \alpha^{|F|-1}\beta^{|V|-1}\lambda T G; 1 + \frac{\alpha}{\beta}\lambda, 1 + \frac{\beta}{\alpha}\lambda). \tag{9}$$

Different weightings yield, of course, different transition polynomials. But there is one remarkable case in which we can assert equality. The proof is left to the exercises.

Lemma 9.21. *Suppose the weightings W and W' differ by an additive constant, that is, $\alpha' = \alpha + m$, $\beta' = \beta + m$, $\gamma' = \gamma + m$. Then $S(\widetilde{G}, W'; -2) = S(\widetilde{G}, W; -2)$.*

Example. Take $W = W_{0,1,-1}$ and $W' = W_{1,2,0}$. The first weighting gives the Penrose polynomial, while the second is covered by the previous theorem. The lemma implies

$$P(G;-2) = S(\tilde{G}, W_{0,1,-1}; -2) = S(\tilde{G}, W_{1,2,0}; -2) = -2^{|V|} T(G;0,-3).$$

Now we know from (3) and Proposition 9.6 that

$$(-1)^{|F|-1} 4\, T(G;0,-3)$$

counts the number of 4-face-colorings of G, which gives another amazing property of the Penrose polynomial.

Corollary 9.22. *We have*

$$P(G;-2) = (-1)^{|F|} 2^{|V|-2} \cdot (\#4\text{-face-colorings of } G).$$

In particular, the 4*-color theorem is equivalent to* $P(G;-2) \neq 0$ *for all connected plane graphs without bridges.*

Next we discuss an interesting connection to 2-in 2-out graphs that will prove useful in the next section. Consider a connected plane graph $G = (V, E, F)$ and the medial graph $\tilde{G}$, and orient the edges of $\tilde{G}$ such that the black face is always on the right, as in the figure:

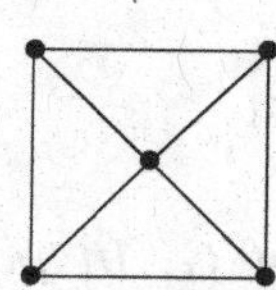

G

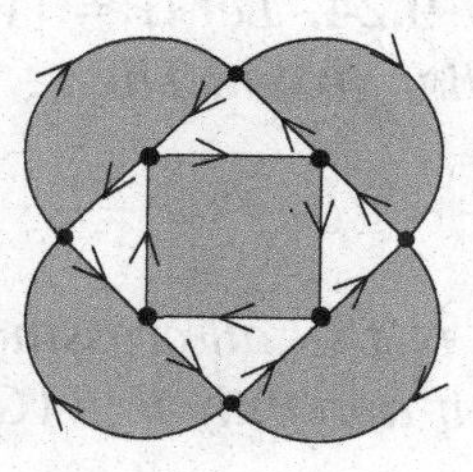

$\tilde{G}$

We call this the *canonical orientation* of $\tilde{G}$, and denote by $\tilde{G}^c$ the resulting 2-in 2-out graph. It follows from the definition of the canonical orientation that with the weighting $\alpha = \beta = 1$, $\gamma = 0$, the transition systems of $\tilde{G}$ correspond bijectively to the Eulerian decompositions of $\tilde{G}^c$. In other words, $S(\tilde{G}, W_{1,1,0}; \lambda) = \sum_{k\geq 1} e_k(\tilde{G}^c)\lambda^k$. Considering (9) we also have

$$S(\tilde{G}, W_{1,1,0}; \lambda) = \lambda\, T(G; 1+\lambda, 1+\lambda),$$

and thus

$$\sum_{k\geq 0} e_{k+1}(\tilde{G}^c)\lambda^k = T(G; 1+\lambda, 1+\lambda). \tag{10}$$

In view of Proposition 9.15 this also gives

$$T(G;\lambda,\lambda) = Q(H;\lambda), \tag{11}$$

where H is any interlace graph of $\widetilde{G}^c$. Now we know that $T(G;1,1)$ counts the number of spanning trees in G, whence (10) gives the following result:

Corollary 9.23. *The number of Eulerian cycles in the 2-in 2-out graph $\widetilde{G}^c$ equals the number of spanning trees of G.*

Consider, finally, the evaluation $\alpha = \beta = 0$, $\gamma = 1$. That is, we take the unique transition system q consisting only of crossing transitions. Subtracting 1, we get the weighting $\overline{\alpha} = \overline{\beta} = -1$, $\overline{\gamma} = 0$. Lemma 9.21 and (9) imply

$$\begin{aligned} S(\widetilde{G}, W_{0,0,1}; -2) = (-2)^{c(q)} &= S(\widetilde{G}, W_{-1,-1,0}; -2) \\ &= (-2)(-1)^{|E|}T(G;-1,-1), \end{aligned}$$

and thus the following result:

Corollary 9.24. *Let $G = (V, E, F)$ be a connected plane graph, and $\widetilde{G}$ its medial graph. Then*

$$T(G;-1,-1) = (-1)^{|E|}(-2)^{c(q)-1}, \tag{12}$$

where q is the all-crossing transition system of $\widetilde{G}$. In particular, $c(q) = 1$ if and only if $|T(G;-1,-1)| = 1$.

Example. We have computed the Tutte polynomial for the graph K_4^- in Section 9.1, obtaining $T(K_4^-;-1,-1) = 2$. Thus $c(q) = 2$, and the figure shows the corresponding decomposition into two cycles.

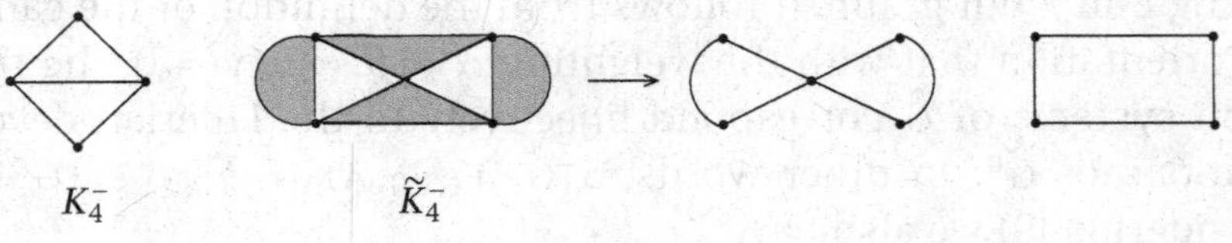

K_4^- $\widetilde{K}_4^-$

Exercises

▷ **9.36** Verify Euler's formula by induction on the number of edges, starting with a tree.

9.37 Prove the equalities in (2).

9.38 Compute the transition polynomial $S(\tilde{G}, W_{\alpha,\beta,\gamma}; \lambda)$ for $G = K_3$, and verify that $S(\tilde{G}, W_{1,1,1}; \lambda) = \sum e_k \lambda^k$, e_k = number of decompositions into k cycles.

9.39 Show that if $G = G_1 \dot{\cup} G_2$ is a disjoint union of plane graphs, then $S(\tilde{G}) = S(\tilde{G}_1)S(\tilde{G}_2)$.

▷ **9.40** Suppose e is a bridge of the connected plane graph G. We have seen that e is then a cut vertex of $\tilde{G}$; let $\tilde{G}_1, \tilde{G}_2$ be the components of $\tilde{G} \smallsetminus e$. Show for $W_{\alpha,\beta,\gamma}$ that $S(\tilde{G}) = \alpha S(\tilde{G}_1)S(\tilde{G}_2) + \frac{\beta+\gamma}{\lambda} S(\tilde{G}_1)S(\tilde{G}_2)$.

9.41 Use the definition of an admissible valuation to deduce a. $P(G;k) \le P(G;k+1)$, b. $P(G;k) \ge \chi(G^*;k)$.

▷ **9.42** Prove that if G has no bridges, then $P(G;\lambda)$ is a polynomial of degree $|F|$.

9.43 Prove Lemma 9.21 by induction on $|E|$.

* * *

▷ **9.44** Let G be a connected plane Eulerian graph, $P(G;\lambda) = \sum_{i=1}^{|F|} a_i \lambda^i$. Show that all coefficients a_i are nonzero and have alternating sign. Prove further: a. $\sum |a_i| = 2^{|E|}$, b. $(-1)^{|F|}P(G;-1) = 2^{|E|}$, c. $|P(G;-1)| < 2^{|E|}$ if G is not Eulerian.

9.45 Suppose $G = (V, E, F)$ (without bridges) has two faces with common boundary edges e, e'. Show that $P(G;\lambda) = 2P(G/e;\lambda)$. Use this to compute the Penrose polynomial of the circuit of length n.

9.46 In continuation of the exercise, suppose $G = (V, E, F)$ has no different faces with two common boundary edges. Prove that the highest coefficient of $P(G;\lambda)$ is 1.

▷ **9.47** Compute the Penrose polynomial of the prism .

9.48 Suppose the Penrose polynomial $P(G;\lambda) = \sum_{i=1}^{|F|} a_i \lambda^i$ has alternating nonzero coefficients. Show that this implies that G is 4-face colorable. Hint: Corollary 9.22.

9.49 Show that an arbitrary graph G has a 2^k-flow if and only if $E(G)$ is the union of k Eulerian subgraphs. Hint: Use the group $A = \mathbb{Z}_2 + \cdots + \mathbb{Z}_2$ with componentwise addition. What is the dual statement when G is plane?

9.50 Consider a connected plane graph $G = (V, E, F)$, and set $N(A) = 1$ if $G_A^* = (V^*, A)$ contains an odd number of spanning trees of G^*, and 0 otherwise. Show that the linear coefficient a_1 of $P(G;\lambda)$ is given by $a_1 = \sum_{A \subseteq E} (-1)^{|A|} N(A)$, and deduce that a_1 is even for $|V| \ge 2$. Hint:

$N(A) \equiv N(A \setminus e) + N(A/e) \pmod 2$. Incidentally, it is an open question whether a_1 is always nonzero.

9.51 Show that $T(G; -1, -1)$ equals $\pm 2^m$ for any graph G.

9.52 Consider the set $\mathcal{V}_k$ of all k-valuations of $\widetilde{G}$ (connected) which have no crossing type, and for $g \in \mathcal{V}_k$ let $t(g)$ be the number of total types (g, e). Prove that $\sum_{g \in \mathcal{V}_k} 2^{t(g)} = kT(G; k+1, k+1)$, where G is the Tait graph of $\widetilde{G}$. Hint: Argue as in the proof of Proposition 9.17.

▷ **9.53** Let G be a connected plane graph, and let $\mathcal{O}$ be the set of all Eulerian orientations of $\widetilde{G}$, that is, those that make $\widetilde{G}$ into a 2-in 2-out graph. For $O \in \mathcal{O}$ let $s(O)$ be the number of *saddle points* e, meaning that at e the edges are oriented alternately in and out. Show that $\sum_{O \in \mathcal{O}} 2^{s(O)} = 2 \cdot T(G; 3, 3)$.

9.4 Knot Polynomials

The counting polynomials considered in the previous sections lead directly to a beautiful field that has been particularly active in recent years—polynomial invariants of knots. A *knot* is a subset of $\mathbb{R}^3$ that is homeomorphic to a circle. A *link* consists of several disjoint knots. In this general setting knots can behave quite strangely. Without going into the topological details we restrict ourselves to so-called *tame* knots, which are ambient isotopic to simple closed polygons.

Knots, Links, and Diagrams.
A knot is usually pictured by means of a regular projection onto a plane, where *regular* means that the projection contains only finitely many multiple points and that these points are all double points v, with the projected pieces *crossing* at v.

Example. The figure shows the (right-handed) *trefoil* T^r and a link L_1 consisting of two components.

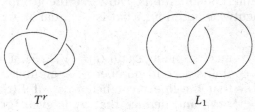

We call such a plane projection a *diagram* D of the knot or link. At a crossing point we speak of an *underpass* or *overpass* with the obvious meaning.

The connection to the last section is immediate. If we regard the crossing points of a diagram D as vertices, and the strings between crossing points as edges, then D becomes a plane 4-regular graph, which can clearly be assumed to be connected. The faces of D are colored black and white as before. Every vertex (crossing point) is thus incident to four faces, alternately colored black and white.

To specify which edge goes over we use the convention "left over right" as seen from the black face. Notice that it makes no difference from which black face the crossing is viewed.

+

−

With this convention, the diagram becomes a connected plane 4-regular graph $\widetilde{D}$ with a signing of the vertices. For example, the graphs $\widetilde{D}$ of the links above are

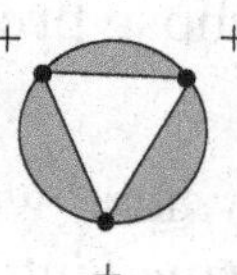

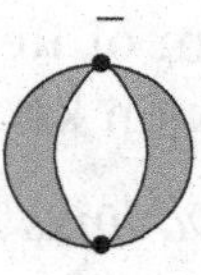

Now we know that $\widetilde{D}$ corresponds to a unique underlying plane graph G, its Tait graph, such that $\widetilde{D} = \widetilde{G}$ is the medial graph of G, with the edges of G corresponding to the vertices of $\widetilde{D}$. In sum, we have proved the following:

Proposition 9.25. *There is a bijection between edge-signed connected plane graphs G and link diagrams D.*

Example. Consider two different edge-signings of K_4^- with the associated diagrams and projections:

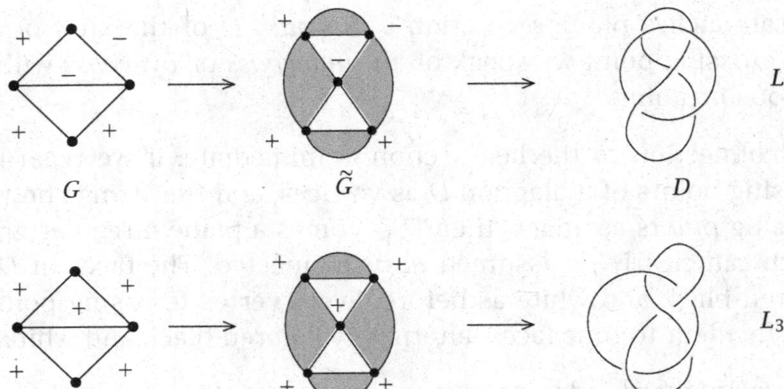

A subclass of diagrams is of particular interest. A link is called *alternating* if it has a diagram in which the crossings are alternating over–under, as we run through the knots. The trefoil T^r and the link L_3 of the last example have alternating diagrams. It is easy to see that an alternating link has a signed Tait graph representation in which all edges have the same sign (Exercise 9.55).

Equivalence of Links and Reidemeister Moves.
Look at the link L_2 of the example. It is immediately seen (by twisting the upper half of the figure eight) that L_2 is topologically equivalent to the link L_1 considered before. Topological equivalence of links is reduced to purely combinatorial conditions by the following famous theorem of Reidemeister, whose proof can be found in any advanced book on knot theory.

Theorem 9.26. *Two links L, L' are equivalent, written $L \cong L'$, if and only if any diagram of L can be transformed into any diagram of L' by a finite sequence of so-called Reidemeister moves:*

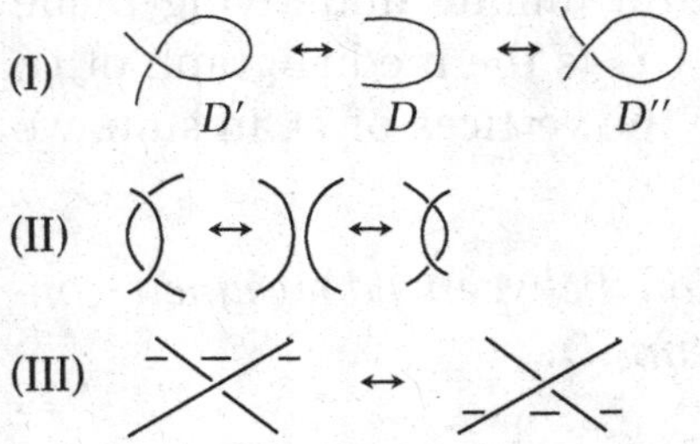

(I) $D' \leftrightarrow D \leftrightarrow D''$

(II) $\leftrightarrow \quad \leftrightarrow$

(III) $\leftrightarrow$

These moves are understood to change only the local configuration depicted; the rest remains unchanged.

The great problem of knot theory is, of course, to decide whether two given knots or links are equivalent, and in particular, when a knot K is equivalent to the "unknot" $L_\circ = \bigcirc$ (K can be unknotted). The traditional approach is to find link invariants. A function f defined on links is called an *invariant* if $L \cong L'$ implies $f(L) = f(L')$. Or turned around: Whenever we can prove $f(L) \neq f(L')$, then L and L' are topologically different links.

A trivial example is the number of component knots of L. Other invariants are known from topology (e.g., the fundamental group), but the complexity status of the equivalence problem is very unclear. Certainly no good algorithm is known.

The great importance of Theorem 9.26 rests on the fact that in order to test whether a function f is an invariant we have only to check that f remains unchanged under the Reidemeister moves. In the sequel we discuss several interesting polynomial invariants that tie in naturally with the previous counting polynomials.

Kauffman Bracket and Jones Polynomial.
The *bracket polynomial* $\langle D\rangle$ of a link diagram is obtained by the following rules. There are constants A, B, and d such that

(K1) $\langle \bigcirc \rangle = 1$, $\bigcirc$ = unknot,

(K2) $\langle D \cup \bigcirc \rangle = d\langle D\rangle$, $D \cup \bigcirc$ disjoint union,

(K3) $\langle \diagup\!\!\!\!\diagdown \rangle = A\langle \asymp \rangle + B\langle)(\rangle$.

Condition (K3) means that we resolve a crossing by going to the right (seen from the underpass), or to the left. The rest is left unchanged. If we apply (K3) to every one of the n crossing points, we see that $\langle D\rangle$ decomposes into 2^n summands (weighted by A or B), where each summand corresponds to a particular *resolution system.*

This looks very much like the transition systems of the last section. Let r be a particular resolution system, which goes right at $i(r)$ crossing points and left at $j(r)$ points, and suppose r decomposes the diagram into $c(r)$ trivial knots. Rules (K1) and (K2) show then that the bracket polynomial is a polynomial in the variables A, B, d, given by

$$\langle D\rangle = \sum_r A^{i(r)} B^{j(r)} d^{c(r)-1}. \tag{1}$$

Before we tackle the problem whether $\langle D \rangle$ is a link invariant, let us look at the smallest examples in which the Tait graph is a bridge or loop. Since the edges are signed we have two cases each. Look at a bridge:

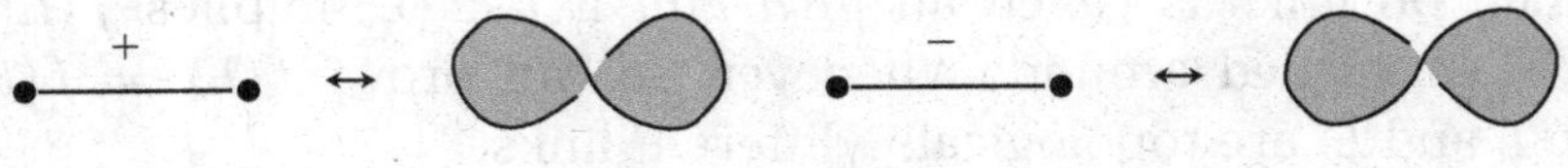

$+$ $\longleftrightarrow$ D_b^+ $-$ $\longleftrightarrow$ D_b^-

By the bracket rules, we get

$$\begin{aligned} \langle D_b^+ \rangle &= A\langle \bigcirc \rangle + B\langle \bigcirc\bigcirc \rangle = A + dB, \\ \langle D_b^- \rangle &= A\langle \bigcirc\bigcirc \rangle + B\langle \bigcirc \rangle = dA + B. \end{aligned} \tag{2}$$

For the loop we obtain analogously

$$\begin{aligned} \langle D_\ell^+ \rangle &= dA + B\,, \\ \langle D_\ell^- \rangle &= A + dB. \end{aligned} \tag{3}$$

Let us study the Reidemeister moves. For a move of type (II) we find by resolving the two crossing points that

$$\begin{aligned} \langle \,)\!\!\!\!\!\;(\, \rangle &= A\,\langle \,\breve{\bigcirc}\, \rangle + B\langle \times \rangle \\ &= A^2\langle \asymp \rangle + AB\langle \,\breve{\bigcirc}\, \rangle + AB\langle\,)(\, \rangle + B^2\langle \asymp \rangle \\ &= (A^2 + B^2 + dAB)\langle \asymp \rangle + AB\langle\,)(\, \rangle\,. \end{aligned}$$

For $\langle \,)\!\!\!\!\!\;(\, \rangle = \langle\,)(\, \rangle$ to hold we must therefore require $AB = 1$, $A^2 + B^2 + dAB = 0$, that is, $B = A^{-1}$, $d = -(A^2 + A^{-2})$.

It is easy to see (Exercise 9.56) that with this choice of variables the bracket polynomial is also invariant under the second case in (II), and under Reidemeister (III). We refer henceforth to the bracket polynomial with this choice of variables $B = A^{-1}$, $d = -(A^2 + A^{-2})$. The bracket polynomial $\langle D \rangle$ is a so-called Laurent polynomial in the variable A. Let us summarize our findings so far:

Proposition 9.27. *The bracket polynomial* $\langle D \rangle$ *is invariant under Reidemeister moves* (II) *and* (III).

Example. For the bridge and loop we get from (2) and (3),

$$\langle D_b^+\rangle = -A^{-3}, \quad \langle D_b^-\rangle = -A^3, \quad \langle D_\ell^+\rangle = -A^3, \quad \langle D_\ell^-\rangle = -A^{-3}. \qquad (4)$$

Let us study next moves of type (I). With the notation there we obtain

$$\begin{aligned} \langle D'\rangle &= \langle \quad \rangle = A\langle \quad \rangle + A^{-1}\langle \quad \rangle = -A^{-3}\langle D\rangle, \\ \langle D''\rangle &= \langle \quad \rangle = A\langle \quad \rangle + A^{-1}\langle \quad \rangle = -A^{3}\langle D\rangle; \end{aligned} \qquad (5)$$

hence $\langle D\rangle$ is not invariant under (I).

To make $\langle D\rangle$ into an invariant we consider *oriented* knots and links. For every component knot we choose one of the two possible orientations as we run through the knot, and say that a crossing has weight 1 or −1 according to the convention left over right:

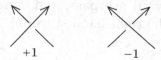

A word of caution: If K is a knot, then the definition is independent of the orientation. Indeed, with the other orientation both arrows point down, and the sign of the weight stays the same. But for links we must fix the orientation, since the sign may change if the arrows belong to different knots.

The *writhe* $w(D)$ is the sum of the weights taken over all crossings.

Theorem 9.28. *The Kauffman polynomial* $f_D(A) = (-A^3)^{-w(D)}\langle D\rangle$ *is an invariant of oriented diagrams. We can therefore uniquely define* $f_L(A)$ *for an oriented link, by setting* $f_L(A) = f_D(A)$ *for any diagram D of L. Clearly,* $f_\circ(A) = 1$ *for the unknot.*

Proof. It is readily checked that the writhe is unchanged under the moves (II) and (III), hence $f_D(A)$ is invariant under (II) and (III) by Proposition 9.27. As to Reidemeister (I), the figure shows that whatever way the curve in D' is oriented we get a negative weight, whereas for D'' it is always positive.

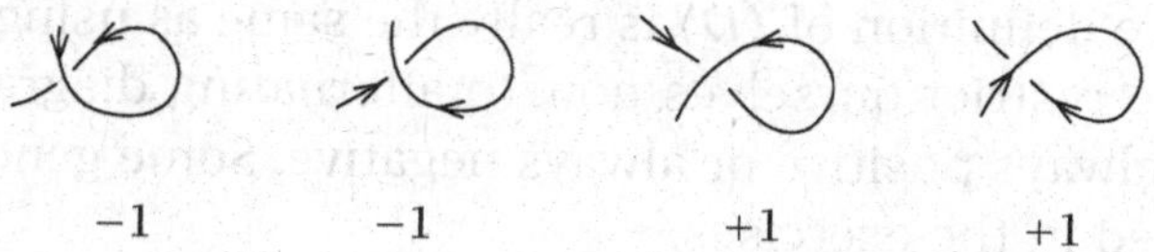

Hence using (5),

$$f_{D'}(A) = (-A^3)^{-w(D')}\langle D'\rangle = (-A^3)^{-w(D)+1}\langle D'\rangle$$
$$= (-A^3)^{-w(D)}(-A^3)(-A^{-3})\langle D\rangle = f_D(A),$$

and similarly $f_{D''}(A) = f_D(A)$. □

The celebrated *Jones polynomial* $V_L(t)$, which was originally defined by an entirely different approach, results from $f_L(A)$ by the substitution $A \mapsto t^{-1/4}$, where L is an oriented link. Hence

$$V_L(t) = (-1)^{w(D)} t^{\frac{3w(D)}{4}} \langle D\rangle_{A=t^{-1/4}}, \tag{6}$$

for any diagram D of L.

Remark. The *mirror image* $\overline{D}$ of a diagram D is obtained by replacing each underpass by an overpass, and conversely. For example, the mirror image of the right-handed trefoil T^r is the left-handed trefoil T^ℓ:

T^r T^ℓ

Looking at the recursive definition of the bracket polynomial one sees that $\langle\overline{D}\rangle$ arises from $\langle D\rangle$ by the substitution $A \mapsto A^{-1}$, that is, $\langle\overline{D}\rangle = \langle D\rangle_{A\mapsto A^{-1}}$. Choosing the same orientation for all component knots, the weights in the mirror image are also exchanged; thus $w(\overline{D}) = -w(D)$, which implies $(-A^3)^{-w(\overline{D})} = (-A^{-3})^{-w(D)}$. This proves the following result.

Corollary 9.29. *Let $\overline{L}$ be the mirror image of an oriented link L. Then*

$$f_{\overline{L}}(A) = f_L(A^{-1}), \quad V_{\overline{L}}(t) = V_L(t^{-1}). \tag{7}$$

Alternating Knots and Links.

We have already remarked that the resolution of crossing points in the recursive definition of $\langle D\rangle$ is really the same as using transition systems. We restrict ourselves now to alternating diagrams, where the sign is always positive or always negative. Some general results are contained in the exercises.

Suppose L is an alternating link, and D a diagram with all crossings positive. The mirror image $\overline{L}$ is then again alternating with all signs negative. For example, the diagram of T^r above is positive, and that of T^ℓ is negative. Let G be the Tait graph of $\widetilde{D}$, which may be assumed to be unsigned since all signs are positive. Recurrence (K3) in the definition of $\langle D \rangle$ says that

$$\langle \times_e \rangle = A\langle \asymp \rangle + A^{-1}\langle)(\rangle,$$

which for G means formally

$$\langle G \rangle = A^{-1}\langle G \setminus e \rangle + A\langle G/e \rangle . \qquad (8)$$

Setting $h(G) = \langle D \rangle$, the equality (8) readily implies that $h(G)$ is a chromatic invariant with $h(\text{bridge}) = -A^{-3}$, $h(\text{loop}) = -A^3$, and $\alpha = A^{-1}$, $\beta = A$. With Theorem 9.5 we arrive at the following result:

Proposition 9.30. *Let D be a positive alternating diagram with Tait graph $G = (V, E, F)$. Then*

$$\langle D \rangle = A^{2|V|-|E|-2}T(G; -A^{-4}, -A^4), \qquad (9)$$

where $T(G)$ is the Tutte polynomial. The mirror image is given by

$$\langle \overline{D} \rangle = A^{-2|V|+|E|+2}T(G; -A^4, -A^{-4}).$$

Note that because of $2|V| - |E| - 2 = |V| - |F|$ (by Euler's formula), we may also write

$$\langle D \rangle = A^{|V|-|F|}T(G; -A^{-4}, -A^4), \langle \overline{D} \rangle = A^{|V^*|-|F^*|}T(G^*; -A^{-4}, -A^4), \qquad (10)$$

where G^* is the dual graph. In other words, mirror images correspond to the duality $G \leftrightarrow G^*$ of plane connected graphs.

Example. The right-handed trefoil T^r has positive alternating diagram D with $G = K_3$. From $T(K_3; x, y) = x^2 + x + y$ we obtain by (9),

$$\langle D \rangle = A(A^{-8} - A^{-4} - A^4) = A^{-7} - A^{-3} - A^5 .$$

The writhe of D is 3, which gives

$$f_{T^r}(A) = -A^{-16} + A^{-12} + A^{-4}, \quad V_{T^r}(t) = -t^4 + t^3 + t .$$

For T^ℓ we have by (7),

$$f_{T^\ell}(A) = -A^{16} + A^{12} + A^4, \quad V_{T^\ell}(t) = -t^{-4} + t^{-3} + t^{-1}.$$

In particular, T^r and T^ℓ are not equivalent, and neither is equivalent to the unknot.

We can now apply all we know about the Tutte polynomial to the polynomials $f_L(A)$ and $V_L(t)$. Let D be a positive alternating diagram, and $G = (V,E,F)$ its Tait graph. The crossing points of D correspond to the edges E. Let E_+ and E_- be the points with positive and negative weight, respectively. Hence

$$|E| = |E_+| + |E_-|, \quad w(D) = |E_+| - |E_-|,$$

which implies $|E| - w(D) \equiv 0 \pmod 2$, that is, $(-1)^{w(D)} = (-1)^{|E|}$. Using (10) and (6), this gives the following result.

Corollary 9.31. *Let L be an alternating link, D a positive alternating diagram, and G = (V,E,F) the Tait graph. Then*

$$\begin{aligned} f_L(A) &= (-1)^{|E|} A^{|V|-|F|-3w(D)} T(G; -A^{-4}, -A^4), \\ V_L(t) &= (-1)^{|E|} t^{\frac{|F|-|V|}{4} + \frac{3w(D)}{4}} T(G; -t, -t^{-1}). \end{aligned} \tag{11}$$

Take $A = t = 1$. We know from Corollary 9.24 that $T(G;-1,-1) = (-1)^{|E|}(-2)^{c(q)-1}$, where $c(q)$ is the number of Eulerian cycles in the all-crossing transition system. But this is just the number $c(L)$ of component knots of L. The formulas in (11) therefore give

$$f_L(1) = V_L(1) = (-2)^{c(L)-1}, \tag{12}$$

and in particular, $f_K(1) = V_K(1) = 1$ for a knot K.

As a final evaluation let us look at $t = -1$ in the Jones polynomial of a positive alternating knot K. The factor in front of (11) is by Euler's formula

$$(-1)^{|E| + \frac{|F|-|V|}{4} + \frac{3w(D)}{4}} = (-1)^{\frac{5|E|}{4} + \frac{3w(D)}{4} - \frac{|V|-1}{2}}.$$

Using E_+, E_- as before, the exponent is

$$2|E_+| + \frac{|E_-| - |V| + 1}{2}.$$

It is readily seen that $|E_-| - |V| + 1$ is always an even number. With (11) we thus arrive at

$$V_K(-1) = (-1)^{\frac{|E_-|-|V|+1}{2}} T(G;1,1),$$

which in turn yields

$$|V_K(-1)| = \#\text{spanning trees of } G. \tag{13}$$

Of course, we could also make use of Corollary 9.23 relating $|V_K(-1)|$ to the number of Eulerian cycles in the canonically oriented 2-in 2-out graph $\vec{\tilde{D}}^c$.

Exercises

9.54 Show that the Tait graph 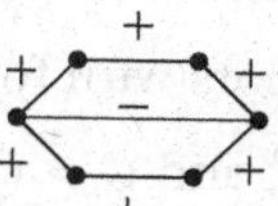

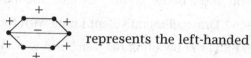

represents the left-handed trefoil by looking at the diagram and performing some Reidemeister moves.

9.55 Verify that an alternating link has a diagram with constant sign on the crossings.

9.56 Show that the bracket polynomial $\langle D\rangle$ is invariant under Reidemeister (III).

▷ **9.57** Verify that the writhe $w(D)$ stays the same under moves (II) and (III).

9.58 Establish as a necessary condition for a knot K to be equivalent to the unknot that $\langle D\rangle$ be $\pm$ a power of A.

9.59 Consider the link L_1 of the text, and compute the Jones polynomial for the four possible orientations of the link.

▷ **9.60** Consider $G = C_{2n+1}$ (circuit), and calculate the Jones polynomial of the corresponding link, where all signs are positive.

9.61 Take $G = K_4$ with all signs positive. The corresponding diagram D decomposes into three knots. Choose orientations of the knots with $w(D) = 0$, and calculate the Jones polynomial. Is $w(D) \neq 0$ possible?

* * *

▷ **9.62** Prove $f_L(1) = V_L(1) = (-2)^{c(L)-1}$ for any oriented link L. Hint: Use (1) with $B = A^{-1}$, $d = -(A^2 + A^{-2})$.

9.63 Suppose L is a link with alternating diagram D, Tait graph G, and all signs negative. Show that the dual graph G^* with all signs positive gives rise to D as well. Hence for an alternating link we can always find a positive alternating diagram.

▷ **9.64** A *nugatory crossing* e decomposes the diagram D into two parts:  . Suppose K is an alternating knot without nugatory crossings. Prove $|V_K(-1)| \geq n = \#$ crossings. Settle the case in which equality holds. Hint: Answer first what nugatory crossing means for the Tait graph, and use then (13) or Exercise 9.33.

9.65 Show that $|V_K(-1)|$ is always an odd integer when K is an alternating knot. Note that this holds for arbitrary knots.

▷ **9.66** We know that $Q(H;1) \leq 2^{n-1}$ for the interlace polynomial of any graph H on n vertices. Use this to show that $|V_K(-1)| \leq 2^{n-1}$, where $n = \#$ crossings, K alternating knot. Can there be equality for $n \geq 2$? Can you find a better bound? Discuss the maximum possible for $n \leq 4$.

9.67 Draw all inequivalent knots with four crossings in the diagram.

9.68 Let D be a link diagram, and $G = (V, E, F)$ its Tait graph. For $X \subseteq E$ let X^+, X^- be the positive and negative edges, respectively. Generalize Proposition 9.30 to

$$\langle D \rangle = A^{|E^-|-|E^+|+2|V|-2} \sum_{X \subseteq E} A^{4(|X^+|-r(X))}(-A^{-4}-1)^{r(G)+|X|-2r(X)},$$

with the notation $r(X)$ as in Theorem 9.3. Hint: Consider the deletion/contraction recurrence in the positive and negative case.

Highlight: The BEST Theorem

We have seen in Corollary 9.23 a somewhat unexpected connection between the number of spanning trees of a plane graph and the number of Eulerian cycles of the associated medial graph, endowed with the canonical orientation.

The following wonderful result generalizes this to arbitrary directed Eulerian graphs. It is called the BEST theorem after its authors de Bruijn, van Ardenne-Ehrenfest, Smith, and Tutte.

Suppose $G = (V, E)$ is a connected directed Eulerian graph, and denote by $e(G)$ the number of Eulerian cycles, and by $d^-(v)$ as usual the out-degree of v. Now recall the highlight to Chapter 5, where we introduced arborescences converging to a vertex s, counted by $t^-(G, s)$.

Theorem. *We have*

$$e(G) = \left(\prod_{v \in V} (d^-(v) - 1)!\right) \cdot t^-(G, s).$$

In particular, $e(G) = t^-(G, s)$ *when* G *is a 2-in 2-out graph.*

Note that this implies the result already proved in Chapter 5 that $t^-(G, s)$ is the same number for all $s \in V$.

Instead of cycle we use in the following the more expressive term *Eulerian walk*, since we will proceed in the cycle step by step.

Choose an arbitrary directed edge $e_1 = (s, \tilde{s})$ as starting edge for all Eulerian walks. We are going to find a bijection between arborescences T converging to s and sets $\mathcal{W}(T)$ of Eulerian walks (starting with e_1). Each set $\mathcal{W}(T)$ will have size $d := \prod_{v \in V} (d^-(v) - 1)!$.

Construction of an Arborescence from an Eulerian Walk.
The walk W gives a numbering $e_1, e_2, \ldots, e_{|E|}$ of the edges, and W touches every vertex at least once. For $|V| = 1$, the theorem is obviously true, so assume $|V| \geq 2$.

For $v \neq s$, let $e(v)$ be the *last* edge leaving v in the walk W. Consider the subgraph $T(W) = (V, F)$, where $F = \{e(v) : v \neq s\}$; $T(W)$ contains $|V| - 1$ edges. Suppose $v_0 \neq s$ with last edge $e(v_0) = (v_0, v_1)$. Note that $v_1 \neq v_0$, since the *last* edge leaving v_0 must come after every loop at v_0. Furthermore, if $v_1 \neq s$, then $e(v_1) = (v_1, v_2)$ has higher number than $e(v_0)$, and thus $v_2 \neq v_0$. Continuing in this way we see that the edges $e(v_0), e(v_1), \ldots$ form a directed path from v_0 to s, and so $T(W)$ is an arborescence converging to s.

Construction of *d* Eulerian Walks from an Arborescence.
Suppose T is an arborescence converging to s. For every vertex u we number the $d^-(u)$ edges leaving u, observing the following two conditions:

A. Among all edges (s, v) leaving s, the edge $(s, \tilde{s})$ receives the number 1.

B. For all $v \neq s$, the *tree-edge* $(v, v') \in E(G)$ receives the highest number $(= d^-(v))$.

Otherwise, the numbering is arbitrary.

Clearly, there are $d = \prod_{v \in V}(d^-(v) - 1)!$ possible numberings. Now we take a fixed numbering that satisfies A and B, and construct the Eulerian walk W as follows:

Start with $(s, \tilde{s})$. Given $W_i = (s, \tilde{s}, \ldots, v_i)$ choose, among all edges that leave v_i and have not been used, the edge with the smallest number. In other words, we leave any vertex in the succession given by the numbering. Continue this process as long as possible and denote by W the resulting walk.

We note the following facts:

1. W terminates at s.
Indeed, G is Eulerian, which means that whenever we enter a vertex, we can also leave it.

2. W contains all edges.
Suppose to the contrary that there is an edge $e = (v_0, v_1')$ that does not appear in W. Then by condition B the tree-edge $(v_0, v_1) \in E(T)$ has also not been used. This, in turn, implies that some outgoing edge from v_1 has not been used, and therefore neither the tree-edge $(v_1, v_2) \in E(T)$. Continuing in this way we obtain a directed path of unused edges (all in $E(T)$) from v_0 to s. But this implies that we could continue the walk at s, contradiction.

The Final Step.
Collect all Eulerian walks constructed from T in this way in the set $\mathcal{W}(T)$. The following three assertions will finish the proof.

1. $|\mathcal{W}(T)| = d$.
Just note that different numberings lead to different Eulerian walks.

2. If T_1, T_2 are different arborescences, then $\mathcal{W}(T_1) \cap \mathcal{W}(T_2) = \emptyset$.
Clearly, different arborescences yield different numberings (condition B).

3. Every Eulerian walk W is in one of the classes $\mathcal{W}(T)$.
By the construction rule, $W \in \mathcal{W}(T(W))$.

In sum,

$$e(G) = \sum_{T \to s} |\mathcal{W}(T)| = d \cdot t^{-}(G,s)\,,$$

and the theorem follows.

The Alexander Polynomial.
One of the earliest invariants for oriented knots and links was the *Alexander polynomial*. Take a diagram D of a link L, and consider the set of all *strands*, where a strand is the portion of the diagram from one underpass to the next (in the given orientation). Clearly, the number of strands equals the number of crossing points.

Example. In the diagram of the link there are four strands:

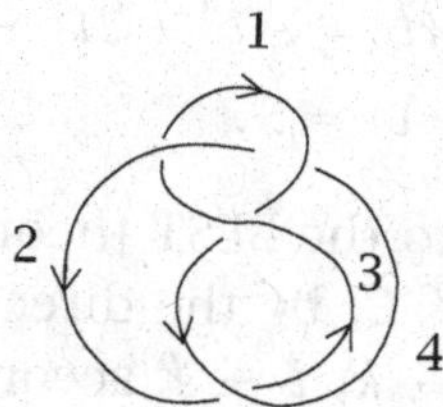

Number the strands $1, \ldots, n$. At any crossing we have the following situation:

k ℓ i

We capture the information in the $n \times n$-matrix $M_D(t) = (m_{ij})$, where t is a variable:

$$m_{ij} = \begin{cases} 1-t & j = i\,, \\ t & j = k\,, \\ -1 & j = \ell\,, \\ 0 & \text{otherwise}\,. \end{cases}$$

In the example above we get the matrix

$$M_D(t) = \begin{pmatrix} 1-t & -1 & 0 & t \\ -1 & 1-t & t & 0 \\ t & 0 & 1-t & -1 \\ 0 & t & -1 & 1-t \end{pmatrix}.$$

The Alexander polynomial $\Delta_D(t)$ is defined as the cofactor

$$\Delta_D(t) = \det M_D(t)_{i,i}\,,$$

where as usual $M_D(t)_{i,i}$ is the matrix with the i-th row and i-th column deleted.

A classical result says that $\Delta_D(t)$ is well defined up to a power $\pm t^m$. In other words, factoring out the highest power of t, any two cofactors agree (apart from possibly the sign). In particular, $|\Delta_D(-1)|$ is well defined. Furthermore, $\Delta_D(t)$ is an invariant of oriented links; we can thus write $\Delta_L(t)$. The invariant $|\Delta_L(-1)|$ is usually called the *determinant* of L.

In our example we compute

$$\begin{aligned} \Delta_L(t) &= -t^3 + 2t^2 - 2t\,, \\ \Delta_L(-1) &= 5\,. \end{aligned}$$

To show the connection to the BEST theorem we consider alternating diagrams. Let $\widetilde{G} = (\widetilde{V}, \widetilde{E})$ be the directed graph on the strands $\widetilde{V} = \{1, \ldots, n\}$, where $i \to k$, $i \to \ell$ according to the figure above. Since D is alternating, $\widetilde{G}$ is a 2-in 2-out graph, with $M_D(-1) = L(\widetilde{G})$, where $L(\widetilde{G})$ is the Laplace matrix defined in Chapter 5.

Tutte's matrix-tree theorem and the BEST theorem imply

$$\Delta_L(-1) = \det(M_D(-1))_{s,s} = t^-(\widetilde{G}, s) = e(\widetilde{G})\,,$$

and $\Delta_L(-1)$ is indeed well defined.

The Alexander polynomial is well understood. For example, the coefficients form a palindromic sequence, that is, $\Delta(t) = t^{\deg \Delta}\Delta(t^{-1})$, and $|\Delta(-1)|$ is always an odd integer.

Notes and References

Chromatic polynomials were invented by Birkhoff (1912) in his attempt to prove the 4-color theorem, that is, finding algebraic reasons that the chromatic polynomial of a planar graph never has 4 as a root. The theory presented in this chapter begins with the work of Whitney and Tutte. For a detailed account of the Tutte polynomial see the article by Brylawski and Oxley, and the book by Tutte. The first evaluation at negative integers (Proposition 9.8) is due to Stanley. The 6-flow theorem was proved by Seymour. The material in Section 9.2 follows in large part the papers by Aigner–van der Holst, and Arratia, Bollobás, and Sorkin. A more general setup in the context of isotropic systems was proposed by Bouchet. Transition polynomials were studied in detail by Jaeger. The Penrose polynomial was introduced in the paper by Penrose. A very readable account of knot polynomials is the lecture notes by Welsh. The references give also the original sources of the Jones polynomial and the Kauffman bracket. For the BEST theorem see Chapter 6 of Tutte's book.

1. M. Aigner and H. van der Holst (2004): Interlace polynomials. *Linear Algebra Appl.* 377, 11–30.
2. R. Arratia, B. Bollobás, and G. Sorkin (2000): The interlace polynomial: a new graph polynomial. In: *Proc. 11th Annual ACM-SIAM Symp. on Discrete Math.*, 237–245.
3. A. Bouchet (1991): Tutte–Martin polynomials and orienting vectors of isotropic systems. *Graphs and Combin.* 7, 235–252.
4. T.H. Brylawski and J. Oxley (1992): The Tutte polynomial and its applications. In: *Matroid Applications*, White, ed., 123–225. Cambridge Univ. Press, Cambridge.
5. F. Jaeger (1990): On transition polynomials of 4-regular graphs. In: *Cycles and Rays*, Hahn et al., eds., 123–150. Kluwer.
6. V.F.R. Jones (1985): A polynomial invariant for knots via von Neumann algebras. *Bull. Amer. Math. Soc.* 12, 103–111.
7. L.H. Kauffman (1987): State models and the Jones polynomial. *Topology* 26, 345–407.

8. R. Penrose (1971): Applications of negative dimensional tensors. In: *Combinatorial Mathematics and Its Applications*, Welsh, ed., 221–244. Academic Press, London.
9. P.D. Seymour (1981): Nowhere-zero 6-flows. J. *Combinatorial Theory B* 30, 130–135.
10. R.P. Stanley (1973): Acyclic orientations of graphs. *Discrete Math.* 5, 171–178.
11. W.T. Tutte (1984): *Graph Theory*. Addison-Wesley, Reading.
12. D.J.A. Welsh (1993): *Complexity: Knots, Colourings and Counting*. London Math. Soc. Lecture Notes Series 186.

10 Models from Statistical Physics

Certain questions from statistical physics give rise to fascinating problems in enumerative combinatorics. The basic setup is as follows. One considers a set of *sites* (occupied by atoms, say), and a set of *bonds* connecting certain pairs of sites that carry an interaction between the corresponding atoms.

Thus we have a (simple) graph, with the sites as vertices, and edges weighted with their interaction. The classical examples are certain lattice graphs, such as the rectangular 2-dimensional grid, which we will consider in detail. A *state* is now an arrangement under the conditions posed by the model at hand. The goal is to compute the so-called *partition function*, which is the generating function of the states and which carries the most important informations about the physical model.

10.1 The Dimer Problem and Perfect Matchings

A *dimer* is a diatomic molecule that occupies two adjacent sites. The dimer problem is to determine the number of ways to cover all sites with dimers such that every site is occupied by exactly one dimer.

In graph-theoretic terms a dimer configuration is a *perfect matching* of the given graph $G = (V, E)$, that is, a set of disjoint edges covering all of V, and the goal is to compute the number $M(G)$ of perfect matchings of G.

Example. An old and famous problem calls for the number $M(n, n)$ of ways to cover an $n \times n$-chessboard with dominoes. Thus $M(n, n) = M(L_{n,n})$, where $L_{n,n}$ is the usual $n \times n$-lattice graph with n^2 vertices. We will derive a formula later which will yield, in particular, the precise answer for the ordinary 8×8-board: $M(8, 8) = 12{,}988{,}816$.

The Pfaffian Approach.
Let $G = (V, E)$ be a simple graph with N vertices, where N is even, since otherwise a perfect matching cannot exist. Now we make use of the Pfaffian discussed in Section 5.3. Let $V = \{1, 2, \ldots, N\}$, and orient G in any way. The (oriented) adjacency matrix $A = (a_{ij})$ is a skew-symmetric matrix defined as

$$a_{ij} = \begin{cases} 1 & \text{if } i \to j\,, \\ -1 & \text{if } j \to i\,, \\ 0 & \text{if } \{i,j\} \notin E\,. \end{cases}$$

Consider now the Pfaffian $\mathrm{Pf}(A) = \sum_\mu (\text{sign } \mu) a_\mu$ over all matchings of V, where the perfect graph matchings are those for which $a_\mu = a_{i_1 j_1} \cdots a_{i_{N/2} j_{N/2}} = \pm 1$. If we can find an orientation such that $(\text{sign } \mu) a_\mu$ is always $+1$ or always -1, then $|\mathrm{Pf}(A)| = M(G)$, and so $M(G)^2 = \mathrm{Pf}(A)^2 = \det A$, or

$$M(G) = \sqrt{\det A}\,, \tag{1}$$

by Theorem 5.6.

The significance of (1) lies in the fact that we have an arsenal of methods at hand to compute a determinant, while in general, the computation of $M(G)$ is an intractable problem.

The proof of Theorem 5.6 rested on the bijection $(\mu_1, \mu_2) \longrightarrow \sigma \in S_e(N)$ with

$$(\text{sign } \mu_1) a_{\mu_1} \cdot (\text{sign } \mu_2) a_{\mu_2} = (\text{sign } \sigma) a_\sigma\,, \tag{2}$$

where σ runs through the set S_e of permutations all of whose cycles have even length. Looking at (2) we have to find an orientation such that $(\text{sign } \sigma) a_\sigma \in \{0, 1\}$ for all $\sigma \in S_e$. Suppose $\sigma = \sigma_1 \sigma_2 \cdots \sigma_t$ is the cycle decomposition with $a_\sigma \neq 0$; then $\text{sign}\, \sigma = (-1)^t$. Hence we require $a_\sigma = a_{\sigma_1} \cdots a_{\sigma_t} = (-1)^t$, and this certainly holds if $a_{\sigma_i} = -1$ for all i.

In sum, we are led to the following definition. An orientation of G is *Pfaffian* if for any $\sigma = \sigma_1 \cdots \sigma_t \in S_e$, $a_\sigma = a_{\sigma_1} \cdots a_{\sigma_t} \neq 0$ implies $a_{\sigma_i} = -1$ for all i. Our task is then to find a Pfaffian orientation. This poses grave difficulties in general, and a characterization of graphs that admit a Pfaffian orientation is not known. But for planar graphs such an orientation always exists as we now show.

Let $G = (V, E, F)$ be a simple connected plane graph in the notation of Section 9.3, with an orientation on the edges. We know that any non-bridge lies on the boundary of two different faces. Consider a face f, and denote by B_f the set of its boundary edges that are not bridges. We say that $e \in B_f$ is oriented *clockwise* if the face f lies on the right of e as we move along e in the direction given by the orientation. In the figure

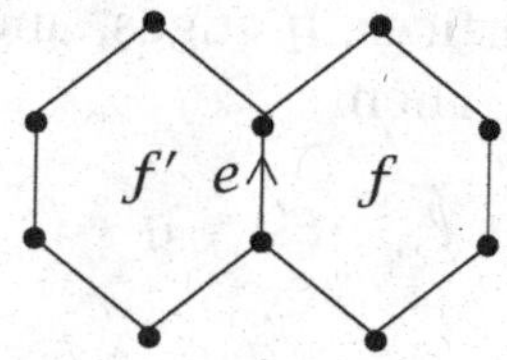

e is oriented clockwise in B_f, and counterclockwise in $B_{f'}$. Bridges are not taken into account.

Lemma 10.1. *Let $G = (V, E, F)$ be a simple connected plane graph. Then there exists an orientation such that for any face f (except possibly the outer face) the number of clockwise oriented edges is odd.*

Proof. We use induction on the number of faces. If $|F| = 1$, then all edges are bridges, and the assertion is vacuously satisfied. Suppose $|F| > 1$. Then there exists a non-bridge e on the boundary of the outer face. Let f be the other face incident with e. In the graph $G \setminus e$, the face f merges with the outer face. By induction there exists an orientation of $G \setminus e$ that meets the requirement of the lemma, and we can now orient e in such a way that the condition also holds for f. The bridges are then oriented arbitrarily. □

Example. The following graph has a required orientation:

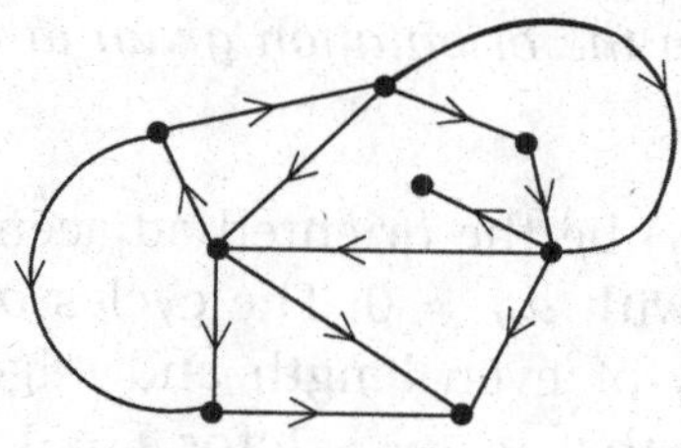

Lemma 10.2. *Let $G = (V, E, F)$ be a simple connected plane graph without bridges, oriented according to the previous lemma, and let*

C be a circuit. Then the number of edges oriented clockwise in C has opposite parity to the number of vertices in the interior of C.

Proof. Let $H = (V', E', F')$ be the subgraph of G consisting of the interior of C, together with C. By Euler's formula (disregarding the outer face),

$$|V'| - |E'| + |F'| = 1. \tag{3}$$

Suppose there are p vertices, q edges, and r faces in the interior, and let C have length ℓ. Then

$$|V'| = p + \ell, \quad |E'| = q + \ell, \quad |F'| = r,$$

which gives by (3),

$$p - q + r = 1. \tag{4}$$

Suppose that the i-th face has c_i edges oriented clockwise, where c_i is odd by assumption, and that c_0 edges of C are oriented clockwise. Since any edge in the interior appears once in clockwise fashion, and the edges on C have the same orientation character on C and on the incident face in the interior, we obtain

$$\sum_{i=1}^{r} c_i = q + c_0,$$

and thus (since all c_i are odd)

$$r \equiv q + c_0 \pmod 2.$$

Appealing to (4), this gives $p + c_0 \equiv 1 \pmod 2$, which means that p and c_0 have opposite parity. □

Theorem 10.3. *Let $G = (V, E, F)$ be a simple connected plane graph without bridges. Then the orientation given in Lemma 10.1 is Pfaffian.*

Proof. Let $A = (a_{ij})$ be the oriented adjacency matrix of G, and $\sigma = \sigma_1 \cdots \sigma_t \in S_e$ with $a_\sigma \neq 0$. The cycles σ_i induce a partition of V into circuits C_i of even length and edges (if σ_i has length 2). We want to show that $a_{\sigma_i} = -1$ for $i = 1, \ldots, t$. If $\sigma_i = (b, c)$ has length 2, then $a_{\sigma_i} = a_{bc}a_{cb} = -1$ holds trivially. Suppose $\sigma_i = (j_1, j_2, \ldots, j_\ell)$ has length $\ell \geq 4$. Running around the circuit

C_i in the direction $j_1, j_2, \ldots, j_\ell$, we see that $a_{\sigma_i} = (-1)^k$, where k is the number of edges oriented in the opposite way. For example, if $\sigma_i = (j_1, \ldots, j_6)$ with orientation as in the figure, then $k = 3$ and therefore $a_{\sigma_i} = -1$.

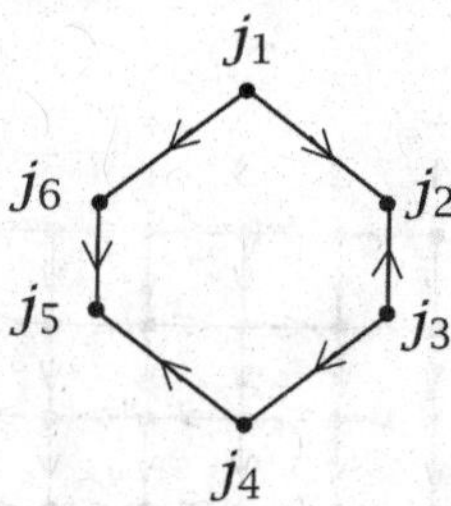

Note that it makes no difference whether we run in clockwise or counterclockwise fashion, since ℓ is even. Hence

$$a_{\sigma_i} = (-1)^k, \quad k = \text{\# edges oriented clockwise},$$

and it remains to show that k is odd. But this is easy. Since $C_1, \ldots, C_t$ is a partition of V, the number of vertices in the interior of C_i must be even (Jordan curve theorem), and so k is odd by Lemma 10.2. □

Corollary 10.4. *Let G be a simple connected plane graph without bridges and A its oriented adjacency matrix according to a Pfaffian orientation. Then*

$$M(G) = \sqrt{\det A}\,. \tag{5}$$

Example. Consider the prism P embedded in the plane with the Pfaffian orientation given in the figure.

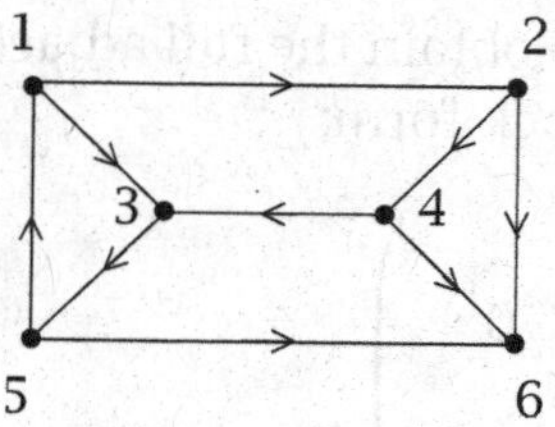

$$A = \begin{pmatrix} 0 & 1 & 1 & 0 & -1 & 0 \\ -1 & 0 & 0 & 1 & 0 & 1 \\ -1 & 0 & 0 & -1 & 1 & 0 \\ 0 & -1 & 1 & 0 & 0 & 1 \\ 1 & 0 & -1 & 0 & 0 & 1 \\ 0 & -1 & 0 & -1 & -1 & 0 \end{pmatrix}$$

One computes $\det A = 16$, and thus $M(P) = 4$.

Domino Tilings.
For the remainder of this section we look at the lattice graph $L_{m,n}$, and compute the number $M(m, n) = M(L_{m,n})$ of domino tilings of

the $m \times n$-chessboard, where m is the number of rows and n the number of columns. Since $|V| = mn$ must be even, we assume that n is even. The following figure shows a Pfaffian orientation of $L_{m,n}$, which by our theorem must exist.

Example. $m = 5$, $n = 6$

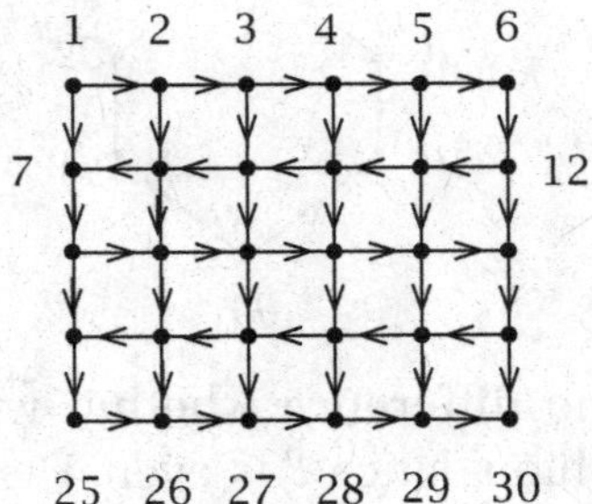

In general, we number the vertices row by row from left to right, orient all vertical edges downward, and the horizontal edges alternately as shown, beginning from left to right. Any interior face has 1 or 3 clockwise oriented edges; hence the orientation is Pfaffian as required.

Let B be the oriented adjacency matrix of the first row; thus B is the $n \times n$-matrix

$$B = \begin{pmatrix} 0 & 1 & 0 & \dots & 0 \\ -1 & 0 & 1 & & \\ & -1 & 0 & 1 & \\ & & & \ddots & 1 \\ 0 & & & -1 & 0 \end{pmatrix}.$$

Denoting by I the $n \times n$-identity matrix, we obtain the full adjacency matrix of $L_{m,n}$ in the following $m \times m$-block form:

$$A = \begin{pmatrix} B & I & & & \\ -I & -B & I & & 0 \\ & -I & B & & \\ 0 & & & \ddots & I \\ & & & -I & \pm B \end{pmatrix}.$$

By suitably multiplying rows and columns by -1 it is readily seen that

$$\det A = \det \begin{pmatrix} B & -I & & & \\ -I & B & -I & & 0 \\ & -I & B & -I & \\ 0 & & \ddots & & -I \\ & & & -I & B \end{pmatrix}. \tag{6}$$

Check this, using the fact that n is even!

Now we can go to work. Let L be the $m \times m$-matrix

$$L = \begin{pmatrix} \lambda & -1 & & & \\ -1 & \lambda & -1 & & 0 \\ & -1 & \lambda & & \\ 0 & & & \ddots & -1 \\ & & & -1 & \lambda \end{pmatrix}.$$

Performing the same row and column operations in L and the block matrix in (6), one obtains

$$\det L = \det \begin{pmatrix} \lambda - \lambda_1 & & & \\ & \lambda - \lambda_2 & & 0 \\ 0 & & \ddots & \\ & & & \lambda - \lambda_m \end{pmatrix},$$

$$\det A = \det \begin{pmatrix} B - \lambda_1 I & & & \\ & B - \lambda_2 I & & 0 \\ 0 & & \ddots & \\ & & & B - \lambda_m I \end{pmatrix},$$

and hence

$$\det A = \prod_{k=1}^{m} \det(B - \lambda_k I).$$

Now $\det L = p_m(\lambda) = \prod_{k=1}^{m}(\lambda - \lambda_k)$ is the characteristic polynomial of the $m \times m$-matrix C,

$$C = \begin{pmatrix} 0 & 1 & & & \\ 1 & 0 & 1 & & 0 \\ & 1 & 0 & & \\ & 0 & & \ddots & 1 \\ & & & 1 & 0 \end{pmatrix}.$$

Thus

$$\det A = \prod_{k=1}^{m} \det(B - \lambda_k I),$$

where the λ_k's are the eigenvalues of C.

Finally, let $q_n(\lambda)$ be the characteristic polynomial of the matrix B with eigenvalues $\mu_1, \ldots, \mu_n$, that is,

$$q_n(\lambda) = \det(\lambda I - B) = (\lambda - \mu_1) \cdots (\lambda - \mu_n).$$

This gives (remember that n is even)

$$\begin{aligned}\det A &= \prod_{k=1}^{m} \det(B - \lambda_k I) = \prod_{k=1}^{m} \det(\lambda_k I - B) \\ &= \prod_{k=1}^{m} q_n(\lambda_k) = \prod_{k=1}^{m} \prod_{\ell=1}^{n} (\lambda_k - \mu_\ell),\end{aligned}$$

and we obtain as our first result

$$M(m,n)^2 = \det A = \prod_{k=1}^{m} \prod_{\ell=1}^{n} (\lambda_k - \mu_\ell), \tag{7}$$

where the λ_k's are the eigenvalues of C, and the μ_ℓ's those of B.

It remains to compute these eigenvalues. Those of C we already know from Section 3.1:

$$\lambda_k = 2\cos\frac{k\pi}{m+1} \quad (k = 1, \ldots, m),$$

and the eigenvalues of B are similarly computed as

$$\mu_\ell = 2i\cos\frac{\ell\pi}{n+1} \quad (\ell = 1, \ldots, n),$$

where $i = \sqrt{-1}$.

By (7), this gives

$$\det A = \prod_{k=1}^{m} \prod_{\ell=1}^{n} \left(2\cos\frac{k\pi}{m+1} - 2i\cos\frac{\ell\pi}{n+1}\right).$$

Since n is even, we have $\cos\frac{(n+1-\ell)\pi}{n+1} = -\cos\frac{\ell\pi}{n+1}$ for $\ell = 1, \ldots, \frac{n}{2}$; hence by grouping the factors into pairs,

$$(\det A)^2 = \prod_{k=1}^{m}\prod_{\ell=1}^{n}\left(4\cos^2\frac{k\pi}{m+1} + 4\cos^2\frac{\ell\pi}{n+1}\right). \tag{8}$$

Then $M(m,n)$ is the fourth root of this expression, so we want to simplify further.

Case A. m even.
Observing that $\cos\frac{(m+1-k)\pi}{m+1} = -\cos\frac{k\pi}{m+1}$, $\cos\frac{(n+1-\ell)\pi}{n+1} = -\cos\frac{\ell\pi}{n+1}$ for $k = 1,\dots\frac{m}{2}$, $\ell = 1,\dots,\frac{n}{2}$ as before, we get

$$(\det A)^2 = 4^{mn}\prod_{k=1}^{m/2}\prod_{\ell=1}^{n/2}\left(\cos^2\frac{k\pi}{m+1} + \cos^2\frac{\ell\pi}{n+1}\right)^4,$$

and thus

$$M(m,n) = 4^{\frac{mn}{4}}\prod_{k=1}^{m/2}\prod_{\ell=1}^{n/2}\left(\cos^2\frac{k\pi}{m+1} + \cos^2\frac{\ell\pi}{n+1}\right). \tag{9}$$

Case B. m odd.
Here we have $\cos\frac{\frac{m+1}{2}\pi}{m+1} = \cos\frac{\pi}{2} = 0$, and you are asked in the exercises to verify the analogous result

$$M(m,n) = 4^{\frac{(m-1)n}{4}}\prod_{k=1}^{(m-1)/2}\prod_{\ell=1}^{n/2}\left(\cos^2\frac{k\pi}{m+1} + \cos^2\frac{\ell\pi}{n+1}\right). \tag{10}$$

In summary we have proved the following astounding formula.

Theorem 10.5 (Fisher–Kasteleyn–Temperley). *For the lattice graph $L_{m,n}$, n even,*

$$M(m,n) = 4^{\lfloor\frac{m}{2}\rfloor\frac{n}{2}}\prod_{k=1}^{\lfloor m/2\rfloor}\prod_{\ell=1}^{n/2}\left(\cos^2\frac{k\pi}{m+1} + \cos^2\frac{\ell\pi}{n+1}\right). \tag{11}$$

It is hard to believe at first sight that the transcendental expression on the right-hand side is a positive integer, let alone that it should be the number of domino tilings. For example, we know from Exercise 3.5 that $M(m,2) = F_{m+1}$, Fibonacci number. Formula (11) thus states with $\cos\frac{\pi}{3} = \frac{1}{2}$,

$$F_{m+1} = 4^{\lfloor \frac{m}{2} \rfloor} \prod_{k=1}^{\lfloor m/2 \rfloor} \left(\cos^2 \frac{k\pi}{m+1} + \frac{1}{4} \right).$$

A Determinant Formula.

As beautiful as the expresssion of Theorem 10.5 is, it is unwieldy to actually compute $M(m,n)$. A closer look at the polynomials $p_m(\lambda)$ and $q_n(\lambda)$ suggests another approach. The polynomials $p_m(\lambda)$ are the Chebyshev polynomials treated in Exercise 3.1,

$$p_m(\lambda) = \lambda^m - \binom{m-1}{1}\lambda^{m-2} + \binom{m-2}{2}\lambda^{m-4} \mp \cdots .$$

Similarly, it is readily seen that

$$q_n(\lambda) = \lambda^n + \binom{n-1}{1}\lambda^{n-2} + \binom{n-2}{2}\lambda^{n-4} + \cdots .$$

Case A. m even.

Consider the associated polynomials

$$\widetilde{p}_m(\lambda) = \lambda^{\frac{m}{2}} - \binom{m-1}{1}\lambda^{\frac{m}{2}-1} + \binom{m-2}{2}\lambda^{\frac{m}{2}-2} \mp \cdots ,$$

$$\widetilde{q}_n(\lambda) = \lambda^{\frac{n}{2}} + \binom{n-1}{1}\lambda^{\frac{n}{2}-1} + \binom{n-2}{2}\lambda^{\frac{n}{2}-2} + \cdots .$$

It follows that α is a root of $p_m(\lambda)$ if and only if α^2 is a root of $\widetilde{p}_m(\lambda)$, and analogously β is a root of $q_n(\lambda)$ if and only if β^2 is a root of $\widetilde{q}_n(\lambda)$. The roots are therefore

$$\widetilde{p}_m(\lambda): \ 4\cos^2 \frac{k\pi}{m+1}, \quad k = 1, \dots, \frac{m}{2},$$

$$\widetilde{q}_n(\lambda): -4\cos^2 \frac{\ell\pi}{n+1}, \quad \ell = 1, \dots, \frac{n}{2}.$$

Looking at (9) we have

$$M(m,n) = \prod_{k=1}^{m/2} \prod_{\ell=1}^{n/2} \left(4\cos^2 \frac{k\pi}{m+1} - \left(-4\cos^2 \frac{\ell\pi}{n+1} \right) \right).$$

Hence if we denote by $\alpha_1, \dots, \alpha_{m/2}$ the roots of $\widetilde{p}_m(\lambda)$ and by $\beta_1, \dots, \beta_{n/2}$ those of $\widetilde{q}_n(\lambda)$, then

$$M(m,n) = \prod_{k=1}^{m/2}\prod_{\ell=1}^{n/2}(\alpha_k - \beta_\ell)\,. \tag{12}$$

Case B. m odd.
It is readily seen that

$$M(m,n) = \prod_{k=1}^{(m-1)/2}\prod_{\ell=1}^{n/2}(\alpha_k - \beta_\ell)\,, \tag{13}$$

where $\alpha_1,\dots,\alpha_{\frac{m-1}{2}}$ are the roots of the polynomial $\tilde{p}_m(\lambda) = \lambda^{\frac{m-1}{2}} - \binom{m-1}{1}\lambda^{\frac{m-3}{2}} \pm \cdots$, and $\tilde{q}_n(\lambda)$ as before.

Now we quote an important result from linear algebra, the so-called resultant theorem.

Fact. *Let $p(x) = x^s + a_1x^{s-1} + \cdots + a_s$, $q(x) = x^t + b_1x^{t-1} + \cdots + b_t$ be polynomials with roots $\alpha_1,\dots,\alpha_s,\beta_1,\dots,\beta_t$, respectively. Then*

$$\prod_{k=1}^{s}\prod_{\ell=1}^{t}(\alpha_k - \beta_\ell) = \det\left(\begin{array}{cccccccc} 1 & a_1 & \dots & a_s & 0 & \dots & & 0 \\ 0 & 1 & a_1 & \dots & a_s & \dots & & 0 \\ & & & \dots & & & & \\ 0 & & \dots & & 1 & a_1 & \dots & a_s \\ \hline 1 & b_1 & \dots & b_t & 0 & \dots & & 0 \\ & & & \dots & & & & \\ 0 & \dots & & & 1 & b_1 & \dots & b_t \end{array}\right)\begin{array}{l} \left.\vphantom{\begin{array}{c}1\\0\\ \\0\end{array}}\right\}t \\ \left.\vphantom{\begin{array}{c}1\\ \\0\end{array}}\right\}s \end{array}.$$

The matrix has $s+t$ rows and columns, where in the top half the coefficients of $p(x)$ are shifted to the right, and in the lower half those of $q(x)$.

Example. $s = 2$, $t = 3$. The resultant matrix is

$$\left(\begin{array}{ccccc} 1 & a_1 & a_2 & 0 & 0 \\ 0 & 1 & a_1 & a_2 & 0 \\ 0 & 0 & 1 & a_1 & a_2 \\ \hline 1 & b_1 & b_2 & b_3 & 0 \\ 0 & 1 & b_1 & b_2 & b_3 \end{array}\right).$$

In our situation this gives a new formula for $M(m,n)$.

Theorem 10.6. *For n even,*

$$M(m,n) = \det \begin{pmatrix} \binom{m}{0} & -\binom{m-1}{1} & \binom{m-2}{2} & \cdots \\ 0 & \binom{m}{0} & -\binom{m-1}{1} & \cdots \\ & & \cdots & \\ \hdashline \binom{n}{0} & \binom{n-1}{1} & \binom{n-2}{2} & \cdots \\ 0 & \binom{n}{0} & \binom{n-1}{1} & \cdots \\ & & \cdots & \end{pmatrix} \begin{matrix} \left.\vphantom{\begin{matrix} a\\a\\a \end{matrix}}\right\} \frac{n}{2} \\ \left.\vphantom{\begin{matrix} a\\a\\a \end{matrix}}\right\} \lfloor \frac{m}{2} \rfloor \end{matrix} . \tag{14}$$

This looks promising (there are lots of zeros), and the result is now certainly an integer.

The Square Lattice.
Let us finally look at the case $m = n$. The resultant matrix in (14) is an $n \times n$-matrix. We perform three operations: First add row $\frac{n}{2} + i$ to row i ($i = 1, \ldots, \frac{n}{2}$), secondly factor out the 2's, and thirdly subtract row i from row $\frac{n}{2} + i$. The matrices should make clear what happens:

$$\begin{pmatrix} \binom{n}{0} & -\binom{n-1}{1} & \binom{n-2}{2} & \cdots \\ & \cdots & & \\ \hdashline \binom{n}{0} & \binom{n-1}{1} & \binom{n-2}{2} & \cdots \\ & \cdots & & \end{pmatrix} \longrightarrow \begin{pmatrix} 2\binom{n}{0} & 0 & 2\binom{n-2}{2} & \cdots \\ & \cdots & & \\ \hdashline \binom{n}{0} & \binom{n-1}{1} & \binom{n-2}{2} & \cdots \\ & \cdots & & \end{pmatrix} \longrightarrow$$

$$2^{\frac{n}{2}} \begin{pmatrix} \binom{n}{0} & 0 & \binom{n-2}{2} & \cdots \\ 0 & \binom{n}{0} & 0 & \binom{n-2}{2} \\ & \cdots & & \\ \hdashline \binom{n}{0} & \binom{n-1}{1} & \binom{n-2}{2} & \cdots \\ & \cdots & & \end{pmatrix} \longrightarrow 2^{\frac{n}{2}} \begin{pmatrix} \binom{n}{0} & 0 & \binom{n-2}{2} & 0 & \cdots \\ 0 & \binom{n}{0} & 0 & \binom{n-2}{2} & \cdots \\ & & \cdots & & \\ \hdashline 0 & \binom{n-1}{1} & 0 & \binom{n-3}{3} & \\ 0 & 0 & \binom{n-1}{1} & 0 & \\ & & \cdots & & \end{pmatrix} \begin{matrix} \left.\vphantom{\begin{matrix} a\\a\\a \end{matrix}}\right\} \frac{n}{2} \\ \left.\vphantom{\begin{matrix} a\\a\\a \end{matrix}}\right\} \frac{n}{2} \end{matrix} .$$

Case A. $n \equiv 2 \pmod 4$.
The last column consists of zeros except the lower right-hand corner $\binom{n/2}{n/2}$. Hence developing the determinant of the matrix according to the first and last columns, we get

$$
M(n,n) = 2^{\frac{n}{2}} \det \left(\begin{array}{ccccc}
\binom{n}{0} & 0 & \binom{n-2}{2} & 0 & \ldots \\
0 & \binom{n}{0} & 0 & \binom{n-2}{2} & \ldots \\
 & & \ldots & & \\
\hline
\binom{n-1}{1} & 0 & \binom{n-3}{3} & 0 & \ldots \\
0 & \binom{n-1}{1} & 0 & \binom{n-3}{3} & \ldots \\
 & & \ldots & &
\end{array} \right)
\begin{array}{l}
\left.\vphantom{\begin{array}{c}1\\1\\1\end{array}}\right\} \frac{n}{2}-1 \\
\left.\vphantom{\begin{array}{c}1\\1\\1\end{array}}\right\} \frac{n}{2}-1
\end{array} . \quad (15)
$$

Case B. $n \equiv 0 \pmod 4$.
One similarly obtains

$$
M(n,n) = 2^{\frac{n}{2}} \det \left(\begin{array}{ccccc}
\binom{n}{0} & 0 & \binom{n-2}{2} & 0 & \ldots \\
0 & \binom{n}{0} & 0 & \binom{n-2}{2} & \ldots \\
 & & \ldots & & \\
\hline
\binom{n-1}{1} & 0 & \binom{n-3}{3} & 0 & \ldots \\
 & & \ldots & &
\end{array} \right)
\begin{array}{l}
\left.\vphantom{\begin{array}{c}1\\1\\1\end{array}}\right\} \frac{n}{2}-2 \\
\left.\vphantom{\begin{array}{c}1\\1\end{array}}\right\} \frac{n}{2}
\end{array} . \quad (16)
$$

These matrices have a very regular form. The rows of the two halves are shifted to the right, and every other element is 0. It is an easy exercise to show that in general, the following formula holds:

$$
\det \left(\begin{array}{l}
a_1 \; 0 \; a_2 \; 0 \; \ldots a_s \; \ldots \\
0 \; a_1 \; 0 \; a_2 \ldots \\
\quad \ldots \\
\hline
b_1 \; 0 \; b_2 \; 0 \; \ldots b_t \; \ldots \\
0 \; b_1 \; 0 \; b_2 \ldots \\
\quad \ldots
\end{array} \right)
\begin{array}{l}
\left.\vphantom{\begin{array}{c}1\\1\\1\end{array}}\right\} 2t-2 \\
\left.\vphantom{\begin{array}{c}1\\1\\1\end{array}}\right\} 2s-2
\end{array}
$$

$$
= \left[\det \left(\begin{array}{l}
a_1 \ldots a_s \; 0 \ldots 0 \\
0 \; a_1 \ldots a_s \ldots \\
\quad \ldots \\
\hline
b_1 \ldots b_t \; 0 \ldots 0 \\
0 \; b_1 \ldots b_t \ldots \\
\quad \ldots
\end{array} \right) \right]^2
\begin{array}{l}
\left.\vphantom{\begin{array}{c}1\\1\\1\end{array}}\right\} t-1 \\
\left.\vphantom{\begin{array}{c}1\\1\\1\end{array}}\right\} s-1
\end{array} . \quad (17)
$$

As an example for $s = t = 2$ we have

$$\det\begin{pmatrix} a_1 & 0 & a_2 & 0 \\ 0 & a_1 & 0 & a_2 \\ b_1 & 0 & b_2 & 0 \\ 0 & b_1 & 0 & b_2 \end{pmatrix} = \left[\det\begin{pmatrix} a_1 & a_2 \\ b_1 & b_2 \end{pmatrix}\right]^2 .$$

Summarizing our findings in (15), (16), (17) we can state the final result.

Theorem 10.7. *Let n be even. Then for $n \equiv 0 \pmod 4$,*

$$M(n,n) = 2^{\frac{n}{2}} \left[\det\begin{pmatrix} \binom{n}{0} & \binom{n-2}{2} & \cdots & \cdots \\ 0 & \binom{n}{0} & \binom{n-2}{2} & \cdots \\ & \cdots & & \\ \hline \binom{n-1}{1} & \binom{n-3}{3} & \cdots & \\ & \cdots & & \end{pmatrix}\right]^2 \begin{matrix} \Big\} \frac{n}{4} - 1 \\ \\ \Big\} \frac{n}{4} \end{matrix} ,$$

and for $n \equiv 2 \pmod 4$,

$$M(n,n) = 2^{\frac{n}{2}} \left[\det\begin{pmatrix} \binom{n}{0} & \binom{n-2}{2} & \cdots & \cdots \\ 0 & \binom{n}{0} & \binom{n-2}{2} & \cdots \\ & \cdots & & \\ \hline \binom{n-1}{1} & \binom{n-3}{3} & \cdots & \\ & \cdots & & \end{pmatrix}\right]^2 \begin{matrix} \Big\} \frac{n-2}{4} \\ \\ \Big\} \frac{n-2}{4} \end{matrix} .$$

Examples. For small n we get

$$M(2,2) = 2,\ M(4,4) = 2^2[\det(3)]^2 = 4 \cdot 9 = 36,$$

$$M(6,6) = 2^3\left[\det\begin{pmatrix} 1 & 6 \\ 5 & 1 \end{pmatrix}\right]^2 = 8 \cdot 29^2 = 6728,$$

$$M(8,8) = 2^4\left[\det\begin{pmatrix} 1 & 15 & 1 \\ 7 & 10 & 0 \\ 0 & 7 & 10 \end{pmatrix}\right]^2 = 16 \cdot 901^2 = 12{,}988{,}816 .$$

Note that $M(n,n)$ is always a square or twice a square. A direct combinatorial argument for this remarkable fact was found only recently.

Asymptotic Growth.
For physical considerations it is important to estimate the growth of $M(m,n)$ as m and n go to infinity. Our discussion suggests that $M(m,n)$ grows exponentially in mn, so we look at

$$c = \lim_{m,n\to\infty} \frac{\log M(m,n)}{mn}.$$

By (11),

$$\frac{\log M(m,n)}{mn} = \frac{1}{mn} \sum_{k=1}^{\lfloor m/2 \rfloor} \sum_{\ell=1}^{n/2} \log\left(4\cos^2 \frac{k\pi}{m+1} + 4\cos^2 \frac{\ell\pi}{n+1}\right).$$

The right-hand side is a discrete mean value, which tends to the continuous mean value, since the functions are continuous. More precisely, setting $x = \frac{k\pi}{m+1}$, $y = \frac{\ell\pi}{n+1}$, the variables x and y range from 0 (since $\frac{1}{m+1}, \frac{1}{n+1} \to 0$) to $\frac{\pi}{2}$ (since $\frac{\lfloor m/2 \rfloor}{m+1}\pi, \frac{n/2}{n+1}\pi \to \frac{\pi}{2}$). Finally, $dx = \frac{\pi}{m+1}$, $dy = \frac{\pi}{n+1}$, where the denominators can be replaced by m and n in the limiting process. Altogether we obtain

$$c = \frac{1}{\pi^2} \int_0^{\pi/2}\int_0^{\pi/2} \log(4\cos^2 x + 4\cos^2 y)\,dx\,dy\,.$$

The last integral is evaluated to

$$c = \frac{G}{\pi}, \text{ where } G = 1 - \frac{1}{3^2} + \frac{1}{5^2} - \frac{1}{7^2} \pm \cdots$$

is called the *Catalan constant*. This gives $c = 0.29156$, and therefore

$$M(m,n) \approx e^{0.29156mn} \approx 1.34^{mn}\,.$$

Exercises

10.1 Let B be an $n\times n$-matrix, $A = \begin{pmatrix} 0 & B \\ -B^T & 0 \end{pmatrix}$. Prove $\mathrm{Pf}(A) = (-1)^{\binom{n}{2}} \det B$.

10.2 Show that the graph $K_{3,3}$ has no Pfaffian orientation, which proves again that $K_{3,3}$ is not planar.

▷ **10.3** Let G be a connected bipartite plane graph without bridges such that all faces have boundary length congruent to 2 (mod 4). Suppose that

V_1, V_2 are the two color classes of G. Show that directing every edge from V_1 to V_2 gives a Pfaffian orientation.

10.4 Prove assertion (16) of the text.

10.5 Use Theorem 10.6 to re-prove $F_{m+1} = \sum_{k\geq 0} \binom{m-k}{k}$.

▷ **10.6** Show that $M(3,n) = \sum_{k\geq 0} 2^{\frac{n}{2}-k}\binom{n-k}{k}$ for even n, and compare it to the result obtained in Exercise 3.15.

10.7 Prove the result (17) mentioned in the text.

* * *

10.8 Let μ be a fixed perfect matching of G. A circuit C is called *alternating* if the edges of C alternate between μ and another matching μ'. Suppose that G is oriented in such a way that all alternating circuits with respect to μ have an odd number of clockwise oriented edges. Prove that the orientation is Pfaffian.

▷ **10.9** Verify formula (10) for $M(m,n)$, when m is odd.
Hint: Show that $\prod_{\ell=1}^{n/2} \cos\frac{\ell\pi}{n+1} = 2^{-n/2}$, using the addition theorem for the cosine.

▷ **10.10** Let G be any simple graph with q edges. For an orientation ε let A^{ε} be the oriented adjacency matrix. Prove that $M(G) = 2^{-q}\sum_{\varepsilon}\det A^{\varepsilon}$, where ε runs through all 2^q orientations. Hint: Show that permutations with a fixed point or cycles of length greater than 2 cancel out, similar to the argument used in Section 5.3.

10.11 Let $L_{2,2,n}$ be the $2\times2\times n$-lattice graph. Compute $M(2,2,n) = M(L_{2,2,n})$ with the recurrence method of Section 3.1. Derive from this and Exercise 10.6 the curious identity $M(3,2n) = M(2,2,n) + M(2,2,n-1)$.

10.12 For a graph G denote by $m(G,r)$ the number of r-matchings in G, that is, of r disjoint edges. Suppose G is plane with a Pfaffian orientation, A the oriented adjacency matrix. Show that the coefficient of λ^{n-2r} in $\det(\lambda I - A)$ is equal to $\sum_H m(H,r)^2$ over all subgraphs H of G with $2r$ vertices.

10.13 Let $G = (V,E)$ be any graph. The *matchings polynomial* $\mu(G;x)$ is given by $\sum_{k\geq 0}(-1)^k m(G,k)x^{n-2k}$ with $m(G,k)$ defined as in the previous exercise. Prove: a. $\mu(G\cup H;x) = \mu(G;x)\mu(H;x)$ for disjoint unions, b. $\mu(G;x) = \mu(G\setminus e;x) - \mu(G\setminus\{u,v\};x)$, $e = \{u,v\}\in E$, c. $\mu(G;x)' = \sum_{u\in V}\mu(G\setminus u;x)$.

▷ **10.14** Compute the matchings polynomial for the paths P_n, complete graphs K_n, and complete bipartite graphs $K_{n,n}$. They are related to classical polynomials; which?

10.2 The Ising Problem and Eulerian Subgraphs

In the general Ising problem we are given a graph $G = (V, E)$, where each vertex i is assigned a *spin* σ_i, which can be either $+1$ or -1. An assignment of spins $\sigma = (\sigma_i)_{i \in V}$ is called a *state*. In addition, each edge $e = \{i, j\}$ has an interaction J_{ij} that is constant on the edge, but may vary from edge to edge. For each state σ the *Hamiltonian* $H(\sigma)$ is defined as

$$H(\sigma) = -\sum_{\{i,j\} \in E} J_{ij}\sigma_i\sigma_j - \sum_i M\sigma_i, \tag{1}$$

where M represents the energy from the external field. The *partition function* $Z(G; \beta, J, M)$ is then

$$Z(G) = \sum_\sigma e^{-\beta H(\sigma)}, \tag{2}$$

where the sum is over all $2^{|V|}$ states, and $\beta = \frac{1}{kT}$ is a parameter associated with the temperature T, and where k is Boltzmann's constant.

The main problem of the Ising model is to find a closed expression for $Z(G)$, and in particular, to determine

$$\lim_{n \to \infty} \frac{\log Z(L_{n,n})}{n^2}$$

for the infinite square lattice.

Eulerian Subgraphs and Bipartitions.
Recall that an *Eulerian subgraph* $G_U = (V, U)$, $U \subseteq E$, is a subgraph in which all vertices have even degree. Note that the vertex set is always assumed to be V, so we may identify G_U with its set U of edges. A *bipartition* (or cut set) $B \subseteq E$ of G corresponds to a partition $V = V_1 \dot\cup V_2$ such that B consists precisely of the edges between V_1 and V_2. The empty set $\emptyset$ is a bipartition by definition.

Let $\mathcal{E}(G) \subseteq 2^E$, $\mathcal{B}(G) \subseteq 2^E$ be the sets of Eulerian subgraphs and bipartitions, respectively. We are interested in the generating functions $\mathcal{E}(G; z) = \sum_{U \in \mathcal{E}(G)} z^{|U|}$, $\mathcal{B}(G; z) = \sum_{B \in \mathcal{B}(G)} z^{|B|}$.

Example. For K_3 and K_4 we find by inspection

$\mathcal{E}(K_3;z) = 1 + z^3$, $\mathcal{B}(K_3;z) = 1 + 3z^2$, $\mathcal{E}(K_4;z) = \mathcal{B}(K_4;z) = 1 + 4z^3 + 3z^4$.

Proposition 10.8. *For* $G = (V,E)$ *we have*

a. $\mathcal{E}(G;z) = (1-z)^{|E|-|V|+k(G)} z^{|V|-k(G)} T(G; \frac{1}{z}, \frac{1+z}{1-z})$,

b. $\mathcal{B}(G;z) = z^{|E|-|V|+k(G)} (1-z)^{|V|-k(G)} T(G; \frac{1+z}{1-z}, \frac{1}{z})$, (3)

c. $\mathcal{B}(G;z) = \frac{(1+z)^{|E|}}{2^{|E|-|V|+k(G)}} \mathcal{E}(G; \frac{1-z}{1+z})$,

where $T(G)$ *is the Tutte polynomial.*

Proof. We show that $\mathcal{E}(z)$ is a chromatic invariant and use the Recipe Theorem 9.5. For a bridge or loop we have

$$\mathcal{E}(\text{bridge}; z) = 1, \quad \mathcal{E}(\text{loop}; z) = 1 + z\,.$$

If e is a bridge of G, then no Eulerian subgraph contains e (clear?); hence $\mathcal{E}(G;z) = \mathcal{E}(G\setminus e;z)$. Similarly, if e is a loop, then e may be added to any Eulerian subgraph of $G\setminus e$; thus $\mathcal{E}(G;z) = (1+z)\mathcal{E}(G\setminus e;z)$. Suppose e is not a bridge or loop; then we claim that

$$\mathcal{E}(G;z) = (1-z)\mathcal{E}(G\setminus e;z) + z\mathcal{E}(G/e;z)\,. \tag{4}$$

Let $e = \{u,v\}$ and E_u the edges emanating from u (apart from e), and E_v those of v (without e):

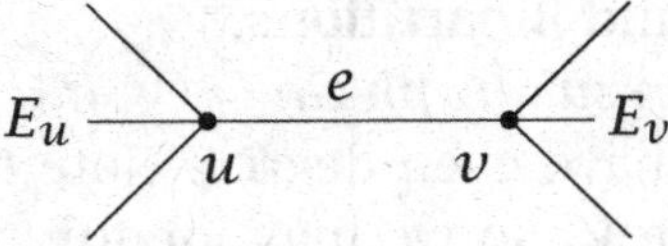

The Eulerian subgraphs of G that do not contain e are in one-to-one correspondence with the Eulerian subgraphs of $G\setminus e$, which accounts for $\mathcal{E}(G\setminus e;z)$ in (4). Suppose $e \in U \in \mathcal{E}(G)$; then $U\setminus e \in \mathcal{E}(G/e)$, since the degree of the contracted vertex is $d(uv) = (d(u)-1) + (d(v)-1) \equiv 0 \pmod 2$. Conversely, if $U \subseteq \mathcal{E}(G/e)$, then $U \cup e \in \mathcal{E}(G)$ precisely when $|E_u|$ and $|E_v|$ are both odd in U. Thus we have to subtract from $\mathcal{E}(G/e)$ those Eulerian subgraphs U for which $|E_u|$ and $|E_v|$ are even. But these are in one-to-one correspondence with $\mathcal{E}(G \setminus e)$. Taking the edge e into account, we get (4) and thus assertion (a) from Theorem 9.5 with $A = 1$, $B = 1 + z$, $\alpha = 1 - z$, $\beta = z$.

Formula (b) is proved in analogous fashion, and (c) follows by considering $\mathcal{E}(G; \frac{1-z}{1+z})$ in (a). □

Example. For $z = 2$ we obtain new evaluations of the Tutte polynomial:

$$\sum_{U \in \mathcal{E}(G)} 2^{|U|} = (-1)^{|E|-|V|+k(G)|} 2^{|V|-k(G)} T\left(G; \frac{1}{2}, -3\right)$$

$$\sum_{B \in \mathcal{B}(G)} 2^{|B|} = (-1)^{|V|-k(G)} 2^{|E|-|V|+k(G)} T\left(G; -3, \frac{1}{2}\right).$$

Let us return to the Ising problem. We make the assumption $M = 0$ (no external field) and $J_{ij} = J$ for all edges. Setting $K = \beta J$, the expression (2) for the partition function becomes

$$Z(G;K) = \sum_{\sigma} e^{K \sum_{\{i,j\} \in E} \sigma_i \sigma_j}. \tag{5}$$

For a state σ let $V^+ = \{i \in V : \sigma_i = 1\}$, $V^- = \{i \in V : \sigma_i = -1\}$, and let E^+, E^- be the edges with both ends in V^+ and in V^-, respectively. Denote by $B_\sigma \subseteq E$ the bipartition induced by the partition $V^+ \cup V^-$; then (1) can be written as

$$H(\sigma) = -J(|E^+| + |E^-|) + J|B_\sigma| = -J|E| + 2J|B_\sigma|.$$

Each bipartition corresponds to exactly $2^{k(G)}$ spin configurations (two for each component by exchanging $+1$ and -1), which all have the same Hamiltonian. Hence we obtain

$$\begin{aligned} Z(G;K) &= \sum_{\sigma} e^{-\beta H(\sigma)} = 2^{k(G)} \sum_{B \in \mathcal{B}(G)} e^{\beta J|E| - 2\beta J|B|} \\ &= 2^{k(G)} e^{K|E|} \sum_{B \in \mathcal{B}(G)} e^{-2K|B|} \\ &= 2^{k(G)} e^{K|E|} \mathcal{B}(G; e^{-2K}), \end{aligned} \tag{6}$$

and with Proposition 10.8b,

$$Z = 2^{k(G)} e^{K|E|} (e^{-2K})^{|E|-|V|+k(G)} \times (1 - e^{-2K})^{|V|-k(G)} T\left(G; \frac{1+e^{-2K}}{1-e^{-2K}}, e^{2K}\right). \quad (7)$$

We can further simplify this expression, using the hyperbolic functions

$$\cosh x = \frac{e^x + e^{-x}}{2}, \quad \sinh x = \frac{e^x - e^{-x}}{2},$$
$$\tanh x = \frac{\sinh x}{\cosh x}, \quad \coth x = \frac{\cosh x}{\sinh x}.$$

Writing $(e^K)^{|E|} = (e^K)^{|E|-|V+k(G)} \cdot (e^K)^{|V|-k|G|}$, and combining the parts with the third and fourth factors above, we obtain the factors

$$(e^{-K})^{|E|-|V|+k(G)}, \quad (e^K - e^{-K})^{|V|-k(G)} = 2^{|V|-k(G)} \sinh(K)^{|V|-k(G)}. \quad (8)$$

Finally, in the Tutte polynomial,

$$\frac{1+e^{-2K}}{1-e^{-2K}} = \frac{e^K + e^{-K}}{e^K - e^{-K}} = \coth(K). \quad (9)$$

Using Proposition 10.8 and (7) we may express Z also in terms of $\mathcal{E}(G;z)$; see Exercise 10.19. Thus with (8) and (9) we obtain three expressions for the partition function.

Theorem 10.9. *Let $G = (V,E)$, $T(G)$ the Tutte polynomial, and $Z(G;K)$ the partition function of the Ising model with constant interaction J, $K = \beta J$, and no outside field. Then*

a. $Z(G;K) = 2^{|V|}(e^{-K})^{|E|-|V|+k(G)} \times (\sinh(K))^{|V|-k(G)} T(G; \coth(K), e^{2K})$,

b. $Z(G;K) = 2^{k(G)} e^{K|E|} \mathcal{B}(G; e^{-2K})$, (10)

c. $Z(G;K) = 2^{|V|}(\cosh(K))^{|E|} \mathcal{E}(G; \tanh(K))$.

Reduction to a Dimer Problem.

The last result shows that we can compute the partition function if we succeed in finding a closed formula for any of the three expressions on the right. Easiest is the Eulerian generating function $\mathcal{E}(G;z)$, and to be specific we look at the usual lattice graph $L_{m,n}$, but the procedure can obviously be generalized.

Set $G = L_{m,n}$. We construct a new graph G^t, called the *terminal graph*, as follows. First, replace each vertex by a complete graph K_4, and join these K_4's as indicated in the figure for $m = 3$, $n = 4$:

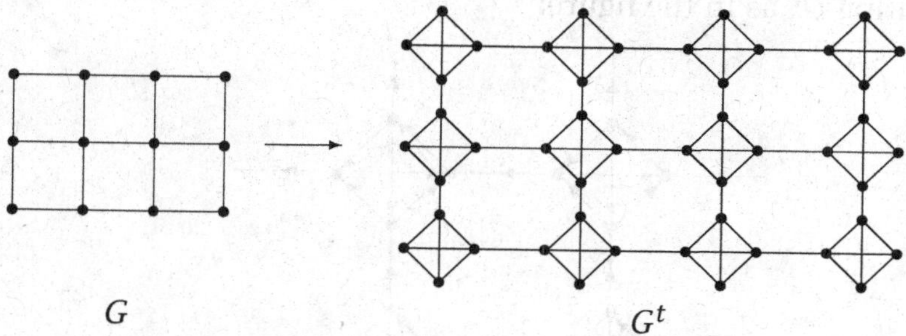

The edges within the K_4's are called *internal edges*, the others *external edges*. The external edges correspond therefore precisely to the original edges of G.

We show next how the Eulerian subgraphs of G are related to perfect matchings in G^t. Let $U \in \mathcal{E}(G)$. The corresponding edges U^t in G^t are disjoint; thus they form a (partial) matching. If they touch a K_4 in G^t with two or four edges, then there exists a unique internal edge that extends the matching (or none). However, when a vertex u has degree 0 in U, then there are three possibilities to complete the matching. The following figure should make the situation clear. The edges of U and U^t are drawn in bold type, and the added internal edges dashed. For the lower left-hand corner there are three possibilities.

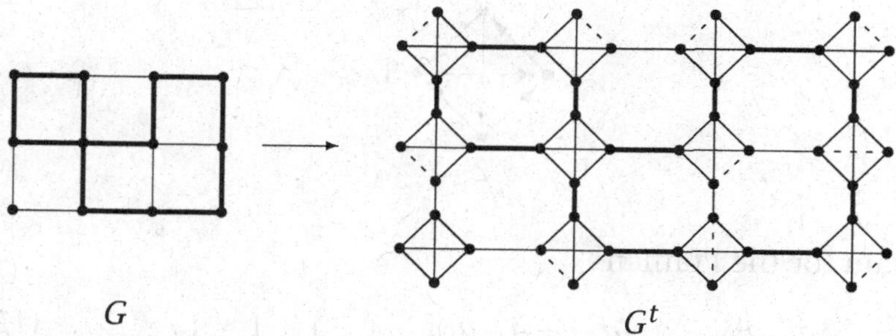

In summary, we see that:

A. To every perfect matching M of G^t there exists a unique Eulerian subgraph U of G, by taking the external edges.

B. To $U \in \mathcal{E}(G)$, there are three possibilities to complete the perfect matching for U^t, for every isolated vertex in U.

Since we want to compute $\mathcal{E}(G;z) = \sum_{U\in\mathcal{E}(G)} z^{|U|}$, and relate $\mathcal{E}(G;z)$ to the Pfaffian of G^t, we have to assign weights to the edges of G^t, and try to get rid of the ambiguity in B. This is accomplished by orienting G^t as in the figure:

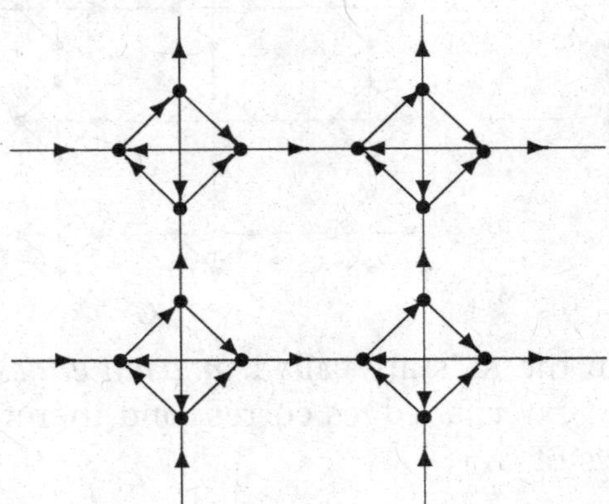

Now we assign the weight ± 1 to the internal edges, and $\pm z$ to the external edges according to the orientation, and obtain thus a skew-symmetric matrix A of G^t.

Theorem 10.10. *Let $G = L_{m,n}$ and G^t the terminal graph with adjacency matrix A according to the orientation above. Then*

$$\mathcal{E}(G;z) = |\mathrm{Pf}(A)|. \tag{11}$$

Proof. We number any K_4 by

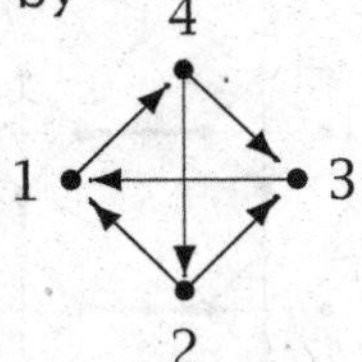

and obtain for the Pfaffian

$$a_{12}a_{34} - a_{13}a_{24} + a_{14}a_{23} = 1 - 1 + 1 = 1. \tag{12}$$

For an Eulerian subgraph U of G, denote by $\mu \to U$ the fact that μ is a perfect matching of G^t belonging to U. Hence we may write

$$\mathrm{Pf}(A) = \sum_{\mu}(\mathrm{sign}\,\mu)a_\mu = \sum_{U\in\mathcal{E}(G)}\sum_{\mu\to U}(\mathrm{sign}\,\mu)a_\mu .$$

Suppose u is an isolated vertex of U, and μ_1, μ_2, μ_3 the three local matchings of the associated K_4 in G^t. Then by (12),

$$\sum_{\mu \to U} (\operatorname{sign} \mu) a_\mu = \sum_{i=1}^{3} (\operatorname{sign} \mu_i) a_{\mu_i} \cdot (\operatorname{sign} \nu) a_\nu = (\operatorname{sign} \nu) a_\nu ,$$

where ν is the remaining matching of G^t. Now we take the next isolated vertex, and so on, and arrive at

$$\sum_{\mu \to U} (\operatorname{sign} \mu) a_\mu = (\operatorname{sign} \nu_U) a_{\nu_U} = \pm z^{|U|} , \tag{13}$$

where ν_U is the set of external matching edges (corresponding to U), all of which have weight $\pm z$.

It remains to prove that all expressions $\pm z^{|U|}$ in (13) have the same sign. For $U \in \mathcal{E}(G)$ extend ν_U to the *unique* perfect matching μ_U of G^t, by always choosing the local matching $\mu_1 : 12, 34$ for every K_4 belonging to an isolated vertex in U. Since $(\operatorname{sign} \mu_1) a_{\mu_1} = 1$, we have

$$\operatorname{Pf}(A) = \sum_{U \in \mathcal{E}(G)} (\operatorname{sign} \mu_U) a_{\mu_U} . \tag{14}$$

To prove that all signs in (14) (and thus in (13)) are the same we proceed as in the plane case treated in the last section. Combining two such perfect matchings B, B' to a permutation $\sigma \in S_e(V(G^t))$ with even cycles, it suffices to show that $a_\sigma = -z^\ell$, where σ corresponds to a circuit C in G^t with edges alternating between B and B', and where ℓ is the number of external edges of C.

Case A. C contains no crossing edges in a K_4.
In this case we note that every face of the "bathroom tiling" contains an odd number of clockwise oriented edges, and this remains true if one of the diagonals is inserted; see the figure. Hence Lemma 10.2 applies, and $a_\sigma = -z^\ell$.

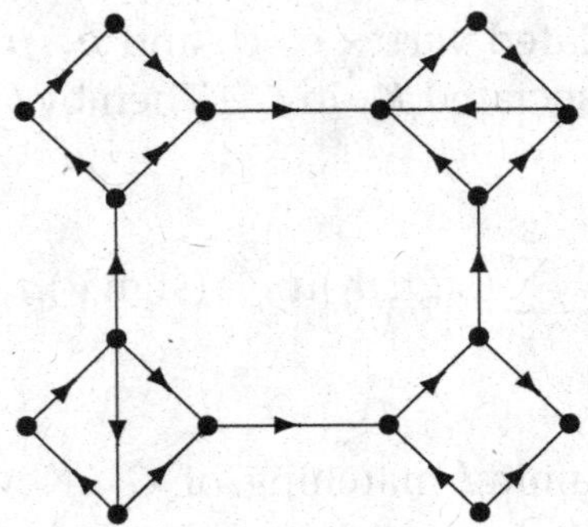

Case B. C contains internal crossing edges.
We proceed by induction on the number c of crossing diagonals, where the induction start $c = 0$ is just case A. By our assumption on the matchings μ_U the diagonals belong to different matchings B and B', whence we have the following situation:

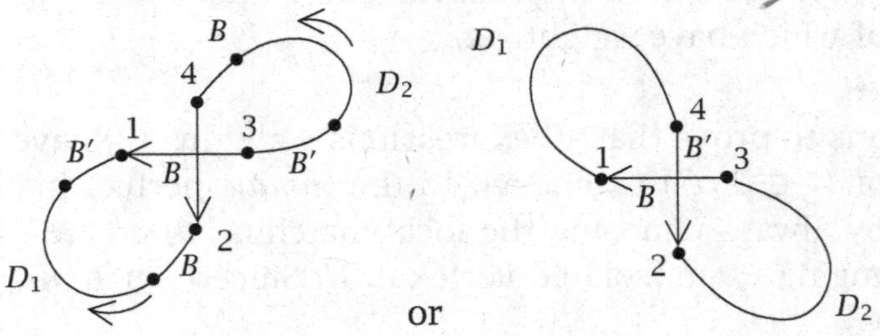

We treat the first case (the other being analogous). Run through C in the given direction, $C = (2, D_1, 1, 3, D_2, 4)$. We have to show that there is an odd number of edges oriented in this direction, and an odd number oriented oppositely. Insert the edges $\{1, 4\}$, respectively $\{2, 3\}$, to produce the circuits C_1 and C_2, which again are made up of two matchings:

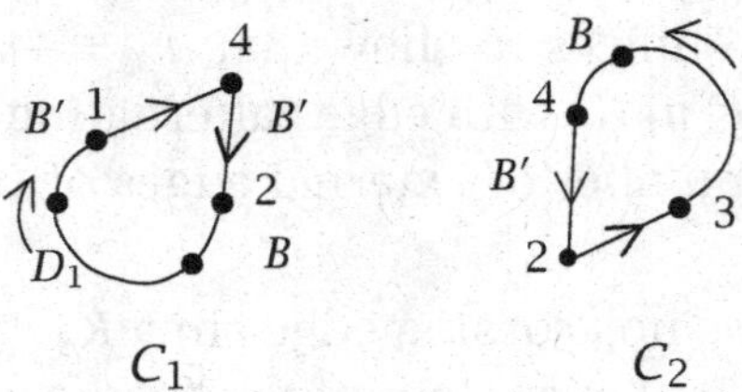

C_1 and C_2 have fewer crossings than C, and so by induction, they agree in an odd number of edges with the given direction. Looking at the picture, D_1 and D_2 also have an odd number in each direction, thus $D_1 \cup D_2$ an even number. Taking the edge $3 \to 1$ into account we conclude that C has an odd number, and the proof is complete. □

Theorem 10.9 and the last result $\mathcal{E}(G;z) = \sqrt{\det A}$ give the partition function Z of the lattice graph $L_{m,n}$, provided we can compute $\det A$. Unfortunately, no one has succeeded in doing that, and the determination of $\det A$ remains one of the great open problems. To overcome this difficulty, we expand $L_{m,n}$ to the toroidal lattice graph $\widetilde{L}_{m,n}$ by wrapping the horizontal and vertical edges around. That is, $\{i,j\}$ and $\{k,\ell\}$ are adjacent if $|i-k| = 1 \pmod{m}$, $j = \ell$, or $i = k$, $|j-\ell| = 1 \pmod{n}$. The picture shows $\widetilde{L}_{3,3}$:

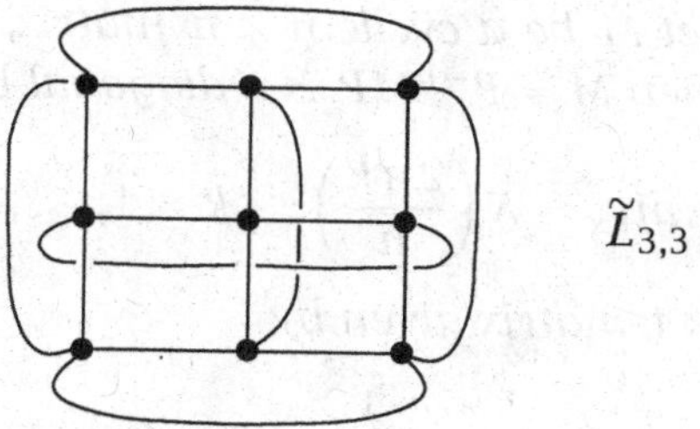

In physics one speaks of a lattice with *periodic boundary*.

Now the state of affairs changes. Owing to the "cyclic" structure of the matrix $\widetilde{A}$ of the corresponding terminal graph $\widetilde{L}^t_{m,n}$, we will be able to compute the determinant. Of course, $\widetilde{L}^t_{m,n}$ is no longer planar, even without crossing diagonals, but the result of Theorem 10.10 can still be used in the limit on the rationale that the contribution of the wrap-around edges is negligible as m and n tend to infinity.

Cyclic Matrices.
Let $M = (m_{k\ell})$ be a $n \times n$-block matrix whose entries $m_{k\ell}$ are $t \times t$-matrices. We call M *cyclic* if there are $t \times t$-matrices $a(0), a(1), \ldots, a(n-1)$ such that $m_{k\ell} = a(\ell - k)$, where $\ell - k$ is taken modulo n.

A cyclic 3×3-matrix looks therefore as follows:

$$M = \begin{pmatrix} a(0) & a(1) & a(2) \\ a(2) & a(0) & a(1) \\ a(1) & a(2) & a(0) \end{pmatrix}.$$

Cyclic matrices can be easily diagonalized, as we now show. From now on, i is not an index but $\sqrt{-1}$. Consider the $n \times n$-block matrix $P = (p_{k\ell})$ whose entries are

$$p_{k\ell} = \frac{1}{\sqrt{n}} \exp\left(k\ell \frac{2\pi i}{n}\right) I_t, \tag{15}$$

where I_t is the $t \times t$-identity matrix. Then P is a unitary matrix, that is, $P^{-1} = \overline{P}^T$, where $\overline{x}$ is the conjugation over $\mathbb{C}$. Just note that

$$\sum_{j=1}^{n} p_{kj}\overline{p_{\ell j}} = I_t \frac{1}{n} \sum_{j=1}^{n} \exp\left(j(k-\ell)\frac{2\pi i}{n}\right) = I_t \delta_{k\ell},$$

since $\sum_{j=1}^{n} \exp(jb\frac{2\pi i}{n}) = 0$ for any integer $b \neq 0$.

Lemma 10.11. *Let M be a cyclic $n \times n$ matrix, with $a(k)$ and P as defined above. Then $\widetilde{M} = P^{-1}MP$ is a diagonal block matrix, with*

$$\widetilde{m}_{kk} = \lambda\left(\frac{2\pi k}{n}\right), \quad k = 1, \ldots, n, \tag{16}$$

where λ is the $t \times t$-matrix given by

$$\lambda(\varphi) = \sum_{j=0}^{n-1} a(j) \exp(ji\varphi). \tag{17}$$

In particular,

$$\det M = \det \widetilde{M} = \prod_{k=1}^{n} \det \lambda\left(\frac{2\pi k}{n}\right).$$

Proof. We compute

$$\begin{aligned}
\widetilde{m}_{k\ell} &= \sum_{r,s=1}^{n} \overline{p_{kr}} m_{rs} p_{s\ell} \\
&= \frac{1}{n} I_t \sum_{r=1}^{n} \exp\left(-kr\frac{\pi i}{n}\right) \sum_{s=1}^{n} a(s-r) \exp\left(s\ell\frac{2\pi i}{n}\right) \\
&= \frac{1}{n} \sum_{r=1}^{n} \exp\left((\ell-k)r\frac{2\pi i}{n}\right) \cdot \sum_{s=1}^{n} a(s-r) \exp\left((s-r)\ell\frac{2\pi i}{n}\right) \\
&= \delta_{k,\ell} \cdot \sum_{j=0}^{n-1} a(j) \exp\left(j\ell\frac{2\pi i}{n}\right) = \delta_{k,\ell} \cdot \lambda\left(\frac{2\pi \ell}{n}\right). \qquad \square
\end{aligned}$$

Now, if the matrices $a(k)$ are themselves cyclic block matrices, then we repeat this process. In general, we have the following result, whose proof is identical to the one just given.

Definition. The matrix $M = (m_{\alpha\beta})$ whose entries are $t \times t$-matrices is called a *general cyclic matrix* if the rows and columns can be indexed by vectors $\alpha = (\alpha_1, \dots, \alpha_d)$, $\alpha_i = 1, 2, \dots, n_i$, such that $m_{\alpha\beta} = a(\beta - \alpha)$, where $\beta_k - \alpha_k$ is taken componentwise mod n_k, $k = 1, \dots, d$, $0 \le \beta_k - \alpha_k \le n_k - 1$.

Proposition 10.12. *Let $M = (m_{\alpha\beta})$ be a general cyclic matrix. Then*

$$\det M = \prod_{\alpha} \det \lambda \left(\frac{2\pi\alpha_1}{n_1}, \dots, \frac{2\pi\alpha_d}{n_d} \right), \tag{18}$$

where λ is the $t \times t$-matrix given by

$$\lambda(\varphi_1, \dots, \varphi_d) = \sum_{\alpha} a(\alpha) \exp(i\alpha \cdot \varphi), \tag{19}$$

and $i\alpha \cdot \varphi = i \sum_{j=1}^{d} \alpha_j \varphi_j$.

Solution of the Ising Problem.
Now let us go to work to compute $\det \widetilde{A}$ for the square terminal graph $\widetilde{L}^t_{n,n}$ with the orientation and weighting given as in Theorem 10.10. Let us verify that $\widetilde{A}$ is a generalized cyclic matrix with $t = 4$, $d = 2$. We number the K_4's from bottom up, and from left to right. Choosing the suggestive notation R (right), L (left), U (up), D (down),

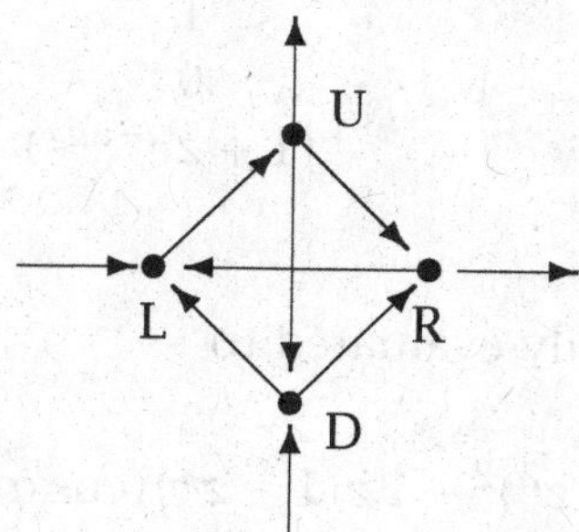

we have

$$a(0,0) = \begin{array}{c} \\ R \\ L \\ U \\ D \end{array} \begin{array}{c} \begin{array}{cccc} R & L & U & D \end{array} \\ \begin{pmatrix} 0 & 1 & -1 & -1 \\ -1 & 0 & 1 & -1 \\ 1 & -1 & 0 & 1 \\ 1 & 1 & -1 & 0 \end{pmatrix} \end{array}.$$

Passing to the right or left, we see from the figure (since R is joined to L, and nothing else) that

$$a(1,0) = \begin{matrix} & R & L & U & D \\ R & 0 & z & 0 & 0 \\ L & 0 & 0 & 0 & 0 \\ U & 0 & 0 & 0 & 0 \\ D & 0 & 0 & 0 & 0 \end{matrix}\,, \quad a(-1,0) = \begin{matrix} & R & L & U & D \\ R & 0 & 0 & 0 & 0 \\ L & -z & 0 & 0 & 0 \\ U & 0 & 0 & 0 & 0 \\ D & 0 & 0 & 0 & 0 \end{matrix}\,.$$

Similarly, moving in vertical direction we obtain

$$a(0,1) = \begin{matrix} & R & L & U & D \\ R & 0 & 0 & 0 & 0 \\ L & 0 & 0 & 0 & 0 \\ U & 0 & 0 & 0 & z \\ D & 0 & 0 & 0 & 0 \end{matrix}\,, \quad a(0,-1) = \begin{matrix} & R & L & U & D \\ R & 0 & 0 & 0 & 0 \\ L & 0 & 0 & 0 & 0 \\ U & 0 & 0 & 0 & 0 \\ D & 0 & 0 & -z & 0 \end{matrix}\,.$$

All other local 4×4-matrices are zero. Since $\widetilde{L}_{n,n}^t$ is wrapped around at the ends it is easily seen that $\widetilde{A}$ is indeed cyclic. We conclude that $\lambda(\varphi_1,\varphi_2) = a(0,0) + a(1,0)e^{i\varphi_1} + a(-1,0)e^{-i\varphi_1} + a(0,1)e^{i\varphi_2} + a(0,-1)e^{-i\varphi_2} =$

$$\begin{pmatrix} 0 & 1+ze^{i\varphi_1} & -1 & -1 \\ -(1+ze^{-i\varphi_1}) & 0 & 1 & -1 \\ 1 & -1 & 0 & 1+ze^{i\varphi_2} \\ 1 & 1 & -(1+ze^{-i\varphi_2}) & 0 \end{pmatrix}.$$

The determinant is readily evaluated to

$$\det\lambda(\varphi_1,\varphi_2) = (1+z^2)^2 - 2z(1-z^2)(\cos\varphi_1 + \cos\varphi_2)\,, \qquad (20)$$

and we obtain from (18) the expression

$$\det\widetilde{A} = \prod_{k=1}^{n}\prod_{\ell=1}^{n}\left[(1+z^2)^2 - 2z(1-z^2)\left(\cos\frac{2\pi k}{n} + \cos\frac{2\pi\ell}{n}\right)\right]. \qquad (21)$$

Combining Theorems 10.9, 10.10 and (21), and noting that $|\widetilde{E}| = 2n^2$ (since $\widetilde{L}_{n,n}$ is 4-regular), we thus obtain for large n,

$$Z(\widetilde{L}_{n,n}; K) \approx 2^{n^2} \left[\prod_{k=1}^{n} \prod_{\ell=1}^{n} \cosh(K)^4 \right. \\ \left. \times \left((1+z^2)^2 - 2z(1-z^2)(\cos\frac{2\pi k}{n} + \cos\frac{2\pi \ell}{n}) \right) \right]^{\frac{1}{2}} \tag{22}$$

evaluated for $z = \tanh(K)$. It is easily seen that

$$(1+\tanh^2(K))^2 = \frac{\cosh(2K)^2}{\cosh(K)^4},$$

$$2\tanh(K)(1-\tanh^2(K)) = 2\frac{\sinh(K)}{\cosh^3(K)}.$$

Multiplying by $\cosh(K)^4$, this gives $\cosh(2K)^2, 2\sinh(K)\cosh(K) = \sinh(2K)$, and therefore

$$Z(\widetilde{L}_{n,n}; K) \approx \\ 2^{n^2} \left[\prod_{k=1}^{n} \prod_{\ell=1}^{n} \left(\cosh(2K)^2 - \sinh(2K) \left(\cos\frac{2\pi k}{n} + \cos\frac{2\pi \ell}{n} \right) \right) \right]^{\frac{1}{2}}.$$

Now we consider $\frac{\log Z}{n^2}$ and let n go to infinity. As in the dimer problem, the right-hand side becomes a double integral with bounds 0 and 2π, and we have finally proved the celebrated solution of the 2-dimensional Ising problem, first derived by Onsager.

Theorem 10.13 (Onsager). *We have*

$$\lim_{n\to\infty} \frac{\log(Z(\widetilde{L}_{n,n}; K)}{n^2} = \log 2 + \frac{1}{2(2\pi)^2} \int_0^{2\pi}\int_0^{2\pi} Q(K, \varphi_1, \varphi_2) d\varphi_1\, d\varphi_2 \,,$$

where

$$Q = \log[\cosh(2K)^2 - \sinh(2K)(\cos\varphi_1 + \cos\varphi_2)] \,.$$

Exercises

10.15 For a graph $G = (V, E)$ prove: a. $|\mathcal{B}(G)| = 2^{|V|-k(G)}$, b. $|\mathcal{E}(G)| = 2^{|E|-|V|+k(G)}$.

▷ **10.16** Prove directly that $\mathcal{B}(G; -1) = |\mathcal{B}(G)|$ if G is Eulerian, and 0 otherwise, and similarly $\mathcal{E}(G; -1) = |\mathcal{E}(G)|$ if G is bipartite, and 0 otherwise.

10.17 Deduce from the previous exercises that $T(G; -1, 0) = (-1)^{|V|-k(G)}$ if G is bipartite, $T(G; 0, -1) = (-1)^{|E|-|V|+k(G)}$ if G is Eulerian.

10.18 Verify assertions (b) and (c) in Proposition 10.8.

10.19 Express $Z(G; K)$ in terms of $\mathcal{E}(G; z)$ (Theorem 10.9c).

▷ **10.20** Compute the partition function $Z(C_n; K)$ for the circuit C_n, $K > 0$. What do you get for $\frac{\log Z}{n}$ as $n \to \infty$?

* * *

▷ **10.21** Consider the ladder graph $L_{2,n}$. All Eulerian subgraphs are unions of circuits, and hence have even size. Let $\mathcal{F}_n(z) = \mathcal{E}(L_{2,n}; \sqrt{z}) = \sum_{k \ge 0} a_k z^k$, $\mathcal{F}_0(z) = 1$. Show that the generating function $H(x) = \sum_{n \ge 0} \mathcal{F}_n(z) x^n$ is given by $\frac{1-xz}{1-(1+z)x+(z-z^2)x^2}$. Deduce $\mathcal{F}_n(1) = 2^{n-1}$, $\mathcal{F}_n(-1) = 2^{\lfloor \frac{n}{2} \rfloor}$, $\mathcal{F}_n(-2) = \frac{1}{5}(2^{n+2} + (-3)^n)$. Show further that $a_2 = n - 1$, $a_3 = n - 2$, and $a_n = F_{n-1}$ (Fibonacci).

10.22 In the *Potts model* on $G = (V, E)$ every vertex can assume $q \ge 2$ spins, and the partition function is $Z_P(G; K) = \sum_\sigma e^{K \sum_{\{i,j\} \in E} \delta(\sigma_i, \sigma_j)}$ over all $q^{|V|}$ states, $\delta(\sigma_i, \sigma_j)$ = Kronecker delta. Let $B(G; \lambda, s)$ be the monochromial introduced in Exercise 9.10. Prove $Z_P(G; K) = B(G; q, e^K)$.

10.23 Show that $Z_P(G; K) = q^{k(G)}(e^K - 1)^{|V|-k(G)} T\left(G; \frac{e^K+q-1}{e^K-1}, e^K\right)$, $T(G)$ Tutte polynomial.

10.24 Set $K = -\infty$ in the Potts model. What do you get for $Z(G; -\infty)$?

▷ **10.25** Show that $Z_P(G; 2K) = e^{K|E|} Z_{\text{Ising}}(G; K)$ for $q = 2$.

10.26 Express the Tutte polynomial in terms of the monochromial $B(G; \lambda, s)$.

▷ **10.27** Suppose $G = (V, E, F)$ is a plane connected graph, $\widetilde{G}$ the medial graph, and $S(\widetilde{G}; W_{\alpha,\beta}, \lambda)$ the crossing-free transition polynomial as in Section 9.3. Prove $Z_P(G; K) = q^{1-\frac{|F|}{2}} S(\widetilde{G}, W_{\sqrt{q}, e^K - 1}; \sqrt{q})$.
Hint: Express x, y in the Tutte polynomial of Exercise 10.23 in the form $1 + \frac{\alpha}{\beta}\lambda$, $1 + \frac{\beta}{\alpha}\lambda$.

10.28 Generalize Theorem 10.13 as follows. Assign weight z_1 to the horizontal edges of $\widetilde{L}_{n,n}$, and z_2 to the vertical edges, with K_1, K_2 the corresponding parameters replacing K. Prove that $\lim_{n\to\infty} \frac{\log Z}{n^2} = \log 2 + \frac{1}{2(2\pi)^2} \int_0^{2\pi} \int_0^{2\pi} Q\, d\varphi_1\, d\varphi_2$, where $Q = \log[\cosh(2K_1)\cosh(2K_2) - \sinh(2K_1) \cdot \cos\varphi_1 - \sinh(2K_2)\cos\varphi_2]$.

10.3 Hard Models

In statistical physics *hard models* describe a collection of particles on a lattice with the restriction that particles must not be on adjacent sites. Interpreted in terms of graphs we have the following problem: Given the simple graph $G = (V, E)$, determine the generating function $\mathcal{U}(G; z) = \sum_{k\geq 0} u_k z^k$, where u_k is the number of independent sets of size k (including the empty set). Physicists usually call the variable z the *activity*.

Of particular interest are the cases $z = 1$ and $z = -1$, enumerating the total number of independent sets, and the difference between the number of even-sized and odd-sized sets. Given $G = (V, E)$ and a vertex v we may partition the independent sets U according to whether U does or does not contain v. Denoting by $N(v)$ the neighborhood of v (including v), we thus have the following recurrence.

Lemma 10.14. *Consider $G = (V, E)$ and $v \in V$. Then*

$$\mathcal{U}(G; z) = \mathcal{U}(G \setminus v; z) + z\mathcal{U}(G \setminus N(v); z). \tag{1}$$

Example. For the path P_n with n vertices the independent sets correspond to subsets of $\{1, \ldots, n\}$ without consecutive entries. We have computed these numbers in Exercise 1.8 and obtain

$$\mathcal{U}(P_n; z) = \sum_{k\geq 0} \binom{n-k+1}{k} z^k, \tag{2}$$

and in particular,

$$\mathcal{U}(P_n; 1) = F_{n+2} \quad \text{(Fibonacci number)}. \tag{3}$$

Picking v as an end vertex of P_n, Lemma 10.14 implies

$$\mathcal{U}(P_n; -1) = \mathcal{U}(P_{n-1}; -1) - \mathcal{U}(P_{n-2}; -1) = -\mathcal{U}(P_{n-3}; -1).$$

With $\mathcal{U}(P_0;-1) = 1$ (by definition), $\mathcal{U}(P_1;-1) = 0$, $\mathcal{U}(P_2;-1) = -1$, this gives

$$\mathcal{U}(P_n;-1) = \begin{cases} 1 & n \equiv 0,5 \pmod 6, \\ 0 & n \equiv 1,4 \pmod 6, \\ -1 & n \equiv 2,3 \pmod 6. \end{cases}$$

Application of (1) to the circuit C_n yields

$$\begin{aligned} \mathcal{U}(C_n;z) &= \mathcal{U}(P_{n-1};z) + z\mathcal{U}(P_{n-3};z) \\ &= \sum_{k\geq 0} \left[\binom{n-k}{k} + \binom{n-k-1}{k-1} \right] z^k , \end{aligned}$$

and in particular,

$$\mathcal{U}(C_n;1) = F_{n+1} + F_{n-1} = L_n \,. \tag{4}$$

The number L_n is called the n-th *Lucas number*. The first Lucas numbers are $(L_0 = 1, 1, 3, 4, 7, 11, 18, \ldots)$.

The Lattice Graph.
As before, we concentrate on the rectangular lattice graph $L_{m,n}$. It is too much to expect that a closed formula for $\mathcal{U}(L_{m,n};z)$ can be obtained, or even an expression for the total number $\mathcal{U}(L_{m,n};1)$ of independent sets. Some exact results for small fixed m are contained in the exercises. But we can say something about the asymptotic growth.

Since $L_{m,n}$ is a bipartite graph, any subset of either color class is independent, which gives

$$2^{\frac{mn}{2}} \leq \mathcal{U}(L_{m,n};1) \leq 2^{mn} ,$$

or

$$\sqrt{2} \leq \liminf_{m,n\to\infty} \mathcal{U}(L_{m,n};1)^{\frac{1}{mn}} \leq \limsup_{m,n\to\infty} \mathcal{U}(L_{m,n};1)^{\frac{1}{mn}} \leq 2 .$$

The main question we want to tackle is therefore whether $\lim_{m,n\to\infty} \mathcal{U}(L_{m,n};1)^{\frac{1}{mn}}$ exists, and what its value ξ is, where $\sqrt{2} \leq \xi \leq 2$.

For notational reasons it is convenient to consider the function

$$f(m,n) = \mathcal{U}(L_{m+1,n+1};1) ,$$

where, of course, $\lim_{m,n\to\infty} f(m,n)^{\frac{1}{mn}} = \xi$ if it exists. Note that $f(m,n) = f(n,m)$.

Example. For $m = 2$, $n = 1$, $f(2,1) = 17$ is easily computed:

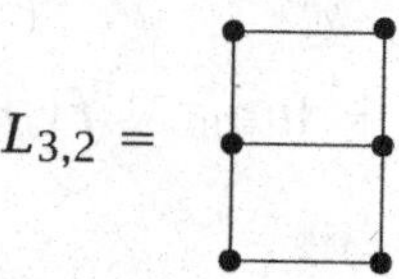

Let us fix m. Then clearly

$$f(m, n_1 + n_2) \le f(m, n_1) f(m, n_2), \tag{5}$$

since any independent set in L_{m+1,n_1+n_2+1} gives rise to a pair of such sets on L_{m+1,n_1+1} and L_{m+1,n_2+1}, as in the figure:

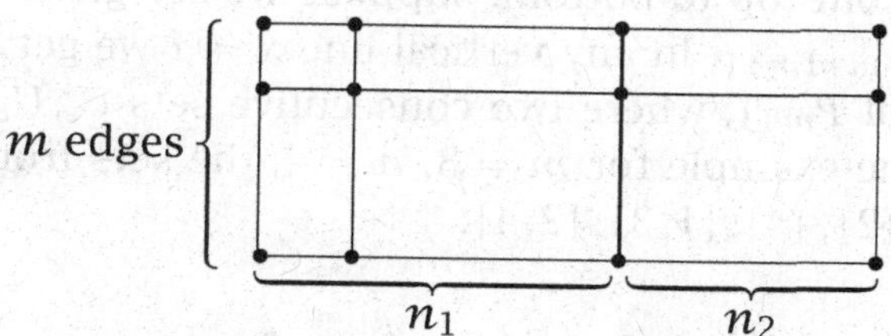

The following result, called the lemma of Fekete, comes in handy.

Lemma 10.15. *Suppose the function $g : \mathbb{N} \to \mathbb{R}$ satisfies $g(r+s) \le g(r)g(s)$ for all r and s. Then $\lim_{n\to\infty} g(n)^{\frac{1}{n}}$ exists, and it is, in fact, equal to* $\inf g(n)^{\frac{1}{n}}$.

Proof. Fix r and ℓ with $\ell \le r$. The inequality implies by induction $g(\ell + kr) \le g(\ell) g(r)^k$, and thus

$$\limsup_{k\to\infty} g(\ell + kr)^{\frac{1}{\ell + kr}} \le g(r)^{\frac{1}{r}}.$$

Since any n can be written in the form $\ell + kr$, this gives

$$\limsup_{n\to\infty} g(n)^{\frac{1}{n}} \le g(r)^{\frac{1}{r}} \text{ for all } r. \tag{6}$$

Letting r go to infinity, we obtain

$$\limsup_{n\to\infty} g(n)^{\frac{1}{n}} \le \liminf_{r\to\infty} g(r)^{\frac{1}{r}},$$

and so $\lim_{n\to\infty} g(n)^{\frac{1}{n}}$ exists. The last assertion follows now immediately from (6). □

If we set $g(n) = f(m,n)$, then the lemma and (5) imply the following:

Corollary 10.16. *The limit $\Theta_m = \lim_{n\to\infty} f(m,n)^{\frac{1}{n}}$ exists for every fixed m.*

Example. For $m = 0$, $L_{1,n+1} = P_{n+1}$ and $f(0,n) = F_{n+3}$ according to (3), which gives $\Theta_0 = \lim_{n\to\infty} F_{n+3}^{\frac{1}{n}} = \tau = \frac{1+\sqrt{5}}{2}$, the golden ratio.

The Transfer Matrix.

The following idea, well known to physicists, provides an alternative method to compute Θ_m. We make use of the symmetric structure of the lattice graph. Number the vertices of P_{m+1} in each column $1, 2, \ldots, m+1$ from top to bottom. Suppose we are given an independent set of $L_{m+1,n+1}$. In any vertical line $x = i$ we get an independent set U_i of P_{m+1}, where two consecutive sets U_i, U_{i+1} must be disjoint. In the example for $m = 3$, $n = 4$, the sets from left to right are $\{1,3\}, \{2\}, \{4\}, \{1,3\}, \{2,4\}$:

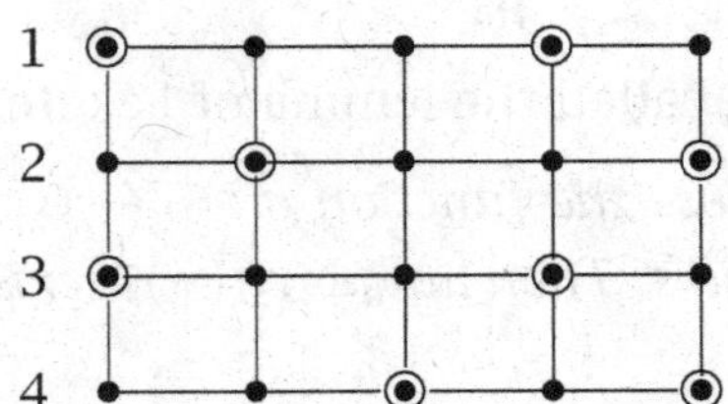

Now we associate to $L_{m+1,n+1}$ the following graph $G_m = (V_m, E_m)$. The vertices are the F_{m+3} independent sets of P_{m+1}, and we join two such sets U_i, U_j if and only if they are disjoint. A moment's thought shows that the independent sets of $L_{m+1,n+1}$ correspond bijectively to *walks* of length n in G_m, where a walk from u to v of length n in any graph is a sequence $u = v_0, v_1, \ldots, v_n = v$ with $\{v_i, v_{i+1}\} \in E$, for all i.

An easy result of graph theory (see Exercise 10.34) states that for any graph G with adjacency matrix A, we have

$$A^n(u,v) = \#\ \textit{walks of length } n \textit{ from } u \textit{ to } v.$$

Accordingly, we consider the adjacency matrix $T_m = (t_{ij})$ of G_m, called the *transfer matrix*. Number the F_{m+3} independent sets of P_{m+1} in any way $U_1, \ldots, U_{F_{m+3}}$; then

$$t_{ij} = \begin{cases} 1 & \text{if } U_i \cap U_j = \emptyset\,, \\ 0 & \text{if } U_i \cap U_j \neq \emptyset\,. \end{cases}$$

In our case this means that $f(m,n)$ is the sum of *all* entries of T_m^n, that is,

$$f(m,n) = \mathbf{1}^T T_m^n \mathbf{1}\,, \tag{7}$$

where $\mathbf{1}$ is the all ones vector of length F_{m+3}.

Example. For $m = 2$, $P_3 = \begin{matrix} 1 \bullet \\ 2 \bullet \\ 3 \bullet \end{matrix}$; let $U_1 = \emptyset$, $U_2 = \{1\}$, $U_3 = \{2\}$, $U_4 = \{3\}$, $U_5 = \{1,3\}$. Then

$$T_2 = \begin{pmatrix} 1 & 1 & 1 & 1 & 1 \\ 1 & 0 & 1 & 1 & 0 \\ 1 & 1 & 0 & 1 & 1 \\ 1 & 1 & 1 & 0 & 0 \\ 1 & 0 & 1 & 0 & 0 \end{pmatrix},$$

and $\mathbf{1}^T T_2 \mathbf{1} = 17$, agreeing with $f(2,1) = 17$ as noted above.

Now we turn to linear algebra. The matrix T_m is real and symmetric and thus diagonizable. Note that the diagonal consists of a 1 (corresponding to $\emptyset$), and 0's otherwise; hence the trace is $\operatorname{tr}(T_m) = 1$. Denote by Λ_m the largest eigenvalue of T_m, $\Lambda_m = \lambda_1 \geq \lambda_2 \geq \cdots \geq \lambda_{F_{m+3}}$. Because of $\operatorname{tr}(T_m) = \sum_{i \geq 1} \lambda_i = 1$, we have $\Lambda_m > 0$, and the theorem of Perron–Frobenius states further that $|\lambda_i| \leq \Lambda_m$ for all i.

Let us finally quote two other easy theorems from linear algebra. Suppose A is a real symmetric matrix with eigenvalues $\Lambda = \lambda_1 \geq \lambda_2 \geq \cdots \geq \lambda_t$. Then the eigenvalues of A^n are $\lambda_1^n, \lambda_2^n, \ldots, \lambda_t^n$. Furthermore,

$$x^T A x \leq \Lambda(x^T x) \quad \text{for all vectors } x\,. \tag{8}$$

With these preparations we can prove the following result.

Proposition 10.17. *For all* m,

$$\Theta_m = \lim_{n \to \infty} f(m,n)^{\frac{1}{n}} = \Lambda_m\,. \tag{9}$$

Proof. Since $|\lambda| \le \Lambda_m$ for any eigenvalue λ of T_m, $\Lambda_m^n > 0$ is the largest eigenvalue of T_m^n. Now by (8),

$$f(m,n) = \mathbf{1}^T T_m^n \mathbf{1} \le \Lambda_m^n(\mathbf{1}^T\mathbf{1}) = \Lambda_m^n F_{m+3},$$

and hence

$$f(m,n)^{\frac{1}{n}} \le \Lambda_m F_{m+3}^{\frac{1}{n}}.$$

Letting $n \to \infty$, this implies $\Theta_m \le \Lambda_m$.

In the other direction we obtain for $p \ge 0$,

$$\Lambda_m^{2p} \le \operatorname{tr}(T_m^{2p}) \le \mathbf{1}^T T_m^{2p}\mathbf{1} = f(m,2p), \tag{10}$$

since $\operatorname{tr}(T_m^{2p}) = \sum \lambda_i^{2p}$, and $\lambda_i^{2p} \ge 0$ for all i. We conclude that $\Lambda_m \le f(m,2p)^{\frac{1}{2p}}$, and thus with $p \to \infty$, $\Lambda_m \le \Theta_m$. □

Remark. Lemma 10.15 implies $\Lambda_m = \inf f(m,n)^{\frac{1}{n}}$ and $\Lambda_m \le f(m,n)^{\frac{1}{n}}$ for all n.

Example. Let us again look at the smallest case $m = 0$. Here $T_0 = \begin{pmatrix} 1 & 1 \\ 1 & 0 \end{pmatrix}$, $\det(xI - T_0) = x^2 - x - 1$ with roots $\frac{1\pm\sqrt{5}}{2}$. Hence $\Theta_0 = \Lambda_0 = \tau$ as noted above.

The proposition immediately implies

$$\begin{aligned} \liminf_{m\to\infty} \Lambda_m^{\frac{1}{m}} &= \lim_{m,n\to\infty} \inf f(m,n)^{\frac{1}{mn}} \le \lim_{m,n\to\infty} \sup f(m,n)^{\frac{1}{mn}} \\ &= \limsup_{m\to\infty} \Lambda_m^{\frac{1}{m}}. \end{aligned} \tag{11}$$

Theorem 10.18. *The limit* $\lim\limits_{m,n\to\infty} f(m,n)^{\frac{1}{mn}}$ *exists, and we have*

$$\xi = \lim_{m,n\to\infty} f(m,n)^{\frac{1}{mn}} = \lim_{m\to\infty} \Lambda_m^{\frac{1}{m}}. \tag{12}$$

Moreover, $\xi = \inf\limits_{m\to\infty} \Lambda_m^{\frac{1}{m}}$.

Proof. According to (11) it remains to show that $\lim_{m\to\infty} \Lambda_m^{\frac{1}{m}}$ exists, but this follows immediately from the lemma of Fekete. Indeed, $f(r+s,n) \le f(r,n)f(s,n)$ implies

$$f(r+s,n)^{\frac{1}{n}} \le f(r,n)^{\frac{1}{n}} f(s,n)^{\frac{1}{n}},$$

and with $n \to \infty$,

$$\Lambda_{r+s} \le \Lambda_r \Lambda_s .$$

Lemma 10.15 finishes the proof. □

Bounds for the Limit.

We know that $\xi = \inf \Lambda_m^{\frac{1}{m}}$, but the behavior of the sequence $(\Lambda_m^{\frac{1}{m}})$ is not clear. It may be monotonically decreasing, but it could also have a minimum. The exercises contain a few results.

We already know the trivial lower bound $\xi \ge \sqrt{2} = 1.414$. Using (8) this can be improved in the following way. Let $p, q \in \mathbb{N}$. Then (recall $f(m,n) = f(n,m)$)

$$\Lambda_m^p (T_m^q \mathbf{1})^T (T_m^q \mathbf{1}) \ge (T_m^q \mathbf{1})^T T_m^p (T_m^q \mathbf{1}),$$

or

$$\Lambda_m^p (\mathbf{1}^T T_m^{2q} \mathbf{1}) = \Lambda_m^p f(2q, m) \ge \mathbf{1}^T T_m^{p+2q} \mathbf{1} = f(p+2q, m).$$

Taking the m-th root and letting $m \to \infty$ yields

$$\xi^p \Lambda_{2q} \ge \Lambda_{p+2q},$$

that is,

$$\xi \ge \left(\frac{\Lambda_{p+2q}}{\Lambda_{2q}} \right)^{\frac{1}{p}} \quad \text{for } p, q \in \mathbb{N}. \tag{13}$$

In particular, for $p = 1$ we get

$$\xi \ge \frac{\Lambda_{2q+1}}{\Lambda_{2q}}. \tag{14}$$

Example. For $p = 1$, $q = 0$ we find that Λ_1 is the largest root of $x^3 - x^2 - 3x - 1 = 0$ with $2.41 < \Lambda_1 < 2.42$. We know that $\Lambda_0 = \tau = \frac{1+\sqrt{5}}{2}$, and so

$$\xi \ge \frac{\Lambda_1}{\Lambda_0} = 1.492066.$$

Of course, we expect that larger values of q might give better bounds. For $q = 4$ one computes

$$\xi \ge \frac{\Lambda_9}{\Lambda_8} = 1.50304808, \tag{15}$$

and it may be that $\frac{\Lambda_{2q+1}}{\Lambda_{2q}}$ is an increasing subsequence yielding successively better bounds. You are asked in the exercises to show that $\frac{\Lambda_{2q+1}}{\Lambda_{2q}} \le \frac{\Lambda_{2q+2}}{\Lambda_{2q+1}}$, and it is conjectured that $\xi \le \frac{\Lambda_{2q+2}}{\Lambda_{2q+1}}$ for all $q \ge 0$. If this is true, then $\xi = 1.50304808$ up to 8 decimals since $\frac{\Lambda_8}{\Lambda_7} = 1.50304808$.

For an upper bound we look at (10),

$$\Lambda_m^{2p} \le \operatorname{tr}(T_m^{2p}).$$

The trace of T_m^{2p} counts the number of all *closed* walks of length $2p$ in G_m from left to right. The figure shows the situation for $m = 3$, $2p = 6$:

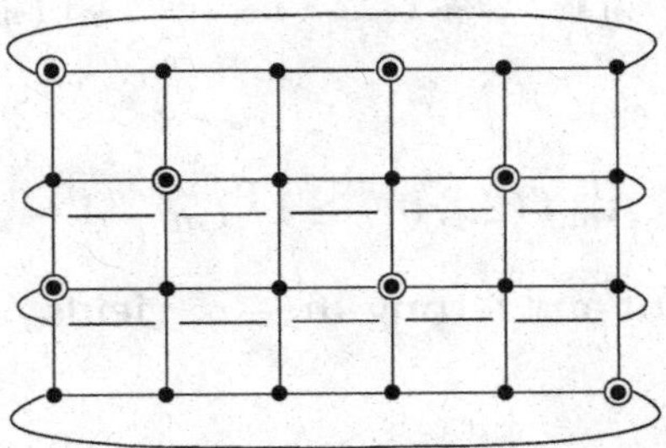

But we can also look from top to bottom. The rows give independent sets of the *circuit* C_{2p}, with consecutive sets being disjoint. We therefore define the graph H_{2p} whose vertices are the independent sets of C_{2p}, and edges corresponding to disjoint pairs. Note that $|V(H_{2p})|$ is the Lucas number L_{2p}.

We conclude that the number of closed walks of length $2p$ in G_m equals the number of walks of length m in H_{2p}. Denoting by S_{2p} the adjacency matrix of H_{2p} we thus have

$$\operatorname{tr}(T_m^{2p}) = \mathbf{1}^T S_{2p}^m \mathbf{1}. \tag{16}$$

By the same argument as before,

$$\lim_{m\to\infty} (\mathbf{1}^T S_{2p}^m \mathbf{1})^{\frac{1}{m}} = \Gamma_{2p},$$

where Γ_{2p} is the largest eigenvalue of S_{2p}. By (10) and (16),

$$\Lambda_m \le (\mathbf{1}^T S_{2p}^m \mathbf{1})^{\frac{1}{2p}},$$

and taking the m-th root and letting $m \to \infty$,

$$\xi \le \Gamma_{2p}^{\frac{1}{2p}} . \tag{17}$$

Example. For $p = 1$, the graph H_2 is a 2-circuit and thus has the same independent sets as $G_1 = K_2$, that is, $\Gamma_2 = \Lambda_1 < 2.42$, which gives

$$\xi < \sqrt{2.42} < 1.556 .$$

Again we expect that larger values of p will give better bounds. For example, for $p = 3$ one obtains

$$\xi < 1.503514809 ,$$

which agrees with the lower bound in (15) in the first three decimals.

Exercises

▷ **10.29** Show for the Fibonacci numbers that $F_{m+k} = F_{k+1}F_m + F_kF_{m-1}$ for $m, k \ge 0$. Deduce for the Lucas numbers that $L_nF_n = F_{2n}$, and derive $L_n = \tau^n + \hat{\tau}^n$ $(n \ge 1)$.

10.30 Let G and H be disjoint graphs, and denote by $G * H$ the graph with $V(G * H) = V(G) \cup V(H)$, $E(G * H) = E(G) \cup E(H) \cup \{\{u, v\} : u \in V(G), v \in V(H)\}$. An example is the wheel $W_n = K_1 * C_n$. Prove $\mathcal{U}(G * H; z) = \mathcal{U}(G; z) + \mathcal{U}(H; z) - 1$.

10.31 Use the previous exercise to compute $\mathcal{U}(K_n; z)$, $\mathcal{U}(K_{n_1,\ldots,n_t}; z)$, and $\mathcal{U}(W_n; z)$.

10.32 Compute $\mathcal{U}(K_{n_1,\ldots,n_t}; -1)$ and $\mathcal{U}(C_n; -1)$.

▷ **10.33** Let T be a tree on n vertices. Prove the bounds $F_{n+2} \le \mathcal{U}(T; 1) \le 2^{n-1} + 1$, and determine the trees that achieve equality in the bounds.

10.34 Let A be the adjacency matrix of the graph G. Show that $A^n(i, j)$ equals the number of walks of length n from i to j in G.

10.35 The n-dimensional cube Q_n has $\{0, 1\}^n$ as vertex set with $u = (u_i)$ and $v = (v_i)$ adjacent if they differ in only one coordinate. Compute $\mathcal{U}(Q_3; z)$ and $\mathcal{U}(Q_4; z)$.

* * *

10.36 Let G be a graph and $\mathcal{I} = \{U_1, \ldots, U_t\}$ the family of independent sets. Define the transfer matrix $T_G = (t_{ij})$ as usual, $t_{ij} = 1$ if $U_i \cap U_j = \emptyset$, $t_{ij} = 0$ otherwise. Prove $\det T_G = (-1)^\ell$, $\ell = \#$ odd-sized sets in $\mathcal{I}$. Hint: Exercise 5.47.

▷ **10.37** Consider $L_{m,n}$, $\mathcal{I}$ the family of independent sets of P_m, $|\mathcal{I}| = F_{m+2}$. Let $D_m(z)$ be the diagonal matrix with $D_m(z)_{U,U} = z^{|U|}$, and $T_m(z)$ the matrix with $T_m(z)_{U,V} = z^{|V|}$ for $U \cap V = \emptyset$, $= 0$ otherwise. Prove $\mathcal{U}(L_{m,n}; z) = \mathbf{1}^T D_m(z) T_m(z)^{n-1} \mathbf{1}$.

10.38 Calculate the generating function of $\mathcal{U}(L_{2,n}; z)$ using the approach of Section 3.1 and evaluate $\mathcal{U}(L_{2,n}; 1)$ and $\mathcal{U}(L_{2,n}; -1)$.

10.39 Consider the matrix $T_m(-1)$ as in Exercise 10.37. Show that $\det T_m(-1) = 1$, and that 1 is an eigenvalue of $T_m(-1)$ for $m \geq 2$.

▷ **10.40** Let the eigenvalues Λ_k be defined as in the text. Prove $\Lambda_{k+\ell}^2 \leq \Lambda_{2k}\Lambda_{2\ell}$ for $k, \ell \geq 0$, and deduce $\frac{\Lambda_{2n+1}}{\Lambda_{2n}} \leq \frac{\Lambda_{2n+2}}{\Lambda_{2n+1}}, \frac{\Lambda_{2n}}{\Lambda_{2n-2}} \leq \frac{\Lambda_{2n+2}}{\Lambda_{2n}}$. Hint: Use the Cauchy–Schwarz inequality from linear algebra.

10.41 Prove $\Lambda_k^{\frac{1}{k}} \leq \Lambda_{2q}^{\frac{1}{2q}}$ for $k > 2q$.

▷ **10.42** Consider the periodic lattice $\widetilde{L}_{m,n}$ as in the section on the Ising problem, and let $\widetilde{T}_m$ be the transfer matrix of C_m. Show that $\mathcal{U}(\widetilde{L}_{m,n}; 1) = \operatorname{tr}(\widetilde{T}_m^n)$ and compute the generating function of $\mathcal{U}(\widetilde{L}_{2,n}; z)$. Evaluate $U(\widetilde{L}_{2,n}; 1)$ and $\mathcal{U}(\widetilde{L}_{2,n}; -1)$.

10.43 We have considered the half-periodic case in the text. Let $g(m,n)$ be the number of independent sets in $L'_{m,n}$ where the columns are periodic (that is, C_m). Does $\lim_{m,n\to\infty} g(m,n)^{\frac{1}{mn}}$ exist? Is it equal to ξ?

10.44 Now let $h(m,n) = \mathcal{U}(\widetilde{L}_{m,n}; 1)$, and answer the same questions for $\lim_{m,n\to\infty} h(m,n)^{\frac{1}{mn}}$.

10.4 Square Ice

We close this chapter and the book with one of the most spectacular successes of enumerative combinatorics, the exact solution of the square ice model due to Zeilberger and Kuperberg. In the general case, an ice model concerns the number of ways of orienting a 4-regular graph G such that G becomes a 2-in 2-out graph. As before, we consider the $n \times n$-lattice graph, which is 4-regular if we add periodic boundary conditions as we did before. For the usual graph $L_{n,n}$ we thus have to proceed a little differently.

The Problem.
Square ice consists of an $n \times n$-lattice arrangement of oxygen atoms. Between any two adjacent O-atoms lies one hydrogen atom, and there are also H-atoms at the left and right boundaries. The task is to count all possible configurations in which every O-atom is attached to exactly two of its surrounding H-atoms, forming H_2O. The figure shows a possible configuration for $n = 5$, where $\circ$ refers to the O-atoms and simple dots to H-atoms.

Example.

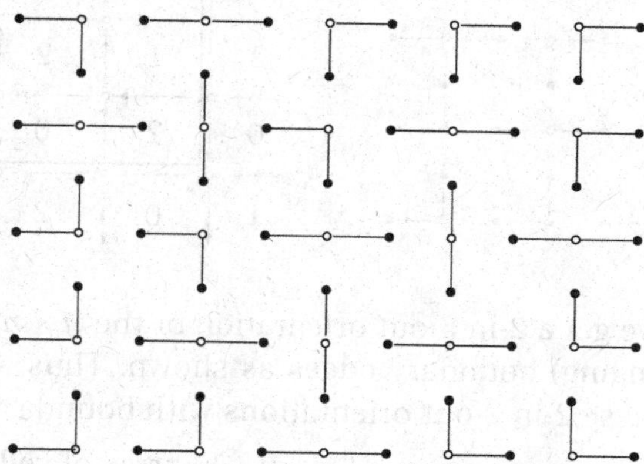

Let A_n be the number of all $n \times n$-square ice configurations. As first values we have $A_1 = 1$, $A_2 = 2$ as shown in the figure:

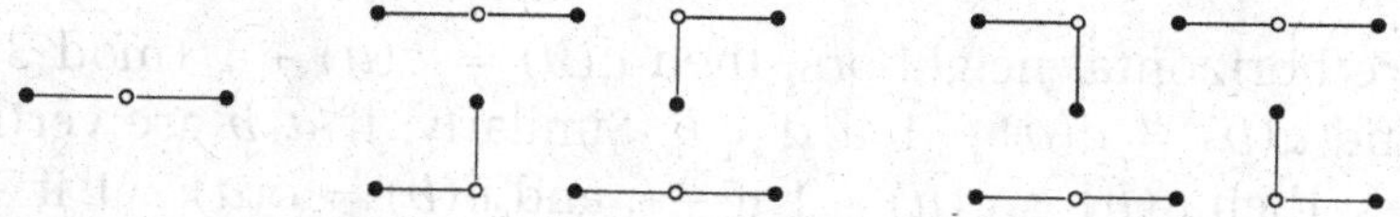

The table shows the next values:

n	1	2	3	4	5	6	7
A_n	1	2	7	42	429	7436	218348

The determination of A_n for arbitrary n combines several beautiful ideas into a glorious finish, masterly explained by David Bressoud in his book *Proofs and Confirmations*, whose exposition we follow closely.

Connection to 2-in 2-out Graphs and 3-Colorings.
There is an obvious bijection between $n \times n$-ice configurations and 2-in 2-out graphs on the lattice graph of the O-atoms, with boundary conditions. Let C be a configuration, and u, v be two adjacent O-atoms. Orient the edge $u \to v$ if the H-atom between u and v is attached to v. The figure explains the correspondence:

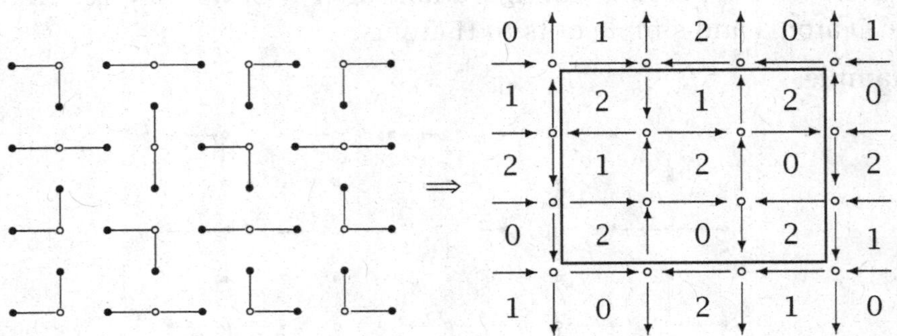

In this way we get a 2-in 2-out orientation of the $n \times n$-lattice graph with the (hanging) boundary edges as shown. Thus A_n equals the number of these 2-in 2-out orientations with boundary conditions.

The graph gives rise to an $(n+1) \times (n+1)$-array of cells, where each O-vertex is surrounded by four cells. Next we color the cells with the additive group $\mathbb{Z}_3 = \{0, 1, 2\}$ in the following way. We color the cell in the upper left-hand corner with 0, and proceed by the following rule. Suppose a, b are neighboring cells, and $c(a), c(b)$ their colors:

If a, b are horizontal neighbors, then $c(b) = c(a) + 1 \pmod 3$ if $a \uparrow b$, and $c(b) = c(a) - 1$ if $a \downarrow b$. Similarly, if a, b are vertical neighbors, then $c(b) = c(a) + 1$ if $\overset{a}{\underset{b}{\longrightarrow}}$, and $c(b) = c(a) - 1$ if $\overset{a}{\underset{b}{\longleftarrow}}$. Observing this rule, we get the 3-coloring of the cells in the above example.

From the setup of the 2-in 2-out graph we see that the boundary cells on top are colored $0, 1, 2, 0, \ldots$ from left to right, and similarly for the other boundaries. Note that the cell in the lower right-hand corner is again colored 0. Altogether we obtain in this way a proper 3-coloring of the $(n+1) \times (n+1)$-array of cells with adjacent cells receiving different colors, observing the boundary conditions. The (easy) proofs of this and the following lemma are left to the exercises.

Lemma 10.19. *The coloring rule gives a bijection between the $n \times n$-ice configurations and the proper 3-colorings of the $(n+1) \times (n+1)$-array of cells, observing the boundary conditions.*

Example. For $n = 3$ we have the outer array as shown,

$$\begin{array}{cccc} 0 & 1 & 2 & 0 \\ 1 & \multicolumn{2}{c}{\boxed{}} & 2 \\ 2 & & & 1 \\ 0 & 2 & 1 & 0 \end{array},$$

which can be extended as follows:

$$\begin{array}{ccccccc} 01 & 01 & 20 & 20 & 21 & 21 & 21 \\ 10 & 12 & 02 & 12 & 02 & 10 & 12 \end{array}.$$

Thus $A_3 = 7$.

Classification of the O-Atoms.

As a first step toward determining A_n let us classify the O-atoms according to the way they are connected to the H-atoms. There are six possibilities, which we denote by H = horizontal, V = vertical, and so on, as shown. It is easily seen that the colors of the four surrounding cells observe the relations as shown in the figure (always mod 3):

$$\begin{array}{lccc} H & \begin{array}{ccc} i & & i+1 \\ & \rightarrow \circ \leftarrow & \\ i+1 & & i \end{array} & NW \begin{array}{ccc} i & & i+1 \\ & \circ \leftarrow & \\ i-1 & \uparrow & i \end{array} & NE \begin{array}{ccc} i & & i+1 \\ & \rightarrow \circ & \\ i+1 & \uparrow & i-1 \end{array} \\[2ex] V & \begin{array}{ccc} i & \downarrow & i-1 \\ & \circ & \\ i-1 & \uparrow & i \end{array} & SE \begin{array}{ccc} i & \downarrow & i-1 \\ & \rightarrow \circ & \\ i+1 & & i \end{array} & SW \begin{array}{ccc} i & \downarrow & i-1 \\ & \circ \leftarrow & \\ i-1 & & i+1 \end{array} \end{array} \qquad (1)$$

Suppose we are given an ice configuration C. Then we observe the following two facts:

A. In every row, the number of H-vertices is one more than the number of V-vertices. Thus altogether,

$$\#H = \#V + n\,. \tag{2}$$

For the proof look at a row of the lattice graph, and consider the difference $c(a) - c(b)$ of the colors of the cells a above and b below. From (1) we find that the difference changes from -1 to $+1$ when we pass an H-vertex, from $+1$ to -1 at a V-vertex, and stays the same otherwise. Since it is -1 at the left boundary and $+1$ at the right boundary, we must have $\#H = \#V + 1$.

B. We have

$$\#NE = \#SW \quad \text{and} \quad \#NW = \#SE\,. \tag{3}$$

The 2-in 2-out representation implies that the total number of edges directed to the right equals the total number of edges directed to the left. Looking at (1), this means that $\#NW + \#SW = \#NE + \#SE$. Considering the vertical edges, it follows similarly that $\#NE + \#NW = \#SE + \#SW$. Subtracting we get $\#NE - \#SW = \#SW - \#NE$; hence $\#NE = \#SW$, and analogously $\#NW = \#SE$.

Finding the Right Weights.
Now we assign weights $w(t)$ to every type $H, V, \ldots, SW$, and set as usual $w(C) = \prod_{t \in C} w(t)$ for an ice configuration C. The partition function is then given by

$$Z_n(w) = \sum_C w(C)\,.$$

It was Baxter's great discovery that the following weights will do the job. Let a be a number $\neq 0, 1, -1$, and set for any $s \neq 0$, $[s] = \frac{s - s^{-1}}{a - a^{-1}}$. We define the weights as follows:

$$\begin{aligned} w(H) &= z, \quad w(V) = z^{-1}, \\ w(NW) &= w(SE) = [z] = \frac{z - z^{-1}}{a - a^{-1}}, \\ w(NE) &= w(SW) = [az] = \frac{az - (az)^{-1}}{a - a^{-1}}\,. \end{aligned} \tag{4}$$

For the corresponding partition function we write $Z_n(z; a)$.

Let us first try $z = a$. Here $w(H) = z$, $w(V) = z^{-1}$, $w(NW) = w(SE) = 1$, $w(NE) = w(SW) = \frac{z^2-z^{-2}}{z-z^{-1}} = z + z^{-1}$, and we conclude by (2) and (3) that

$$\begin{aligned} Z_n(z;z) &= \sum_C z^{\#H}(z^{-1})^{\#V}(z+z^{-1})^{\#NE+\#SW} \\ &= \sum_C z^n(z+z^{-1})^m, \end{aligned}$$

where m is an even integer. In order to make the second factor equal to 1, we consider $z + z^{-1} = -1$, that is, $z^2 + z + 1 = 0$. A solution is the third root of unity $\omega = \frac{-1+\sqrt{-3}}{2}$. In sum, we obtain for $z = a = \omega$,

$$A_n = \frac{1}{\omega^n} Z_n(\omega;\omega), \tag{5}$$

and the problem reduces to finding an expression for $Z_n(\omega;\omega)$.

The Triangle-to-Triangle Relation.

The key to determining $Z_n(z;a)$ lies in specifying the weights even further. Assign to the O-atom in the i-th row and j-th column the variable $z \mapsto \frac{x_i}{y_j}$. Hence (4) becomes

$$\begin{aligned} w(H) &= \frac{x_i}{y_j}, \quad w(V) = \frac{y_j}{x_i}, \\ w(NW/SE) &= \left[\frac{x_i}{y_j}\right], \quad w(NE/SW) = \left[\frac{ax_i}{y_j}\right]. \end{aligned} \tag{6}$$

Theorem 10.20 (Baxter). *The partition function $Z_n(x_1,\ldots,x_n, y_1,\ldots,y_n;a)$ is symmetric in the variables x_i and the y_j.*

Proof. Consider rows i and $i+1$ of the $n \times n$-grid of O-atoms. We insert an additional triangle at the left boundary as in the figure and color the cells appropriately. By the boundary condition the only possible color for the new cell is i.

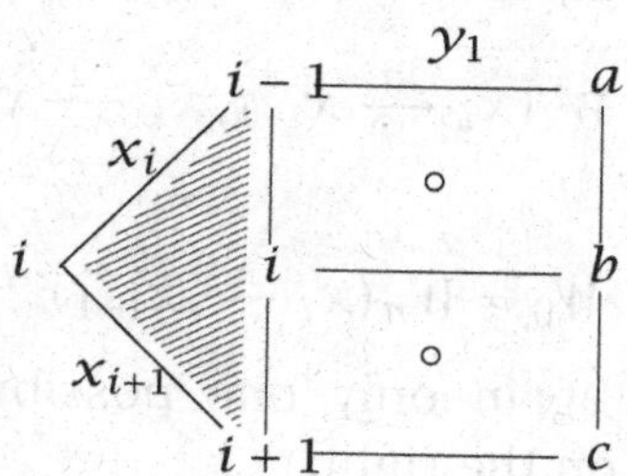

The triangle gets weight $z = \frac{x_{i+1}}{x_i}$, and its type is determined by rotating it counterclockwise by 45°,

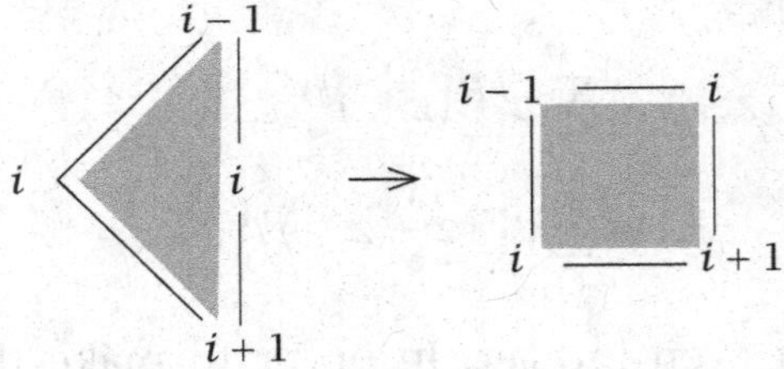

Looking at (4), the weight is $\left[\frac{ax_{i+1}}{x_i}\right]$.

Let W_0 be the partition function of this enlarged lattice; thus

$$W_0 = \left[\frac{ax_{i+1}}{x_i}\right] Z_n .$$

Here is the crucial step, which we shall justify in a moment. Push the triangle across column 1 to the right. Then we claim that

$$w\left(\begin{array}{ccc} i-1 & & a \\ & \frac{x_i}{y_1} & \\ i \;\; \frac{x_{i+1}}{x_i} \;\; i & & b \\ & \frac{x_{i+1}}{y_1} & \\ i+1 & & c \end{array}\right) = w\left(\begin{array}{ccc} i-1 & & a \\ & \frac{x_{i+1}}{y_1} & \\ i & & x \;\; \frac{x_{i+1}}{x_i} \;\; b \\ & \frac{x_i}{y_1} & \\ i+1 & & c \end{array}\right)$$

where the right-hand side is summed over all possible colors x. Note that in the regular cells x_i and x_{i+1} are interchanged. Thus if W_1 is the partition function with the triangle in the first column, then

$$W_0 = W_1(x_i \overset{y_1}{\longleftrightarrow} x_{i+1}) ,$$

meaning that in all expressions involving regular cells and y_1, the variables x_i and x_{i+1} are interchanged. Next we push the triangle into the second column, and obtain $W_1 = W_2(x_i \overset{y_2}{\longleftrightarrow} x_{i+1})$, that is,

$$W_0 = W_2(x_i \overset{y_1}{\longleftrightarrow} x_{i+1}, x_i \overset{y_2}{\longleftrightarrow} x_{i+1}) ,$$

and finally,

$$W_0 = W_n(x_i \overset{y}{\longleftrightarrow} x_{i+1}) .$$

But in W_n we have again only one possibility, considering the boundary conditions on the right;

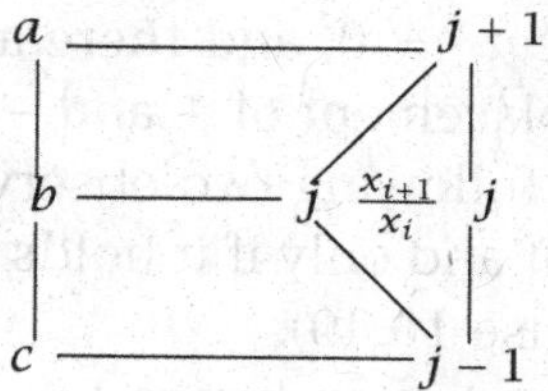

and we obtain $W_n = \left[\frac{ax_{i+1}}{x_i}\right] Z_n$. But this gives

$$W_0 = W_n = W_0(x_i \overset{y}{\longleftrightarrow} x_{i+1}),$$

and therefore $Z_n = Z_n(x_i \longleftrightarrow x_{i+1})$. Thus Z_n is invariant under all transpositions $x_i \longleftrightarrow x_{i+1}$, and therefore symmetric in the variables x_i. The proof for the y_j's proceeds analogously, pushing the triangle down.

It remains to prove the following lemma. Set $z_0 = \frac{x_{i+1}}{x_i}$, $z_1 = \frac{x_i}{y_j}$, $z_2 = \frac{x_{i+1}}{y_j}$; thus $z_0 z_1 = z_2$.

Lemma 10.21. *Let $z_0 z_1 = z_2$, and suppose $a_0, a_1, \ldots, a_5$ is an admissible 3-coloring. Then*

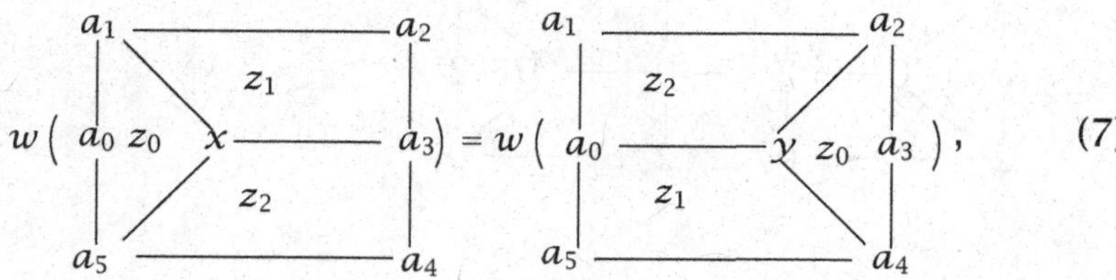

where we sum over all admissible colorings x and y.

Proof. Let us collect a few facts about the a_i's. Go around clockwise from a_0, and write + if $a_{i+1} - a_i = 1$, and − if $a_{i+1} - a_i = -1$. Either we have three + and three −, or all signs are the same, but the second case cannot happen (why?). Next we observe that we may add a fixed m to all a_i (mod 3), and the situation remains the same, since the possible colorings and the types do not change.

Hence we may assume $a_0 = 0$, and there are $\binom{6}{3} = 20$ possibilities corresponding to the placement of + and −. We reduce this number further by making the following two observations:

A. Equality (7) holds if and only if it holds when the configurations are turned 180° (Exercise 10.49).

B. If we keep all 0's, and interchange $1 \leftrightarrow 2$, then the types $H \leftrightarrow V$ are interchanged, and the other types remain the same (according to (1)). Hence making the replacements $z_i \mapsto z_i^{-1}$, $a \mapsto a^{-1}$, we get $z_0^{-1} z_1^{-1} = z_2^{-1}$, and the weights stay the same. In other words, (7) holds if and only if it holds for the configuration with 1 and 2 interchanged.

We may thus assume $a_0 = 0$, $a_1 = 1$, with 10 cases remaining. Using a combination of (A) and (B) and another symmetry (see Exercise 10.51), we finally arrive at the following five cases:

(1)
1 0 1 0
z_1 z_2
0 z_0 0 1 = 0 1/2 z_0 1
z_2 z_1
2 0 2 0

(2)
1 0 1 0
0 0 2 = 0 1/2 2
2 0 2 0

(3)
1 0 1 0
0 0 2 = 0 2 2
2 1 2 1

(4)
1 2 1 2
0 0 1 = 0 1 1
2 0 2 0

(5)

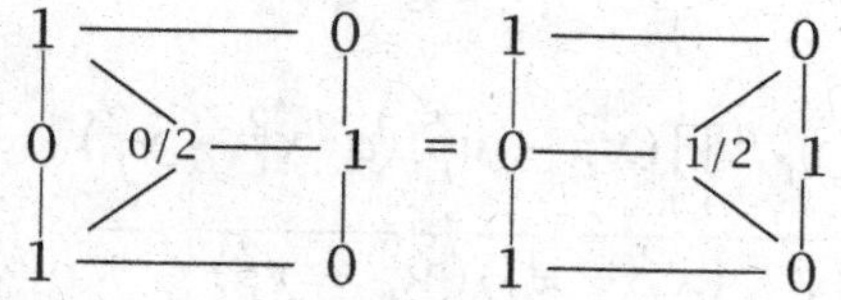

1 — 0   1 — 0
0   0/2 — 1   =   0 — 1/2   1
1 — 0   1 — 0

Looking at the types, we have to verify the following five equalities:

1. $[az_0]z_1^{-1}[z_2] = z_0[z_1]z_2^{-1} + [z_0]z_1^{-1}[az_2]$;
2. $[az_0][az_1]z_2^{-1} = [z_0][z_1]z_2^{-1} + z_0^{-1}z_1^{-1}[az_2]$;
3. $[az_0][az_1][az_2] = [az_0][az_1][az_2]$;
4. $[az_0][z_1][z_2] = [az_0][z_1][z_2]$;
5. $z_0^{-1}z_1^{-1}z_2 + [z_0][z_1][az_2] = z_0z_1z_2^{-1} + [z_0][z_1][az_2]$.

Equalities (3), (4), and (5) are obviously true (recall that $z_0z_1 = z_2$). Multiplying (1) by z_1 and (2) by z_2, we get

$$[az_0][z_2] = [z_1] + [z_0][az_2], \quad [az_0][az_1] = [z_0][z_1] + [az_2],$$

and both equalities are immediately verified. □

The Determinant Formula.
We know that $Z_n(x_1,\dots,x_n,y_1,\dots,y_n;a)$ is symmetric in the x_i's and y_j's. So we expect that Z_n might be expressible via a determinant, and this is indeed the case, as the following result due to Izergin shows.

Theorem 10.22 (Izergin). *We have*

$$Z_n(x,y;a) = \frac{\prod\limits_{i=1}^{n} \frac{x_i}{y_i} \prod\limits_{i,j=1}^{n} [\frac{x_i}{y_j}][\frac{ax_i}{y_j}]}{\prod\limits_{1\le i<j\le n} [\frac{x_i}{x_j}][\frac{y_j}{y_i}]} \det\left(\frac{1}{[\frac{x_i}{y_j}][\frac{ax_i}{y_j}]}\right)_{i,j=1}^{n} . \tag{8}$$

Proof. Let $F_n(x,y;a)$ be the right-hand side. Using the definition $[\frac{x_i}{y_j}] = \frac{x_i/y_j - y_j/x_i}{a-a^{-1}} = \frac{(x_i^2-y_j^2)a}{x_iy_j(a^2-1)}$, and similarly for the other [ ]-expressions, we obtain (check it!)

$$F_n(x,y;a) =$$

$$\frac{1}{(a^2-1)^{n(n-1)}} \frac{\prod_{i=1}^{n} x_i^{2-n} y_i^{-n} \prod_{i,j}(x_i^2-y_j^2)(a^2x_i^2-y_j^2)}{\prod_{1\le i<j\le n} (x_i^2-x_j^2)(y_j^2-y_i^2)}$$

$$\times \det\left(\frac{1}{(x_i^2-y_j^2)(a^2x_i^2-y_j^2)}\right)_{i,j=1}^{n}.$$

The product

$$\prod_{i,j}(x_i^2-y_j^2)(a^2x_i^2-y_j^2)\det\left(\frac{1}{(x_i^2-y_j^2)(a^2x_i^2-y_j^2)}\right)$$

is an alternating polynomial in the x_i^2 and also the y_j^2. Hence dividing this product by $\prod_{1\le i<j\le n}(x_i^2-x_j^2)(y_j^2-y_i^2)$ gives a *symmetric* polynomial in x_i^2 and y_j^2. To get rid of the factor x_1^{2-n} we multiply F_n by x_1^{n-2}, and conclude that

$$Q_n = x_1^{n-2}F_n(x,y;a) \tag{9}$$

is a polynomial of degree $n-1$ in x_1^2 that is symmetric in the y_j's.

Now consider $Z_n(x,y;a)$. The first row contains exactly one O-vertex of type H with weight $\frac{x_1}{y_{j_0}}$ for some j_0. The other $n-1$ vertices are of type NW or NE, that is,

$$\left[\begin{matrix}x_1\\ y_j\end{matrix}\right] = \frac{a}{x_1y_j}\frac{x_1^2-y_j^2}{a^2-1}, \quad \left[\begin{matrix}ax_1\\ y_j\end{matrix}\right] = \frac{a^2x_1^2-y_j^2}{x_1y_j(a^2-1)}.$$

Multiplying the weights, we see that x_1 appears as x_1^{2-n}, and x_1^2 to the power $n-1$. Hence

$$P_n = x_1^{n-2}Z_n(x,y;a) \tag{10}$$

is a polynomial in x_1^2 of degree $n-1$, which is symmetric in the y_j's according to Theorem 10.20.

Since P_n, Q_n are polynomials of degree $n-1$ in x_1^2, it remains to show that they agree at n points. We claim that

$$P_n\left(x_1=\frac{y_k}{a}\right) = Q_n\left(x_1=\frac{y_k}{a}\right) \quad \text{for } k=1,\dots,n,$$

which will prove the theorem. Let us check this for $k = 1$, that is, for $x_1 = \frac{y_1}{a}$. The remaining cases follow then easily by the symmetry of the y_j's.

Let $x_1 = \frac{y_1}{a}$. If the vertex in the upper left-hand corner has type NE, then the weight is $[\frac{ax_1}{y_1}] = [1] = \frac{1-1}{a-a^{-1}} = 0$. So we may assume it has type H with weight $\frac{x_1}{y_1} = \frac{1}{a}$. In this case, the other vertices in the top row are of type NW, and the vertices in column 1 of type SE as shown in the figure:

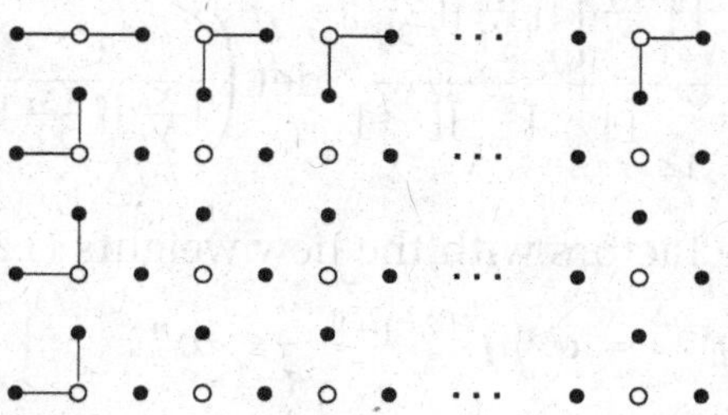

The remaining array is then counted by $Z_{n-1}(x, y; a)$, and we obtain from (10), with $x_1 = \frac{y_1}{a}$,

$$P_n(\frac{y_1}{a}) = \frac{y_1^{n-2}}{a^{n-1}} \prod_{j \neq 1} \left[\frac{y_1}{ay_j}\right]\left[\frac{x_j}{y_1}\right] Z_{n-1}(x_2, \dots, x_n, y_2, \dots, y_n; a).$$

You are asked in the exercises to check that similarly,

$$Q_n(\frac{y_1}{a}) = \frac{y_1^{n-2}}{a^{n-1}} \prod_{j \neq 1} \left[\frac{y_1}{ay_j}\right]\left[\frac{x_j}{y_1}\right] F_{n-1}(x_2, \dots, x_n, y_2, \dots, y_n; a), \tag{11}$$

and the theorem follows by induction. □

Recall that $A_n = \frac{1}{\omega^n} Z_n(\omega; \omega)$. The simplest idea would be to try $x_i = \omega$, $y_j = 1$, $a = \omega$, which gives $\frac{x_i}{y_j} = \omega$ for all i, j, and use Theorem 10.22. Unfortunately, this will not work, since in the expression (8) we have $[\frac{x_i}{x_j}] = [1] = 0$ in the denominator. But the following wonderful idea due to Kuperberg will indeed succeed. Let q be a variable, and set

$$x_i = \omega q^{\frac{i-1}{2}}, \quad y_j = q^{-\frac{j}{2}}, \quad a = \omega, \tag{12}$$

and denote by $Z_{n,q}$ the corresponding partition function. Then

$$A_n = \frac{1}{\omega^n} \lim_{q\to 1} Z_{n,q}, \tag{13}$$

and we are faced with the task to determine $Z_{n,q}$.

Computing $Z_{n,q}$.
According to Theorem 10.22,

$$Z_{n,q} = \frac{\prod_{i=1}^{n} \frac{x_i}{y_i} \prod_{i,j} [\frac{x_i}{y_j}][\frac{ax_i}{y_j}]}{\prod_{1\le i<j\le n} [\frac{x_i}{x_j}][\frac{y_j}{y_i}]} \det\left(\frac{1}{[\frac{x_i}{y_j}][\frac{ax_i}{y_j}]}\right).$$

Let us compute the factors with the new weights (12):

1. $\prod_{i=1}^{n} \frac{x_i}{y_i} = \omega^n \prod_{i=1}^{n} q^{i-\frac{1}{2}} = \omega^n q^{\binom{n+1}{2}-\frac{n}{2}} \underset{q\to 1}{\longrightarrow} \omega^n$.
2. $\prod_{i,j}[\frac{x_i}{y_j}][\frac{ax_i}{y_j}] = \prod_{i,j}[\omega q^{\frac{i+j-1}{2}}][\omega^2 q^{\frac{i+j-1}{2}}]$. Using $\omega + \omega^2 = -1$, you can easily check that as $q \to 1$ this goes to

$$\prod_{i,j}(1 + q^{i+j-1} + q^{2(i+j-1)}) \frac{\omega^{2n^2}}{(\omega^2-1)^{2n^2}} \underset{q\to 1}{\longrightarrow} \frac{\omega^{2n^2}}{(\omega^2-1)^{2n^2}} \cdot 3^{n^2}.$$

3. A similar computation for the denominator gives

$$\prod_{1\le i<j\le n} [\frac{x_i}{x_j}][\frac{y_j}{y_i}]$$

$$= \prod_{1\le i<j\le n} [q^{\frac{i-j}{2}}][q^{\frac{i-j}{2}}] \underset{q\to 1}{\longrightarrow} \prod_{i<j}(1-q^{j-i})^2 \frac{\omega^{n(n-1)}}{(\omega^2-1)^{n(n-1)}}.$$

4. For the determinant we get

$$\det\left(\frac{1}{[\frac{x_i}{y_j}][\frac{ax_i}{y_j}]}\right) \underset{q\to 1}{\longrightarrow} \det\left(\frac{1}{1+q^{i+j-1}+q^{2(i+j-1)}}\right) \cdot \frac{(\omega^2-1)^{2n}}{\omega^{2n}}.$$

Next we get rid of the ω-factor. Collecting all factors in (1) to (4) we obtain, using $\omega^2 = \omega^{-1}$,

$$\omega^n \frac{1}{(1+2\omega)^{n^2-n}} = \omega^n \frac{1}{(-3)^{\binom{n}{2}}},$$

since $\omega = \frac{-1+\sqrt{-3}}{2}$, and hence $1 + 2\omega = \sqrt{-3}$.

Altogether we have arrived at the following expression (note that the ω^n-factor cancels out),

$$A_n = (-1)^{\binom{n}{2}} 3^{n^2-\binom{n}{2}} \lim_{q\to 1} \frac{\det\left(\frac{1-q^{i+j-1}}{1-q^{3(i+j-1)}}\right)}{\prod\limits_{1\le i<j\le n} (1-q^{j-i})^2}, \tag{14}$$

and we turn to the determinant.

Evaluation of the Determinant.

To compute the determinant in (14), Exercise 8.8 comes to our help. With $t = q^3$ we obtain

$$q^{n^3-\frac{3}{2}n^2+\frac{n}{2}} \prod_{1\le i<j\le n} (1-q^{3(j-i)})^2 \prod_{i,j=1}^{n} \frac{1-q^{3(j-i)+1}}{1-q^{3(i+j-1)}}. \tag{15}$$

Look at the numerator of the second product in (15). For $i = j$ we get $1 - q$, and for $i > j$ we multiply $1 - q^{3(j-i)+1}$ by $-q^{3(i-j)-1}/ - q^{3(i-j)-1}$, which gives $-\frac{1-q^{3(i-j)-1}}{q^{3(i-j)-1}}$. As $q \to 1$ the expression (15) therefore goes to

$$(-1)^{\binom{n}{2}} (1-q)^n \frac{\prod\limits_{i<j} (1-q^{3(j-i)+1})(1-q^{3(j-i)})^2(1-q^{3(j-i)-1})}{\prod\limits_{i,j} (1-q^{3(i+j-1)})}.$$

Together with (14) this gives for A_n,

$$3^{n^2-\binom{n}{2}} \lim_{q\to 1} (1-q)^n \prod_{i<j} \frac{(1-q^{3(j-i)+1})(1-q^{3(j-i)})^2(1-q^{3(j-i)-1})}{(1-q^{j-i})^2} \times \prod_{i,j=1}^{n} \frac{1}{1-q^{3(i+j-1)}}. \tag{16}$$

Finale.

To go with $q \to 1$ we use, of course,

$$\frac{1-q^m}{1-q} = 1 + q + \cdots + q^{m-1} \underset{q\to 1}{\longrightarrow} m.$$

Hence, when we multiply numerator and denominator in (16) by $(1-q)^{2n(n-1)}$ we get

$$A_n = 3^{n^2-\binom{n}{2}} \prod_{1\le i<j\le n} \frac{(3(j-i)+1)(3(j-i))^2(3(j-i)-1)}{(j-i)^2} \times \prod_{i,j=1}^{n} \frac{1}{3(i+j-1)}$$

$$= \prod_{1\le i<j\le n} \frac{(3(j-i)+1)(3(j-i))(3(j-i)-1)}{j-i} \prod_{i,j=1}^{n} \frac{1}{i+j-1}$$

$$= \prod_{k=1}^{n-1} [\frac{(3k+1)3k(3k-1)}{k}]^{n-k} \prod_{k=1}^{n-1} \frac{1}{k^k} \prod_{k=n}^{2n-1} \frac{1}{k^{2n-k}}$$

$$= \frac{[(3n-2)(3n-3)(3n-4)][(3n-5)(3n-6)(3n-7)]^2 \cdots [4.3.2]^{n-1}}{[(n-1)^n(n-2)^n \cdots 1^n][(2n-1)(2n-2)^2 \cdots n^n]},$$

and out comes the following astounding formula for the number of $n \times n$-ice configurations:

$$A_n = \frac{(3n-2)!(3n-5)! \cdots 4!1!}{(2n-1)!(2n-2)! \cdots (n+1)!n!}.$$

Exercises

10.45 For a 4-regular graph G let ice(G) be the number of 2-in 2-out orientations of G. Compute ice(G) for $G = K_5$, $G = Q_4$ and $G = \text{Oct}$, where Q_4 is the 4-dimensional cube graph and Oct the octahedral graph, which is K_6 minus a perfect matching.

10.46 Let $\overline{C}_n$ be the n-circuit with all edges doubled. Compute ice$(\overline{C}_n)$.

▷ **10.47** Denote by $\widetilde{L}_{2,n}$ the periodic $2 \times n$-lattice graph considered in the last section. Show that ice$(\widetilde{L}_{2,n}) = 3^n + 2^{n+1} + 1$, $n \ge 1$.

10.48 Prove Lemma 10.19.

▷ **10.49** Verify assertion (A) in Lemma 10.21.

10.50 Check the points (2) to (4) in the computation of $Z_{n,q}$, including the ω-factor.

* * *

10.51 Find another symmetry to complete the proof of Lemma 10.21.

10.52 Verify the equality (11) in the proof of Theorem 10.22.

▷ **10.53** Let G be a 4-regular graph on n vertices. Show that $\text{ice}(G) \geq (\frac{3}{2})^n$. Hint: Look at the transitions at a vertex and proceed by induction.

10.54 What is the growth of A_n? Hint: Use the formula for A_n to get an expression for A_n/A_{n-1}.

▷ **10.55** Consider the periodic $n \times n$-lattice graph $\widetilde{L}_{n,n}$. Construct the following $2^n \times 2^n$-transfer matrix $T_n = (t_{ij})$. The rows and columns of T_n correspond to the 2^n ± 1-vectors v_i of length n. Now set $t_{ij} = 2$ if $v_i = v_j$, $t_{ij} = 1$ if $v_i \neq v_j$, and the coordinates k with $v_{ik} \neq v_{jk}$ alternate, meaning that $v_{ik} = 1$, $v_{jk} = -1$ is followed by $v_{i\ell} = -1$, $v_{j\ell} = 1$, and $t_{ij} = 0$ otherwise. Prove that $\text{ice}(\widetilde{L}_{n,n}) = \text{tr}(T_n^n)$.
Hint: Look at the arrows between the first and second columns from top to bottom, and choose $+1$ if the arrow points right, and -1 if it points left. Now consider two adjacent such collections of horizontal arrows.

10.56 Let Λ_n be the maximal eigenvalue of T_n. Deduce as in the previous section that $\lim_{n\to\infty} \text{ice}(\widetilde{L}_{n,n})^{\frac{1}{n^2}}$ exists and equals $\lim_{n\to\infty} \Lambda_n^{\frac{1}{n}}$. Remark: The limit is known to be $(\frac{4}{3})^{3/2} = 1.5396007$, compare this to Exercise 10.53.

Highlight: The Rogers–Ramanujan Identities

Of the many beautiful formulas due to Ramanujan, the Rogers–Ramanujan identities are probably the most famous. What's more, they have the touch of the extraordinary because of the mysterious appearance of the number 5. Here they are:

$$\sum_{k\geq 0}\frac{q^{k^2}}{(1-q)\cdots(1-q^k)}=\prod_{i\geq 1}\frac{1}{(1-q^{5i-1})(1-q^{5i-4})}, \qquad (1)$$

$$\sum_{k\geq 0}\frac{q^{k^2+k}}{(1-q)\cdots(1-q^k)}=\prod_{i\geq 1}\frac{1}{(1-q^{5i-2})(1-q^{5i-3})}. \qquad (1')$$

We noted in Exercise 3.76 that the left-hand side of (1) is the generating function of the partitions of n whose parts differ by at least 2, while the right-hand side counts, of course, the partitions of n with all summands congruent to 1 or 4 (mod 5). Similarly, the left-hand side of (1′) is the generating function of partitions of n whose parts differ at least by 2, and where, in addition, 1 is not a summand, while the right-hand side counts all partitions with all parts congruent to 2 or 3 (mod 5). So, surprisingly, these numbers are equal for all n.

Example. For $n = 11$, we have in (1) the partitions

$$11, 10\,1, 92, 83, 74, 731, 641,$$

respectively

$$11, 911, 641, 611111, 44111, 41111111, 11111111111,$$

and in (1′)

$$11, 92, 83, 74 \quad \text{respectively} \quad 83, 722, 3332, 32222\,.$$

Some of the greatest names have contributed after Rogers (1894) and Ramanujan (1917). Issai Schur (1917) proved a finite version of the identities that we will see later, Garsia and Milne (1981) provided an ingenious bijection proof of the corresponding sets of partitions.

Perhaps the most unexpected route was taken by Rodney Baxter in 1980 when he discovered and re-proved the Rogers–Ramanujan identities in his work on the hard hexagon model. We follow his beautiful argument with some refinements due to J. Cigler.

Hard Hexagon Model.
Consider the usual lattice graph $L_{n,n}$ with diagonals; the figure shows the graph for $n = 5$:

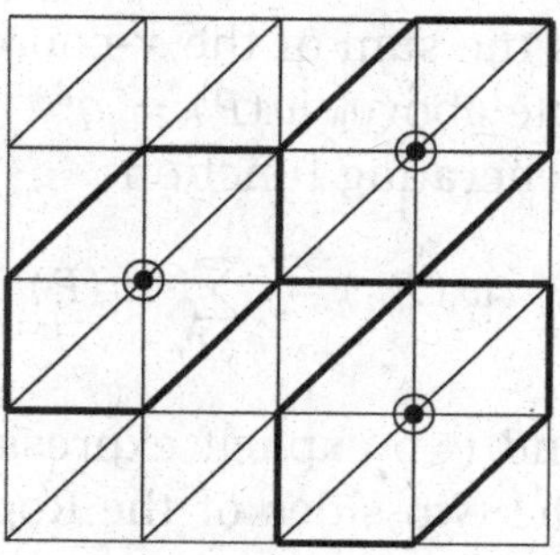

As in Section 10.3 we are interested in the independent sets. The term "hexagon" refers to the fact that the neighborhood of a site forms a hexagon, that is, independent sets are equivalent to non-overlapping hexagons.

On his way to determining the behavior of the generating function for $n \to \infty$, Baxter reduced the 2-dimensional situation to a 1-dimensional problem on lattice paths, and this is the setup we are going to study.

Lattice Paths.
Consider the independent sets U of the (graph) path on $\{1, \ldots, n-1\}$. We may code U by setting $y_i = 1$ for $i \in U$, and $y_i = 0$ when $i \notin U$. In addition, we have the boundary conditions $y_0 = y_n = 0$. Every independent set corresponds in this way to a sequence $(y_0 = 0, y_1, \ldots, y_{n-1}, y_n = 0)$ with no two adjacent 1's. The number of these sequences is, of course, F_{n+1} as noted in Section 10.3.

To every sequence $(y_0, \ldots, y_n)$ corresponds a lattice path by joining (i, y_i) to $(i+1, y_{i+1})$ by a straight line. As an example for $(0, 0, 1, 0, 1, 0, 0, 1, 0, 0)$ we get the path

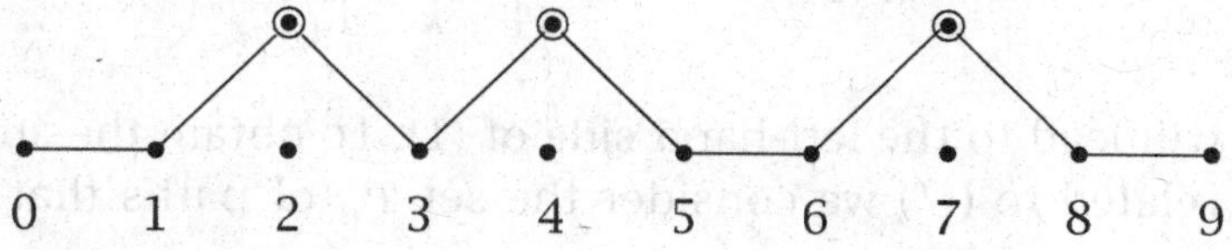

Let $\mathcal{P}_n$ be the set of these lattice paths from $(0,0)$ to $(n,0)$. The points with $y_i = 1$ are called *extremal points.*

Next, we assign to every path $P = (y_0, y_1, \ldots, y_n)$ the weight

$$w(P) = q^{\sum_{i=0}^{n} i y_i}\,. \tag{2}$$

The exponent is thus the sum of the x-coordinates of the extremal points. In the example above, $w(P) = q^{2+4+7} = q^{13}$. What we are interested in is the generating function

$$w(\mathcal{P}_n) = \sum_{P \in \mathcal{P}_n} w(P)\,, \tag{3}$$

and our goal is to find two explicit expressions for $w(\mathcal{P}_n)$, which will correspond to the two sides of the Rogers–Ramanujan identities.

First Evaluation.
Let $\mathcal{P}_{n,k}$ be the set of paths in $\mathcal{P}_n$ with exactly k extremal points. Clearly, the minimum exponent q^m arises for the unique path P_0 with the extremal points at $x = 1, 3, \ldots, 2k-1$; thus $w(P_0) = q^{k^2}$. All other paths in $\mathcal{P}_{n,k}$ are obtained by moving the extremal points to the right. Suppose the i-th extremal point is moved λ_i steps to the right. Notice that

$$0 \le \lambda_1 \le \lambda_2 \le \cdots \le \lambda_k \le n - 1 - (2k-1) = n - 2k\,.$$

We conclude that the paths $P \in \mathcal{P}_{n,k}$ correspond bijectively to the number partitions λ with at most k parts and largest summand less than or equal to $n - 2k$. Furthermore, if P corresponds to λ, then $w(P) = w(P_0)q^{|\lambda|} = q^{k^2 + |\lambda|}$. Invoking Proposition 1.1, this gives

$$w(\mathcal{P}_{n,k}) = \sum_{P \in \mathcal{P}_{n,k}} w(P) = q^{k^2} \begin{bmatrix} n-k \\ k \end{bmatrix}_q\,.$$

Summing over k, we arrive at the first evaluation:

$$w(\mathcal{P}_n) = \sum_{k \ge 0} q^{k^2} \begin{bmatrix} n-k \\ k \end{bmatrix}_q\,, \tag{4}$$

which will lead to the left-hand side of (1). To obtain the analogous result related to (1′) we consider the set $\mathcal{P}'_n$ of paths that start at

$(0,1)$, terminate at $(n,0)$, and otherwise observe the same rules. The extremal points are defined as before, except that the starting point is not considered an extremal point. Consider again the set $\mathcal{P}'_{n,k}$. The unique path $P'_0 \in \mathcal{P}_{n,k}$ of minimal weight has the extremal points at $x = 2,4,\ldots,2k$ with $w(\mathcal{P}'_0) = q^{k^2+k}$, and an analogous argument leads to

$$w(\mathcal{P}'_n) = \sum_{k\geq 0} q^{k^2+k} \begin{bmatrix} n-1-k \\ k \end{bmatrix}_q . \tag{4'}$$

Second Evaluation.

To obtain a second expression we first give a new interpretation of the paths in $\mathcal{P}_n$. Consider the set $\mathcal{Q}_n$ of all lattice paths starting at $(0,0)$ with $\lfloor \frac{n}{2} \rfloor$ *diagonal* up-steps and $\lfloor \frac{n+1}{2} \rfloor$ *diagonal* down-steps, and which stay within the strip $-2 \leq y \leq 1$. Thus every $Q \in \mathcal{Q}_n$ terminates in $(n,0)$ or $(n,-1)$ depending on whether n is even or odd.

The following correspondence maps $\mathcal{Q}_n$ bijectively onto $\mathcal{P}_n$. Replace every diagonal step of Q between $y = 0$ and $y = -1$ by a horizontal step $(y = 0)$, and every *peak* $y = 1$ or *valley* $y = -2$ by an up-down step $y = 1$. In this way, we obtain a path $P \in \mathcal{P}_n$, and the mapping is easily seen to be bijective.

Example. The path $P \in \mathcal{P}_9$ above corresponds to $Q \in \mathcal{Q}_9$;

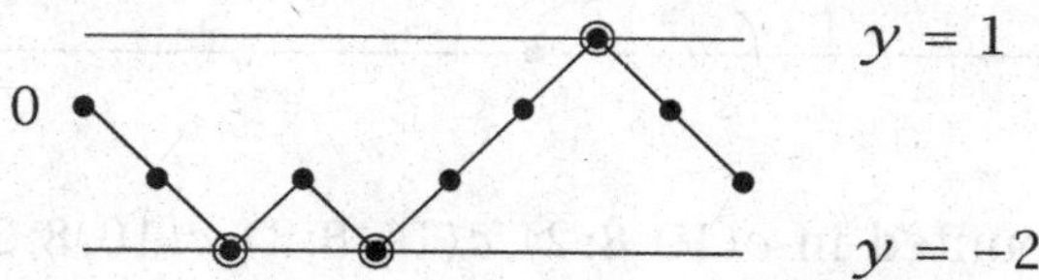

From now on we regard $\mathcal{P}_n$ as the set of these paths with diagonal steps staying within the strip $-2 \leq y \leq 1$. The *extremal points* are then the peaks and valleys as defined above, and the weight of P retains its meaning.

In order to apply induction we consider a more general class of paths. Let $\mathcal{P}(k,\ell)$ be the set of lattice paths that start at $(0,0)$

and have k diagonal up-steps and ℓ diagonal down-steps. Any $P \in \mathcal{P}(k,\ell)$ ends in the point $(k+\ell, k-\ell)$, and $|\mathcal{P}(k,\ell)| = \binom{k+\ell}{k}$. An extremal point of $P \in \mathcal{P}(k,\ell)$ is a peak with $y \geq 1$ or a valley with $y \leq -2$, where the weight $w(P)$ is defined as before.

Let $\mathcal{P}^1_{-2}(k,\ell) \subseteq \mathcal{P}(k,\ell)$ be the subset of paths that stay within the strip $-2 \leq y \leq 1$. In this notation, $\mathcal{P}_n = \mathcal{P}^1_{-2}(\lfloor \frac{n}{2} \rfloor, \lfloor \frac{n+1}{2} \rfloor)$. Our goal is to find an explicit expression for $w(\mathcal{P}^1_{-2}(k,\ell))$, and the method we are going to use is inclusion–exclusion. Consider $\mathcal{P}(k,\ell)$, and let $r_1 r_2 \ldots r_i$ be one of the two possible sequences of length i, alternating between 2 and 3. Thus for $i = 5$, we get the sequences 2 3 2 3 2 and 3 2 3 2 3.

Denote by $c(k,\ell; r_1, \ldots, r_i)$ the number of paths P in $\mathcal{P}(k,\ell)$ that touch (or cross) alternately the bounds $y = 2$ or $y = -3$ as given by $r_1 \ldots r_i$. Note that P may touch or cross more often; what is required is a sequence of i alternating times, prescribed by $r_1, \ldots, r_i$. Hence for $i = 0$, $c(k,\ell) = |\mathcal{P}(k,\ell)| = \binom{k+\ell}{k}$.

Example. $k = 10$, $\ell = 8$

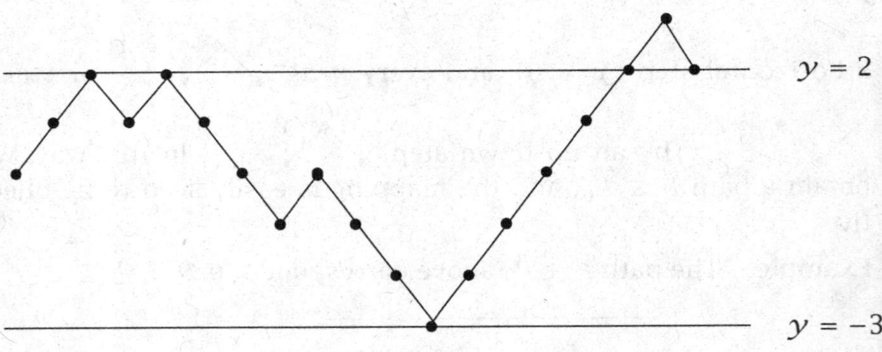

This path is counted in $c(10,8;2), c(10,8;3), c(10,8;2,3), c(10,8;3,2)$, and $c(10,8;2,3,2)$.

Claim 1. *We have*

$$\left|\mathcal{P}^1_{-2}(k,\ell)\right| = \binom{k+\ell}{k} + \sum_{i \geq 1} (-1)^i [c(k,\ell; r_1, \ldots, r_i = 2) + c(k,\ell; r_1, \ldots, r_i = 3)]. \quad (5)$$

If P is in $\mathcal{P}^1_{-2}(k,\ell)$, then P is counted once on either side. Now suppose P has a *maximal* sequence of $m \ge 1$ alternate touches (or crossings).

If $m = 2j$, then P is counted on the right-hand side

$$1 - \underset{i=1}{2} + \underset{i=2}{2} \mp \cdots - \underset{i=2j-1}{2} + 1 = 0$$

times, and if $m = 2j+1$, then it is counted

$$1 - \underset{i=1}{2} + \underset{i=2}{2} \mp \cdots + \underset{i=2j}{2} - 1 = 0$$

times, and the claim follows.

If instead of the cardinalities we use the weights of the sets involved, the inclusion-exclusion formula (5) carries over. Thus if $w(k,\ell;r_1,\ldots,r_i)$ denotes the weight of the paths counted by $c(k,\ell;r_1,\ldots,r_i)$, then

$$\begin{aligned} w(\mathcal{P}^1_{-2}(k,\ell)) &= w(\mathcal{P}(k,\ell)) \\ &+ \sum_{i\ge1}(-1)^i[w(k,\ell;r_1,\ldots,r_i=2) + w(k,\ell;r_1,\ldots,r_i=3)], \end{aligned} \tag{6}$$

and it remains to compute $w(k,\ell;r_1,\ldots,r_i)$ for $i \ge 0$.

Claim 2. *Set* $a_i = \sum_{j=1}^{i} r_j$, $b_i = r_i + 3r_{i-1} + 5r_{i-2} + \cdots + (2i-1)r_1$. *Then the following holds:*

$$\begin{aligned} &\text{a. If } k-\ell \le 1, \text{ then } w(k,\ell;r_1,\ldots,r_i=2) = q^{b_i}\begin{bmatrix} k+\ell \\ k-a_i \end{bmatrix}_q. \\ &\text{b. If } k-\ell \ge -2, \text{ then } w(k,\ell;r_1,\ldots,r_i=3) = q^{b_i}\begin{bmatrix} k+\ell \\ k+a_i \end{bmatrix}_q. \end{aligned} \tag{7}$$

The definitions of a_i and b_i readily imply

$$a_i = a_{i-1} + r_i, \quad b_i = b_{i-1} + a_{i-1} + a_i, \quad a_0 = b_0 = 0. \tag{8}$$

Next we recall the recurrences for the Gaussian coefficients (leaving out the suffix q for ease of notation):

$$\begin{bmatrix} n \\ m \end{bmatrix} = \begin{bmatrix} n-1 \\ m-1 \end{bmatrix} + q^m \begin{bmatrix} n-1 \\ m \end{bmatrix}, \qquad (9)$$
$$\begin{bmatrix} n \\ m \end{bmatrix} = \begin{bmatrix} n-1 \\ m \end{bmatrix} + q^{n-m} \begin{bmatrix} n-1 \\ m-1 \end{bmatrix}.$$

We prove (a) and (b) simultaneously by induction on i, and within i by induction on $k+\ell$. Note first that $w(k,\ell;r_1,\dots,r_i) = 1$ for $k+\ell \le 1$ and any i, since in this case there is only one path (of weight 1). Furthermore, $w(k,\ell;r_1,\dots,r_i = 2) > 0$ implies $k \ge a_i$, since we need at least k up-steps to reach the line $y = 2$ as prescribed by $r_1,\dots,r_i = 2$. Similarly, $w(k,\ell;r_1,\dots,r_i = 3) > 0$ implies $\ell \ge a_i$.

Let us prove (a) (the proof of (b) being analogous). The proof consists of two parts. First we derive a recurrence for $w(k,\ell;r_1,\dots,r_i = 2)$ by looking at the last step, and second we verify that $q^{b_i}[{k+\ell \atop k-a_i}]$ satisfies the recurrence.

To start the induction we look at $i = 0$, that is, we want to show that $w(\mathcal{P}(k,\ell)) = w(k,\ell) = [{k+\ell \atop k}]$, where we may assume $k+\ell \ge 2$ as noted above.

Suppose $k - \ell \ge 0$. Look at the last step. If it goes up, then we obtain $w(k-1,\ell)$. If it goes down, then we have the following two possibilities:

$y = k - \ell$ $\qquad$ $y = k - \ell$

In the first case there is an extremal point at position $k+\ell-1$ accounting for $q^{k+\ell-1}w(k-1,\ell-1)$. In the second case the contribution is $w(k,\ell-1) - w(k-1,\ell-1)$. In sum, we obtain the recurrence

$$w(k,\ell) = w(k-1,\ell) + q^{k+\ell-1}w(k-1,\ell-1) + w(k,\ell-1) - w(k-1,\ell-1).$$

Using induction on $k+\ell$ and repeatedly (9), it is easily shown that

$$\begin{bmatrix} k+\ell-1 \\ k-1 \end{bmatrix} + q^{k+\ell-1} \begin{bmatrix} k+\ell-2 \\ k-1 \end{bmatrix} + \begin{bmatrix} k+\ell-1 \\ k \end{bmatrix} - \begin{bmatrix} k+\ell-2 \\ k-1 \end{bmatrix}$$

equals $[{k+\ell \atop k}]$.

When $k - \ell \le 1$, we get the analogous recurrence with k and ℓ interchanged, so (a) holds for $i = 0$.

Now assume $i \geq 1$. Here we have three cases. We just give the recurrences and consider in detail only the most interesting third case.

Case 1. $k - \ell \leq -1$. Looking at the last step we obtain

$$\begin{aligned} w(k,\ell;r_1,\dots,r_i = 2) = & w(k,\ell-1;r_1,\dots,r_i = 2) \\ & + q^{k+\ell-1} w(k-1,\ell-1;r_1,\dots,r_i = 2) \\ & + w(k-1,\ell;r_1,\dots r_i = 2) \\ & - w(k-1,\ell-1;r_1,\dots,r_i = 2) \end{aligned}$$

with solution $q^{b_i}\left[{k+\ell \atop k-a_i}\right]$.

Case 2. $k - \ell = 0$. We get the same recurrence as in case 1 with k and ℓ interchanged, and thus again the solution $q^{b_i}\left[{k+\ell \atop k-a_i}\right]$, since $k = \ell$.

Case 3. $k-\ell = 1$. If the last step is up, then $w(k-1,\ell;r_1,\dots,r_i = 2)$ results. On the other hand, if it goes down, then the path hits the line $y = 2$ at $k+\ell-1$, which means that these paths are in the class counted by $w(k,\ell-1;r_1,\dots,r_{i-1} = 3)$. Taking the possible extremal point into account we arrive at

$$\begin{aligned} w(k,\ell;r_1,\dots,r_i = 2) = & w(k-1,\ell;r_1,\dots,r_i = 2) \\ & + q^{k+\ell-1} w(k-1,\ell-1;r_1,\dots,r_{i-1} = 3) \\ & + w(k,\ell-1;r_1,\dots,r_{i-1} = 3) \\ & - w(k-1,\ell-1;r_1,\dots,r_{i-1} = 3)\,. \end{aligned}$$

With induction on i and $k+\ell$, the right-hand side is equal to

$$\begin{aligned} q^{b_i}\begin{bmatrix} k+\ell-1 \\ k-1-a_i \end{bmatrix} + q^{k+\ell-1} q^{b_{i-1}} & \begin{bmatrix} k+\ell-2 \\ k-1+a_{i-1} \end{bmatrix} \\ + q^{b_{i-1}} & \left(\begin{bmatrix} k+\ell-1 \\ k+a_{i-1} \end{bmatrix} - \begin{bmatrix} k+\ell-2 \\ k-1+a_{i-1} \end{bmatrix} \right). \end{aligned}$$

According to (8), $b_i = b_{i-1} + a_i + a_{i-1}$, $a_i = a_{i-1} + 2$. Factoring out q^{b_i} and using $\ell = k-1$, this gives

$$q^{b_i}\left(\begin{bmatrix} k+\ell-1 \\ k-1-a_i \end{bmatrix} + q^{k+\ell+1-2a_i}\begin{bmatrix} k+\ell-2 \\ k-a_i \end{bmatrix} + q^{k-a_i}\begin{bmatrix} k+\ell-2 \\ k-1-a_i \end{bmatrix}\right)$$
$$= q^{b_i}\left(\begin{bmatrix} k+\ell-1 \\ k-1-a_i \end{bmatrix} + q^{k-a_i}\left(\begin{bmatrix} k+\ell-2 \\ k-1-a_i \end{bmatrix} + q^{k-a_i}\begin{bmatrix} k+\ell-2 \\ k-a_i \end{bmatrix}\right)\right)$$
$$= q^{b_i}\left(\begin{bmatrix} k+\ell-1 \\ k-1-a_i \end{bmatrix} + q^{k-a_i}\begin{bmatrix} k+\ell-1 \\ k-a_i \end{bmatrix}\right) = q^{b_i}\begin{bmatrix} k+\ell \\ k-a_i \end{bmatrix},$$

as claimed.

Schur's Identity.
Now we set $k = \lfloor \frac{n}{2} \rfloor$, $\ell = \lfloor \frac{n+1}{2} \rfloor$ and make use of (6) and (7). It is an easy matter to compute a_i and b_i. We have

$$r_i = 2: \quad a_i = \begin{cases} \frac{5i}{2} \\ \frac{5i-1}{2} \end{cases}, \qquad b_i = \begin{cases} \frac{i(5i+1)}{2} & i \text{ even} \\ \frac{i(5i-1)}{2} & i \text{ odd} \end{cases},$$

$$r_i = 3: \quad a_i = \begin{cases} \frac{5i}{2} \\ \frac{5i+1}{2} \end{cases}, \qquad b_i = \begin{cases} \frac{i(5i-1)}{2} & i \text{ even} \\ \frac{i(5i+1)}{2} & i \text{ odd} \end{cases}.$$

Hence for even i we get

$$w\left(\left\lfloor\frac{n}{2}\right\rfloor, \left\lfloor\frac{n+1}{2}\right\rfloor; r_1, \ldots, r_i = 2\right) + w\left(\left\lfloor\frac{n}{2}\right\rfloor, \left\lfloor\frac{n+1}{2}\right\rfloor; r_1, \ldots, r_i = 3\right)$$
$$= q^{\frac{i(5i+1)}{2}}\begin{bmatrix} n \\ \lfloor\frac{n-5i}{2}\rfloor \end{bmatrix} + q^{\frac{i(5i-1)}{2}}\begin{bmatrix} n \\ \lfloor\frac{n+5i}{2}\rfloor \end{bmatrix},$$

and replacing i by $-i$ in the second summand we obtain $q^{\frac{i(5i+1)}{2}}$ $\begin{bmatrix} n \\ \lfloor\frac{n-5i}{2}\rfloor \end{bmatrix}$ again. For odd i the same expression results.
Summing over all $i \in \mathbb{Z}$ we thus obtain the identity first established by Schur, by equating the two expressions for $w(\mathcal{P}_n)$:

$$\sum_{k \ge 0} q^{k^2}\begin{bmatrix} n-k \\ k \end{bmatrix}_q = \sum_{i \in \mathbb{Z}} (-i)^i q^{\frac{i(5i+1)}{2}}\begin{bmatrix} n \\ \lfloor\frac{n-5i}{2}\rfloor \end{bmatrix}_q. \tag{10}$$

Entirely analogous reasoning for the set $\mathcal{P}'_n$ leads to the second identity

$$\sum_{k \ge 0} q^{k^2+k}\begin{bmatrix} n-1-k \\ k \end{bmatrix}_q = \sum_{i \in \mathbb{Z}} (-1)^i q^{\frac{i(5i+3)}{2}}\begin{bmatrix} n \\ \lfloor\frac{n-1-5i}{2}\rfloor \end{bmatrix}_q \tag{10'}$$

Going to Infinity.

The rest proceeds on familiar ground. From Section 1.6 we know that

$$\begin{bmatrix} n-k \\ k \end{bmatrix} = \sum_{j\geq 0} p(j; \leq k; \leq n-2k)q^j .$$

Letting n go to ∞, we have no restriction on the summands, and so

$$\begin{bmatrix} n-k \\ k \end{bmatrix} \longrightarrow \sum_{j\geq 0} p(j; \leq k)q^j = \frac{1}{(1-q)\cdots(1-q^k)},$$

and similarly

$$\begin{bmatrix} n-1-k \\ k \end{bmatrix} \longrightarrow \frac{1}{(1-q)\cdots(1-q^k)} .$$

The left-hand sides of (10) and (10′) therefore go to

$$\sum_{k\geq 0} \frac{q^{k^2}}{(1-q)\cdots(1-q^k)} \text{ respectively } \sum_{k\geq 0} \frac{q^{k^2+k}}{(1-q)\cdots(1-q^k)},$$

which are precisely the left-hand sides of the Rogers–Ramanujan identities.

Similarly,

$$\begin{bmatrix} n \\ \lfloor \frac{n-5i}{2} \rfloor \end{bmatrix} = \sum_{j\geq 0} p\left(j; \leq \left\lfloor \frac{n-5i}{2} \right\rfloor; \leq \left\lceil \frac{n+5i}{2} \right\rceil\right) q^j \longrightarrow \sum_{j\geq 0} p(j)q^j$$
$$= \frac{1}{\prod_{i\geq 1}(1-q^i)},$$

and also

$$\begin{bmatrix} n \\ \lfloor \frac{n-5i-1}{2} \rfloor \end{bmatrix} \longrightarrow \frac{1}{\prod_{i\geq 1}(1-q^i)} .$$

The right-hand sides of (10) and (10′) therefore go to

$$\frac{1}{\prod_{i\geq 1}(1-q^i)} \sum_{i\in\mathbb{Z}} (-1)^i q^{\frac{i(5i+1)}{2}} \text{ respectively } \frac{1}{\prod_{i\geq 1}(1-q^i)} \sum_{i\in\mathbb{Z}} (-1)^i q^{\frac{i(5i+3)}{2}} . \tag{11}$$

The coup de grâce is now delivered by Jacobi's triple product theorem:

$$\prod_{k\geq 1}(1+zq^k)(1+z^{-1}q^{k-1})(1-q^k) = \sum_{i\in\mathbb{Z}} q^{\frac{i(i+1)}{2}} z^i .$$

With the substitutions $z \mapsto -q^{-2}$, $q \mapsto q^5$ respectively $z \mapsto -q^{-1}$, $q \mapsto q^5$, we obtain

$$\prod_{k\geq 1}(1-q^{5k-2})(1-q^{5k-3})(1-q^{5k}) = \sum_{i\in\mathbb{Z}} (-1)^i q^{\frac{i(5i+1)}{2}} ,$$

respectively (12)

$$\prod_{k\geq 1}(1-q^{5k-1})(1-q^{5k-4})(1-q^{5k}) = \sum_{i\in\mathbb{Z}} (-1)^i q^{\frac{i(5i+3)}{2}} .$$

Comparing (11) and (12), we have to divide (12) by $\prod_{k\geq 1}(1-q^k)$, which shows that the right-hand sides of (10) and (10′) go to

$$\prod_{i\geq 1}\frac{1}{(1-q^{5i-1})(1-q^{5i-4})} \text{ respecively } \prod_{i\geq 1}\frac{1}{(1-q^{5i-2})(1-q^{5i-3})} ,$$

and we are done.

And the mystery of the number 5? Well, 5 equals 2 + 3 after all ...

Notes and References

There exist several excellent texts on physical models, in particular lattice statistics, such as the survey article by Kasteleyn and the book by Percus. For an advanced treatment see the book by Baxter. The solution of the dimer problem via Pfaffians was accomplished independently (and at the same time) by Kasteleyn and Fisher-Temperley. The combinatorial argument that $M(n,n)$ is always a square or twice a square is due to Jockusch. Welsh is a good source for the connection of the Ising problem to Eulerian subgraphs (originally observed by van der Waerden), and to the Tutte polynomial. The solution of the Ising model by means of the terminal graph is due to Kasteleyn. The references list also the famous paper of Onsager. The presentation of hard models follows in part the paper by Calkin and Wilf. The square ice problem is intimately connected to the alternating sign matrix conjecture. For the fascinating story of this conjecture and its solution the book of Bressoud is highly recommended. The references contain the papers of Izergin, and of the proofs due to Kuperberg and Zeilberger. The highlight follows the ideas of Baxter and the recent paper of Cigler.

1. R.J. Baxter (1982): *Exactly Solved Models in Statistical Mechanics.* Academic Press, London.
2. D.M. Bressoud (1999): *Proofs and Confirmations. The Story of the Alternating Sign Matrix Conjecture.* Cambridge Univ. Press, Cambridge.
3. N.J. Calkin and H.S. Wilf (1998): The number of independent sets in a grid graph. *SIAM J. Discrete Math.* 11, 54–60.
4. J. Cigler (2005): Fibonacci-Zahlen, Gitterpunktwege und die Identitäten von Rogers–Ramanujan. *Math. Semesterber.* 52, 97–125.
5. W. Jockusch (1994): Perfect matchings and perfect squares. *J. Combinatorial Theory A* 67, 100–115.
6. P.W. Kasteleyn (1967): Graph theory and crystal physics. In: *Graph Theory and Theoretical Physics*, Harary, ed., 43–110. Academic Press, London.
7. A.G. Izergin (1987): Partition function of a six-vertex model in a finite volume (Russian). *Dokl. Akad. Nauk SSR* 297, 331–333.

8. G. Kuperberg (1996): Another proof of the alternating sign matrix conjecture. *International Math. Research Notes* 139–150.
9. L. Onsager (1944): Crystal statistics I. A two-dimensional model with an order-disorder transition. *Phys. Rev.* 65, 117–149.
10. J.K. Percus (1971): *Combinatorial Methods.* Springer, New York.
11. H.N.V. Temperley and M.E. Fisher (1961): Dimer problem in statistical mechanics – an exact result. *Phil. Mag.* 6, 1061–1063.
12. D, Zeilberger (1996): Proof of the refined alternating sign matrix conjecture. *New York J. Math.* 2, 59–68.

Solutions to Selected Exercises

Chapter 1

1.2 The conditions $1 \le i \le 75$, $76 - i \le j \le 75$ give as number $\sum_{i=1}^{75} i = 2850$.

1.5 Classify the triples in $\{1, \ldots, n+1\}$ according to the middle element respectively the last element.

1.8 Classify the sets according to whether 1 is in the set or not. This gives the recurrence $f(n,k) = f(n-1,k) + f(n-2,k-1)$. Now use induction and Pascal's recurrence. The sum satisfies the Fibonacci recurrence.

1.11 Suppose this is false. Look at the incidence system $(A, N \smallsetminus A)$ with iIj if $|i-j| = 9$. For $a \in A$, $d(a) = 2$ for $10 \le a \le 91$; otherwise $d(a) = 1$; hence $\sum_{a \in A} d(a) \ge 18 + 2 \cdot 37 = 92$. On the other hand, $\sum_{b \in N \smallsetminus A} d(b) \le 2 \cdot 45 = 90$, contradiction. For $|A| = 54$, take six blocks of 9 consecutive numbers, the blocks 9 apart.

1.13 We have $\binom{2n}{2k}\binom{2n-2k}{n-k}\binom{2k}{k} = \binom{2n}{2n-2k}\binom{2n-2k}{n-k}\binom{2k}{k} = \binom{2n}{n+k}\binom{n+k}{n-k}\binom{2k}{k} = \binom{2n}{n+k}\binom{n+k}{2k}\binom{2k}{k} = \binom{2n}{n+k}\binom{n+k}{n}\binom{n}{k} = \binom{2n}{n}\binom{n}{k}^2$.

1.17 Suppose $n_1 \ge n_2 + 2$; then $(n_1-1)!(n_2+1)! < n_1!n_2!$; hence $\binom{n}{n_1 \cdots n_k} < \binom{n}{n_1-1\, n_2+1 \cdots n_k}$. For the last assertion use induction.

1.19 Let $k \ge 4$ be even; then the element in position $(\frac{k}{2}, k)$ is 1, which is not circled. Suppose k is odd, not a prime, and p a prime divisor. Then $p \mid \frac{k-p}{2}$, and for $n = \frac{k-p}{2}$, $n \nmid \binom{n}{k-2n} = \binom{(k-p)/2}{p}$. For $k = p$ it is easily verified that $n \mid \binom{n}{p-2n}$ for all n.

1.21 The runs can be placed in $\binom{n+1}{k}$ ways (Exercise 1.8), and the runs can be filled in $\binom{m-1}{k-1}$ ways. The answer is therefore $\binom{n+1}{k}\binom{m-1}{k-1}$.

1.24 If there are k diagonal steps, then there are $m + n - k$ steps altogether. Contracting the diagonal steps gives $m + n - 2k$ $(1,0)$- and $(0,1)$-steps. Hence $D_{m,n} = \sum_k \binom{m+n-k}{k}\binom{m+n-2k}{n-k} = \sum_k \binom{m+n-k}{n-k}\binom{m}{m-k} = \sum_k \binom{m+n-k}{m}\binom{m}{m-k} = \sum_k \binom{n+k}{m}\binom{m}{k}$ by the index shift $k \mapsto m-k$.

1.27 From $(x+1)^{n+1} = \sum_k S_{n+1.k+1}(x+1)^{\underline{k+1}} = (x+1)\sum_k S_{n+1,k+1} x^{\underline{k}}$ follows $(x+1)^n = \sum_k S_{n+1,k+1} x^{\underline{k}}$. On the other hand, $(x+1)^n = \sum_i \binom{n}{i} x^i =$

$\sum_i \binom{n}{i} \sum_k S_{i,k} x^{\underline{k}} = \sum_k (\sum_i \binom{n}{i} S_{i,k}) x^{\underline{k}}$, hence $S_{n+1,k+1} = \sum_i \binom{n}{i} S_{i,k}$. For a combinatorial proof classify the partitions according to the size of the block containing $n+1$. Summation over k yields the recurrence for the Bell numbers.

1.32 Both sides count the pairs (P_1, P_2), where P_1 is a $(k+r)$-partition of $\{1, \ldots, n\}$ and P_2 is a subfamily of k blocks of P_1.

1.35 Given $P \in \Pi(n-1)$ construct $\phi P \in \Pi(n)$ as follows. If $A = \{i_1 < \cdots < i_k\}$ is a block of P, then the pairs $\{i_1, i_2+1\}, \{i_2, i_3+1\}, \ldots \{i_{k-1}, i_k+1\}$ are in blocks of ϕP. Now merge intersecting blocks, and add remaining elements as singletons. This is a bijection. Example: $13|246|5 \to 147|25|36$.

1.40 Classify the permutations in $S(n+1)$ according to the length $n+1-k$ of the cycle containing 1, where $1 \le n+1-k \le r$, that is, $n+1-r \le k \le n$.

1.42 Part (a) is clear, and (b) follows by looking at the reverse permutation $\sigma^* = a_n \ldots a_1$ of $\sigma = a_1 \ldots a_n$. c. The number of $\sigma = a_1 \ldots a_n$ with $\mathrm{inv}(\sigma) = k$, $a_n = n$ is $I_{n-1,k}$. Suppose $n = a_i$ $(i < n)$. Then interchanging a_i, a_{i+1} gives $\sigma' \in S(n)$ with $\mathrm{inv}(\sigma') = k-1$. Hence we get $I_{n,k-1}$, since no $\sigma' \in S(n)$ with $\mathrm{inv}(\sigma') = k-1$ can have n in first place (because of $n > k$). The example $I_{3,3} = 1, I_{3,2} = 2$ shows that $n = k$ no longer works. d. Interchange of the first two elements produces a bijection between even and odd permutations.

1.46 Looking at the cycles of σ as directed graph circuits, one readily finds that $\pi\sigma = \sigma\pi$ holds if and only if any k-cycle $(i_1, \ldots, i_k)$ of σ is mapped under π onto a k-cycle of σ. Since we may choose the image of i_1 in k ways, the result follows.

1.48 A permutation $\sigma \in S(n)$ with k runs is created either by inserting n into $\sigma' \in S(n-1)$ with k runs at the end of a run, resulting in $kA_{n-1,k}$, or into $\sigma'' \in S(n-1)$ with $k-1$ runs not at the end of a run, which gives $(n-k+1)A_{n-1,k-1}$. For the second assertion write $\sum_k A_{n,k}\binom{x+n-k}{n} = \sum_k A_{n-1,k}\left[k\binom{x+n-k}{n} + (n-k)\binom{x+n-1-k}{n}\right]$. The expression in parentheses is easily seen to be $x\binom{x+n-1-k}{n-1}$, and we obtain $x \cdot x^{n-1} = x^n$ by induction. For the last assertion we have $(k-i)^n = \sum_{j=0}^{k-i} A_{n,j}\binom{n+k-i-j}{k-i-j} = \sum_{j=0}^{k-i} A_{n,j}(-1)^{k-i-j}\binom{-n-1}{k-i-j}$ for $i = 0, \ldots, k$. Multiply both sides by $(-1)^i\binom{n+1}{i}$ and sum over i.

1.51 The expected value is $\frac{1}{n!}\sum_k k s_{n,k}$. Differentiation of $x^{\overline{n}} = \sum_k s_{n,k} x^k$ gives $\sum_k k s_{n,k} x^{k-1} = \sum_{i=0}^{n-1} \frac{x^{\overline{n}}}{x+i}$, and with $x = 1$, $\sum_k k s_{n,k} = n! H_n$. The expected value is thus the harmonic number H_n.

1.53 Decompose a self-conjugate partition into hooks.

Example:  53211 → 93. This is a bijection.

1.59 Every positive solution of $x_1 + \cdots + x_k = m$ corresponds to an ordered partition of m. The number of positive solutions of $x_1 + \cdots + x_k \le n$ is therefore $\sum_{m=k}^{n} \binom{m-1}{k-1} = \sum_{m=k-1}^{n-1} \binom{m}{k-1} = \binom{n}{k}$. For nonnegative solutions use multisets to obtain $\binom{n+k}{k}$.

1.62 Deleting m from λ, we get the recurrence $\sum_{|\lambda|=n} f_m(\lambda) = \sum_{|\lambda|=n-m} f_m(\lambda) + p(n-m)$, and so $\sum_{|\lambda|=n} f_m(\lambda) = p(n-m) + p(n-2m) + \cdots$. Suppose λ contains t distinct parts k that appear at least m times. Delete m occurences of each k to obtain t partitions. This results in the same expression $\sum_{|\lambda|=n} g_m(\lambda) = p(n-m) + p(n-2m) + \cdots$.

1.64 Let λ be a perfect partition, and suppose 1 appears a_1 times. The next part must be $a_1 + 1$. Suppose it appears a_2 times. Then the next part is $a_2(a_1+1) + a_1 + 1 = (a_1+1)(a_2+1)$. Altogether we obtain $n + 1 = (a_1+1)(a_2+1)\cdots(a_t+1)$, where t is the number of distinct parts. For a given factorization reverse the procedure.

1.66 Use $\left[{n \atop k}\right] = \left[{n \atop n-k}\right]$ and apply recurrence (1) to $\left[{n \atop n-k}\right]$.

1.70 Classify the paths in the $(k+1)\times(n-k)$-grid according to the height $y = i - k$ when the path first hits the line $x = k+1$. The contribution is $\left[{i \atop k}\right] q^{(k+1)(n-i)}$ to the sum $\left[{n+1 \atop k+1}\right]$.

1.73 We have $G_{n+1} = \sum_k \left[{n+1 \atop k}\right] = \sum_k \left[{n \atop k-1}\right] + \sum_k q^k \left[{n \atop k}\right] = 2G_n - \sum_k (1 - q^k)\left[{n \atop k}\right] = 2G_n - (1-q^n)G_{n-1}$, since $(1-q^k)\left[{n \atop k}\right] = (1-q^n)\left[{n-1 \atop k-1}\right]$.

1.75 Use repeatedly the recurrence in (1) and Exercise 1.66.

Chapter 2

2.2 We have $A(z)^2 = \frac{1}{1-z}$; hence $A(z) = (1-z)^{-1/2}$ with $a_n = \binom{2n}{n}/4^n$ by Exercise 1.16.

2.6 Convolution gives $F(z) = (1+z)^r(1+z^2)^r = (1+z+z^2+z^3)^r$.

2.8 Write $\binom{2n+1}{n} = 2\binom{2n}{n} - \frac{1}{n+1}\binom{2n}{n}$ and use the same approach as for $\binom{2n}{n}$ to compute $\sum_n \binom{2n+1}{n} z^n = \frac{1-4z-\sqrt{1-4z}}{-2z+8z^2}$. For $\binom{n}{\lfloor n/2 \rfloor}$ distinguish the cases n even or odd, and deduce $\sum \binom{n}{\lfloor n/2 \rfloor} z^n = \sum \binom{2n}{n} z^{2n} + z \sum \binom{2n+1}{n} z^{2n}$. Plugging in the generating functions for $\binom{2n}{n}, \binom{2n+1}{n}$, you obtain $\sum_n \binom{n}{\lfloor n/2 \rfloor} z^n = \frac{1-2z-\sqrt{1-4z^2}}{-2z+4z^2}$.

2.11 Set $F(z) = 1 + G(z)$, then $\log F = \sum \frac{(-1)^{n-1}G^n}{n}$. From this we get $(\log F)' = \sum(-1)^{n-1}G^{n-1}G' = \frac{G'}{1+G}$, thus $(\log F)' = \frac{F'}{F}$.

2.16 $D_qF(cz) = \frac{F(cz)-F(qcz)}{(1-q)z}$, and $(D_qF)(cz) = \frac{F(cz)-F(qcz)}{(1-q)cz}$.

2.18 We have

$$D_qH_n(x) = \sum\left[{n \atop k}\right]\frac{1-q^k}{1-q}x^{k-1} = \sum\left[{n-1 \atop k-1}\right]\frac{1-q^n}{1-q}x^{k-1} = \frac{1-q^n}{1-q}H_{n-1}(x).$$

The recurrence follows from (1) in Section 1.6.

2.21 We have

$$\begin{aligned}\prod_{k=1}^{m}\frac{1}{1-zq^k} &= \sum_{\ell_1} q^{\ell_1}z^{\ell_1}\cdot\sum_{\ell_2} q^{2\ell_2}z^{\ell_2}\cdots\sum_{\ell_m} q^{m\ell_m}z^{\ell_m}\\ &= \sum_n\left(\sum_i p(i;n;\le m)q^i\right)z^n = \sum_n\left(\left[{m+n \atop n}\right]-\left[{m+n-1 \atop n-1}\right]\right)z^n\\ &= \sum_n\left[{m+n-1 \atop n}\right](qz)^n.\end{aligned}$$

Now substitute $qz \mapsto z$. Alternatively,

$$D_q\sum_n\left[{m+n-1 \atop n}\right]z^n = \frac{1-q^m}{1-q}\sum\left[{m+n \atop n}\right]z^n.$$

Hence, setting $F(z,m)$ for the sum we obtain $D_qF(z,m) = \frac{1-q^m}{1-q}F(z,m+1)$. Now prove the same recurrence for the right-hand side using (7).

2.24 Using (7), $D_qf_n(x) = q\frac{1-q^n}{1-q}(1+xq^2)\cdots(1+xq^n)$; hence

$$D_q^{(k)}f_n(x)\,|_{x=0} = q^{\binom{k+1}{2}}\frac{(1-q^n)\cdots(1-q^{n-k+1})}{(1-q)^k}.$$

On the other hand,

$$D_q^{(k)}\sum_{k=0}^{n} q^{\binom{k+1}{2}}a_{n,k}x^k\,|_{x=0} = q^{\binom{k+1}{2}}a_{n,k}\frac{(1-q^k)\cdots(1-q)}{(1-q)^k}.$$

Comparison of coefficients gives the result $a_{n,k} = \left[{n \atop k}\right]$.

2.28 We have

$$\deg\left(\prod_{i=1}^{k}A_i - \prod_{i=1}^{k-1}A_i\right) = \deg\left((A_k-1)\prod_{i=1}^{k-1}A_i\right) = \deg(A_k-1),$$

since $\deg A_i = 0$.

2.31 From $F' = \lim_{k\to\infty}\left(\prod_{i=1}^{k}F_i\right)' = \lim_{k\to\infty}\sum_{i=1}^{k}\left(\prod_{j\ne i}F_j\right)F_i' = \lim_{k\to\infty}\sum_{i=1}^{k}\left(\frac{F_i'}{F_i}\right)\prod_{j=1}^{k}F_j = \left(\sum_{i\ge1}\frac{F_i'}{F_i}\right)F$ follows $\frac{F'}{F} = \sum_{i\ge1}\frac{F_i'}{F_i}$. For $F(z) = \prod_{i\ge1}\frac{1}{1-z^i}$ this gives $\frac{F'}{F} = \sum_{i\ge1}\frac{iz^{i-1}}{1-z^i} = \sum_{i\ge1}iz^{i-1}\sum_{k\ge0}z^{ki}$. Hence $[z^n]\frac{F'}{F} = \sum_{i\ge1}i$ over all i such that $i(k+1) = n+1$ for some k; that is $[z^n]\frac{F'}{F} = \sigma(n+1)$, where $\sigma(n+1)$ is the sum of divisors of $n+1$.

2.32 Since $\frac{z^k}{(1-z)^k} = \sum_i \binom{k+i-1}{i} z^{k+i}$ we have

$$e^{z/1-z} = 1 + \sum_{k\ge1}\sum_{i\ge0}\binom{k+i-1}{i}\frac{z^{k+i}}{k!} = 1 + \sum_{n\ge1}\sum_{k=1}^{n}\binom{n-1}{n-k}\frac{z^n}{k!}$$
$$= 1 + \sum_{n\ge1}\left(\sum_{k\ge1}\frac{n!}{k!}\binom{n-1}{k-1}\right)\frac{z^n}{n!}\,.$$

On the other hand, $[\frac{z^n}{n!}]\prod_{i\ge1} e^{z^i}$ is $\sum \frac{n!}{k_1!\cdots k_n!}$ over all $(k_1,\ldots,k_n)$ with $k_1 + 2k_2 + \cdots + nk_n = n$, which clearly counts the partitions in question.

2.34 We have

$$\log\prod_n (1-z^n)^{-\overline{\mu}(n)/n} = \sum_n -\frac{\overline{\mu}(n)}{n}\log(1-z^n) = \sum_n \frac{\overline{\mu}(n)}{n}\sum_k \frac{z^{nk}}{k}$$
$$= \sum_m \sum_{d|m}\frac{\overline{\mu}(d)}{m} z^m = z\,.$$

Hence $F(z) = e^z$.

2.40 Vandermonde gives $\binom{x+n-1}{n} = \sum_k \binom{n-1}{k-1}\binom{x}{k}$.
Thus $x^{\overline{n}} = \sum_{k=0}^{n}\frac{n!}{k!}\binom{n-1}{k-1}x^{\underline{k}}$, and so $L_{n,k} = \frac{n!}{k!}\binom{n-1}{k-1}$.

2.43 Binomial inversion turns $n! = \sum_k \binom{n}{k}a_k k!$ into the equality $n!a_n = \sum_k (-1)^{n-k}\binom{n}{k}k!$. Hence $a_n = \sum_{k=0}^{n}\frac{(-1)^{n-k}}{(n-k)!} = \frac{D_n}{n!}$, D_n derangement number.

2.47 Substitute in the q-binomial formula of Corollary 1.2 x by $-\frac{x}{q^n}$ and multiply by $q^{\binom{n}{2}}$. This gives $(q^{n-1}-x)(q^{n-2}-x)\cdots(1-x) = \sum_{k=0}^{n}\left[{n \atop k}\right]q^{\binom{n-k}{2}}(-1)^k x^k$; thus $g_n(x) = \sum_{k=0}^{n}(-1)^{n-k}q^{\binom{n-k}{2}}\left[{n \atop k}\right]x^k$. The inversion formula follows.

2.49 Set $A(z) = \sum a_n z^n$, $B(z) = \sum b_n z^n$. Then $A(z) = \frac{B(z)}{(1-z)^{m+1}}$, $B(z) = (1-z)^{m+1}A(z)$, and therefore $b_n = \sum_{k=0}^{n}(-1)^k\binom{m+1}{k}a_{n-k}$.

2.51 The result is $P_X(z) = \frac{1}{n}\frac{1-z^n}{1-z}$.

2.54 We have $\text{Prob}(X = k) = \binom{k-1}{n-1}/2^k$, since there must be $n-1$ heads among the first $k-1$ throws. Hence $P_X(z) = \frac{z}{2}\sum_{k\ge1}\binom{k-1}{n-1}(\frac{z}{2})^{k-1} = (\frac{z}{2})^n\frac{1}{(1-z/2)^n} = \frac{z^n}{(2-z)^n}$. Using (2) and (3), this gives $\mathbb{E}X = \text{Var}X = 2n$.

2.58 When k numbers have been seen, the probability of rolling a new one is equivalent to tossing a coin with success probability $\frac{6-k}{6}$. From this follows easily that $P_X(z) = \prod_{k=0}^{5}\frac{(6-k)z}{6-kz}$, thus $\mathbb{E}X = P_X'(1) = \sum_{k=0}^{5}\frac{6}{6-k} = 6H_6 = \frac{147}{10}$, H_6 = harmonic number.

2.61 The distance between the particles during the process is 0, 1, or 2, initially 1. Let $P_X(z) = A_0(z)$, and $A_1(z), A_2(z)$ the probability generating functions of the particles being at distance 1 respectively 2 after n

moves. Then $A_0(z) = \frac{1}{4}zA_2(z)$, $A_1(z) = 1 + \frac{3}{4}zA_1(z) + \frac{1}{4}zA_2(z)$, $A_2(z) = \frac{1}{4}zA_1(z) + \frac{1}{2}zA_2(z)$. Solving for $A_0(z)$ we obtain $A_0(z) = \frac{z^2}{16-20z+5z^2}$, and from this $\mathbb{E} = 12$, $\mathrm{Var} = 100$. For initially two apart the result is $\mathbb{E} = 8$, $\mathrm{Var} = 88$.

Chapter 3

3.4 We have $f(0) = 1$, $f(1) = 2$. Suppose $A \subseteq \{1,\dots,n\}$ is fat. If $n \notin A$, then A is fat in $\{1,\dots,n-1\}$; if $n \in A$, then $\{i-1 : i \in A \setminus n\}$ is fat in $\{1,\dots,n-2\}$ or $\emptyset$. It follows that $f(n) = f(n-1) + f(n-2)$; thus $f(n) = F_{n+2}$. The fat k-sets in $\{1,\dots,n-1\}$ are precisely the k-subsets of $\{k,\dots,n-1\}$. which proves (b). Assertion (c) is by (a) equivalent to $(n+1) + \sum_{k=1}^{n-1} \binom{n+1}{n-k} f(k-1) = f(2n)$. Let $X = \{1,\dots,n-1\}$, $Y = \{n,\dots,2n\}$; then (c) follows by classifying the fat sets $A \subseteq \{1,\dots,2n\}$ according to $|A \cap Y| = n-k$.

3.6 Considering the first entry we get the recurrence $f(n) = 2(f(n-1) + f(n-2))$, $n \ge 2$, with $f(0) = 1$. For the generating function this means $F(z) = 2zF(z) + 2z^2F(z) + 1 + z$; hence $F(z) = \frac{1+z}{1-2z-2z^2}$. The usual method gives $f(n) = \sum_{i\ge 0} \left[\binom{n+1}{2i+1} + \binom{n}{2i+1}\right] 3^i$.

3.11 The Catalan recurrence $C = zC^2 + 1$ implies $C' = C^2 + 2zCC'$, thus $C' = \frac{C^2}{1-2zC} = \frac{C^2}{\sqrt{1-4z}}$. Since $\frac{1}{\sqrt{1-4z}} = \sum_n \binom{2n}{n} z^n$ we obtain with $C^2 = \frac{C-1}{z}$, $(n+1)C_{n+1} = \sum_{k=0}^{n} \binom{2k}{k} C_{n-k+1}$.

3.15 Following the hint we get the recurrences $A_n = 2B_{n-1} + A_{n-2} + [n=0]$, $B_n = A_{n-1} + B_{n-2}$. Elimination of B gives $A(z) = \frac{1-z^2}{1-4z^2+z^4}$. It follows that $A_{2n+1} = 0$, which is clear, and $A_{2n} = \frac{(2+\sqrt{3})^n}{3-\sqrt{3}} + \frac{(2-\sqrt{3})^n}{3+\sqrt{3}}$ by our four steps. Using the binomial theorem we may also write this as $A_{2n} = \sum_{k=0}^{n} \binom{n}{k} 2^{n-k} 3^{\lfloor k/2 \rfloor}$.

3.18 The conditions yield the recurrences $a_n = b_n + a_{n-2}$, $b_n = a_{n-1} + a_{n-2} + 1$ $(n \ge 1)$ with $a_0 = b_0 = 0$. The usual method gives $a_n = \frac{2^{n+2}-3+(-1)^{n+1}}{6}$, $b_n = 2^{n-1}$ $(n \ge 1)$.

3.20 We have

$$\begin{aligned}
\sum_n \left(\sum_k \binom{n}{k}\binom{2k}{k} x^k\right) z^n &= \sum_k \binom{2k}{k} x^k \sum_n \binom{n}{k} z^n = \sum_k \binom{2k}{k} x^k \frac{z^k}{(1-z)^{k+1}} \\
&= \frac{1}{1-z} \sum_k \binom{2k}{k} \left(\frac{xz}{1-z}\right)^k = \frac{1}{1-z} \frac{1}{\sqrt{1-4xz/(1-z)}} \\
&= [(1-z)(1-z-4xz)]^{-1/2}.
\end{aligned}$$

For $x = -\frac{1}{2}$ we get $(1-z^2)^{-1/2}$, thus $p_n(-\frac{1}{2}) = \frac{1}{4^{n/2}}\binom{n}{n/2}[n \text{ even}]$, and for $x = -\frac{1}{4}$, $(1-z)^{-1/2}$ results, hence $p_n(-\frac{1}{4}) = p_{2n}(-\frac{1}{2})$.

3.24 Set $q_m = p_{-m}$ $(m \geq 1)$. Then $q_m = \frac{1}{2}(q_{m+1} + q_{m-2})$, $q_0 = 1$, $q_1 = a$, $q_2 = b$, where $2a = b + (\tau - 1)$, since $p_1 = \tau - 1$. This gives for $Q(z) = \sum_{m\geq 0} q_m z^m = \frac{1+(a-2)z+(b-2a)z^2}{1-2z+z^3} = \frac{1+(a-2)z+(b-2a)z^2}{(1-z)(1-\tau z)(1-\hat{\tau}z)}$, with $q_m = \alpha + \beta\tau^m + \gamma\hat{\tau}^m$ for constants α, β, γ. It follows that $\beta = 0$, since otherwise $q_m \to \infty$, which cannot be. With the initial conditions one obtains $q_m = 3\tau - 4 + (5 - 3\tau)\hat{\tau}^m$, and $\hat{\tau}^m = F_{m+1} - F_m\tau$ by induction. Thus $\lim_{m\to\infty} q_m = 3\tau - 4 \sim 0.83$.

3.25 We have

$$z\sum_n \binom{2n+k+1}{n} z^n = \sum_n \binom{2n+k-1}{n-1} z^n = \sum_n \binom{2n+k}{n} z^n - \sum_n \binom{2n+k-1}{n} z^n,$$

and hence by induction on k,

$$z\sum_{n\geq 0} \binom{2n+k+1}{n} z^n = \frac{C^k}{\sqrt{1-4z}} - \frac{C^{k-1}}{\sqrt{1-4z}} = \frac{C^{k-1}}{\sqrt{1-4z}}(C-1) = z\frac{C^{k+1}}{\sqrt{1-4z}}.$$

3.27 The recurrence implies $D(y,z) = yD + zD + yzD + 1$, and so $D_{m,n} = \sum_k 2^k \binom{m}{k}\binom{n}{k}$ by the previous exercise with $a = 2$.

3.31 Pascal gives for the sum $s_{n+1} = 2s_n$, hence $s_n = 2^n$. Now use snake oil to get $\sum_k \binom{n+k}{k} 2^{-k} = 2^{n+1}$, and thus $\sum_{k>n} \binom{n+k}{k} 2^{-k} = 2^n$.

3.35 Show first that $\sum_k (-1)^k \binom{n}{k} F_k = -F_n$ by induction. With $\hat{F}(z) = \sum F_n \frac{z^n}{n!}$ the wanted sum corresponds to $(z\hat{F} - \hat{F})\hat{D} = (z-1)\hat{F}\frac{e^{-z}}{1-z} = (-\hat{F})e^{-z}$. The result follows from above.

3.40 We have

$$\begin{aligned}\sum_n \sum_k \binom{m}{k}\binom{n+k}{m} x^{m-k} z^n &= \sum_k \binom{m}{k} x^{m-k} z^{-k} \sum_n \binom{n+k}{m} z^{n+k} \\ &= \sum_k \binom{m}{k} x^{m-k} z^{-k} \frac{z^m}{(1-z)^{m+1}} \\ &= \frac{(xz)^m}{(1-z)^{m+1}} \sum_k \binom{m}{k} \frac{1}{(xz)^k} = \frac{(1+xz)^m}{(1-z)^{m+1}}.\end{aligned}$$

Now check that the right-hand sum has the same generating function.

3.41 Fix x; then

$$\begin{aligned}F(z)^x &= \exp\left(x \log(1 + \textstyle\sum_{k\geq 1} a_k z^k)\right) \\ &= \exp\left(x \textstyle\sum_{\ell\geq 1} \frac{(-1)^{\ell+1}}{\ell} (\sum_{k\geq 1} a_k z^k)^\ell\right) \\ &= \textstyle\sum_{m\geq 0} \frac{x^m}{m!} \left(\sum_{\ell\geq 1} \frac{(-1)^{\ell+1}}{\ell} (\sum_{k\geq 1} a_k z^k)^\ell\right)^m.\end{aligned}$$

Hence $[z^n]F(z)^x$ is a polynomial $p_n(x)$ of degree n with $p_n(0) = [n = 0]$. The first convolution follows from $F(z)^x F(z)^y = F(z)^{x+y}$, and the second by comparing the coefficients $[z^{n-1}]$ in $F'F^{x-1}F^y = F'F^{x+y-1}$, since $F'F^{x-1} = x^{-1}(F^x)' = x^{-1}\sum_n np_n(x)z^{n-1}$.

3.43 Following the hint, $\sum_{k=0}^{n-1} k^m = \sum_{k=0}^{n-1}\sum_j S_{m,j}k^{\underline{j}} = \sum_j S_{m,j}\sum_{k=0}^{n-1} k^{\underline{j}} = \sum_j S_{m,j}\frac{n^{\underline{j+1}}}{j+1} = \sum_j \frac{S_{m,j}}{j+1}\sum_k s_{j+1,k}(-1)^{j+1-k}n^k$. Now compare this with formula (6) of the text. For $k = 1$ we obtain $B_m = \sum_j S_{m,j}\frac{(-1)^j j!}{j+1}$.

3.46 Here $\hat{G}(z) = \frac{1}{1-z}$, $\hat{F}(z) = \frac{z}{1-z}$, hence $\hat{H}(z) = \frac{1-z}{1-2z}$, and so $k_n = n!2^{n-1}$. For a direct proof, take a permutation and insert up to $n-1$ bars to create the blocks.

3.48 Set $H(z) = e^z - 1 = z\frac{e^z-1}{z}$. Then $[z^n]\log(1+z) = \frac{(-1)^{n-1}}{n} = \frac{1}{n}[z^{n-1}](\frac{z}{e^z-1})^n$. It follows that $(-1)^{n-1} = \sum_{(k_1,\ldots,k_n)}\frac{B_{k_1}\cdots B_{k_n}}{k_1!\cdots k_n!}$, where the sum is taken over all $(k_1,\ldots,k_n)$ with $k_1+\cdots+k_n = n-1$.

3.53 We have $\hat{A}(z) = \exp\left(\sum\frac{z^{2n+1}}{2n+1}\right) = \exp\left(\frac{1}{2}(-\log(1-z)+\log(1+z))\right) = \sqrt{\frac{1+z}{1-z}}$; similarly $\hat{B}(z) = \frac{1}{\sqrt{1-z^2}}$. It follows that $\hat{A}(z) = (1+z)\hat{B}(z)$, that is, $a_n = b_n + nb_{n-1}$. Since clearly $b_n = 0$ for n odd, we obtain $a_n = b_n$ (n even) and $a_n = nb_{n-1}$ (n odd).

3.55 For $n \geq 1$, set $f(n) = \sum_i c_{n,i}\alpha^i$, where $c_{n,i}$ is the number of connected graphs on n vertices and i edges. Set $g(k) = \beta^k$; then $\hat{H}(z) = \hat{G}(\hat{F}(z))$ is the desired exponential generating function. Now $\hat{G}(z) = e^{\beta z}$, thus $\hat{H}(z) = (e^{\hat{F}(z)})^\beta$. Observe, finally, that $e^{\hat{F}(z)}$ is the generating function of *all* graphs, enumerated according to the number of vertices and edges. Thus $e^{\hat{F}(z)} = \sum_n(1+\alpha)^{\binom{n}{2}}\frac{z^n}{n!}$, and the result follows.

3.58 By the composition formula the desired function is $\frac{\hat{T}(z)^k}{k!}$. By Exercise 3.56, $[z^n]\hat{T}(z)^k = \frac{k}{n}[z^{n-k}]e^{nz} = \frac{k}{n}\frac{n^{n-k}}{(n-k)!}$. The number of rooted forests is therefore $\frac{(n-1)!n^{n-k}}{(k-1)!(n-k)!} = \binom{n-1}{k-1}n^{n-k}$.

3.61 Considering alternating forests, one easily obtains $\hat{H} = e^{\frac{z}{2}(\hat{H}+1)}$. Set $A(z) = z(\hat{H}(z)+1) = 2z + \sum_{n\geq 1} h_n\frac{z^{n+1}}{n!}$. Then $\frac{A}{z} - 1 = e^{\frac{A}{2}}$, $A = z(1+e^{\frac{A}{2}})$, and thus $A^{\langle -1\rangle} = \frac{z}{1+e^{z/2}}$. By Lagrange $[z^{n+1}]A = \frac{h_n}{n!} = \frac{1}{n+1}[z^n](1+e^{\frac{z}{2}})^{n+1} = \frac{1}{n+1}[z^n]\sum_k\binom{n+1}{k}e^{\frac{kz}{2}} = \frac{1}{n+1}[z^n]\sum_k\binom{n+1}{k}\sum_i\frac{k^i}{2^i}\frac{z^i}{i!} = \frac{1}{n+1}\frac{1}{2^n n!}\sum_{k=1}^{n+1}\binom{n+1}{k}k^n = \frac{1}{2^n n!}\sum_{k=0}^{n}\binom{n}{k}(k+1)^{n-1}$. Thus we obtain $h_n = \frac{1}{2^n}\sum_{k=0}^{n}\binom{n}{k}(k+1)^{n-1}$.

3.63 With $y = \hat{T}$, $\hat{F} = e^y$, $y = ze^y$ as in the text, we get $(e^y)' = e^y y'$, hence $f_{n+1} = \sum_k\binom{n}{k}f_k t_{n-k+1}$. Using $f_n = T_{n+1}$, this gives $(n+2)^n = \sum_k\binom{n}{k}(k+1)^{k-1}(n-k+1)^{n-k}$.

3.65 The parts arising from a fixed λ_i are clearly distinct. Now if $2^k\lambda_i = 2^\ell\lambda_j$, then $k = \ell$ and $\lambda_i = \lambda_j$, since the λ_i's are odd. For the converse, write a part μ_i as $\mu_i = 2^\ell \nu_i$ where 2^ℓ is the highest power of 2, and collect the powers $(2^{\ell_1} + \cdots + 2^{\ell_t})\nu_i$, creating $2^{\ell_1} + \cdots + 2^{\ell_t}$ summands equal to ν_i.

3.68 The generating function for the first set of partitions is

$$\prod_{i\geq 1} \frac{(1-z^{6i-2})(1-z^{6i-4})}{(1-z^{3i-1})(1-z^{3i-2})} = \prod_{i\geq 1}(1+z^{3i-1})(1+z^{3i-2}),$$

which also counts the second set of partitions.

3.71 We have $\prod_{i\geq 1}(1+qz^i) = \sum_{n,k} p_d(n;k)q^k z^n$. Now set $q = -1$.

3.75 By the same idea as in Exercise 3.71 we have $\sum_n (e(n) - o(n))z^n = \prod_{i\geq 1} \frac{1}{1-z^{2i-1}} \cdot \prod_{i\geq 1} \frac{1}{1+z^{2i}}$. Now we know that $\prod_{i\geq 1}(1+z^i) = \prod_{i\geq 1} \frac{1}{1-z^{2i-1}}$; thus $\prod_{i\geq 1}(1+z^{2i}) = \prod_{i\geq 1} \frac{1}{1-z^{4i-2}}$. It follows that $\sum_{n\geq 0} (e(n) - o(n))z^n = \prod_{i\geq 1} \frac{1-z^{4i-2}}{1-z^{2i-1}} = \prod_{i\geq 1}(1+z^{2i-1}) = \sum_{n\geq 0} sc(n)z^n$.

3.77 We know that $\frac{P'}{P} = \sum_{n\geq 0} \sigma(n+1)z^n$ from Exercise 2.31, that is, $P' = \sum \sigma(n+1)z^n \cdot P$. Convolution gives for the coefficient $[z^{n-1}]$, $p(n) = \frac{1}{n}\sum_{i=1}^n \sigma(i)p(n-i)$.

3.81 Write the triple product theorem in the form $\prod_{k\geq 1}(1+zq^k)(1+z^{-1}q^k)(1-q^k) = \frac{z}{z+1}\sum_{n\in\mathbb{Z}} q^{\frac{n(n+1)}{2}} z^n$. The substitution $z \mapsto y-1$ gives $\prod_{k\geq 1}(1+(y-1)q^k)(1+\frac{1}{y-1}q^k)(1-q^k) = \frac{y-1}{y}\sum_{n\in\mathbb{Z}} q^{\frac{n(n+1)}{2}}(y-1)^n$. For $y = 0$ we obtain $\prod_{k\geq 1}(1-q^k)^3$ on the left. Since $q^{\frac{(-n-1)(-n)}{2}} = q^{\frac{(n+1)n}{2}}$ for $n \geq 0$, the sum on the right has constant coefficient 0; hence $[y^0] = (-1)(\sum_{n\in\mathbb{Z}} q^{\frac{n(n+1)}{2}}(y-1)^n)'\,|_{y=0} = \sum_{n\in\mathbb{Z}}(-1)^n n q^{\frac{n(n+1)}{2}}$. By the observation $q^{\frac{(-n-1)(-n)}{2}} = q^{\frac{(n+1)n}{2}}$ above, this finally gives $\prod_{k\geq 1}(1-q^k)^3 = \sum_{n\geq 0}(-1)^n(2n+1)q^{\frac{n(n+1)}{2}}$.

3.82 The substitution $x \mapsto q^{-N}z$ in the q-binomial formula for $2N$ gives

$$\begin{aligned}&(1+q^{-(N-1)}z)\cdots(1+q^{-1}z)(1+z)(1+qz)\cdots(1+q^N z)\\ &\quad = \sum_{k=0}^{2N}\left[{2N \atop k}\right] q^{\frac{(k+1)k}{2}-kN} z^k = \sum_{i=-N}^{N}\left[{2N \atop N+i}\right] q^{\frac{(N+i)(-N+i+1)}{2}} z^{N+i}.\end{aligned}$$

Now multiply both sides by $q^{1+2+\cdots+(N-1)}z^{-N}$. Then we obtain

$$\begin{aligned}&(1+q^{N-1}z^{-1})\cdots(1+qz^{-1})(1+z^{-1})(1+qz)\cdots(1+q^N z)\\ &\quad = \sum_{i=-N}^{N}\left[{2N \atop N+i}\right] q^{\frac{i(i+1)}{2}} z^i.\end{aligned}$$

With $N \to \infty$ the left-hand side goes to $\prod_{k\geq 1}(1+zq^k)(1+z^{-1}q^{k-1})$. Since $\left[{2N \atop N+i}\right] = \sum_j p(j; \leq N+i; \leq N-i)q^j \overset{N\to\infty}{\longrightarrow} \sum_j p(j)q^j = \frac{1}{\prod_{k\geq 1}(1-q^k)}$, the right-hand side goes to $\frac{1}{\prod_{k\geq 1}(1-q^k)}\sum_{n\in\mathbb{Z}} q^{\frac{n(n+1)}{2}} z^n$, and this is Jacobi's theorem.

Chapter 4

4.3 Write $P = NK - g(n)K - f(n)$ as $P = (K - I)(N - g(n)) + N - f(n) - g(n)$. Hence $S(N,n) = N - f(n) - g(n)$ is a recurrence operator for the sum $s(n)$, that is, $s(n+1) = (f(n) + g(n))s(n)$.

4.5 Cross-multiplication gives the recurrence $s(n) = \frac{n}{n+m}s(n-1)$ for the sum, hence $s(n) = \frac{1}{m}\binom{m+n}{n}^{-1}$ as $s(0) = \frac{1}{m}$.

4.8 Set $n = m = p$ in Dixon's identity; then $\sum_{i=-n}^{n}(-1)^i\binom{2n}{n+i}^3 = \frac{(3n)!}{(n!)^3}$ and with $k = n + i$, $\sum_{k=0}^{2n}(-1)^k\binom{2n}{k}^3 = (-1)^n\frac{(3n)!}{(n!)^3}$.

4.12 We have $\frac{a(n)}{a(n-1)} = \frac{n(n-2)}{(n-1)(n+1)}$, $p(n) = n$, $q(n) = n - 2$, $r(n) = n + 1$. The equation is $(n-1)f(n) - (n+1)f(n-1) = n$ with solution $f(n) = -n - \frac{1}{2}$. Hence $S(n+1) = \frac{n-1}{(n^2-1)n}(-n - \frac{1}{2}) = -\frac{2n+1}{2n(n+1)}$, and $\sum_{i=2}^{n}\frac{1}{i^2-1} = S(n+1) - S(3) + a(2) = \frac{3n^2-n-2}{4n(n+1)}$. The second sum is not a closed form.

4.15 Here $\frac{a(n,k)}{a(n,k-1)} = \frac{\binom{n}{k}}{\binom{n}{k-1}} = \frac{n-k+1}{k}$, $q(n) = n - k + 1$, $r(k) = k$, and $(n-k)f(k) - kf(k-1) = 1$ has no solution. For $a(n+1,k) - a(n,k)$ one obtains $S(n+1) = -\binom{n}{k}/2^{n+1}$.

4.18 We have $\frac{a(n)}{a(n-1)} = \frac{(4n-2)}{n}a$, $p(n) = 1$, $q(n) = a(4n-2)$, $r(n) = n$, $a(4n+1)f(n) - nf(n-1) = 1$. For this to be solvable, the highest coefficients must cancel, which forces $a = \frac{1}{4}$. For $a = \frac{1}{4}$ we obtain $f(n) = 2$ as solution, and thus $S(n+1) = \frac{2n+1}{4^n}\binom{2n}{n}$. Since $S(1) = a(0) = 1$, $\sum_{k=0}^{n}\binom{2n}{k}(\frac{1}{4})^k = \frac{2n+1}{4^n}\binom{2n}{n}$.

4.20 The result is $\frac{2^n}{n^2} - 2$.

4.23 The value is $n + 1$.

4.25 Using the approach of the last example we obtain $t_0 = z$, $t_1 = z - 1$, $S(N,n) = z + (z-1)N$, $H(n,k) = S(N,n)\binom{k}{n}z^k = \binom{k}{n}z^{k+1} + \binom{k}{n+1}z^k(z-1) = G(n,k+1) - G(n,k)$. Proceeding as usual, one gets $G(n,k+1) = \binom{k+1}{n+1}z^{k+1}$. Now sum $H(n,k) = G(n,k+1) - G(n,k)$ from $k = 0$ to $k = 2n+2$. This gives $zs_n(z) + \binom{2n+1}{n}z^{2n+2} + \binom{2n+2}{n}z^{2n+3} + (z-1)s_{n+1}(z) = \binom{2n+3}{n+1}z^{2n+3}$, or $s_n(z) = \frac{z}{1-z}[s_{n-1}(z) + (1-2z)z^{2n-1}\binom{2n-1}{n}]$. Unwrap the recurrence to get the result, which also proves the assertion for $z = \frac{1}{2}$.

4.28 For the combinatorial argument classify the mappings according to the largest k such that $\{1,\ldots,k+1\}$ is mapped bijectively, $k = 0,\ldots,n-1$. This gives $\sum_{k=0}^{n+1}\binom{n}{k+1}(k+1)!(k+1)n^{n-2-k} = n^n$. Now divide by n^{n-1}.

4.33 With index shift $\sum_{k=0}^{n-1}(k+1)\binom{m-1-k}{m-n}$, $t_0 = \binom{m-1}{m-n}$, $\frac{t_{k+1}}{t_k} = \frac{k+2}{k+1}\frac{n-k-1}{m-k-1}$. Hence the sum equals $\binom{m-1}{m-n}F\left({2,-(n-1) \atop 1-m};1\right) = \binom{m-1}{m-n} \cdot \frac{(m+1)^{\overline{n-1}}}{(m-1)^{\overline{n-1}}} = \binom{m-1}{m-n} \cdot \frac{(m+1)m}{(m-n+2)(m-n+1)} = \binom{m+1}{m-n+2} = \binom{m+1}{n-1}$.

4.35 We have $\frac{t_{k+1}}{t_k} = -\frac{m-k}{n+k+1}\frac{p-k}{m+k+1}\frac{n-k}{p+k+1}$, $t_0 = \binom{n+m}{n}\binom{m+p}{m}\binom{n+p}{p}$. Hence $F\left({-m,-n,-p,1 \atop m+1,n+1,p+1};1\right) = \frac{(m+n+p)!m!n!p!}{(n+m)!(m+p)!(n+p)!}$.

4.37 Proceeding as usual, $\frac{t_{k+1}}{t_k} = -\frac{r-k}{k+1}\frac{n-k}{r+n-k}\frac{s+n-k}{s+n-k-1}$, $t_0 = \binom{r+n}{n}\frac{s}{s+n}$. The sum is therefore $\binom{r+n}{n}\frac{s}{s+n}F\left({-r,-(s+n),-n \atop -(r+n),-(s+n-1)};1\right)$, which by Saalschütz equals $\frac{(s-r)^{\overline{n}}}{(s+1)^{\overline{n}}}$.

4.41 With index shift the sum becomes $\sum_{k\geq 0}\binom{2n+1+k}{n+1+k}2^{-k} = 2^{2n+1}$. Evaluating the sum as before, we obtain $F\left({2n+2,1 \atop n+2};\frac{1}{2}\right) = 2^{2n+1}/\binom{2n+1}{n}$. Now apply the preceding exercise with $a = n+1$, $b = \frac{1}{2}$ to obtain $F\left({n+1,\frac{1}{2} \atop n+2};1\right) = 2^{2n+1}/\binom{2n+1}{n}$.

4.43 The identity follows by differentiating the equation of the previous exercise n times. With the index shift $k \mapsto m-n-k$ and the usual method the left-hand side becomes $\binom{m+r}{m-n}F\left({n+1,-(m-n) \atop r+n+1};-x\right)$, and the right-hand side $(-1)^{m-n}\binom{-r}{m-n}F\left({n+1,-(m-n) \atop -r-m+n+1};1+x\right)$. Now set $z = 1+x$, $a = n+1$, $c = r+n+1$, and $m-n \mapsto n$ to obtain the formula.

4.45 a. One of $\frac{N+1}{2}$, $\frac{N}{2}$ is an integer; suppose $\frac{N}{2} = m$. Then by (3),

$$F = \frac{(n+m+1)^{\overline{m}}}{(n+\frac{1}{2})^{\overline{m}}} = \frac{(n+m+1)\cdots(n+2m)2^m}{(2n+1)(2n+3)\cdots(2n+2m-1)}.$$

Multiply numerator and denominator by $(2n+2)\cdots(2n+2m)$. This gives

$$F = 2^N\frac{(n+m+1)\cdots(n+2m)(n+1)\cdots(n+m)}{(2n+1)\cdots(2n+2m)} = 2^N\frac{(n+N)!(2n)!}{n!(2n+N)!}.$$

A similar argument settles the case in which N is odd, and the other assertions are proved analogously.

Chapter 5

5.3 The number of permutations with k (or more) cycles of length ℓ is $S_k = \sum_{T:|T|=k} N_{\supseteq T} = \binom{n}{\ell}\binom{n-\ell}{\ell}\cdots\binom{n-(k-1)\ell}{\ell}(\ell-1)!^k(n-k\ell)!\frac{1}{k!} = \frac{n!}{\ell^k k!}$. Hence $N_{=\emptyset} = n!\sum_{k\leq n/\ell}\frac{(-1)^k}{\ell^k k!}$. To use the composition formula set $f(\ell) = 0$, $f(i) = 1$ for $i \neq \ell$; then the exponential generating function is $\exp(\sum f(n)\frac{z^n}{n}) = \exp(-\log(1-z) - \frac{z^\ell}{\ell}) = \frac{1}{1-z}e^{-z^\ell/\ell}$ with $[z^n] = n!\sum_{k\leq n/\ell}\frac{(-1)^k}{\ell^k k!}$.

5.6 Let e_n and o_n denote the number of permutations in $S(n)$ with an even respectively odd number of fixed points. Set $f(1) = -1$, $f(n) = 1$ for $n > 1$ and apply Theorem 3.5. Then $h(n) = e_n - o_n$, and $\widehat{H}(z) = \exp(-z + \frac{z^2}{2} + \frac{z^3}{3} + \cdots) = \exp(-\log(1-z) - 2z) = \frac{e^{-2z}}{1-z}$. Hence $h(n) = n! \sum_{k=0}^{n} \frac{(-2)^k}{k!}$, and this is easily seen to be greater than 0 for $n \geq 4$.

5.9 We have

$$\begin{aligned}\textstyle\sum_{p=0}^{n} N_p x^p &= \textstyle\sum_{p=0}^{n} \left(\sum_{k=p}^{n} (-1)^{k-p} \binom{k}{p} S_k\right) x^p \\ &= \textstyle\sum_{k=0}^{n} \left(\sum_{p=0}^{k} (-1)^{k-p} \binom{k}{p} x^p\right) S_k = \sum_{k=0}^{n} (x-1)^k S_k .\end{aligned}$$

5.11 Let e_i be the property that i is red and $i-1$ blue, $i = 2, \ldots, 2n$. Two red positions must be two places apart; hence $\sum_{|T|=k} N_{\supseteq T} = \binom{2n-k}{k} 2^{2n-2k}$, and so $N_{=\emptyset} = \sum_k (-1)^k \binom{2n-k}{k} 2^{2n-2k}$. On the other hand, we clearly have $N_{=\emptyset} = 2n+1$, since these colorings correspond to a string of red followed by a string of blue. For the generating function use snake oil. For the last assertion color the integers $1, \ldots, n$ such that if i is not blue, then $i-1$ is also not blue.

5.14 Seat the women first in alternate places ($2n!$ possibilities), and label the woman at the end 1, the next 2, and so on. The condition for the men calls for a permutation $a_1 \ldots a_n$ with $a_1 \neq 1$, $a_i \neq i-1, i$ $(i \geq 2)$. Consider properties $e_1, \ldots, e_{2n-1}$ where e_{2i} means $a_{i+1} = i$, and e_{2i+1} means $a_{i+1} = i+1$. Properties e_{2i}, e_{2i+1} and e_{2i}, e_{2i-1} are mutually exclusive. Thus $\sum_{|T|=k} N_{\supseteq T} = \binom{2n-k}{k}(n-k)!$, and # seatings $= 2n! \sum_{k=0}^{n} (-1^k \binom{2n-k}{k} (n-k)!$.

5.18 Let $x_1, x_2, \ldots, x_{n-k+1}$ be the lengths of the runs between the $n-k$ missing integers. Then $0 \leq x_i < s$, $\sum x_i = k$, and so $C(n,k,s) = \sum_i (-1)^i \binom{n-k+1}{i} \binom{n-is}{n-k}$ by (6).

5.21 Let X be the set of all collections of r packages, $|X| = \binom{sn}{r}$, and e_i the property that coupon i is not in the package. Clearly, $N_{\geq k} = \binom{s(n-k)}{r}$; hence $N_{=0} = \sum_k (-1)^k \binom{n}{k} \binom{s(n-k)}{r}$. The probability of getting a complete set is $N_{=0} / \binom{sn}{r}$, $S_1 = n\binom{s(n-1)}{r}$; hence $\mathbb{E} = n(1 - \binom{s(n-1)}{r} / \binom{sn}{r})$.

5.23 Set $w(m) = m^2$ for $m \in \{1, \ldots, n\}$, and let e_i be the property that $p_i \mid m$. Suppose $T = \{e_1, \ldots, e_k\}$, $\prod = p_1 \cdots p_k$. Then $W_{\supseteq T} = \Pi^2(1^2 + 2^2 + \cdots + (n/\Pi)^2) = \frac{n(n+\Pi)(2n+\Pi)}{6\Pi} = \frac{n}{6}\left[\frac{2n^2}{\Pi} + 3n + \Pi\right]$. The signed first sum over all T is $\frac{n^2}{3} \sum_T (-1)^k \frac{n}{p_1 \cdots p_k} = \frac{n^2}{3}\varphi(n)$. The second sum is 0, and the third $\frac{1}{6}(-1)^t p_1 \cdots p_t \varphi(n)$. For cubes one obtains $\frac{n^3}{4}\varphi(n) + (-1)^t \frac{n}{4} p_1 \cdots p_t \varphi(n)$.

5.27 From $\zeta = \delta + \eta$ follows $\mu = \zeta^{-1} = \frac{1}{\delta+\eta} = \sum_k (-1)^k \eta^k$. Now $\eta^k(\mathbb{C}_n) = \binom{n-1}{k-1}$ (ordered number partitions), $\eta^k(\mathbb{B}(n)) = k! S_{n,k}$ (ordered set-partitions). Hence $\sum_k (-1)^k \binom{n-1}{k-1} = -1$ for $n = 1$, $= 0$ for $n \geq 2$, and $\sum_k (-1)^k k! S_{n,k} = (-1)^n$.

5.31 We have $\varphi(n) = \sum_{d|n} \overline{\mu}(\frac{n}{d})d$; hence by Möbius inversion $n = \sum_{d|n} \varphi(d)$.

5.34 We have $\sum_1^n \frac{H_x}{(x+1)(x+2)} = \sum_1^n H_x x^{\underline{-2}} = -x^{\underline{-1}} H_x \big|_1^n + \sum_1^n \frac{1}{x+1}(x+1)^{\underline{-1}} = -\frac{H_x}{x+1}\big|_1^n + \sum_1^n x^{\underline{-2}} = -\frac{H_x}{x+1}\big|_1^n - \frac{1}{x+1}\big|_1^n = 1 - \frac{H_n+1}{n+1}$.

5.36 Since $\mu = \frac{1}{\delta+\eta}$ we have $\mu^r = \sum_{k\geq 0} \binom{r+k-1}{k}(-1)^k \eta^k$. Now $\mu^r = \mu^{r-1} * \mu$ and induction yields $\mu^r(\mathbb{C}_n) = (-1)^n \binom{r}{n}$ and $\mu^r(\mathbb{B}(n)) = (-1)^n r^n$. The identities follow from Exercise 5.27.

5.38 Let $g(X)$ be the number of linear maps $h : V^n \to V^r$ with $\mathrm{im}(h) = X \subseteq V^r$. Then $f(A) = \sum_{X\subseteq A} g(X) = \#\{h : \mathrm{im}(h) \subseteq A\} = q^{(\dim A)n}$. For the number of surjective maps we get $g(V^r) = \sum_X f(X)\mu(X, V_r) = \sum_{k=0}^r \left[{r \atop k}\right]_q q^{kn}(-1)^{r-k} q^{\binom{r-k}{2}}$ by the previous exercise.

5.40 Following the hint, $f(a) = \sum_{\pi \geq a} g(\pi)$ counts all maps h with $\ker(h) \geq a$. Hence $f(a) = x^{b(a)}$, where $b(a)$ is the number of blocks. By Möbius inversion, $g(0) = \sum_y \mu(0,y) f(y) = \sum_y \mu(0,y) x^{b(y)}$. Since $g(0)$ counts all injective maps, we have $g(0) = x^{\underline{n}} = x(x-1)\cdots(x-n+1)$. Now, $\mu(0,1)$ is the linear coefficient of the polynomial, and so $\mu(0,1) = (-1)^{n-1}(n-1)!$.

5.44 We have $nQ_n = n(n-1) + 2\sum_{k=0}^{n-1} Q_k$ $(n \geq 1)$, $(n-1)Q_{n-1} = (n-1)(n-2) + 2\sum_{k=0}^{n-2} Q_k$ $(n \geq 2)$. Subtraction gives $nQ_n = (n+1)Q_{n-1} + 2(n-1)$ for $n \geq 2$, and this holds for $n = 1$ as well. Unwrapping the recurrence, we get $Q_n = 2(n+1)\sum_{k=0}^{n-1} \frac{k}{(k+1)(k+2)}$. Now use the difference calculus to obtain the result $Q_n = 2(n+1)H_n - 4n$.

5.47 Since $\mathcal{D}$ is a down-set, it is also a lower semilattice. Set $f(\emptyset) = 1$ and $f(U) = 0$ for $U \neq \emptyset$, then $M = (m_{ij})$ satisfies $m_{ij} = f(U_i \cap U_j)$. Clearly, $f(U) = \sum_{X\subseteq U} g(X)$ with $g(X) = (-1)^{|X|}$; hence $\det M = (-1)^\ell$, where ℓ is the number of odd-sized sets in $\mathcal{D}$.

5.51 For odd k, $\varphi : \{a_1, \ldots, a_k\} \mapsto \{n+1-a_1, \ldots, n+1-a_k\}$ is an alternating involution without fixed points; hence $|S^+| - |S^-| = 0$. Suppose k is even. We denote the places by $1, 1', 2, 2', \ldots, \frac{k}{2}, \frac{k'}{2}$, and write $A \in S$ as $a_1, a_1', \ldots, a_{k/2}, a_{k/2}'$. Call a pair (a_i, a_i') good if $a_i' = a_i + 1$, a_i odd; otherwise, bad. Let a_j, a_j' be the last bad pair; then $\varphi A = \{a_1, a_1', \ldots, a_{j-1}, a_{j-1}', a_j, b_j, \ldots, a_{k/2}, a_{k/2}'\}$, where $b_j = a_j' - 1$ if a_j' is even, and $b_j = a'_j + 1$ if a'_j is odd. If there is no bad pair, set $\varphi A = A$. Check that this involution works.

5.53 Consider $S = \{(\lambda, \mu) : \lambda \in \text{Par}, \mu \in \text{Par}_d(; \geq k+1)\}$ with $w(\lambda, \mu) = x^{|\lambda|+|\mu|}$, sign $(\lambda, \mu) = (-1)^{b(\mu)}$, where $b(\mu)$ is the number of parts in μ. To define φ take the largest part m of λ and μ. If $m \in \mu$ move it to λ, if $m \in \lambda \setminus \mu$ and $m \geq k+1$ move it to μ. Otherwise $\varphi(\lambda, \mu) = (\lambda, \mu)$. Clearly, $\text{Fix}_\varphi S = \{\lambda, \emptyset); \lambda \in \text{Par}(; \leq k)\}$, $w(\text{Fix}_\varphi S) = \frac{1}{(1-z)\cdots(1-z^k)}$. Finally, observe that $w(S^+) - w(S^-) = w(\text{Par}) \cdot \sum (p_{d,e}(n; \geq k+1) - p_{d,o}(n; \geq k+1))z^n = \frac{1}{\prod_{i\geq 1}(1-z^i)} \cdot (1-z^{k+1})(1-z^{k+2})\cdots$, and the result follows.

5.55 For $n \geq 2$ define on $S(n)$ the involution $\sigma \mapsto \sigma^*$ as in the hint. It follows that $i \in D(\sigma^*) \iff n+1-\sigma^*(i) \in D(\sigma)$; hence $|D(\sigma^*)| = |D(\sigma)|$. This gives $P(\sigma) + P(\sigma^*) = \frac{1}{|D(\sigma)|} \sum_{i \in D^*(\sigma)} (i + \sigma^*(i) + (n+1-\sigma^*(i)) + (n+1-i)) = 2(n+1)$. Summation over the pairs (σ, σ^*) gives the result.

5.60 If i and $i+1$ are not next to each other in a cycle of σ, then $e(\hat{\sigma}) = e(\sigma)$; otherwise $e(\hat{\sigma}) = e(\sigma) \pm 1$. In the first case it is easily seen that $\#\hat{\mu}_1 + \#\hat{\mu}_2 = \#\mu_1 + \#\mu_2$ or $\#\mu_1 + \#\mu_2 \pm 2$, and in the second case $\#\hat{\mu}_1 + \#\hat{\mu}_2 = \#\mu_1 + \#\mu_2 \pm 1$.

5.62 Note first that the number of 1's in λ of a pair $\lambda \times \mu$ in the course of the algorithm has the same parity as the number of 1's in the starting pair $\lambda_0 \times \emptyset$. Hence when we run the algorithm on $1 \ldots 1 \in \text{Par}_o(2n)$ and $1 \ldots 11 \in \text{Par}_o(2n+1)$, the steps $\lambda \times \mu \to \lambda' \times \mu'$ and $\lambda 1 \times \mu \to \lambda' 1 \times \mu'$ run in parallel. It follows that $h(2n) = h(2n+1)$. Furthermore, all parts that appear are powers of 2; hence the mate is $2^{k_1} \ldots 2^{k_t}$, where $2^{k_1} + \cdots + 2^{k_t}$ is the binary expansion of n.

5.65 Suppose without loss of generality that the n-th row is linearly dependent on the other rows, $A_n = \lambda_1 A_1 + \cdots + \lambda_{n-1} A_{n-1}$. Construct a graph with A_n on top, arrows with weight λ_i to A_i, and otherwise m_{ij} as before $(1 \leq i \leq n-1, 1 \leq j \leq n)$. Then M is the path matrix, but there is no vertex-disjoint path system. Hence $\det M = 0$.

5.68 Construct the lattice graph as in the Catalan example with diagonal steps up-down, and horizontal steps all with weight 1. Then H_n is the path matrix, and there is only one vertex-disjoint path system $P_i : A_i \to B_i$ of weight 1.

5.71 Replace the Ferrers diagram by a lattice graph with vertices in the cells and all steps up and to the right of weight 1. Furthermore, add vertices $A_1, \ldots, A_k$ below the k columns of the Durfee square D, and $B_1, \ldots, B_k$ to the right of the k rows of D. Then M is the path matrix, and there is clearly only one system $\mathcal{P} : A_i \to B_i$ of weight 1.

5.73 Consider the lattice graph with diagonal up-steps to the right of weight 1 and $(1,0)$-steps with weight c_k when $y = k$. Let $A_i = (-i, 0)$, $B_j = (m, j)$, $i, j = 1, \ldots, n$. The recurrence shows that $(a_{m+i,j})$ is the path matrix. The only vertex-disjoint path systems are $\mathcal{P} : A_i \to B_i$. Classify these paths according to the number i_j of horizontal steps of the path $A_j \to B_j$ at $y = j - 1$. Since $\mathcal{P}$ is vertex-disjoint, we have $m \geq i_1 \geq i_2 \geq \cdots \geq i_n \geq 0$, and $w(\mathcal{P}) = c_0^{i_1} c_1^{m-i_1} c_1^{i_2} c_2^{m-i_2} \cdots c_{n-1}^{i_n} c_n^{m-i_n} = (c_1 \cdots c_n)^m s_1^{i_1} \cdots s_n^{i_n}$. The summation formula follows. For $c_i = q^{-i}$ we get $\det = q^{-\binom{n+1}{2}} \left[{m+n \atop n} \right]_q$.

5.76 Consider the lattice graph with $(1,0)$- and $(0,1)$-steps of weight 1. Let $A_0 = (0,0)$, $B_0 = (r,s)$, and $A_i = B_i = (a_i, b_i)$, $i = 1, \ldots, n$. The vertex-disjoint path systems from $\mathcal{A}$ to $\mathcal{B}$ correspond precisely to the paths from $(0,0)$ to (r,s) that avoid S. The path matrix $M = (m_{ij})$ has entries $m_{00} = \binom{r+s}{r}$, $m_{0j} = \binom{a_j+b_j}{a_j}$, $m_{i0} = \binom{r+s-a_i-b_i}{r-a_i}$, $m_{ii} = 1$ $(i > 0)$, $m_{ij} = \binom{a_j+b_j-a_i-b_i}{a_j-a_i}$ $(i < j)$, $m_{ij} = 0$ $(i > j)$. Hence # paths = $\det M$. For inclusion–exclusion use properties e_i, meaning that A_i is hit, $i = 1, \ldots, n$.

Chapter 6

6.3 Let X be the set of colorings. Under C_6 we have $|X_{\rho^0}| = 2^6 = 64$, $|X_\rho| = |X_{\rho^5}| = 2$, $|X_{\rho^2}| = |X_{\rho^4}| = 4$, $|X_{\rho^3}| = 8$. Hence by Burnside–Frobenius $|\mathcal{M}| = \frac{1}{6}(64 + 4 + 8 + 8) = 14$. For D_6 one obtains $|\mathcal{M}| = 13$.

The two patterns are

```
   ●         ●
 ○   ●     ○   ●
 ○   ○     ●   ○
   ●         ○     .
```

6.5 The lattice paths correspond to sequences of n 0's and n 1's, where 0 represents a $(1,0)$-step and 1 a $(0,1)$-step. The group consists of the identity, the reverse $\sigma : a_1 \ldots a_{2n} \to a_{2n} \ldots a_1$, the exchange $\varepsilon : a_1 \ldots a_{2n} \to \overline{a}_1 \ldots \overline{a}_{2n}$, where $\overline{a}_i \neq a_i$, and $\sigma\varepsilon$. The fixed-point sets X_g are easily seen to have sizes $|X_{\text{id}}| = \binom{2n}{n}$, $|X_\sigma| = \binom{n}{n/2}[n \text{ even}]$, $|X_\varepsilon| = 0$, $|X_{\sigma\varepsilon}| = 2^n$. Hence $|\mathcal{M}| = \frac{1}{4}\left(\binom{2n}{n} + 2^n + \binom{n}{n/2}[n \text{ even}]\right)$.

6.8 The group is C_3 with $|X_{\rho^0}| = 3^{\binom{n+1}{2}}$, $|X_\rho| = |X_{\rho^2}| = 3^{\lceil \frac{n(n+1)}{6} \rceil}$, thus $|\mathcal{M}| = \frac{1}{3}\left[3^{\binom{n+1}{2}} + 2 \cdot 3^{\lceil \frac{n(n+1)}{6} \rceil}\right]$.

6.11 Clearly, $A \in X_g$ if and only if A contains all elements of a cycle of g or none. It follows that $|X_g| = 2^{c(g)}$, $c(g)$ = # cycles in g. Hence $m = \frac{1}{|G|} \sum_{g \in G} 2^{c(g)}$.

6.13 The group consists of id and $g : x \to \overline{x}$, where $\overline{x} = 1 \iff x = 0$. The multiplication table $x \cdot y$ is fixed by g iff $\overline{x \cdot y} = \overline{x} \cdot \overline{y}$. Suppose $0 \cdot 0 = a$, then $\overline{a} = \overline{0 \cdot 0} = \overline{0} \cdot \overline{0} = 1 \cdot 1$, and if $0 \cdot 1 = b$ then $\overline{b} = \overline{0 \cdot 1} = 1 \cdot 0$. Hence the tables fixed by g are of the form $\begin{array}{c|cc} & 0 & 1 \\ \hline 0 & a & b \\ 1 & \overline{b} & \overline{a} \end{array}$, that is, $|X_g| = 4$. By Burnside-Frobenius $|\mathcal{M}_2| = \frac{1}{2}(2^4 + 4) = 10$. For three elements one obtains $|\mathcal{M}_3| = 3330$.

6.17 Let $\sigma = \tau^{-1}\pi\tau$, and suppose $(1, 2, \ldots, k)$ is a cycle of σ. Then $\tau^{-1}\pi\tau(i) = i+1$ or $\pi\tau(i) = \tau(i+1)$, and we conclude that $(\tau(1), \ldots, \tau(k))$ is a k-cycle of π. Hence σ has the same type as π. Now if $C(\pi)$ is the set of permutations conjugate to π, then by the previous exercise $|C(\pi)| = n!/|S(n)_\pi|$, where $S(n)_\pi = \{\rho \in S(n) : \rho\pi = \pi\rho\}$. According to Exercise 1.46, $|S(n)_\pi| = 1^{c_1}c_1!2^{c_2}c_2! \cdots n^{c_n}c_n!$, where $1^{c_1} \ldots n^{c_n}$ is the type of π; thus $|C(\pi)| = n!/1^{c_1}c_1! \cdots n^{c_n}c_n!$. But this last expression is precisely the number of permutations of type $1^{c_1} \ldots n^{c_n}$, and the result follows.

6.20 Consider all mappings $f : N \to X$, $|N| = n$, $|X| = x$ under the group $S(n)$. Then $|\mathcal{M}| = \binom{x+n-1}{n} = \frac{x^{\overline{n}}}{n!} = \frac{1}{n!}Z(S(n); x, \ldots, x) = \frac{1}{n!}\sum_\sigma x^{c_1(\sigma)} \ldots x^{c_n(\sigma)} = \frac{1}{n!}\sum_\sigma x^{c(\sigma)}$, where $c(\sigma)$ is the number of cycles of σ. Thus $x^{\overline{n}} = \sum_k s_{n,k}x^k$.

6.24 a. For $G = \{\text{id}\}$ we obtain $\sum_{k=0}^n \binom{n}{k}x^k = (1 + x)^n$. b. For $G = S(n)$ we have $m_k = 1$ for all k. For $x = 1$ this gives $n + 1 = \frac{1}{n!}\sum_\sigma 2^{c(\sigma)}$, thus $(n + 1)! = \sum_\sigma 2^{c(\sigma)}$. c. For the cyclic group C_n we get $\sum_{k=0}^n m_k x^k = \frac{1}{n}\sum_{d|n}\varphi(d)(1 + x^d)^{n/d}$. Now set $x = 0$ to obtain $1 = \frac{1}{n}\sum_{d|n}\varphi(d)$.

6.29 Looking at the cycle type of the reflections one easily obtains

$$Z(D_n) = \frac{1}{2}Z(C_n) + \begin{cases} \frac{1}{4}(z_2^{n/2} + z_1^2 z_2^{n/2-1}) & n \text{ even}, \\ \frac{1}{2}z_1 z_2^{(n-1)/2} & n \text{ odd}. \end{cases}$$

The number of self-complementary patterns under D_n is therefore $\frac{1}{2}|\mathcal{M}^c(C_n)| + 2^{n/2-2}[n \text{ even}]$.

6.33 The coefficient of z^{r+1-m} on the left is $e_{r+1-m}(1, 2, \ldots, r) = \sum k_1 k_2 \cdots k_{r+1-m}$ over all $1 \le k_1 < k_2 < \cdots < k_{r+1-m} \le r$. The right-hand side is $(1 + z)(1 + 2z) \cdots (1 + rz) = z^{r+1}\frac{1}{z}(\frac{1}{z} + 1) \cdots (\frac{1}{z} + r) = z^{r+1}(\frac{1}{z})^{\overline{r+1}} = z^{r+1}\sum_k s_{r+1,k}z^{-k}$. Hence $[z^{r+1-m}] = s_{r+1,m}$. Now set $n = r + 1$.

6.34 We know from (16) that $\sum_{n\ge0} Z(S(n); z_1, -z_2, \ldots, (-1)^{n-1}z_n)y^n = \exp(\sum_{k\ge1}(-1)^{k-1}z_k\frac{y^k}{k})$. Now set $z_k = p_k(x_1, \ldots, x_r) = \sum_{i=1}^r x_i^k$. Then $\sum_{k\ge1}(-1)^{k-1}p_k\frac{y^k}{k} = \sum_{k\ge1}\frac{(-1)^{k-1}}{k}((x_1y)^k + \cdots + (x_ry)^k) = \log(1 + x_1y) + \cdots + \log(1 + x_ry) = \log\prod_{i=1}^r(1 + x_iy)$. It follows that $\sum_{n\ge0} Z(S(n); p_1, -p_2, \ldots, (-1)^{n-1}p_n)y^n = \prod_{i=1}^r(1 + x_iy)$. Now compare coefficients for y^n, using the previous exercise.

6.36 Let the tree T be rooted at $e = \{a, b\}$, and define ϕT as the forest rooted at a and b after deletion of e. Then ϕ is a bijection between the isomorphism classes of edge-rooted trees on n vertices and the unordered pairs of isomorphism classes of rooted trees (including pairs whose trees are isomorphic) whose vertex numbers add to n. Hence $u_n^{(2)} = \frac{1}{2}\sum_{k=0}^{n} u_k u_{n-k}$ for n odd, and $u_n^{(2)} = \frac{1}{2}\sum_{k=0}^{n} u_k u_{n-k} + \frac{1}{2}u_{n/2}$ for n even. This translates into $U^{(2)}(x) = \frac{1}{2}[U(x)^2 + U(x^2)]$.

6.40 The edge group is the dihedral group D_n; hence the number of non-isomorphic subgraphs equals $Z(D_n; 2, \ldots, 2) = \frac{1}{2n}\sum_{d|n} \varphi(d)2^{n/d} + 2^{n/2-2} + 2^{n/2-1}$ for even n, and $\frac{1}{2n}\sum_{d|n}\varphi(d)2^{n/d} + 2^{(n-1)/2}$ for odd n.

6.43 The symmetry group is $S(m)[S(n)]$. The desired number is therefore $Z(S(m)[S(n)]; 2, \ldots, 2) = \frac{1}{m!}\sum_{\sigma \in S(m)} (n+1)^{c(\sigma)} = \frac{1}{m!}\sum_k s_{m,k}(n+1)^k = \frac{1}{m!}(n+1)^{\overline{m}} = \binom{m+n}{m}$.

6.47 Taking complements we have $g_{n,k} = g_{n,\binom{n}{2}-k}$. Now $\binom{n}{2}$ is even for $n \equiv 0, 1 \pmod 4$, and odd for $n \equiv 2, 3 \pmod 4$. Hence when $\binom{n}{2}$ is odd, that is, for $n \equiv 2, 3$, we have $e_n = o_n$. Suppose $n \equiv 0, 1 \pmod 4$; then $e_n - o_n = \sum_k g_{n,k}(-1)^k = Z(\overline{S}(n); 0, 2, 0, 2, \ldots) = \#$ self-complementary graphs, which is greater than or equal to 0.

6.49 Consider graphs on $4n$ vertices; then $s_{4n} = Z(\overline{S}(4n); 0, 2, \ldots, 0, 2)$. Following the analysis $Z(S(4n)) \to Z(\overline{S}(4n))$ in the text one sees that only those monomials $z_1^{c_1} \cdots z_{4n}^{c_{4n}}$ in $Z(S(4n))$ have to be taken into account for which $c_i > 0$ implies $i \equiv 0 \pmod 4$. From this it is easy to derive $s_{4n} = \sum_{(c_1,\ldots,c_n)} \frac{z^{a(c)}}{1^{c_1}c_1! \cdots n^{c_n}c_n!}$, where $\sum ic_i = n$ and $a(c) = 2\sum_{k=1}^{n} c_k(kc_k - 1) + 4\sum_{1 \le i < j \le n} \gcd(i,j)c_ic_j$. Now perform the same analysis for $s_{\vec{2n}}$ to arrive at the same expression.

6.51 We have

$$\begin{aligned}\sum_{n\ge 0} m(S(n), H)z^n &= \frac{1}{|H|}\sum_h Z(S(n); \lambda_1(h), \ldots, \lambda_n(h))z^n \\ &= \frac{1}{|H|}\sum_h \exp\Big(\sum_{k\ge 1}\lambda_k(h)\frac{z^k}{k}\Big).\end{aligned}$$

Now $\lambda_k(h) = \sum_{j|k} jc_j(h)$ and thus

$$\begin{aligned}\sum_{k\ge 1}\lambda_k(h)\frac{z^k}{k} &= \sum_{j\ge 1} jc_j(h)\Big(\frac{z^j}{j} + \frac{z^{2j}}{2j} + \cdots\Big) = \sum_{j\ge 1} c_j(h)(-\log(1 - z^j)) \\ &= \log\prod_{j=1}^{r}\Big(\frac{1}{1-z^j}\Big)^{c_j(h)}.\end{aligned}$$

The sum $\sum_{n\ge 0} m(S(n), H)z^n$ is therefore equal to

$$\frac{1}{|H|}\sum_h \Big(\frac{1}{1-z}\Big)^{c_1(h)}\Big(\frac{1}{1-z^2}\Big)^{c_2(h)} \cdots = Z\Big(H; \frac{1}{1-z}, \ldots \frac{1}{1-z^r}\Big).$$

For $H = \{\text{id}\}$ we get $\sum_{n\geq 0} m(S(n), \{\text{id}\})z^n = \frac{1}{(1-z)^r}$, confirming our old result $m(S(n), \{\text{id}\}) = \binom{r+n-1}{n}$.

6.54 For $G = S(n)$ we have

$$|\mathcal{M}_{\text{Inj}}| = \frac{1}{n!|H|} \sum_h \sum_{(c_1,\ldots,c_n)} \frac{n!}{1^{c_1}c_1!\cdots n^{c_n}c_n!} \cdot \prod_{k=1}^{n} k^{c_k}c_k!\binom{c_k(h)}{c_k},$$

where $(c_1,\ldots,c_n)$ runs through all cycle types in $S(n)$. This gives

$$|\mathcal{M}_{\text{Inj}}| = \frac{1}{|H|} \sum_{h\in H} \sum_{(i_1,\ldots,i_n)} \prod_{k=1}^{n} \binom{c_k(h)}{i_k}$$

over all $i_1 + 2i_2 + \cdots + ni_n = n$. Hence $|\mathcal{M}_{\text{Inj}}| = [x^n]Z(H; 1+x, 1+x^2, \ldots, 1+x^r)$. To see this directly note that $|\mathcal{M}_{\text{Inj}}|$ counts the number of all n-subset patterns of R under H, which are counted by $Z(H, 1+x, \ldots, 1+x^r)$ according to Exercise 6.23.

6.56 Take $g \in G$, $h \in H$ and set $A_h = \prod_{k=1}^{r} e^{kc_k(h)(z_k+z_{2k}+\cdots)}$. Then $(\frac{\partial}{\partial z_1})^{c_1(g)} \cdots (\frac{\partial}{\partial z_n})^{c_n(g)} A_h = c_1(h)^{c_1(g)}(c_1(h)+2c_2(h))^{c_2(g)} \cdots \cdot A_h$, which at $z = 0$ becomes $\prod_{k=1}^{n} \lambda_k(h)^{c_k(g)}$ as in (8). Summing, we obtain the result $\frac{1}{|H|}\sum_h Z(G; \lambda_1(h), \ldots, \lambda_n(h)) = |\mathcal{M}|$.

6.59 We have $Z(D_p; z_1, \ldots, z_p) = \frac{1}{2p}(z_1^p + (p-1)z_p + pz_1z_2^{\frac{p-1}{2}})$, hence $|\mathcal{M}_{\text{Bij}}| = Z(D_p) \cap Z(D_p) = \frac{1}{4p^2}(p! + (p-1)^2p + p^2 2^{\frac{p-1}{2}}(\frac{p-1}{2})!)$. Now 4 divides $2^{\frac{p-1}{2}}$ for $p \geq 5$, and so $4p|(p-1)! + (p-1)^2$.

Chapter 7

7.3 An OPS $(p_n(x))$ is a basis of $\mathbb{C}[x]$; hence L is uniquely determined by $Lp_n(x) = \delta_{n,0}$. Assume, conversely, that $(p_n(x))$ and $(q_n(x))$ are OPS for L. Then $p_0(x) = q_0(x) = 1$. For $n \geq 1$ let $q_n(x) = \sum_{i=0}^{n} c_{n,i}p_i(x)$, then $0 = L(q_n(x)) = \sum_{i=0}^{n} c_{n,i}L(p_i(x)) = c_{n,0}$; thus $c_{n,0} = 0$. Assume $c_{n,0} = \cdots = c_{n,k} = 0$ for $k \leq n-2$. By the same argument $0 = L(x^{k+1}q_n(x)) = \sum_{i=k+1}^{n} c_{n,i}L(x^{k+1}p_i(x)) = c_{n,k+1}\lambda_{k+1}$, which implies $c_{n,k+1} = 0$. Since $c_{n,n} = 1$, $q_n(x) = p_n(x)$ follows.

7.5 Looking at the right half of the triangle we infer $Tr_n = B_n^{\sigma,\tau}$ for $\sigma \equiv 1$, $\tau = (2, 1, 1, \ldots)$. The last assertion follows by expanding $(1 + x + x^2)^n$ and induction.

7.11 The recurrence follows upon developing the determinant according to the last column. Translated into exponential generating functions this gives $\hat{U}' = x\hat{U} - \alpha z\hat{U} + z^2\hat{U}'$. Hence $(\log \hat{U})' = \frac{\hat{U}'}{\hat{U}} = \frac{x-\alpha z}{1-z^2} = \frac{1}{2}\frac{x-\alpha}{1-z} + \frac{1}{2}\frac{x+\alpha}{1+z} =$

$-\frac{x-\alpha}{2}(\log(1-z))' + \frac{x+\alpha}{2}(\log(1+z))'$, that is, $\hat{U}(x,\alpha;z) = (1+z)^{\frac{x+\alpha}{2}}/(1-z)^{\frac{x-\alpha}{2}}$. The convolution formula follows from the multiplicativity of the functions, similarly the reciprocity law.

7.14 Let D_0 be the derivative as in Chapter 2. The ballot recurrence translates into $D_0B_k = B_{k-1} + B_k + \cdots, B_k(0) = 0$ for $k \geq 1$, $D_0B_0 = B_0 + B_1 + \cdots, B_0(0) = 1$. Following the hint we set $B_k = B_0F^k$ with $F(0) = 0$. Then $D_0B_k = D_0(B_0F) = B_0F^{k-1}(D_0F) = B_0F^{k-1}\frac{1}{1-F}$. Canceling B_0F^{k-1} this gives with $D_0F = \frac{F}{z}$ the solution $F(z) = \frac{1-\sqrt{1-4z}}{2} = zC(z)$, $C(z)$ ordinary Catalan series. Now $D_0B_0 = \frac{B_0-1}{z} = B_0\frac{1}{1-F} = \frac{B_0F}{z}$ yields $B_0 = \frac{1}{1-F} = \frac{F}{z} = C(z)$, hence $b_{n,0} = C_n$.

7.17 We know from Exercise 7.14 that $B_k(z) = z^kC(z)^{k+1}$; hence $q_0(z) = 1$, and $B_1(z) = zC^2(z) = C(z) - 1$ implies $q_1(z) = r_1(z) = 1$. Using induction we obtain $B_{k+1} = z^{k+1}C^{k+2} = (zC)B_k = (zC)(q_kC - r_k) = zq_kC^2 - r_kzC = q_k(C-1) - r_kzC = (q_k - zr_k)C - q_k$, which proves the degree conditions. Furthermore, $r_{k+1} = q_k$, $q_{k+1} = q_k - zr_k = q_k - zq_{k-1}$. The recurrence for $q_k(z)$ is easily seen to have the solution $q_k(z) = \sum_i(-1)^i\binom{k-i}{i}z^i$. This yields $[z^n]B_k(z) = b_{n,k} = \sum_{i\geq 0}(-1)^i\binom{k-i}{i}C_{n-i}$, since $q_{k-1}(z)$ has degree $\leq \frac{k-1}{2} < n$.

7.20 Clearly, $\det H_n = 1$. For $\det H_n^{(1)}$ the recurrence is $r_n = r_{n-1} - r_{n-2}$, $r_{-1} = r_0 = 1$ with solution $r_n = \begin{cases} 1 & n \equiv 0,5 \\ 0 & n \equiv 1,4 \pmod 6 \\ -1 & n \equiv 2,3. \end{cases}$

7.23 To apply the lemma of Gessel-Viennot set A_0 and B_0 two apart and consider the strip of width 2. Alternatively, (8) gives $d_n^{(2)}d_n^{(0)} = d_{n+1}^{(0)}d_{n-1}^{(2)} + (d_n^{(1)})^2$, that is, $d_n^{(2)}T_0T_1\cdots T_n = T_0T_1\cdots T_{n+1}d_{n-1}^{(2)} + r_n^2T_0^2\cdots T_n^2$. This yields $d_n^{(2)}/T_0\cdots T_{n+1} = d_{n-1}^{(2)}/T_0\cdots T_n + r_n^2/T_{n+1}$ and therefore $d_n^{(2)}/T_0\cdots T_{n+1} = \sum_{j=1}^n r_j^2/T_{j+1} + d_0^{(2)}/T_1$. With $d_0^{(2)} = s_0^2 + t_1$ we obtain $d_0^{(2)}/T_1 = (s_0^2 + t_1)/t_1 = r_0^2/T_1 + 1$, and the formula follows.

7.25 Start the induction with $n = 0$, $k = 1, 2$, and show that the product satisfies the recurrence (8). Note that a formula for C_k results: $C_k = \det H_0^{(k)} = \prod_{1\leq i\leq j\leq k-1}\frac{i+j+2}{i+j}$.

7.28 It is readily seen that B_{2n} satisfies the recurrence $B_{2n+2} = \alpha B_{2n} + \beta\sum_{k=0}^{n-1}B_{2k}B_{2n-2k}$ (see, e.g., Exercise 7.66). Now we prove the same recurrence for the Catalan paths weighted by the number of peaks. If a Catalan path has $k+1$ peaks, then it has k valleys $\searrow\!\nearrow$. Give such a path P of length $2n$ the weight $w(P) = \alpha^{n-k}\beta^k$, and let $V_n = \sum w(P)$. We classify the paths P of length $2n+2$ according to the first point $(2k, 0)$ when P touches the x-axis, $k = 1, \ldots, n+1$. If $k = n+1$, then we get αV_n as contribution, and for $k \leq n$ $\beta V_{k-1}V_{n+1-k}$. Thus $V_{n+1} =$

$\alpha V_n + \beta \sum_{k=1}^{n} V_{k-1}V_{n+1-k} = \alpha V_n + \beta \sum_{k=0}^{n-1} V_k V_{n-k}$. It suffices therefore to prove the last assertion. Consider all pairs $a_1 + \cdots + a_{k+1}$, $b_1 + \cdots + b_{k+1}$ of ordered partitions of $n+1$ and n, respectively. Now arrange cyclically a_1 A's followed by b_1 B' s, then a_2 A's, and so on. We obtain the same cyclic arrangement by choosing $a_i a_{i+1} \ldots$, $b_i b_{i+1} \ldots$, $i = 1, \ldots, k+1$. Since $n+1$ and n are relatively prime, any two such pairs are distinct. Hence the number of these cyclic arrangements is $\frac{1}{k+1}\binom{n}{k}\binom{n-1}{k} = \frac{1}{n}\binom{n}{k}\binom{n}{k+1}$. Now check that for any arrangement there is exactly one way to break it into a linear word of n A's and n B's that begins with A and after deleting this initial A has always at least as many A's as B's. The result follows.

7.30 The recurrence reads $D_q(Bz^k) = q_k Bz^{k-1} + s_k Bz^k + u_{k+1}Bz^{k+1}$ $(k \geq 1)$, $D_q B = aB + bBz$. We know that $D_q(Bz^k) = q^k z^k (D_q B) + Bq_k z^{k-1}$. Inserting $D_q B = aB + bBz$ into the first equation and canceling gives $aq^k + bq^k z = s_k + u_{k+1}z$, that is, $s_k - aq^k = (bq^k - u_{k+1})z$. It follows that $s_k = aq^k$, $t_k = \frac{1-q^k}{1-q}u_k = \frac{1-q^k}{1-q}bq^{k-1}$ is the only solution.

7.34 Set $H(z) = \sum_{k\geq 0} c_k A_k$, then

$$\begin{aligned} H(\Delta)p_n(x)\big|_{x=0} &= \textstyle\sum_{k\geq 0} c_k A_k(\Delta) p_n(x)\big|_{x=0} \\ &= \textstyle\sum_{k\geq 0} c_k \delta_{n,k} = c_n \,. \end{aligned}$$

7.38 We have $B(z)H(F(z)) = B(z)\sum_{k\geq 0} h_k F(z)^k = \sum_{k\geq 0} h_k A_k(z)$, hence $[z^n]B(z)H(F(z)) = \sum_k h_k a_{n,k}$. For $\widetilde{H}(z)$ we obtain $\widetilde{H}(F(z)) = H(z)$ and so $[z^n]B(z)H(z) = \sum_k \widetilde{h}_k a_{n,k}$.

7.40 From $A_k(z) = B(z)F(z)^k = \frac{z^k}{(1-z)^{k+1}}$ follows $a_{n,k} = \binom{n}{k}$, and $\beta_{n,j} = \sum_{k=0}^{n}\binom{k}{j} = \binom{n+1}{j+1}$. Consider $H(z) = \frac{1}{1-z}$; then $H(F(z)) = \frac{1-z}{1-2z}$, and $\widetilde{H}(z) = H(\frac{z}{1+z}) = 1 + z$. This gives the identities $\sum_{k=0}^{n}[z^k]\frac{1}{1-2z} = \sum_{k=0}^{n} 2^k = \sum_{j=0}^{n}\binom{n+1}{j+1}$, and $\sum_{k=0}^{n}[z^k]\frac{1}{(1-z)^2} = \sum_{k=0}^{n}(k+1) = \binom{n+2}{2}$. For $H(z) = z^m / \prod_{i=1}^{m}(1 - iz)$ one obtains $\sum_{k=0}^{n} S_{k+1,m+1} = \sum_{j=0}^{n} S_{j,m}\binom{n+1}{j+1}$. The choice $\alpha_{n,k} = H_k$ leads to the identities $\sum_{k=0}^{n} 2^k H_k = \sum_{j=0}^{n}\binom{n+1}{j+1}(H_{n+1} - \frac{1}{j+1})$ and $\sum_{k=0}^{n}(k+1)H_k = \binom{n+2}{2}H_{n+1} - \binom{n+2}{2} + \binom{n+1}{2}\frac{1}{2}$.

7.43 We know that $U \leftrightarrow (1/B(F^{\langle -1\rangle}), F^{\langle -1\rangle})$. Let a', b', s', u' be the parameters belonging to U, where $a' = -a$, $s' = -s$ by the previous exercise. The differential equation for $F^{\langle -1\rangle}$ reads therefore $F^{\langle -1\rangle\prime} = 1 - sF^{\langle -1\rangle} + u'(F^{\langle -1\rangle})^2$. Now $F^{\langle -1\rangle\prime} = \frac{1}{F'(F^{\langle -1\rangle}(z))} = \frac{1}{1+sz+uz^2}$; hence $\frac{1}{1+sF+uF^2} = 1 - sz + u'z^2$, that is, $(1 + sF + uF^2)(1 - sz + u'z^2) = 1$. With $F_2 = \frac{s}{2}$ it follows that $u + u' = \frac{s^2}{2}$, and similarly, $b + b' = as$, $au' = \frac{bs}{2}$. Case 1. $\tau' \equiv -\tau$. Then $b' = -b$, $u' = -u$; hence $s = 0$. If $a \neq 0$, then $u' = u = 0$, and we obtain $\sigma \equiv a$, $\tau = (bk)$. If $a = 0$, then a short computation yields again $u = 0$. Case 2. $\tau' \equiv \tau$. Here $b' = b$, $u' = u$; thus $u = \frac{s^2}{4} = (\frac{s}{2})^2$. Set $\frac{s}{2} = m$; then $u = m^2$. Furthermore, $2b = as = 2am$, that is, $b = am$. The sequences are therefore $\sigma = (a + 2mk)$, $\tau = (k(am + m^2(k-1))$.

7.46 We have $C_n^{(x)}(a+n-1) = \sum_{k=0}^{n}\binom{n}{k}(-x)^{n-k}(a+n-1)^{\underline{k}} = \sum_{k=0}^{n}\binom{n}{k}(-x)^{k}(a+n-1)^{\underline{n-k}} = n!\sum_{k=0}^{n}\binom{a+n-1}{n-k}\frac{(-x)^k}{k!} = (-1)^n L_n^{(a)}(x)$. For the second assertion apply the Sheffer identity to $L_n^{(x+1)}(a+b)$.

7.49 The recurrence follows immediately from the recurrence for $p_n(x)$. Translated into generating functions we get $P' = (x-z)(zP)' + xzP' = (x-z)(P+zP') + xzP'$, hence $\frac{P'}{P} = \frac{x-z}{1-2xz+z^2}$. So, $(\log P)' = -\frac{1}{2}(\log(1-2xz+z^2))'$, which gives $P(z) = \sum_{n\ge0} P_n(x)z^n = \frac{1}{\sqrt{1-2xz+z^2}}$.

7.51 The equation $D_qB = aB + bBz$ translates into $B_{n+1} = aB_n + b\frac{1-q^n}{1-q}B_{n-1}$. Define the operator L as in the hint; then we get $Lx^n = \sum_{k=0}^{n}\left[{n \atop k}\right]Lg_k(x) = \sum_{k=0}^{n}\left[{n \atop k}\right]\lambda^k$. Hence we have to prove the recurrence $Lx^{n+1} = aLx^n + b\frac{1-q^n}{1-q}Lx^{n-1}$. From $g_{n+1}(x) = xg_n(x) - q^n g_n(x)$ follows $\lambda^{n+1} = L(xg_n(x)) - q^n\lambda^n$; thus $L(xg_n(x)) = \lambda^n(\lambda+1) - (1-q^n)\lambda^n = a\lambda^n - (1-q^n)\lambda^n$. Furthermore, $LD_qg_n(x) = \frac{1-q^n}{1-q}\lambda^{n-1} = -\frac{1-q^n}{b}\lambda^n$, that is, $L(xg_n(x)) = aLg_n(x) + bLD_qg_n(x)$. By linear extension, $L(xp(x)) = aLp(x) + bLD_qp(x)$ for any polynomial. Now set $p(x) = x^n$ to obtain $Lx^{n+1} = aLx^n + b\frac{1-q^n}{1-q}Lx^{n-1}$.

7.54 If the first step is a diagonal step, then the contribution is Sch_n. Otherwise, the path starts with a $(1,0)$-step. Let $x = k+1$ be the first x-coordinate where the path hits the line $y = x$, $0 \le k \le n$. Then the contribution is $Sch_k Sch_{n-k}$, and the recurrence follows. The formula involving C_{n-k} follows upon classifying the paths according to the number k of diagonal steps (see the argument in Exercise 1.24). The congruence is clear from $Sch_n = a_{n,0} = 2a_{n-1,0} + 2a_{n-1,1}$ and induction.

7.58 If $n+1$ is a fixed point, then the contribution is aB_n. Suppose $n+1$ is in a cycle of length $n-1-k \ge 2$, $0 \le k \le n-1$; then the contribution is $b\binom{n}{k}(n-k)!B_k$.

7.60 By the previous exercise, $C_{n+1} = Fi_n + 2Fi_{n+1}$; hence $C(z) = zFi(z) + 2Fi(z) - 1$, that is, $Fi(z) = \frac{C(z)+1}{z+2} = \frac{1}{1-z^2C(z)^2}$, since $(C+1)(1-z^2C^2) = (C+1)(1-z(C-1)) = C+1-z(C^2-1) = C+1-(C-1)+z = z+2$. From $B_k(z) = z^kC(z)^{k+1}$ we obtain $\sum_{i\ge1}B_{2i-1}(z) = \frac{zC(z)^2}{1-z^2C(z)^2}$. Thus $Fi(z) = z\sum_{k \text{ odd}}B_k(z) + 1$, and so $Fi_{n+1} = \sum_{k \text{ odd}} b_{n,k}$.

7.62 For the two equations use (3). For $a = 1$, $s = 0$ this gives $\binom{n}{\lfloor n/2\rfloor} = 2^n - \sum_{k\ge0}2^{n-1-2k}C_k$, and for $a = 0$, $s = 1$, $B_n^{(0,1)}(z) = \frac{1}{1+z}(1 + zB^{(1,1)}(z))$, thus $R_n = M_{n-1} - M_{n-2} \pm \cdots + (-1)^{n-2}M_1 + (-1)^{n-1}M_0 + (-1)^n = M_{n-1} - M_{n-2} \pm \cdots + (-1)^{n-2}M_1$.

7.65 Using (4) one obtains $B(1-sz) - uz^2B^2 = 1$, and by a short computation $(zB)' = \frac{2-B(1-sz)}{1-2sz+(s^2-4u)z^2}$; hence $(zB)'(1-2sz+(s^2-4u)z^2) = 2 - B(1-sz)$. Comparing coefficients for z^n, this gives $(n+1)B_n - 2snB_{n-1} + (s^2-4u)(n-1)B_{n-2} = -B_n + sB_{n-1}$ $(n \geq 1)$; thus $(n+2)B_n = (2n+1)sB_{n-1} + (4u-s^2)(n-1)B_{n-2}$. For the Motzkin numbers $s = u = 1$; hence $(n+2)M_n = (2n+1)M_{n-1} + 3(n-1)M_{n-2}$, or $\frac{M_n}{M_{n-1}} = \frac{2n+1}{n+2} + \frac{3n-3}{n+2}\frac{1}{M_{n-1}/M_{n-2}}$. Define the function $f : [2,\infty] \to \mathbb{R}$ by $f(x) = \frac{2x+1}{x+2} + \frac{3x-3}{x+2}\frac{1}{f(x-1)}$, $f(2) = 2$. Then f is easily seen to be monotonically increasing with $f(n) = \frac{M_n}{M_{n-1}}$, which gives $\frac{M_n}{M_{n-1}} \leq \frac{M_{n+1}}{M_n}$. Furthermore, $\frac{M_n}{M_{n-1}} \leq \frac{2n+1}{n+2} + \frac{3n-3}{(n+2)2} = \frac{7n-1}{2(n+2)} \leq \frac{7}{2}$. Therefore $\alpha = \lim \frac{M_n}{M_{n-1}}$ exists with $\alpha = \lim \frac{2n+1}{n+2} + \lim \frac{3n-3}{n+2} \cdot \frac{1}{\alpha} = 2 + \frac{3}{\alpha}$, and so $\alpha = 3$.

7.70 We have the recurrence $a_{n,k} = a_{n-1,k-1} + (2d)a_{n-1,k} + a_{n-1,k+1}$, $a_{n,0} = (2d)a_{n-1,0} + a_{n-1,1}$. The number of walks is therefore the Catalan number with $\sigma \equiv 2d$, $\tau \equiv 1$, and the desired generating function is $S^{(2d)}(z) = B^{(2d+1,2d)}(z)$ according to Exercise 7.68. Since $B_n^{(1,0)} = \binom{n}{\lfloor n/2 \rfloor}$, $B_n^{(3,2)} = \binom{2n+1}{n}$, the binomial formula in Proposition 7.20 yields $S_n = \sum_{k=0}^{n}\binom{n}{k}(2d)^{n-k}\binom{k}{\lfloor k/2 \rfloor} = \sum_{k=0}^{n}\binom{n}{k}(2d-2)^{n-k}\binom{2k+1}{k}$. For $d = 2$ this gives $S_n = \sum_{k=0}^{n}\binom{n}{k}2^{n-k}\binom{2k+1}{k}$.

7.73 The differential equations $F' = 1 + (u+1)F + uF^2$, $B' = uB + uBF$ lead to $B = 1 + uF$, and hence to the recurrence $B_{n+1} = uB_n + \sum_{k=0}^{n-1}\binom{n}{k}B_kB_{n-k}$. Define the weight $w(\pi) = u^{\rho(\pi)}$, where $\rho(\pi)$ is the number of runs, and classify according to whether $n+1$ is in first place or not. This gives $B_n = \sum_{k\geq 0} A_{n,k}u^k$.

Chapter 8

8.3 For every i factor $x_i + 1$ out of the i-th row to obtain $\det\binom{x_i+j}{j} = \prod_{i=1}^{n}(x_i+1)\cdot\det\left(\frac{(x_i+j)^{\underline{j-1}}}{j!}\right)$. As a polynomial this determinant has degree $\binom{n}{2}$, hence $\det\left(\frac{(x_i+j)^{\underline{j-1}}}{j!}\right) = c\prod_{i<j}(x_i - x_j)$. Comparing the coefficient of $x_1^0x_2^1\cdots x_n^{n-1}$ gives $1/1!2!\cdots n! = (-1)^{\binom{n}{2}}c$. The desired determinant is therefore equal to $\frac{(x_1+1)\cdots(x_n+1)}{1!\cdots n!}\prod_{i<j}(x_j - x_i)$. For $x_i = i$ this gives $\det = n+1$.

8.5 The polynomial $\prod_{i,j}(1 - x_i y_j)\det(\frac{1}{1-x_i y_j})$ is alternating in the x_i and y_j, hence equal to $c\prod_{i<j}(x_i - x_j)(y_i - y_j)$. Since the degree is $n-1$ in each x_i and y_j, we infer $c = 1$.

8.8 Following the hint, $P(q)$ is a polynomial in q of degree less than or equal to n^2. Since $1 - q$ can be factored out of every row, 1 is a root of $P(q)$ of multiplicity at least n. For $1 \le k < n$, show that each row vector of the matrix $((1 - t^{k(i+j-1)})/(1 - t^{i+j-1}))$ is a linear combination of the k vectors $(1, t^i, t^{2i}, \dots, t^{(n-1)i})$, $0 \le i \le k-1$. Therefore the rank of this matrix is at most k, and thus t^k is a root of $P(q)$ with multiplicity at least $n - k$. The same reasoning applies for t^{-k}, which gives the expression for $P(q)$ as in the hint. Finally $P(0) = \det(\frac{1}{1-t^{i+j-1}}) = \prod_{i<j}(t^i - t^j)(t^{i-1} - t^{j-1})\prod_{i,j}\frac{1}{1-t^{i+j-1}}$ by Exercise 8.5. Now rearrange the terms to obtain the formula.

8.9 One obtains $m_{1111} = e_4$, $m_{211} = e_{31} - 4e_4$, $m_{22} = e_{22} - 2e_{31} + 2e_4$, $m_{31} = e_{211} - 2e_{22} - e_{31} + 4e_4$, $m_4 = e_{1111} - 4e_{211} + 2e_{22} + 4e_{31} - 4e_4$.

8.13 The coefficient of a monomial $x^\lambda y^\mu$ in the product is equal to the number of 0,1-matrices with row sum λ and column sum μ, that is, $M_{\lambda\mu}$.

8.17 We have $e_k \in \langle h_\mu : \mu \in \mathrm{Par}(k)\rangle$. Hence in $e_\lambda = e_{\lambda_1}\cdots e_{\lambda_r}$ the lexicographically largest possible h_μ in the expansion is $h_{\lambda_1}\cdots h_{\lambda_r}$, that is, $\mu \le \lambda$.

8.19 By (3), $\sum_{m\ge 0} h_m(1, q, \dots, q^n)z^m = \prod_{i=0}^n \frac{1}{1-q^i z}$. Exercise 2.21 now yields $\sum_{m\ge 0} h_m(1, q, \dots, q^n)z^m = \sum_{m\ge 0}\left[{m+n \atop m}\right]z^m$; thus $h_m(1, q, \dots, q^n) = \left[{m+n \atop m}\right]$. For the second assertion by (1), $\sum_{m\ge 0} e_m(1, q, \dots, q^{n-1})z^m = (1+z)\prod_{i=1}^{n-1}(1 + q^i z)$, which is equal to $(1+z)\sum_{k=0}^{n-1}\left[{n-1 \atop k}\right]q^{\binom{k+1}{2}}z^k$ by the q-binomial theorem. The coefficient of z^m is therefore

$$\left[{n-1 \atop m}\right]q^{\binom{m+1}{2}} + \left[{n-1 \atop m-1}\right]q^{\binom{m}{2}} = q^{\binom{m}{2}}\left(\left[{n-1 \atop m}\right]q^m + \left[{n-1 \atop m-1}\right]\right) = q^{\binom{m}{2}}\left[{n \atop m}\right].$$

8.21 Waring's formula in Exercise 6.34 expresses e_k as a linear combination of terms $p_1^{c_1}\cdots p_k^{c_k}$ with $c_1 + 2c_2 + \cdots + kc_k = k$, that is, of terms $p_{\lambda_1}\cdots p_{\lambda_r}$, $\lambda \in \mathrm{Par}(k)$.

8.25 For the first assertion use induction on k to show that $\mu_1^* + \cdots + \mu_k^* \le \lambda_1^* + \cdots + \lambda_k^*$. Suppose $\lambda \prec \mu$, and let i be the smallest index with $\lambda_i \ne \mu_i$. Then $\sum_{k=1}^i \lambda_k = \sum_{k=1}^{i-1}\lambda_k + \lambda_i \le \sum_{k=1}^{i-1}\mu_k + \mu_i$, and so $\lambda_i < \mu_i$.

8.29 If $\mu \preceq \lambda$ then $\mu_1 \le \lambda_1$. Now fill up the two rows from left to right with μ_1 1's, μ_2 2's,..., to obtain an SST of shape λ and type μ. Thus $K_{\lambda\mu} > 0$.

8.31 Consider the lattice graph with horizontal steps of weight 1, and diagonal $(1,1)$-steps of weight x_k, when the step goes from $x = k-1$ to $x =$

k. Let $\mathcal{A} = \{A_1, \ldots, A_n\}$, $\mathcal{B} = \{B_1, \ldots, B_n\}$, where $A_i = (0, n-i)$, $B_j = (n - j + \lambda_j)$. For the path matrix $M = (m_{ij})$, $m_{ij} = \sum_{1 \le i_1 < \cdots < i_k \le n} x_{i_1} \cdots x_{i_k}$, $k = \lambda_j - j + i$. Thus $m_{ij} = e_{\lambda_j - j + i}$, and so $\det M = \det(e_{\lambda_i - i + j})$. The vertex-disjoint path systems are of the form $\mathcal{P} : A_i \to B_i$. To every $\mathcal{P}$ associate bijectively a table $T_{\mathcal{P}}$, where the i-th row corresponds to the diagonal steps $x = k-1$ to $x = k$. $T_{\mathcal{P}}$ has shape $\lambda_1 \lambda_2 \ldots \lambda_n$ with rows strictly increasing and $w(T_{\mathcal{P}}) = w(\mathcal{P})$. The transpose T^* is therefore an SST of shape λ^*, and the result follows.

8.34 Dividing $s_\lambda(q^{\mu_1+n-1}, \ldots, q^{\mu_n})$ by $s_\mu(q^{\lambda_1+n-1}, \ldots, q^{\lambda_n})$ we get

$$\frac{\det(q^{(\mu_j+n-j)(n-i+\lambda_i)})}{\det(q^{(\lambda_j+n-j)(n-i+\mu_i)})} \prod_{i<j} \frac{q^{\lambda_i+n-i} - q^{\lambda_j+n-j}}{q^{\mu_i+n-i} - q^{\mu_j+n-j}} \, .$$

The first factor is 1, since the matrix in the denominator is the transpose of the matrix in the numerator. Now it is easily checked that

$$s_\lambda(1, q, \ldots, q^{n-1}) = (-1)^{\binom{n}{2}} \prod_{i<j} \frac{q^{n-i+\lambda_i} - q^{n-j+\lambda_j}}{q^{i-1} - q^{j-1}} \, ,$$

whence the result.

8.37 Regard the symmetric plane partition λ_{ij} as a 3-dimensional arrangement with a stack of height λ_{ij} at (i, j). The blocks at level 1 form a self-conjugate Ferrers diagram F_1, and similarly up to the highest level less than or equal to t. Now rearrange F_1 with self-conjugate partition $\lambda_1 \ldots \lambda_r$ into decreasing stacks of odd length corresponding to the hooks as in Exercise 1.53, and similarly for the other F_i. Arrange the level stacks vertically and decreasing along columns. This gives the desired bijection, since the height of a stack is at most $2r - 1$, there are less than or equal to r rows and less than or equal to t columns that are strictly decreasing. Example:

```
3221      • • • •   • • •   •     • • • • • • •   • • • • •   •      751
222   →   • • •     • • •      →  • • •           • • •           →  33
221       • • •     • •           •                                  1
1         •
```

8.41 For type (r, r) the r 1's must be in the first row, and there may be k 2's in row 1, $k = 0, 1, \ldots, r$. For type (r, r, r) we count first those SSTs that have at most two rows. Classifying them according to the number k of 2's in the second row, we get $\sum_{k=0}^{r/2}(r+1) + \sum_{k=r/2+1}^{r}(2r - 2k + 1) = \frac{3r^2}{4} + \frac{3r}{2} + 1$ SSTs for even r, and similarly $\frac{3r^2}{4} + \frac{3r}{2} + \frac{3}{4}$ for odd r. If the SST has three rows, then the first column must be $\begin{smallmatrix}1\\2\\3\end{smallmatrix}$. Hence we obtain for the total number the recurrence $A_r = A_{r-1} + \frac{3r^2}{4} + \frac{3r}{2} + \begin{cases} 1 & r \text{ even} \\ 3/4 & r \text{ odd} \end{cases}$. Iteration down gives with $A_1 = 4$ the result.

8.43 By Exercise 8.14 and Corollary 8.16, $\sum_\lambda m_\lambda(x)h_\lambda(y) = \prod_{i,j}\frac{1}{1-x_iy_j} = \sum_\lambda s_\lambda(x)s_\lambda(y)$. Now compare the coefficients of $m_\mu(x)$. For e_μ apply the involution ω.

8.47 We have $\omega_x\prod_{i,j}(1+x_iy_j) = \omega_x\sum_\lambda e_\lambda(x)m_\lambda(y) = \sum_\lambda h_\lambda(x)m_\lambda(y) = \prod_{i,j}\frac{1}{1-x_iy_j} = \sum_\lambda s_\lambda(x)s_\lambda(y) = \sum_\lambda s_{\lambda^*}(x)s_{\lambda^*}(y)$. By the previous exercise this is also equal to $\omega_x\sum_\lambda s_\lambda(x)s_{\lambda^*}(y) = \sum_\lambda(\omega_x s_\lambda(x))s_{\lambda^*}(y)$, hence $\omega_x s_\lambda(x) = s_{\lambda^*}(x)$.

8.49 We have

$$\prod_{i\ge1}\frac{1}{1-qx_i}\prod_{i<j}\frac{1}{1-x_ix_j} = \Big(\prod_{i\ge1}\sum_{a_{ii}\ge0}(qx_i)^{a_{ii}}\Big)\cdot\Big(\prod_{i<j}\sum_{a_{ij}\ge0}(x_ix_j)^{a_{ij}}\Big)$$
$$= \sum_{A\text{ symm.}} q^{\mathrm{tr}(A)}x^{\mathrm{row}(A)}\,.$$

The correspondence of the previous exercise and Corollary 8.25 show that the last expression equals $\sum_\lambda q^{o(\lambda^*)}s_\lambda(x)$.

8.52 We know from Exercise 8.37 that symmetric plane partitions correspond to column-strict plane partitions with less than or equal to r rows, less than or equal to t columns, and max $\le 2r-1$, for which all entries are odd. Hence the generating function is $\prod\limits_{i=1}^{r}\frac{1}{1-q^{2i-1}}\prod_{1\le i<j\le r}\frac{1}{1-q^{2i+2j-2}}$. Now apply Exercise 8.49 with the substitutions $q\mapsto 1$, $x_i\mapsto q^{2i-1}$.

8.56 Consider $\pi \xrightarrow{\text{ins}} P$. If $\lambda_1\le m$, $\lambda_1^*\le n$, then $|\lambda|\le mn$, contradiction. For $S(mn)$ take $\pi = n\,n{-}1\dots1\,2n\,2n{-}1\dots n{+}1\dots mn\,mn{-}1\dots(m-1)n+1$; then $is(\pi)=m$, $ds(\pi)=n$.

8.59 Let T be an ST of shape λ, $|\lambda| = n$. If n is in the i-th row, then clearly $f(\lambda,-i)\neq 0$. Thus after deletion of the box containing n we obtain an ST φT of shape $(\lambda,-i)$. Conversely, to any T' of shape $(\lambda,-i)$ we may add n at the end of the i-th row. This proves (a), and (b) is shown similarly. For (c) we have by induction, $\sum_{|\lambda|=n}f(\lambda)^2 = \sum_{|\lambda|=n}f(\lambda)\sum_j f(\lambda,-j) = \sum f(\alpha)f(\beta)$, where the sum extends over all pairs (α,β) with $|\alpha| = n$, $|\beta| = n-1$ and $\alpha = (\beta,+i)$, $\beta = (\alpha,-j)$ for some (i,j). Hence $\sum f(\alpha)f(\beta) = \sum_{|\gamma|=n-1}f(\gamma)\sum_i(\gamma,+i) = \sum_{|\gamma|=n-1}nf^2(\gamma) = n(n-1)! = n!$.

8.61 Let $A(n)$ be the number of STs with at most two rows. Classifying the tableaux according to whether n is in the first or second row we obtain the recurrence $A(n) = 2A(n-1) - [n \text{ odd}]$ with solution $A(n) = \binom{n}{\lfloor n/2\rfloor}$. The number of STs with at most three rows equals the number of involutions $\pi\in S(n)$ with $ds(\pi)\le 3$. Represent an involution π by a diagram $1\,2\dots\,\overset{\frown}{i\,\dots\,j}\,\dots\,n$ with arcs if (ij) is a 2-cycle. Clearly, the forbidden configurations are $i\;j\quad k\;\ell$ (nested arcs) since then in the word representation, $\pi = \dots \ell.k.j.i \dots$ with $\ell > k > j > i$. Classify these permutations according to the number k of 2-cycles. It is easy to see that the number of diagrams with k arcs without the forbidden configurations equals the Catalan number C_k. Hence we get as total number $\sum_k \binom{n}{2k} C_k = M_n$ according to Section 7.4.

8.64 Suppose there is an even hook length. Then it is easy to see that there is a hook length equal to 2. Remove the cells containing 2 and a neighboring cell containing 1. Check that the difference odd minus even remains unchanged in the new diagram. Continue until all hook lengths are odd. The resulting partition must be of the form $\widetilde{\lambda} = m\, m-1 \dots 2\,1$.

Chapter 9

9.2 Classification of the λ-colorings with respect to the number k of colors used in the left color class yields $\chi(K_{m,n};\lambda) = \sum_{k=0}^{m} \binom{\lambda}{k} k! S_{m,k} (\lambda - k)^n$ and $ac(K_{m,n}) = \sum_{k=0}^{m} (-1)^{m-k} k! S_{m,k} (k+1)^n$.

9.6 Recurrence (1) gives with induction $a_1 = -|E|$. For the next coefficient one easily obtains $a_2 = \binom{|E|}{2}$ – # triangles, again by (1).

9.10 We have $\frac{B(\text{bridge})}{\lambda} = \frac{\lambda(\lambda-1)+\lambda s}{\lambda} = \lambda - 1 + s$, $\frac{B(\text{loop})}{\lambda} = \frac{\lambda s}{\lambda} = s$. Let $e = \{u, v\}$ be a bridge, and consider $b_i(G;\lambda)$. Suppose u and v are colored alike, then $b_i(G;\lambda) = \lambda m_{i-1}$, where m_{i-1} is the number of colorings of the rest with $i-1$ bad edges. Since $b_{i-1}(G \setminus e;\lambda) = \lambda^2 m_{i-1}$, we get $\lambda b_i(G;\lambda) = b_{i-1}(G \setminus e;\lambda)$. If u and v are colored differently, then $b_i(G;\lambda) = \lambda(\lambda-1) m_i$ and $b_i(G \setminus e;\lambda) = \lambda^2 m_i$; hence $\lambda b_i(G;\lambda) = (\lambda-1) b_i(G \setminus e;\lambda)$. Altogether, $\lambda b_i(G;\lambda) = (\lambda-1) b_i(G \setminus e;\lambda) + b_{i-1}(G \setminus e;\lambda)$, which gives $\frac{B(G)}{\lambda^{k(G)}} = \frac{(\lambda-1+s) B(G \setminus e)}{\lambda^{k(G)+1}}$. The other recurrences are proved similarly. Now apply the recipe theorem with $A = \lambda - 1 + s$, $B = s$ and $\alpha = 1$, $\beta = s - 1$.

9.13 Let $s(G)$ be the number of strong orientations of G. We have $s(\text{bridge}) = 0$, $s(\text{loop}) = 2$, and the recurrence is $s(G) = s(G \setminus e) + s(G/e)$ by a similar argument as in the proof of Proposition 9.7. The recipe theorem yields $s(G) = T(G;0,2)$.

9.15 Take a fixed orientation of G, and let ϕ be a 3-flow in $\{1,-1\}$. Now turn all edges e with $\phi(e) = -1$ around, and assign them the value 1. The flow condition remains valid, and since all edges have $\phi(e) = 1$, it must be a 2-in 2-out orientation. The converse construction is clear.

9.18 Let $f(n) = \sum_T y^{\text{inv}T}$ over all rooted trees on n vertices, and $t(n) = \sum_T y^{\text{inv}T}$ when the root is 1. As in the proof of Theorem 3.8 we have $t(n+1) = h(n)$, $t(z) = \sum_n t(n+1) \frac{z^n}{n!} = \sum_n h(n) \frac{z^n}{n!} = e^{F(z)}$. Now when we consider all roots, then $f(n+1) = h(n)(1+y+\cdots+y^n) = h(n)\frac{y^{n+1}-1}{y-1}$, and thus $F'(z) = \sum_n f(n+1)\frac{z^n}{n!} = \sum_n h(n)\frac{z^n}{n!}\frac{y^{n+1}-1}{y-1} = \frac{y}{y-1}t(yz) - \frac{1}{y-1}t(z)$. Furthermore, $t'(z) = e^{F(z)}F'(z) = t(z)(\frac{y}{y-1}t(yz) - \frac{1}{y-1}t(z))$, which is precisely the recurrence in Exercise 9.17. The trees with no inversion correspond to $t(z)\,|_{y=0}$. With induction we obtain $t_{n+1}(0) = \sum_{k=0}^{n-1}\binom{n-1}{k}k!(n-1-k)! = n(n-1)! = n!$. The number of n-trees with $\text{inv}T = 0$ is thus $(n-1)!$.

9.22 Using recurrence (3) one easily obtains

$$Q(P_n;x) = \sum_{k\ge1}\left[\binom{n-k}{k-1} + \binom{n-1-k}{k-1}\right]x^k$$

with $Q(P_n;1) = F_{n+1}$, and $Q(K_{m,n};x) = (1+x+\cdots+x^{m-1})(1+x+\cdots+x^{n-1}) + x^m + x^n - 1$ with $Q(K_{m,n};1) = mn+1$.

9.27 We have $Q(C_3;x) = 4x$, $Q(C_4;x) = 3x^2+2x$. For $n \ge 5$ the recurrence gives $Q(C_n;x) = Q(P_{n-1};x) + Q(C_{n-2};x) + xQ(P_{n-3};x)$. The linear coefficient therefore satisfies $q_1(C_n) = 2 + q_1(C_{n-2})$, which gives $q_1(C_n) = n-2$ for n even, and $q_1(C_n) = n+1$ for n odd. If $Q(H;x)$ has $q_1 = 2$, then H is connected and contains no odd induced circuit by the previous exercise. Hence H is bipartite.

9.29 It suffices to consider connected graphs. Let U be an independent set with $|U| = \alpha(G)$, and let $\{u,v\} \in E$, $u \notin U$. Then $Q(G) = Q(G\smallsetminus u) + Q(G^{uv}\smallsetminus v)$. By induction, $\deg Q(G\smallsetminus u) \ge \alpha(G\smallsetminus u) = \alpha(G)$, and so $\deg Q(G) \ge \deg(Q\smallsetminus u) \ge \alpha(G)$. Suppose, conversely, $H_0 \approx G$ with $\alpha(H_0) = \max_{H\approx G}\alpha(H)$. Then $Q(H_0) = Q(H_0\smallsetminus u) + Q(H_0^{uv}\smallsetminus v)$. By induction, $\deg Q(H_0\smallsetminus u) = \alpha(H')$ for some $H' \approx H_0\smallsetminus u$. Now $\alpha(H') = \alpha(H''\smallsetminus u)$ for some $H'' \approx H_0$, and therefore $\alpha(H') \le \alpha(H'') \le \alpha(H_0)$, that is, $\deg Q(H_0\smallsetminus u) \le \alpha(H_0)$. Similarly, $\deg Q(H_0^{uv}\smallsetminus v) \le \alpha(H_0)$, and we conclude that $\deg Q(H_0) \le \alpha(H_0)$, and thus $\deg Q(H_0) = \alpha(H_0)$.

9.33 Let $H_1,\ldots,H_t$ be the connected components of H, with $|V(H_i)| = n_i \ge 2$, $|E(H_i)| = m_i$. Then $Q(H;1) = \prod_{i=1}^t Q(H_i;1) \ge (m_1+1)\cdots(m_t+1) \ge n_1\cdots n_t \ge n$. Equality forces $t = 1$, $m+1 = n$, in which case H is a tree, or $t = 2$ and $H = K_2 \cup K_2$ with $Q(H;x) = 4x^2$, $Q(H;1) = 4 = n$. Suppose H is a tree with $Q(H;1) = n$. Since $Q(H;1) = \#$ matchings, we conclude that there is no matching with two disjoint edges, and $H = K_{1,n-1}$ results. For $K_{1,n-1}$ we have $Q(K_{1,n-1};x) = x^{n-1}+x^{n-2}+\cdots+x^2+2x$, $Q(K_{1,n-1};1) = n$. The 2-word is $12\ldots n1nn-1\ldots2$, and the associated 2-in 2-out graph is an n-circuit with two edges joining adjacent vertices in either direction.

9.35 We have $f_k(n) = \sum_H L(H)$ over all interlace graphs H with $Q(H;1) = k$, and where $L(H)$ is the number of ways to fit H into the chord diagram. For $k = 1$ we get $H = \overline{K}_n$ (n isolated vertices) as the only graph with

$L(H) = C_n$ (Catalan number). Thus $f_1(n) = C_n$. The results for $k = 2, 3$ are $f_2(n) = \binom{2n}{n-2}$, $f_3(n) = 3\binom{2n}{n-3}$.

9.36 For a tree, $|V| = n$, $|E| = n - 1$, $|F| = 1$; hence $|V| - |E| + |F| = 2$. Suppose G is not a tree. Delete a non-bridge and apply induction.

9.40 For the black transition we get $\alpha S(\widetilde{G}_1)S(\widetilde{G}_2)$ with $c(p) = c(p_1) + c(p_2)$. For the white or crossing case $c(p) = c(p_1) + c(p_2) - 1$; hence $S(\widetilde{G}) = \alpha S(\widetilde{G}_1)S(\widetilde{G}_2) + \frac{\beta+\gamma}{\lambda} S(\widetilde{G}_1)S(\widetilde{G}_2)$.

9.42 If G has no bridge, then G^* has no loop, and so $\chi(G^*; k) > 0$ for k large enough. By the previous exercise $P(G; k) \geq \chi(G^*; k) > 0$; hence $P(G; \lambda)$ cannot be the 0 polynomial. The degree condition is easily seen.

9.44 Since G is Eulerian, G^* is bipartite. Let A and B be the color classes of the faces with the outer face in B. In $\widetilde{G}$ orient the edges around faces in A clockwise, around faces in B counterclockwise, and the outer face clockwise. With this orientation the edges around a black face of $\widetilde{G}$ receive alternate directions. Note that the cycles of a transition system p in the Penrose polynomial correspond precisely to directed Eulerian cycles. Let $X \subseteq E$ be the set of crossing vertices in p. By induction it is easy to see that $|X| + c(p) \equiv |F| \pmod 2$. Hence for all p with $c(p) = i$ we have $x(p) \equiv |F| - i \pmod 2$, that is, $(-1)^{x(p)}$ is constant. It follows that $a_{|F|} > 0$ $a_{|F|-1} < 0, \ldots$, and therefore $P(G; -\lambda) = \sum_p (-1)^{x(p)} (-\lambda)^{c(p)} = \sum_p (-1)^{|F|} \lambda^{c(p)} = (-1)^{|F|} \sum_p \lambda^{c(p)}$. We conclude that $(-1)^{|F|} P(G; -1) = 2^{|E|}$, $\sum |a_i| = 2^{|E|}$. The last assertion is readily seen.

9.47 By the previous results $P(G; \lambda)$ is a polynomial of degree 5 with leading coefficient 1, $P(G; 0) = P(G; 1) = P(G; 2) = 0$, $P(G; 3) =$ # 3-edge-colorings $= 16$, $P(G; -2) = (-16) \cdot$ # 4-face-colorings $= (-16) \cdot 24$. From these six conditions one computes $P(G; \lambda) = \lambda(\lambda - 1)(\lambda - 2)^3$.

9.53 Associate to $O \in \mathcal{O}$ the 2-valuation $g : E(\widetilde{G}) \to \{1, 2\}$, where $g(f) = 1$ if in the orientation the black face is to the right of f, otherwise $g(f) = 2$. This gives $g \in \mathcal{V}_2$ as in the previous exercise, where the total types correspond to saddle points. Exercise 9.52 therefore yields $\sum_O 2^{s(O)} = 2T(G; 3, 3)$.

9.57 Check that for move (II) the weights are $+1$ and -1 no matter how the curves are oriented, whereas for (III) the weights of the three corners of the triangle stay the same.

9.60 The Tutte polynomial is $T(C_{2n+1}; x, y) = x^{2n} + x^{2n-1} + \cdots + x + y$. With writhe $w(D) = 2n + 1$ this gives $V(t) = -t^{n+1} \cdot (t^{2n} - t^{2n-1} \pm \cdots - t - t^{-1}) = -t^{3n+1} + t^{3n} \mp \cdots + t^{n+2} + t^n$.

9.62 Using (1) we obtain $\langle D \rangle = \sum_p A^{i(p)-j(p)} (-A^2 - A^{-2})^{c(p)-1}$. With $A = 1$ this gives

$$\langle D\rangle_{A=1} = \sum_p (-2)^{c(p)-1} = -\frac{1}{2}S(\widetilde{G}, W_{1,1,0}; -2) = -\frac{1}{2}S(\widetilde{G}, W_{0,0,-1}; -2)$$
$$= -\tfrac{1}{2}(-2)^{c(L)}(-1)^{|E|} = (-1)^{|E|}(-2)^{c(L)-1}\,.$$

Since $w(D) \equiv |E| \pmod 2$, $f_L(1) = V_L(1) = (-2)^{c(L)-1}$ results.

9.64 The nugatory crossings in D correspond to bridges in the Tait graph G. Hence if G has no bridges, then $|E| = n \geq |V(G)|$. Since G has at least n spanning trees, $|V_K(-1)| = \#$ trees in $G \geq n$. Since the number of spanning trees in G equals the number of Eulerian cycles in $\widetilde{G}^c$ (Corollary 9.22) we find that $|V_K(-1)| = n$ if and only if the interlace graph H corresponding to $\widetilde{G}^c$ is $K_{1,n-1}$ or $K_2 \cup K_2$ (Exercise 9.33), but $K_2 \cup K_2$ cannot happen (why?).

9.66 We have $|V_K(-1)| = T(G;1,1) = e(\widetilde{G}^c) = Q(H;1)$, where H is an interlace graph to $\widetilde{G}^c$ with the canonical orientation. Hence $|V_K(-1)| \leq 2^{n-1}$, $n = |E|$, and equality can hold only for $H = K_n$. Check that for $H \neq K_n$, $Q(H;1) \leq \frac{3}{4}2^{n-1}$. Since $|V_K(-1)|$ is odd (Exercise 9.61), $|V_K(-1)| = 2^{n-1}$ cannot hold. For $n = 2$ we get $|V_K(-1)| \leq \frac{3}{4} \cdot 2$; thus $|V_K(-1)| \leq 1$ with the Tait graph $G = P_3$ achieving equality. For $n = 3$, $|V_K(-1)| \leq 3$ with $G = K_3$ as example, and for $n = 4$, $|V_K(-1)| \leq 6$. But any connected graph G with 4 edges has at most 4 spanning trees. Hence $|V_K(-1)| \leq 4$ with $G = C_4$ achieving the bound.

Chapter 10

10.3 Any face has boundary length $4k+2$ for some k. Since G is bipartite, the directions around a face alternate. So an odd number $2k+1$ of edges are directed clockwise.

10.6 We have $M(3,n) = \det A$, where

$$A = \left(\begin{array}{ccccc} 1 & -2 & 0 & \dots & \\ 0 & 1 & -2 & & \\ & & \dots & & \\ & & & 1 & -2 \\ \binom{n}{0} & \binom{n-1}{1} & \dots & & \binom{n/2}{n/2} \end{array}\right) \left.\begin{array}{c} \\ \\ \\ \\ \end{array}\right\} \frac{n}{2}\,.$$

Developing the determinant according to the last row yields the result. Exercise 3.15 gave $\det A = \sum_k \binom{n/2}{k} 2^{n/2-k} 3^{\lfloor k/2 \rfloor}$.

10.9 As $M(m,n) = 4^{\frac{mn}{4}} \prod_{k=1}^{(m-1)/2} \prod_{\ell=1}^{n/2} (\cos^2 \frac{k\pi}{m+1} + \cos^2 \frac{\ell\pi}{n+1}) \prod_{\ell=1}^{n/2} \frac{\cos \ell\pi}{n+1}$ we have to prove $\prod_{\ell=1}^{n/2} \frac{\cos \ell\pi}{n+1} = 2^{-n/2}$. Let $\omega = e^{\frac{2i\pi}{n+1}}$ be $(n+1)$-st root of unity; then $\omega^\ell = \cos \frac{2\ell\pi}{n+1} + i \sin \frac{2\ell\pi}{n+1}$, and thus $\cos \frac{2\ell\pi}{n+1} = \frac{\omega^\ell + \omega^{-\ell}}{2}$. By the cosine theorem, $2\cos^2\frac{\ell\pi}{n+1}=\cos\frac{2\ell\pi}{n+1}+1$; hence $\cos^2\frac{\ell\pi}{n+1}=\frac{\omega^\ell+\omega^{-\ell}+2}{4}$. This gives $\prod_{\ell=1}^{n/2}\cos^2\frac{\ell\pi}{n+1}=\prod_{\ell=1}^{n/2}(\omega^\ell+\omega^{-\ell}+2)4^{-n/2}$. Now we claim that $\prod_{\ell=1}^{n/2}(\omega^\ell+\omega^{-\ell}+2)=1$, which will finish the proof, taking square roots. Since $\omega^{-\ell}=\omega^{n+1-\ell}$ we have $\prod_{\ell=1}^{n/2}(\omega^\ell+\omega^{-\ell}+2)=\prod_{\ell=1}^{n/2}(\omega^\ell+1)(\omega^{-\ell}+1)=\prod_{\ell=1}^{n}(\omega^\ell+1)$. Finally, $1+x+\cdots+x^n=(x-\omega)(x-\omega^2)\cdots(x-\omega^n)$, and setting $x=-1$ yields $1=(-1-\omega)\cdots(-1-\omega^n)=\prod_{\ell=1}^{n}(\omega^\ell+1)$, since n is even.

10.10 Write $\varepsilon(i,j)=1$ if $i\overset{\varepsilon}{\to}j$ and $\varepsilon(i,j)=-1$ if $i\overset{\varepsilon}{\leftarrow}j$. Then

$$\begin{aligned}\textstyle\sum_\varepsilon \det A^\varepsilon &= \textstyle\sum_\varepsilon\sum_\sigma(\operatorname{sign}\sigma)\prod_{i=1}^n a_{i\sigma(i)}\varepsilon(i,\sigma(i))\\ &= \textstyle\sum_\sigma\left((\operatorname{sign}\sigma)\prod_{i=1}^n a_{i\sigma(i)}\right)\sum_\varepsilon\prod_{i=1}^n\varepsilon(i,\sigma(i))\,.\end{aligned}$$

If σ has i as fixed point, then $a_{ii}=0$. Suppose σ has a cycle of length greater than 2, say, $(i,j,\ldots)$. The number of orientations ε with $\varepsilon(i,j)=1$ equals those with $\varepsilon(i,j)=-1$; hence the last factor $\sum_\varepsilon\prod_{i=1}^n\varepsilon(i,\sigma(i))$ becomes 0. Finally, if σ has $n/2$ cycles of length 2, then $\prod_{i=1}^n\varepsilon(i,\sigma(i))=(-1)^{n/2}$, $\operatorname{sign}\sigma=(-1)^{n/2}$, and so $\sum_\varepsilon\det A^\varepsilon=2^qM(G)$.

10.14 For the path P_n we get $\mu(P_n;x)=x\mu(P_{n-1};x)-\mu(P_{n-2};x)$; hence $\mu(P_n;x)$ is the n-th Chebyshev polynomial. For K_n one easily computes $m(K_n;k)=\binom{n}{2k}(2k-1)(2k-3)\cdots=\binom{n}{2k}\frac{(2k)!}{k!2^k}$, and $\mu(K_n;x)=\sum_{k=0}^n(-\frac{1}{2})^k\frac{n^{\underline{2k}}}{k!}x^{n-2k}=H_n^{(1)}(x)$ is a Hermite polynomial. Finally, $m(K_{n,n};k)=\binom{n}{k}^2k!$, which gives $\mu(K_{n,n};x)=\sum_{k=0}^n(-1)^k\binom{n}{k}^2k!x^{n-2k}$, related to the Laguerre polynomial $L_n^{(1)}(x)=\sum_{k=0}^n(-1)^k\binom{n}{k}^2k!x^{n-k}$.

10.16 Let G be Eulerian, and $U\subseteq E$ a bipartition (V_1,V_2). Then $|U|=\sum_{u\in V_1}d(u)-2|E(G(V_1))|\equiv 0\pmod 2$, since $d(u)$ is even for all u. It follows that $\mathcal{B}(G;-1)=\sum_U(-1)^{|U|}=|\mathcal{B}(G)|$. Suppose G is not Eulerian, and $u\in V$ has odd degree. Consider the pairs of bipartitions $U=(V_1,V_2)$, $u\in V_1$, and $U'=(V_1\smallsetminus u,V_2\cup u)$. Then $|U|+|U'|\equiv 1\pmod 2$, and hence $(-1)^{|U|}+(-1)^{|U'|}=0$. Summing over all pairs U,U' yields $\mathcal{B}(G;-1)=0$. The assertion for $\mathcal{E}(G;-1)$ is proved similarly.

10.20 From $\mathcal{E}(C_n;z)=1+z^n$ follows, with Theorem 10.9, $Z(C_n;K)=2^n(\cosh(K))^n\left(1+\left(\frac{\sinh(K)}{\cosh(K)}\right)^n\right)=2^n(\cosh^n(K)+\sinh^n(K))=(e^K+e^{-K})^n+(e^K-e^{-K})^n$. From $Z(C_n;K)\ge(e^K+e^{-K})^n$ we get $\frac{\log Z}{n}\ge\log(e^K+e^{-K})\ge\log e^K=K$. Conversely, $\log\frac{a+b}{2}\le\frac{\log a+\log b}{2}$ implies $\frac{1}{n}\log\frac{(e^K+e^{-K})^n+(e^K-e^{-K})^n}{2}\le\frac{\log(e^K+e^{-K})+\log(e^K-e^{-K})}{2}=\frac{\log(e^{2K}-e^{-2K})}{2}\le\frac{\log e^{2K}}{2}=K$. Hence $\frac{\log Z}{n}\le K+\frac{\log 2}{n}$, and so $\lim_{n\to\infty}\frac{\log Z}{n}=K$.

10.21 Classify the Eulerian subgraphs according to the leftmost vertical edge. This gives $\mathcal{F}_0=\mathcal{F}_1=1$, $\mathcal{F}_2=1+z^2$, $\mathcal{F}_n=\mathcal{F}_{n-1}+z^2\mathcal{F}_{n-2}+$

$z^3\mathcal{F}_{n-3} + \cdots + z^n\mathcal{F}_0$. Use the same recurrence for $z\mathcal{F}_{n-1}$ and subtract to obtain $\mathcal{F}_n - z\mathcal{F}_{n-1} = \mathcal{F}_{n-1} + (z^2 - z)\mathcal{F}_{n-2}$, and thus $\mathcal{F}_n = (1+z)\mathcal{F}_{n-1} + (z^2 - z)\mathcal{F}_{n-2} + [n = 0] - [n = 1]z$. The generating function is therefore $H(x) = \frac{1-zx}{1-(1+z)x-(z^2-z)x^2}$. For $z = 1$ we obtain $\frac{1-x}{1-2x}$, hence $\mathcal{F}_n(1) = 2^{n-1}$. For $z = -1$ this gives $\frac{1+x}{1-2x^2} = \sum 2^n x^{2n} + \sum 2^n x^{2n+1}$; thus $\mathcal{F}_n(-1) = 2^{\lfloor n/2 \rfloor}$. For $z = -2$, $H(x) = \frac{1+2x}{1+x-6x^2}$, and with the methods of Section 3.1 one computes $\mathcal{F}_n(-2) = \frac{1}{5}(2^{n+2} + (-3)^n)$. The coefficient a_2 equals the number of 4-circuits which is clearly $n-1$, and similarly $a_3 = n-2$ is the number of 6-circuits. Finally, $a_n = F_{n-1}$ is easily seen.

10.25 By Exercise 10.23 we have

$$Z_P(G; 2K) = 2^{k(G)}(e^{2K} - 1)^{|V|-k(G)}T(G; \tfrac{e^{2K}+1}{e^{2K}-1}, e^{2K}),$$

and a short computation gives

$$Z_P(G; 2K) = 2^{|V|}(e^K)^{|V|-k(G)}(\sinh(K))^{|V|-k(G)}T(G; \coth(K), e^{2K}).$$

This last expression is equal to $e^{K|E|}Z_{\text{Ising}}(G; K)$ by Theorem 10.9.

10.27 Set $\alpha = \sqrt{q}$, $\beta = e^K - 1$, $\lambda = \sqrt{q}$, then $S(\widetilde{G}, W_{\alpha,\beta,0}; \sqrt{q}) = q^{\frac{|F|-1}{2}}(e^K - 1)^{|V|-1}\sqrt{q}T(G; 1+\frac{\sqrt{q}}{e^K-1}\sqrt{q}, 1+\frac{e^K-1}{\sqrt{q}}\sqrt{q}) = q^{\frac{|F|}{2}}(e^K-1)^{|V|-1}T(G; 1+\frac{q}{e^K-1}, e^K)$. Hence $q^{1-\frac{|F|}{2}}S(\widetilde{G}, W; \sqrt{q}) = q(e^K - 1)^{|V|-1}T(G; \frac{e^K+q-1}{e^K-1}, e^K) = Z_{\text{Potts}}$ by Exercise 10.23.

10.29 The formula for the Fibonacci numbers is derived by induction on k. It follows that $L_nF_n = F_{n-1}F_n + F_{n+1}F_n = F_{2n}$, and hence $L_n = \frac{\tau^{2n}-\hat{\tau}^{2n}}{\tau^n-\hat{\tau}^n} = \tau^n + \hat{\tau}^n$.

10.33 Let v be a leaf of T. Then by (1) and induction $\mathcal{U}(T; 1) = \mathcal{U}(T \setminus v; 1) + \mathcal{U}(T \setminus N(v); 1) \le 2^{n-2} + 1 + 2^{n-2} = 2^{n-1} + 1$, and the upper bound is attained by $K_{1,n-1}$. For the lower bound consider more generally forests F. If $E = \emptyset$, then $\mathcal{U}(F; 1) = 2^n$, and $2^n \ge F_{n+2}$ holds by induction. If $E \ne \emptyset$, let v be a leaf; then $\mathcal{U}(F; 1) = \mathcal{U}(F \setminus v; 1) + \mathcal{U}(F \setminus N(v); 1) \ge F_{n+1} + F_n = F_{n+2}$. The bound is attained for the path P_n.

10.37 The independent sets correspond to walks of length $n-1$ in G_m: $U, V_1, \ldots, V_{n-1} = V$ with weight $z^{|U|+|V_1|+\cdots+|V_{n-1}|}$. Hence $\mathcal{U}(L_{m,n}; z) = \mathbf{1}^T D_m(z) T_m(z)^{n-1}\mathbf{1}$.

10.40 We have $f(k+\ell, n) = \mathbf{1}^T T_{k+\ell}^n \mathbf{1}$; hence $f(k+\ell, n)^2 = f(n, k+\ell)^2 = (\mathbf{1}^T T_n^{k+\ell}\mathbf{1})^2 = (T_n^k\mathbf{1} \cdot T_n^\ell\mathbf{1})^2$. By the Cauchy-Schwarz inequality $(T_n^k\mathbf{1} \cdot T_n^\ell\mathbf{1})^2 \le (\mathbf{1}^T T_n^{2k}\mathbf{1})(\mathbf{1}^T T_n^{2\ell}\mathbf{1}) = f(n, 2k)f(n, 2\ell) = f(2k, n)f(2\ell, n)$. Taking the n-th root and letting $n \to \infty$, this gives $\Lambda_{k+\ell}^2 \le \Lambda_{2k}\Lambda_{2\ell}$. For $k = n$, $\ell = n+1$, $\Lambda_{2n+1}^2 \le \Lambda_{2n}\Lambda_{2n+2}$ results, and $k = n-1$, $\ell = n+1$ yields $\Lambda_{2n}^2 \le \Lambda_{2n-2}\Lambda_{2n+2}$.

10.42 Clearly, $\mathcal{U}(\widetilde{L}_{m,n};1)$ equals the number of closed walks of length n in the graph G_m; thus $\mathcal{U}(\widetilde{L}_{m,n};1) = \operatorname{tr}(\widetilde{T}_m^n)$. To compute $\mathcal{U}(\widetilde{L}_{2,n};z)$, we have $\widetilde{T}_2(z) = \begin{pmatrix} 1 & z & z \\ 1 & 0 & z \\ 1 & z & 0 \end{pmatrix}$ by Exercise 10.37; thus $\mathcal{U}(\widetilde{L}_{2,n};z) = \operatorname{tr}(\widetilde{T}_2(z)^n)$. The eigenvalues of $\widetilde{T}_2)(z)$ are $-z, \frac{z+1}{2} \pm \sqrt{\frac{z^2+6z+1}{4}}$; thus $\mathcal{U}(\widetilde{L}_{2,n};z) = (-z)^n + (\frac{z+1}{2} + \sqrt{\frac{z^2+6z+1}{4}})^n + (\frac{z+1}{2} - \sqrt{\frac{z^2+6z+1}{4}})^n$. For $z = 1$ one easily computes $\mathcal{U}(\widetilde{L}_{2,n};1) = (-1)^n + \sum_{i\geq 0}\binom{n}{2i}2^{i+1}$. For $z = -1$, $\mathcal{U}(\widetilde{L}_{2,n};-1) = 1 + i^n + (-i)^n$, $i = \sqrt{-1}$; hence

$$\mathcal{U}(\widetilde{L}_{2,n};-1) = \begin{cases} 1 & n \text{ odd}, \\ 3 & n \equiv 0 \pmod 4), \\ -1 & n \equiv 2 \pmod 4). \end{cases}.$$

10.47 We know from Exercise 9.15 that $\operatorname{ice}(\widetilde{L}_{2,n}) = |T(\widetilde{L}_{2,n};0,-2)|$. Since $\widetilde{L}_{2,n}$ is plane, $\operatorname{ice}(\widetilde{L}_{2,n}) = \frac{1}{3}$ (# 3-vertex-colorings of the dual graph $\widetilde{L}_{2,n}^*$). $\widetilde{L}_{2,n}^*$ consists of a path $A_1B_1A_2\ldots A_{n-1}B_{n-1}A_n$ plus two vertices x, y joined to all B_i, and an outer vertex z joined to x, y, A_1, A_n. If x and y are colored differently, then the coloring of z and $B_1, \ldots, B_{n-1}$ is fixed, and for each A_j we have two choices not equal to $c(z)$. Hence the number of these 3-colorings is $6 \cdot 2^n = 3 \cdot 2^{n+1}$. If x and y are colored alike, it is easy to see that $6 \cdot \sum_{i=0}^{n}\binom{n}{2i}2^{n-2i} = 6 \cdot \frac{3^n+1}{2} = 3(3^n+1)$ colorings result. Altogether this gives $\operatorname{ice}(\widetilde{L}_{2,n}) = 2^{n+1} + 3^n + 1$.

10.49 Just note that the types H and V stay the same, while $NW \leftrightarrow SE$ and $NE \leftrightarrow SW$ are interchanged.

10.53 For $n = 1$ the graph G consists of two loops. Hence $\operatorname{ice}(G) = 4 \geq \frac{3}{2}$. For $n > 1$ look at a vertex v and denote by $A, B, \ldots, F$ the number of 2-in 2-out orientations of G where at v we have the following situation:

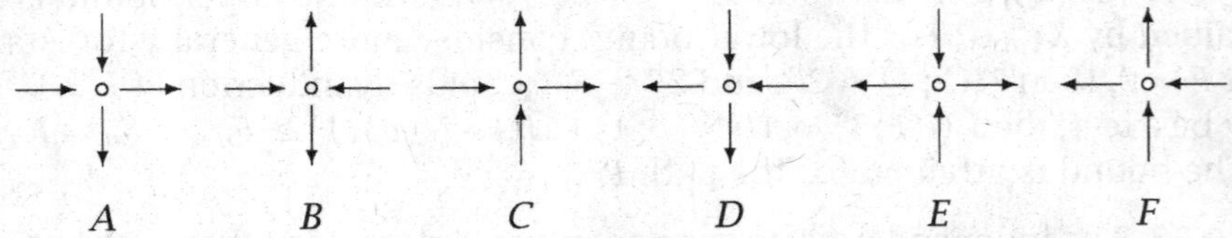

Split G into three 4-regular graphs G_1, G_2, G_3 on $n-1$ vertices by considering the transitions at v:

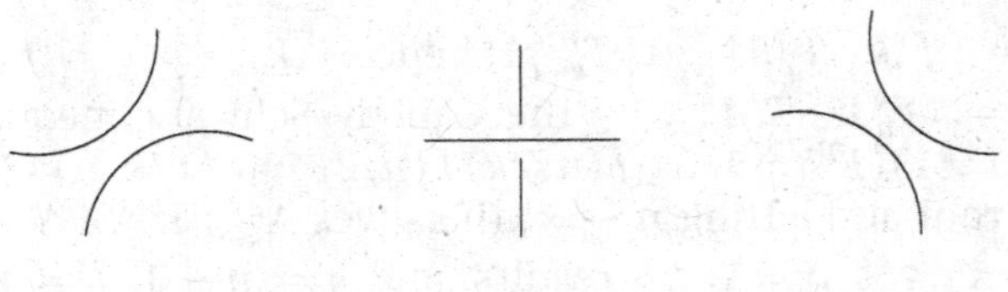

G_1 G_2 G_3

Then $\mathrm{ice}(G_1) \le B+C+D+E$, $\mathrm{ice}(G_2) \le A+C+D+F$, $\mathrm{ice}(G_3) \le A+B+E+F$. Hence $\sum_{i=1}^{3} \mathrm{ice}(G_i) \le 2 \cdot \mathrm{ice}(G)$, and so by induction $\mathrm{ice}(G) \ge \frac{3}{2}(\frac{3}{2})^{n-1} = (\frac{3}{2})^n$.

10.55 Consider columns 1, 2, and 3, and a 2-in 2-out orientation. Let $v_i \in \{1,-1\}^n$ correspond to the horizontal arrows between columns 1 and 2, and v_j to the arrows between columns 2 and 3, as in the hint. If $v_i = v_j$, then all points of column 2 have horizontally out-degree and in-degree equal to 1. Hence there are two possibilities to orient the vertical arrows. In the other cases there is one possibility or none. Hence with this weighting $\mathrm{ice}(\widetilde{L}_{n,n}) = \mathrm{tr}(T_n^n)$.

Notation

Numbers

$A_{n,k}$	Eulerian
B_n	Bernoulli
$\mathrm{Bell}(n)$	Bell
$B_n^{\sigma,\tau}$	Catalan
$b_{n,k}$	ballot
C_n	(ordinary) Catalan
D_n	derangement
$D_{m,n}$	Delannoy
F_n	Fibonacci
Fi_n	Fine
G_n	Galois
H_n	harmonic
$I_{n,k}$	inversion
i_n	involution
$K_{\lambda\mu}$	Kostka
$L_{n,k}$	Lah
M_n	Motzkin
$M(m,n)$	domino tilings
$N(n,k)$	Narayana
$\binom{n}{k}$	binomial
$n^{\underline{k}}$	falling factorial
$n^{\overline{k}}$	rising factorial
$\binom{n}{k_1 \dots k_m}$	multinomial
$[n]_q$	q-integer
$\left[{n \atop k}\right]_q$	Gaussian
$p(n)$	partition
$p(n;k)$	
$p(n;k;m)$	
$pp(n)$	plane partition
R_n	Riordan
Sch_n	Schröder
$S_{n,k}$	Stirling 2nd kind
$s_{n,k}$	Stirling 1st kind
sec_n	secant
$spp(n)$	strict plane partition
Tr_n	central trinomial
tan_n	tangent

Polynomials

$B(G;\lambda,s)$	monochromial
$C_n^{(a)}(x)$	Charlier
$c_n(x)$	Chebyshev
$\langle D\rangle$	bracket
$e_n(x_1,\dots,x_r)$	elementary symmetric
$e(G;x)$	Eulerian
$F(G;\lambda)$	flow
$f_L(A)$	Kauffman
$g_n(x)$	Gaussian
$H_n^{a,b}(x)$	Hermite
$L_n^{(a)}(x)$	Laguerre
$P_m(n)$	power sum
$P_n(x)$	Legendre
$P(G;\lambda)$	Penrose
$p_n(x_1,\dots,x_r)$	symmetric power
$Q(H;x)$	interlace
$R(G;u,v)$	rank generating
$S(\widetilde{G},W;\lambda)$	transition
$T(G;x,y)$	Tutte
$Z(G;z_1,\dots,z_n)$	cycle index
$Z(G)\cap Z(H)$	cap product
$V_L(t)$	Jones
$x^{\underline{n}}$	falling factorial
$x^{\overline{n}}$	rising factorial
$\Delta_L(t)$	Alexander
$\mu(G;x)$	matchings
$\chi(G;\lambda)$	chromatic

Functions

$\det A$	determinant	$\mathrm{Pf}A$	Pfaffian
$e_\lambda(x)$	elementary symmetric	$\mathrm{per}A$	permanent
$F(z)$	generating	$s_\lambda(x)$	Schur
$\hat{F}(z)$	exponential generating	$Z(G;K)$	partition Ising model
$F^{\langle -1\rangle}(z)$	compositional inverse	$Z_P(G;K)$	Potts model
$F\left(\begin{smallmatrix} a_1 \ldots a_m \\ b_1 \ldots b_n \end{smallmatrix}; z\right)$	hypergeometric	$Z_n(z;a)$	partition square ice
$H(\sigma)$	Hamiltonian	$\mathcal{B}(G;z)$	bipartition
$h_\lambda(x)$	complete symmetric	$\mathcal{E}(G;z)$	Eulerian
$m_\lambda(x)$	monomial symmetric	$\mathcal{U}(G;z)$	hard model
		$\Gamma(z)$	gamma
		$\mu(a,b), \overline{\mu}(n)$	Möbius
		$\varphi(n)$	Euler
		$\zeta(P)$	zeta

Sets, Posets, Groups, Graphs

$\mathrm{Bij}(N,R)$	bijection map	$S(n)$	symmetric
$\mathrm{Inj}(N,R)$	injection map	$\overline{S}(n)$	pair group
$\mathrm{Map}(N,R)$	map	$\overline{S}_G$	edge group
Par, $\mathrm{Par}(n)$	number-partition	X_g	fixed-point set
$S(X), S(n)$	permutation	C_n	circuit
$\mathrm{Surj}\,(N,R)$	surjective map	G^*	dual graph
$\binom{X}{k}$	k-subset	$\tilde{G}$	medial graph
$\Pi(X), \Pi(n)$	set-partition	G^t	terminal graph
$\mathcal{T}(n,\lambda)$	semistandard tableau (SST)	$G \setminus e, G/e$	restriction, contraction
$\Lambda(X), \Lambda^n(X)$	homogeneous symmetric	$H(C)$	interlace graph
$\mathbb{A}(P)$	incidence algebra	K_n	complete graph
$\mathbb{B}(n)$	subset lattice	$K_{m,n}$	complete bipartite graph
$\mathbb{C}(n)$	chain	$K_{n_1,\ldots,n_t}$	complete multipartite graph
$\mathbb{D}$	divisor lattice	$L_{m,n}$	lattice graph
$\mathbb{L}(n,q)$	subspace lattice	$\tilde{L}_{m,n}$	periodic lattice graph
C_n	cyclic group	Q_n	cube
D_n	dihedral group	W_n	wheel
G_x	stabilizer		

Symbols, Operators, Parameters

$A^{\sigma,\tau}$	Catalan matrix
$H_n^{(k)}$	Hankel matrix
$L^-(G), L^+(G)$	Laplace matrix
$\overline{M}$	cofactor matrix
$N_{\supseteq T}, N_{=T}$	inclusion-exclusion
$N_{\geq k}, N_{=k}$	
SST	semistandard tableau
ST	standard tableau
T^r, T^ℓ	trefoil
$w(\mathcal{F}; G)$	weight enumerator
$\delta_{m,n}$	Kronecker
π^*	reverse permutation
$\tau, \hat{\tau}$	golden section
$\Delta F(z)$	derivative
D_0F, F', D_qF	
$\Delta f(x)$	forward difference operator
$\sum f(x)$	forward sum operator
N, K	shift operator
$S(n)$	indefinite sum
$S(N, n)$	sum recurrence operator
$ac(G)$	acyclic orientations
$b(P)$	number of blocks
$c(\sigma)$	number of cycles
$d(u)$	degree of vertex
$d^-(u), d^+(u)$	out-degree, in-degree
$ds(\pi)$	decreasing subword
h_{ij}	hook length
$\mathrm{ice}(G)$	2-in 2-out orientation
$\mathrm{inv}(\sigma)$	inversion permutation
$\mathrm{inv}(s)$	inversion multiset
$\mathrm{inv}T$	inversion tree
$is(\pi)$	increasing subword
$M(G)$	perfect matchings
$t(\sigma)$	type of permutation
$t^-(G, u), t^+(G, u)$	arborescences

Index

Graduate Texts in Mathematics

(continued from page ii)

64 EDWARDS. Fourier Series. Vol. I. 2nd ed.
65 WELLS. Differential Analysis on Complex Manifolds. 2nd ed.
66 WATERHOUSE. Introduction to Affine Group Schemes.
67 SERRE. Local Fields.
68 WEIDMANN. Linear Operators in Hilbert Spaces.
69 LANG. Cyclotomic Fields II.
70 MASSEY. Singular Homology Theory.
71 FARKAS/KRA. Riemann Surfaces. 2nd ed.
72 STILLWELL. Classical Topology and Combinatorial Group Theory. 2nd ed.
73 HUNGERFORD. Algebra.
74 DAVENPORT. Multiplicative Number Theory. 3rd ed.
75 HOCHSCHILD. Basic Theory of Algebraic Groups and Lie Algebras.
76 IITAKA. Algebraic Geometry.
77 HECKE. Lectures on the Theory of Algebraic Numbers.
78 BURRIS/SANKAPPANAVAR. A Course in Universal Algebra.
79 WALTERS. An Introduction to Ergodic Theory.
80 ROBINSON. A Course in the Theory of Groups. 2nd ed.
81 FORSTER. Lectures on Riemann Surfaces.
82 BOTT/TU. Differential Forms in Algebraic Topology.
83 WASHINGTON. Introduction to Cyclotomic Fields. 2nd ed.
84 IRELAND/ROSEN. A Classical Introduction to Modern Number Theory. 2nd ed.
85 EDWARDS. Fourier Series. Vol. II. 2nd ed.
86 VAN LINT. Introduction to Coding Theory. 2nd ed.
87 BROWN. Cohomology of Groups.
88 PIERCE. Associative Algebras.
89 LANG. Introduction to Algebraic and Abelian Functions. 2nd ed.
90 BRØNDSTED. An Introduction to Convex Polytopes.
91 BEARDON. On the Geometry of Discrete Groups.
92 DIESTEL. Sequences and Series in Banach Spaces.
93 DUBROVIN/FOMENKO/NOVIKOV. Modern Geometry—Methods and Applications. Part I. 2nd ed.
94 WARNER. Foundations of Differentiable Manifolds and Lie Groups.
95 SHIRYAEV. Probability. 2nd ed.
96 CONWAY. A Course in Functional Analysis. 2nd ed.
97 KOBLITZ. Introduction to Elliptic Curves and Modular Forms. 2nd ed.
98 BRÖCKER/TOM DIECK. Representations of Compact Lie Groups.
99 GROVE/BENSON. Finite Reflection Groups. 2nd ed.
100 BERG/CHRISTENSEN/RESSEL. Harmonic Analysis on Semigroups: Theory of Positive Definite and Related Functions.
101 EDWARDS. Galois Theory.
102 VARADARAJAN. Lie Groups, Lie Algebras and Their Representations.
103 LANG. Complex Analysis. 3rd ed.
104 DUBROVIN/FOMENKO/NOVIKOV. Modern Geometry—Methods and Applications. Part II.
105 LANG. $SL_2(\mathbf{R})$.
106 SILVERMAN. The Arithmetic of Elliptic Curves.
107 OLVER. Applications of Lie Groups to Differential Equations. 2nd ed.
108 RANGE. Holomorphic Functions and Integral Representations in Several Complex Variables.
109 LEHTO. Univalent Functions and Teichmüller Spaces.
110 LANG. Algebraic Number Theory.
111 HUSEMÖLLER. Elliptic Curves. 2nd ed.
112 LANG. Elliptic Functions.
113 KARATZAS/SHREVE. Brownian Motion and Stochastic Calculus. 2nd ed.
114 KOBLITZ. A Course in Number Theory and Cryptography. 2nd ed.
115 BERGER/GOSTIAUX. Differential Geometry: Manifolds, Curves, and Surfaces.
116 KELLEY/SRINIVASAN. Measure and Integral. Vol. I.
117 J.-P. SERRE. Algebraic Groups and Class Fields.
118 PEDERSEN. Analysis Now.
119 ROTMAN. An Introduction to Algebraic Topology.
120 ZIEMER. Weakly Differentiable Functions: Sobolev Spaces and Functions of Bounded Variation.
121 LANG. Cyclotomic Fields I and II. Combined 2nd ed.
122 REMMERT. Theory of Complex Functions. *Readings in Mathematics*
123 EBBINGHAUS/HERMES et al. Numbers. *Readings in Mathematics*

124 DUBROVIN/FOMENKO/NOVIKOV. Modern Geometry—Methods and Applications Part III.
125 BERENSTEIN/GAY. Complex Variables: An Introduction.
126 BOREL. Linear Algebraic Groups. 2nd ed.
127 MASSEY. A Basic Course in Algebraic Topology.
128 RAUCH. Partial Differential Equations.
129 FULTON/HARRIS. Representation Theory: A First Course. *Readings in Mathematics*
130 DODSON/POSTON. Tensor Geometry.
131 LAM. A First Course in Noncommutative Rings. 2nd ed.
132 BEARDON. Iteration of Rational Functions.
133 HARRIS. Algebraic Geometry: A First Course.
134 ROMAN. Coding and Information Theory.
135 ROMAN. Advanced Linear Algebra. 2nd ed.
136 ADKINS/WEINTRAUB. Algebra: An Approach via Module Theory.
137 AXLER/BOURDON/RAMEY. Harmonic Function Theory. 2nd ed.
138 COHEN. A Course in Computational Algebraic Number Theory.
139 BREDON. Topology and Geometry.
140 AUBIN. Optima and Equilibria. An Introduction to Nonlinear Analysis.
141 BECKER/WEISPFENNING/KREDEL. Gröbner Bases. A Computational Approach to Commutative Algebra.
142 LANG. Real and Functional Analysis. 3rd ed.
143 DOOB. Measure Theory.
144 DENNIS/FARB. Noncommutative Algebra.
145 VICK. Homology Theory. An Introduction to Algebraic Topology. 2nd ed.
146 BRIDGES. Computability: A Mathematical Sketchbook.
147 ROSENBERG. Algebraic *K*-Theory and Its Applications.
148 ROTMAN. An Introduction to the Theory of Groups. 4th ed.
149 RATCLIFFE. Foundations of Hyperbolic Manifolds.
150 EISENBUD. Commutative Algebra with a View Toward Algebraic Geometry.
151 SILVERMAN. Advanced Topics in the Arithmetic of Elliptic Curves.
152 ZIEGLER. Lectures on Polytopes.
153 FULTON. Algebraic Topology: A First Course.
154 BROWN/PEARCY. An Introduction to Analysis.
155 KASSEL. Quantum Groups.
156 KECHRIS. Classical Descriptive Set Theory.
157 MALLIAVIN. Integration and Probability.
158 ROMAN. Field Theory.
159 CONWAY. Functions of One Complex Variable II.
160 LANG. Differential and Riemannian Manifolds.
161 BORWEIN/ERDÉLYI. Polynomials and Polynomial Inequalities.
162 ALPERIN/BELL. Groups and Representations.
163 DIXON/MORTIMER. Permutation Groups.
164 NATHANSON. Additive Number Theory: The Classical Bases.
165 NATHANSON. Additive Number Theory: Inverse Problems and the Geometry of Sumsets.
166 SHARPE. Differential Geometry: Cartan's Generalization of Klein's Erlangen Program.
167 MORANDI. Field and Galois Theory.
168 EWALD. Combinatorial Convexity and Algebraic Geometry.
169 BHATIA. Matrix Analysis.
170 BREDON. Sheaf Theory. 2nd ed.
171 PETERSEN. Riemannian Geometry.
172 REMMERT. Classical Topics in Complex Function Theory.
173 DIESTEL. Graph Theory. 3rd ed.
174 BRIDGES. Foundations of Real and Abstract Analysis.
175 LICKORISH. An Introduction to Knot Theory.
176 LEE. Riemannian Manifolds.
177 NEWMAN. Analytic Number Theory.
178 CLARKE/LEDYAEV/STERN/WOLENSKI. Nonsmooth Analysis and Control Theory.
179 DOUGLAS. Banach Algebra Techniques in Operator Theory. 2nd ed.
180 SRIVASTAVA. A Course on Borel Sets.
181 KRESS. Numerical Analysis.
182 WALTER. Ordinary Differential Equations.
183 MEGGINSON. An Introduction to Banach Space Theory.
184 BOLLOBAS. Modern Graph Theory.

185 COX/LITTLE/O'SHEA. Using Algebraic Geometry. 2nd ed.
186 RAMAKRISHNAN/VALENZA. Fourier Analysis on Number Fields.
187 HARRIS/MORRISON. Moduli of Curves.
188 GOLDBLATT. Lectures on the Hyperreals: An Introduction to Nonstandard Analysis.
189 LAM. Lectures on Modules and Rings.
190 ESMONDE/MURTY. Problems in Algebraic Number Theory. 2nd ed.
191 LANG. Fundamentals of Differential Geometry.
192 HIRSCH/LACOMBE. Elements of Functional Analysis.
193 COHEN. Advanced Topics in Computational Number Theory.
194 ENGEL/NAGEL. One-Parameter Semigroups for Linear Evolution Equations.
195 NATHANSON. Elementary Methods in Number Theory.
196 OSBORNE. Basic Homological Algebra.
197 EISENBUD/HARRIS. The Geometry of Schemes.
198 ROBERT. A Course in p-adic Analysis.
199 HEDENMALM/KORENBLUM/ZHU. Theory of Bergman Spaces.
200 BAO/CHERN/SHEN. An Introduction to Riemann–Finsler Geometry.
201 HINDRY/SILVERMAN. Diophantine Geometry: An Introduction.
202 LEE. Introduction to Topological Manifolds.
203 SAGAN. The Symmetric Group: Representations, Combinatorial Algorithms, and Symmetric Functions.
204 ESCOFIER. Galois Theory.
205 FÉLIX/HALPERIN/THOMAS. Rational Homotopy Theory. 2nd ed.
206 MURTY. Problems in Analytic Number Theory.
Readings in Mathematics
207 GODSIL/ROYLE. Algebraic Graph Theory.
208 CHENEY. Analysis for Applied Mathematics.
209 ARVESON. A Short Course on Spectral Theory.
210 ROSEN. Number Theory in Function Fields.
211 LANG. Algebra. Revised 3rd ed.
212 MATOUŠEK. Lectures on Discrete Geometry.
213 FRITZSCHE/GRAUERT. From Holomorphic Functions to Complex Manifolds.
214 JOST. Partial Differential Equations. 2nd ed.
215 GOLDSCHMIDT. Algebraic Functions and Projective Curves.
216 D. SERRE. Matrices: Theory and Applications.
217 MARKER. Model Theory: An Introduction.
218 LEE. Introduction to Smooth Manifolds.
219 MACLACHLAN/REID. The Arithmetic of Hyperbolic 3-Manifolds.
220 NESTRUEV. Smooth Manifolds and Observables.
221 GRÜNBAUM. Convex Polytopes. 2nd ed.
222 HALL. Lie Groups, Lie Algebras, and Representations: An Elementary Introduction.
223 VRETBLAD. Fourier Analysis and Its Applications.
224 WALSCHAP. Metric Structures in Differential Geometry.
225 BUMP: Lie Groups.
226 ZHU. Spaces of Holomorphic Functions in the Unit Ball.
227 MILLER/STURMFELS. Combinatorial Commutative Algebra.
228 DIAMOND/SHURMAN. A First Course in Modular Forms.
229 EISENBUD. The Geometry of Syzygies.
230 STROOCK. An Introduction to Markov Processes.
231 BJÖRNER/BRENTI. Combinatorics of Coxeter Groups.
232 EVEREST/WARD. An Introduction to Number Theory.
233 ALBIAC/KALTON. Topics in Banach Space Theory
234 JORGENSEN. Analysis and Probability
235 SEPANSKI. Compact Lie Groups
236 GARNETT. Bounded Analytic Functions
237 MARTINEZ-AVENDANO/ROSENTHAL. An Introduction to Operators on the Hardy-Hilbert Space
238 AIGNER. A Course in Enumeration
239 COHEN. Number Theory – Volume I: Tools and Diophantine Equations
240 COHEN. Number Theory – Volume II: Analytic and Modern Tools
241 SILVERMAN. The Arithmetic of Dynamical Systems.
242 GRILLET. Abstract Algebra. 2nd ed.